Texts and Monographs in Physics

Series Editors:

R. Balian, Gif-sur-Yvette, France
W. Beiglböck, Heidelberg, Germany
H. Grosse, Wien, Austria
E. H. Lieb, Princeton, NJ, USA
N. Reshetikhin, Berkeley, CA, USA
H. Spohn, München, Germany
W. Thirring, Wien, Austria

T0211064

Springer

Berlin
Heidelberg
New York
Hong Kong
London
Milan
Paris
Tokyo

Masanori Ohya   Dénes Petz

Quantum Entropy and Its Use

 Springer

Professor Masanori Ohya

Tokyo University of Science
Department of Information Sciences
Yamazaki 2641, Noda City
Chiba 278-8510, Japan

Professor Dénes Petz

Budapest University of Technology
and Economics
Department for Mathematical Analysis
Egry József utca 3.
1111 Budapest, Hungary

First Edition 1993
Corrected Second Printing 2004

ISSN 0172-5998

ISBN 3-540-20806-2 Corrected Second Printing
Springer-Verlag Berlin Heidelberg New York

ISBN 3-540-54881-5 First Edition Springer-Verlag Berlin Heidelberg New York

Library of Congress Cataloging-in-Publication Data applied for.
Bibliographic information published by Die Deutsche Bibliothek. Die Deutsche Bibliothek lists this publication
in the Deutsche Nationalbibliografie; detailed bibliographic data is available in the Internet at http://dnb.ddb.de.

This work is subject to copyright. All rights are reserved, whether the whole or part of the material is concerned,
specifically the rights of translation, reprinting, reuse of illustrations, recitation, broadcasting, reproduction on microfilm
or in any other way, and storage in data banks. Duplication of this publication or parts thereof is permitted only under
the provisions of the German Copyright Law of September 9, 1965, in its current version, and permission for use must
always be obtained from Springer-Verlag. Violations are liable for prosecution under the German Copyright Law.

Springer-Verlag is a part of Springer Science+Business Media.

springeronline.com

© Springer-Verlag Berlin Heidelberg 1993
Printed in Germany

The use of general descriptive names, registered names, trademarks, etc. in this publication does not imply, even in the
absence of a specific statement, that such names are exempt from the relevant probreak tective laws and regulations and
therefore free for general use.

Typesetting by the authors
Cover design: *design & production* GmbH, Heidelberg
Printed on acid-free paper   55/3141/tr   5 4 3 2 1 0

Preface to the Corrected Second Printing

Since our monograph was published first in 1993, quantum information theory has developped a lot in connection with the revolutionary achievements in the fields of quantum computing and quantum algorithms. Von Neumann entropy and quantum relative entropy have got several new applications and interpretations and many examples of quantum communication channels have been discovered and discussed.

The second edition keeps the original structure of the first one. At the end of the chapters a short summary is given about some of the new results obtained after the first addition was published. The results covered by the summary sections are selected in a rather arbitrary way and they cannot give an abstract of the huge number of papers published in quantum information theory in the recent years. The bibliography of the book is slightly extended as well. The new items are indicated by a star.

The second edition has benefitted from comments from a number of people and the authors are grateful to many colleagues who helped to correct misprints. The index of the book has been made more comprehensive to help those readers who wants to find some specific subjects.

Masanori Ohya
Dénes Petz

Preface to the First Edition

Entropy is a concept which appears in several fields and it is in the center of interest both in mathematical and physical subjects, sometimes even at other places, for example in communication engineering. The birthplace of quantum entropy was in quantum statistical mechanics. Quantum entropy is not a single concept but rather a family of notions which started with the von Neumann entropy of a density matrix and has developed in a number of directions. The heritage of quantum entropies from quantum mechanics is their strong relation to Hilbert space.

This book begins with the entropy of a state and through the entropy of coarse graining it proceeds to the quantum dynamical entropy. A bunch of topics shows the direct physical relevance of the entropy methods. The mathematical formalism is based on operator theory and mostly the language of operator algebras is used. The use of the concepts and methods of functional analysis not only makes the theory of quantum entropy more uniform and lucid, but it essentially simplifies and extends it. A comprehensive presentation without these elements would not be possible.

Chapters 1–3 are elementary and require of the reader only a basic knowledge of linear algebra and function theory. (Some series expansions may be beyond this level but they are standard in mathematical physics and and exact source indication provides assistance in the few cases they show up.) Most of the results of Chapters 1 and 3 are repeated in Part 2 in a more general setting. Part 1 of the book intends to demonstrate that although the quantum entropy is a more technical subject than the classical one, the basic concepts and properties are accessible without heavy mathematics.

The theory of operator algebras becomes more necessary in Parts 2–4. The very essentials are contained in any standard book on operator algebras but we suggest "Operator Algebras and Quantum Statistical Mechanics 1" by Olla Bratteli and Derek W. Robinson if a reader really would like to see a systematic treatment of this subject. What is badly needed for quantum entropy theory is concisely summarized in Chapter 4. It might be informative to emphasize that the bulk of entropy results are fully understandable without a deep knowledge of functional analysis but the proofs at a general level require sometimes sophisticated tools.

Part 5 is a collection of independent subjects which provide wide space for application of the abstract theory. In our opinion one can get a real insight into these topics without a thorough reading of all the previous chapters but with the understanding of Part 1. Precise references help those readers who have not gone through the preceding chapters.

Almost all the results in this book are accompanied with detailed rigorous proofs. Our attitude is contained in the following motto.

"Even things that are true can be proved."
(O. Wilde)

The end of a proof is indicated by the symbol □. Each chapter is closed by a section "Notes and Remarks". Here one can find the main sources used by the authors in the presentation and references to related results if the authors were aware of those. The literature on quantum entropy is huge and on entropy it is even more tremendous. Our selection must be very subjective and it mostly indicates our sources rather than evaluation of the literature. To find the balance between classical entropy and quantum entropy theory was a delicate point for the authors. Typically the classical theory is neither assumed nor explained in detail in the book. Of course, any slight knowledge of probabilistic entropy concepts is very helpful for the reader and yields a deeper understanding.

We want to thank many colleagues for their kind help in the preparation of the manuscript. We cannot name all of them but we ought to mention Luigi Accardi, Imre Csiszár, András Dévényi, Matthew J. Donald, Mark Fannes, Fumio Hiai, Heide Narnhofer, Walter Thirring and André Verbeure.

Masanori Ohya
Dénes Petz

Table of Contents

Introduction

The task of statistical mechanics is to derive macroscopic or bulk properties of matter from the laws governing a huge number of individual particles. (The word particle is used here for the sake of simplicity, one can think of molecules in a gas, electrons in a plasma, etc.) Entropy relates macroscopic and microscopic aspects and the concept is as old as modern thermodynamics. The name "entropy" is due to Clausius and the main steps towards the concept were taken by Boltzmann and Gibbs. Any study leading to the understanding of entropy must go through probability theory. The Boltzmann-Gibbs-Shannon entropy of a finite probability distribution $(p_1, p_2, \ldots, p_k)$ is the quantity

$$H(p_1, p_2, \ldots, p_k) = -\sum_{i=1}^{k} p_i \log p_i \,, \tag{1}$$

which is strongly related to the asymptotics of certain probabilities. This is the very reason why the same quantity appears in different subjects, statistical mechanics, information theory, ergodic theory, for example.

$$W_N(p_1, p_2, \ldots, p_k) = \frac{N!}{(p_1 N)!(p_2 N)! \ldots (p_k N)!} \tag{2}$$

is the number of ways one can put N objects (say, particles) into k boxes (say, cells of the phase space) so that the different boxes contain objects with proportions $p_1, p_2, \ldots, p_k$. Suppose we are interested in the behavior of W_N for large N. Using Stirling bounds for the factorials one can see that

$$\left| \tfrac{1}{N} \log W_N(p_1, p_2, \ldots, p_k) - H(p_1, p_2, \ldots, p_k) \right| = O(n^{-1} \log n) \,. \tag{3}$$

Boltzmann interpreted the number $\log W_N(p_1, p_2, \ldots, p_k)$ as the "thermodynamical probability" of the macro-state $(p_1, p_2, \ldots p_k)$. He had to divide the phase space into cells in order to speak of the number of "micro-states". In quantum mechanics such discretization is not necessary because contained in the theory.

Before we turn to entropy in quantum systems we summarize the abstract logical foundations of quantum theory. The primary mathematical structure associated with the set $\mathcal{P}$ of all propositions on a physical system is an ordered structure. (In place of proposition "question", "event" and "yes-no experiment" are used in the literature.) On the set $\mathcal{P}$ a reflexive antisymmetric transitive relation (that is, an order relation) is given, which is customarily denoted by $\leq$. The relation $p \leq q$ is interpreted as stating that q is true whenever p is proven to be true. Any (finite) collection $(p_i)_i$ of elements of $\mathcal{P}$ admits a greatest lower bound $\wedge_i p_i$ and a least upper bound $\vee_i p_i$. We write $p \wedge q$ and $p \vee q$ if we have to do with two elements. $\mathcal{P}$ is called a lattice in the language of algebra and it has a least element and a greatest element, which are denoted by O and I respectively. Another algebraic operation on $\mathcal{P}$ is the orthocomplementation. This is a mapping $p \mapsto p^{\perp}$ of $\mathcal{P}$ onto itself such that

(i) $p^{\perp\perp} = p$,
(ii) $p \leq q$ implies $p^{\perp} \geq q^{\perp}$,
(iii) $p \wedge p^{\perp} = 0$ and $p \vee p^{\perp} = I$.

Two elements p and q of $\mathcal{P}$ are said to be disjoint or orthogonal if $p \leq q^{\perp}$ (or equivalently $q \leq p^{\perp}$). Without deeply entering quantum logics we recall that the logical differences between quantum and classical behavior concerns the distributivity of the lattice $\mathcal{P}$.

In this book we always suppose that $\mathcal{P}$ is the set of all orthoprojections (i.e., idempotent self-adjoint elements: $p = p^2 = p^*$) of a *-algebra $\mathcal{A}$. In usual quantum mechanics $\mathcal{A}$ is the set of all bounded operators on a Hilbert space. On the set of projections of $\mathcal{A}$ the ordering is defined by

$$p \leq q \iff pq = p$$

and orthocomplementation is defined by

$$p^{\perp} = I - p$$

where I is the unit element of $\mathcal{A}$. (I is the largest projection.) The lattice operations are not directly expressed by product and sum. In the Hilbert space formulation of quantum mechanics $\mathcal{P}$ is the set of all orthogonal projections of a Hilbert space $\mathcal{H}$. Let $p, q \in \mathcal{P}$ be projections onto the closed subspaces $\mathcal{H}_1$ and $\mathcal{H}_2$, respectively. Then the join $p \vee q$ is the orthogonal projection onto the closed subspace generated by $\mathcal{H}_1$ and $\mathcal{H}_2$ and the meet $p \wedge q$ projects onto the intersection of $\mathcal{H}_1$ and $\mathcal{H}_2$.

Given an orthocomplemented partially ordered set $\mathcal{P}$ we say that a real-valued function μ on $\mathcal{P}$ is a probability measure on $\mathcal{P}$ if

(i) $0 \leq \mu(p) \leq 1$ for all $p \in \mathcal{P}$, $\mu(0) = 0$ and $\mu(I) = 1$,
(ii) for every sequence (p_i) of pairwise orthogonal elements in $\mathcal{P}$ the additivity $\sum_i \mu(p_i) = \mu(\vee_i p_i)$ holds.

The additivity property (ii) may be called finite or countable additivity depending on the sequence required. When $\mathcal{P}$ is the set of projections of a *-algebra and φ is a linear functional on $\mathcal{A}$ such that $0 \leq \varphi(p) \leq 1$ for every projection and $\varphi(I) = 1$, then φ restricted to $\mathcal{P}$ gives rise to a finitely additive measure on $\mathcal{P}$. This is a consequence of the simple fact that for disjoint projections p and q the equality $p \vee q = p + q$ holds. In order to obtain a measure on $\mathcal{P}$ by the restriction of a linear functional φ we need to pose some further conditions on φ. A linear functional φ will be called a state if $\varphi(I) = 1$ and $\varphi(x^*x) \geq 0$ for every $x \in \mathcal{A}$. The states of a *-algebra supply an important class of measures on the set of projections of the algebra but certainly not all the measures may be obtained in this way. A pair $(\mathcal{A}, \varphi)$ consisting of a *-algebra $\mathcal{A}$ and of a state φ will be called a probability algebra. Given a probability space $(X, \mathcal{F}, \mu)$ the set of all complex-valued bounded measurable functions form a *-algebra with pointwise conjugation as involution, on which integration with respect to μ provides a state. On the basis of this example, we are tempted to borrow terminology from classical probability theory when dealing with probability algebras.

In Hilbert space quantum mechanics states are given by statistical operators. A positive operator $D \in B(\mathcal{H})$ is called a statistical operator or density if its spectrum consists only of eigenvalues and the sum of eigenvalues is 1. When the physical system is in a state described by the density D, the expected value of the bounded (self-adjoint) observable $A \in B(\mathcal{H})$ is

$$\text{Trace } DA = \sum_i \lambda_i \langle f_i | \, A \, | f_i \rangle \tag{4}$$

if $D = \sum_i \lambda_i | f_i \rangle \langle f_i |$ in an orthonormal basis $| f_i \rangle$. In physics the underlying Hilbert space is often infinite dimensional and so is the *-algebra $B(\mathcal{H})$. However, because of their for their technical simplicity it is useful to consider finite dimensional *-algebras which readily admit an abstract trace functional. An abstract trace functional is a linear functional Tr taking the value 1 on each minimal projection (that is, on the atoms of the corresponding logic). So $\text{Tr } I$ measures the size of the algebra; it is the number of pairwise orthogonal minimal projections the sum of which is the unity I. By means of Tr the states are described by density matrices in the form

$$\varphi(a) = \text{Tr } D_\varphi a \qquad (a \in \mathcal{A}).$$

Here the density matrix D_φ is positive (semidefinite) and $\text{Tr } D_\varphi = 1$ comes from the normalization $\varphi(I) = 1$. In this formalism for the case of the commutative algebra $\mathcal{A} = \mathbb{C}^n$ we have

$$\text{Tr }(c_1, c_2, \ldots, c_n) = \sum_{i=1}^n c_i \,.$$

In this case the density "matrix" or "statistical operator" becomes a probability distribution on the n-point-space. It is certainly an advantage of the *-algebraic approach to the foundations of quantum theory that it includes traditional probability theory and on the other hand the "matrix mechanics" of quantum systems is covered by the same formalism.

The set of all states, the state space, is a convex set. In the example $\mathcal{A} = \mathbb{C}^n$ the state space is nothing else but the standard n-simplex

$$\mathbf{S}_n = \{(c_1, c_2, \ldots, c_n) \in \mathbb{R}^n \ : \ c_i \geq 0, \ \textstyle\sum_{i=1}^n c_i = 1\}.$$

The extreme points of the state space are called pure states. (Pure states can not be mixtures of other states.) The simplex $\mathbf{S}_n$ has n extreme points and every element is uniquely represented by their convex combination. (A convex set is called "Choquet simplex" if all points are unique mixtures of the extremal boundary.) The state space of a typical quantum system is far from being a simplex. Consider the algebra $\mathcal{A}$ of a single particle of spin $1/2$ which is simply $\mathcal{A} = M_2(\mathbb{C})$, the algebra of 2×2 complex matrices. The state space with an appropriate parameterization may be identified with the three dimensional unit ball. The pure states correspond to surface points and all the inner points are mixtures of pure states in many ways.

Turning to thermodynamical entropy we repeat a gedanken experiment of John von Neumann. Let us have a gas of $N(\gg 1)$ molecules in a rectangular box K. Suppose that the gas behaves as a quantum system and it is described by a probability algebra $(\mathcal{A}, \varphi)$ in a certain physical state. If φ is a mixture $\lambda\varphi_1 + (1 - \lambda)\varphi_2$, then we may think that λN molecules are in state φ_1 and $(1 - \lambda)N$ ones are in the state φ_2. On the ground of phenomenological thermodymanics we assume that if φ_1 and φ_2 are disjoint, then there is a wall which is completely permeable for the φ_1-molecules and isolating for the φ_2-molecules. (If the states φ_1 and φ_2 are disjoint, then this should be demonstrated by a certain filter. Mathematically the disjointness of φ_1 and φ_2 is expressed in the orthogonality of the eigenvectors corresponding to nonzero eigenvalues of the density matrices of the states φ_1 and φ_2.) We add an equally large empty rectangular box K' to the left of the box K and we replace the common wall with two new walls. Wall (a), the one to the left is impenetrable, whereas the one to the right, wall (b), lets through the φ_1-molecules but keeps back the φ_2-molecules. We add a third wall (c) opposite to (b) which is semi-permeable, transparent for the φ_2-molecules and impenetrable for the φ_1-ones. Then we push slowly (a) and (c) to the left, keeping their distance. During this process the φ_1-molecules are pressed through (b) into K' and the φ_2-molecules diffuse through wall (c) and remain in K. No work is done against the gas pressure, no heat is developed. Replacing the walls (b) and (c) with a rigid absolutely impenetrable wall and removing (a) we restore the boxes K and K' and succeed in the separation of the φ_1-molecules from the φ_2-ones without any work being done, without any temperature change and without evolution of heat. The entropy of the original φ-gas (with density N/V) must be the sum of the entropies of the φ_1- and φ_2-gases (with densities $\lambda N/V$ and $(1-\lambda)N/V$, respectively.) If we compress the gases in K and K' to the volumes λV and $(1 - \lambda)V$, respectively, keeping the temperature constant by means of a heat reservoir, the entropy change amounts to $\kappa_{\mathrm{B}}\lambda N \log \lambda$ and $\kappa_{\mathrm{B}}(1 - \lambda)N \log(1 - \lambda)$, respectively. Finally, mixing the φ_1- and φ_2-gases of identical density we obtain a φ-gas of N molecules in a volume V at the original temperature. If $S_0(\psi, N)$ denotes the entropy of a ψ-gas of N molecules (in a volume V and at the given temperature), we conclude that

$$S_0(\varphi_1, \lambda N) + S_0(\varphi_2, (1 - \lambda)N)$$
$$= S_0(\varphi, N) + \kappa_B \lambda N \log \lambda + \kappa_B (1 - \lambda) N \log(1 - \lambda)$$

must hold, where κ_B is Boltzmann's constant. Assuming that $S_0(\psi, N)$ is proportional to N and dividing by N we have

$$\lambda S(\varphi_1) + (1 - \lambda)S(\varphi_2)$$
$$= S(\varphi) + \kappa_B \lambda \log \lambda + \kappa_B (1 - \lambda) \log(1 - \lambda), \tag{5}$$

where S is certain thermodynamical entropy quantity (relative to the fixed temperature and molecule density). We arrived at the mixing property of entropy but we should not forget about the starting assumption: φ_1 and φ_2 are supposed to be disjoint. In order to get the entropy of a φ-gas with statistical operator D_φ we need one more postulate: We shall assume that the entropy of pure states vanishes. If $D_\varphi = \sum_i \lambda_i p_i$ is the spectral decomposition of the statistical operator (p_i's are pairwise orthogonal minimal projections), then

$$S(\varphi) = -\kappa_B \sum \lambda_i \log \lambda_i \tag{6}$$

is easily inferred by repeated application of the mixing equation (5). In this book Boltzmann's constant κ_B will be taken 1. This choice makes the entropy dimensionless (and correspondingly, the temperature should be measured in erg instead of Kelvin). The above gedanken experiment is a heuristical motivation for von Neumann's entropy formula (6). During the deduction we overlooked a problem arising from the ambiguity of the decomposition $\varphi = \lambda \varphi_1 + (1 - \lambda)\varphi_2$. Nevertheless, a closer inspection shows that a different decomposition $\varphi = \mu \psi_1 + (1 - \mu)\psi_2$ would not yield a contradicting conclusion.

The statistical operator is fundamental in quantum theory, but one should be aware of the fact that not all states of any quantum system can be described by a statistical operator. For example, type III von Neumann algebras appearing typically in quantum field theory do not have pure normal states. Notwithstanding this, the von Neumann entropy of a density matrix is right way to understand quantum entropy. Entropy is not an observable like position or angular momentum. There does not exist self-adjoint operator such that its expectation value in a state would yield the entropy of this state. The von Neumann entropy is a functional on the states of the quantum system which can not be represented by an operator.

The von Neumann entropy of a state expresses the degree of "mixing". Pure states have vanishing entropy. Since the state space of a quantum system is never a Choquet simplex, a state φ may have several extremal decompositions like

$$\varphi = \sum_i \lambda_i \varphi_i \qquad (\sum_i \lambda_i = 1, \quad \lambda_i \geq 0), \tag{7}$$

where every φ_i is a pure state by definition. Then

$$S(\varphi) = \inf\{H(\lambda_1, \lambda_2, \ldots) : \varphi = \sum_i \lambda_i \varphi_i\}$$

where the inf is over all (countable) extremal decompositions and H is given by (1). The infimum is attained on orthogonal decompositions (that is, if φ_i and φ_j are orthogonal whenever $i \neq j$). These are decompositions without any interference effects. When φ_i and φ_j are not orthogonal, λ_i and λ_j represent probabilities of non-disjoint events. As we mentioned above, it may happen that φ does not admit an extremal decomposition (7). Then one assigns the value $+\infty$ to the von Neumann entropy. The von Neumann entropy is often infinite and this fact creates a need for other entropy concepts, relative entropy, Rényi entropy, entropy density, etc.

It was Shannon who interpreted the quantity (1) as information. He showed that, up to a constant factor, (1) is the only function of the probabilities $p_1, p_2, \ldots, p_k$ that satifies some postulates natural for an information quantity. Information and uncertainty are very close concepts. The quantity (1) may be thought of as uncertainty as well because we are unable to predict exactly the outcome of an experiment having different outcomes with probabilities $p_1, p_2, \ldots, p_k$. For example, when one of the probabilities is near to 1, our uncertainty in the prediction is small, and so is (1). On the other hand, having performed the experiment, the amount of information we have gained (or more exactly, what we are expected to gain) is expressed by (1). Hence we shall regard information and uncertainty as complementary notions. Uncertainty is simply considered negative information or the lack of information.

A possible reason for uncertainty is the large number of alternatives. Suppose that a source produces N long messages from an "alphabet" $\{1, 2, \ldots, k\}$ and $1 \leq i \leq k$ appears with probability p_i. The number of all possible messages is k^N, but for large N most of them will show up with very small probability. (This is a consequence of the law of large numbers.) If we consider a set L of messages with total probability $1 - \varepsilon$, then this set contains at least $\exp(N H(p_1, p_2, \ldots, p_k))$ messages,

$$\#(L) \geq \exp(N H(p_1, p_2, \ldots, p_k)) . \tag{8}$$

Thus, asymptotically (as $N \to \infty$, ε fixed) there is a simple exponential relation between the Shannon entropy $H(p_1, p_2, \ldots, p_k)$ and the number of alternatives in a set of large probability. A similar result is true in quantum probability. Assume that we are in a position to investigate a quantum system $\mathcal{A}$ composed of N independent and identical subsystems $\mathcal{B}$. Imagine, that a stistical operator D is given on the Hilbert space $\mathcal{H}$ associated with the individual systems. The composite system can be in the state described on $\mathcal{H}_N = \mathcal{H} \otimes \mathcal{H} \otimes \ldots \otimes \mathcal{H}$ by the density $D_N = D \otimes D \otimes \ldots \otimes D$ (both tensor products are N-fold). A subspace $\mathcal{K}$ of the tensor product $\mathcal{H}_N$ carries most of the probability if for the corresponding orthogonal projection P

$$\operatorname{Tr} D P \geq 1 - \varepsilon$$

holds. The number of alternatives within $\mathcal{K}$ is the dimension of $\mathcal{K}$, or equivalently, the rank of P. Corresponding to the above probabilistic result we have

$$\operatorname{Tr} P = \dim \mathcal{K} \geq \exp(N\,S(D))\,. \qquad (9)$$

Although the Hilbert space probability is very different from classical probability, the analogy between (8) and (9) is striking.

The Shannon entropy (1) measures the amount of information carried by the probability distribution $(p_1, p_2, \ldots, p_k)$. Suppose again that p_i is the probabilty of an outcome i in an experiment. If the experiment is observed roughly, for example, if we do not distinguish between the outcomes 1 and 2, then the observation provides less information. In this example, the relevant probability distribution is $(p_1 + p_2, p_3, \ldots, p_k)$, and correspondingly

$$H(p_1 + p_2, p_3, \ldots, p_k) < H(p_1, p_2, p_3, \ldots, p_k)\,.$$

The partition $\{\{1, 2\}, \{3\}, \ldots \{k\}\}$ is coarser than $\{\{1\}, \{2\}, \{3\}, \ldots \{k\}\}$, and indeed a coarser partition always has smaller Shannon entropy. In statistical language, grouped data yields less information. The following analysis of the quantum mechanical analogue leads to a delicate point in the theory of quantum entropy.

Let D_{12} be a density matrix of a composite system $\mathcal{H}_1 \otimes \mathcal{H}_2$. The measurement of observables of the form like $A \otimes I$ gives only partial information on the state D_{12} of the composite system. The observer does not get complete information about the linear functional

$$\varphi_{12} : X \mapsto \operatorname{Tr} D_{12} X \qquad (X \in B(\mathcal{H}_1 \otimes \mathcal{H}_2) \equiv B(\mathcal{H}_1) \otimes B(\mathcal{H}_2))\,,$$

instead, his information is confined to the restriction φ_1 of φ_{12} to the first subsystem $B(\mathcal{H}_1)$. (This is what we shall call coarse graining.) The restriction φ_1 possesses a density matrix D_1 and according to our intuition D_1 carries less information then D_{12}. However, it is by no means necessarily true that the von Neumann entropy of D_1 is less than that of D_{12}. For example, if

$$D_{12} = \frac{1}{2} \begin{pmatrix} 0 & 0 & 0 & 0 \\ 0 & 1 & 1 & 0 \\ 0 & 1 & 1 & 0 \\ 0 & 0 & 0 & 0 \end{pmatrix} \quad \text{then} \quad D_1 = \frac{1}{2} \begin{pmatrix} 1 & 0 \\ 0 & 1 \end{pmatrix}$$

and $S(D_1) = \log 2$ while $S(D_{12}) = 0$ because D_{12} corresponds to a pure state. In fact, much more is known. Any state of $B(\mathcal{H})$ is the restriction of a pure state of $B(\mathcal{H}) \otimes B(\mathcal{H})$. This is a pure quantum phenomena, and this problem with density D_{12} has already caused a lot of problems to philosophers of quantum theory. If given a larger total system $\mathcal{A}$ (the composite system $B(\mathcal{H}_1 \otimes \mathcal{H}_2)$ in the above situation) and a part of the observables is neglected for certain reasons and only a subalgebra $\mathcal{B} \subset \mathcal{A}$ is considered, we shall speak of a coarse graining. When coarse graining is performed, our uncertainty is decreased since prediction is easier if a part of the complicated circumstances is neglected. Let the total system $\mathcal{A}$ be in a state φ. Then the coarse grained state is $\varphi|\mathcal{B}$. Since there is no relation between the von Neumann entropies $S(\varphi)$ and $S(\varphi|\mathcal{B})$, this concept is not suitable to measure the information contained in the restriction to the subalgebra $\mathcal{B}$ in the state φ of $\mathcal{A}$. Hence there is a need for the entropy of a coarse graining, say $H_\varphi(\mathcal{B})$, which must possess the following obvious properties.

(i) $H_\varphi(\mathcal{C}) \leq H_\varphi(\mathcal{B})$ whenever $\mathcal{C} \subset \mathcal{B} \subset \mathcal{A}$.

(ii) $H_\varphi(\mathcal{A}) = S(\varphi)$.

From the viewpoint of communication theory any coarse graining gives rise to a channeling transformation. The state φ of $\mathcal{A}$ is transformed into $\varphi|\mathcal{B}$ and this application is an affine mapping from the state space of $\mathcal{A}$ into that of $\mathcal{B}$. (In communication theory one would speak of mutual information instead of the entropy of coarse graining.) The entropy of coarse graining is a very important concept. An observation on a system is a coarse graining in a wide sense, and our knowledge on a quantum system is based on observations.

In this book the relative entropy of two states is regarded as the fundamental notion and the different other kinds of entropies will be deduced from the relative entropy (like the entropy of coarse graining discussed above). This approach does not follow the path of history, since quantum relative entropy appeared in the work of Umegaki only in 1962. In the simplest case the relative entropy of ω with respect to φ will be defined as

$$S(\omega, \varphi) = \operatorname{Tr} D_\omega (\log D_\omega - \log D_\varphi), \tag{10}$$

provided that the densities D_ω and D_φ have positive eigenvalues. The origin of probabilistic relative entropy is in mathematical statistics, where it is used to measure how much the state φ differs from ω in the sense of statistical distinguishability. Briefly, the larger $S(\omega, \varphi)$ is, the more information for discriminating between the hypotheses ω and φ can be obtained from an observation. There $S(\omega, \varphi)$ is called information for discrimination. The same aspect of relative entropy is present in the quantum statistical setting as well. Perhaps relative entropy is not the most appropriate terminology, but it is rather widespread. For a state ω of the matrix algebra $M_n(\mathbb{C})$ one has

$$S(\omega) = \log n - S(\omega, \tau), \tag{11}$$

where τ is the tracial state Tr/n. The negative sign in (11) implies that the von Neumann entropy and the relative entropy are rather different, for example concerning convexity. The von Neumann entropy as a real uncertainty quantity is a concave functional of the state, while the relative entropy is (jointly) convex (as it is an information quantity, in fact). The von Neumann entropy may be expressed in terms of the relative entropy without the use of a trace functional. Namely,

$$S(\varphi) = \sup \left\{ \sum_i \lambda_i S(\varphi_i, \varphi) : \varphi = \sum_i \lambda_i \varphi_i \right\}, \tag{12}$$

where the greatest lower bound is over all finite convex decompositions in the state space. Formula (12) is the key to the definition of the entropy of coarse graining (satisfying the desiderata (i) and (ii)).

From the point of view of physics the most transparent path to relative entropy starts from a simple thermodynamical model. Consider a finite quantum system $\mathcal{A}$ of spins that is described by a Hamiltonian $H = H^* \in \mathcal{A}$. The free energy functional (also called Helmholtz free energy) is defined on the state space of $\mathcal{A}$ as

$$F : \omega \mapsto \omega(H) - \frac{1}{\beta} S(\omega) \quad \text{where} \quad \beta = \frac{1}{\kappa_B T}$$

is the inverse temperature and $\omega(H)$ is the internal energy term. This functional is minimized by the canonical state (also called Gibbs state)

$$\varphi^c(a) = \frac{\operatorname{Tr} a\, e^{-\beta H}}{\operatorname{Tr} e^{-\beta H}} \qquad (a \in \mathcal{A}). \tag{13}$$

The minimal value of the free energy functional is given by

$$F_0 = -\frac{1}{\beta} \log \operatorname{Tr} e^{-\beta H} \tag{14}$$

and is often called Gibbs free energy. In this example the relative entropy $S(\omega, \varphi^c)$ is the increment of free energy, $\beta(F(\omega) - F_0)$ and in the thermodynamical limit (as as the size of the system tends to infinity) one meets the most important use of quantum entropy in statistical thermodynamics. The gap between finite systems and infinite systems is very serious in the behavior of entropy. To indicate what we have in mind, take von Neumann entropy. On the state space of a finite quantum system this is a continuous real functional. In infinite dimension, or in the thermodynamical limit discontinuities occur (and become dominant) which are frequently related to interesting physical phenomenon and it is by no means an irritating pathology.

In a common quantum statistical model the states of a quantum mechanical system correspond to positive normalized functionals of an algebra of operators. The observer of the physical system obtains information by measurements. In the mathematical model measurements are measures taking positive operator values in the operator algebra. If the system is in a state φ and $E(H)$ is the measure of the set $H \subset \mathbb{R}^n$ then $\varphi(E(H))$ is interpreted as the probability that the measured quantity has its value in the set H. Suppose that we have to decide between two states φ and ω on the algebra. Performing a measurement we receive two probability distributions μ_φ and μ_ω. The decision between φ and ω must be done on the basis of μ_φ and μ_ω. The result may be satisfactory only in the case in which the measurement performed has extracted all the information on the mutual relation of the two states. The information contained in the measures μ_φ and μ_ω is typically less than the information carried by the couple φ and ω. The formulation in terms of relative entropy is the following.

$$S(\mu_\omega, \mu_\varphi) \leq S(\omega, \varphi) \tag{15}$$

In the fortunate case the two information quantities coincide and we shall call such a measurement sufficient. Sufficient measurements do not exist for "non-commuting" states and this is nothing else but a form of the quantum mechanical uncertainty. Although the uncertainty principle is conventionally expressed by non-commuting observables, something similar appears in connection with two non-commuting states as well. Loss of relative entropy of two states is a form of uncertainty during a measurement. A large number of identical and independent systems makes the mean uncertainty small when the measurement is chosen properly.

Due to its importance we devote some more discussion to the position-momentum uncertainty. Recall that P and Q are self-adjoint operators on $\mathcal{H} = L^2(\mathbb{R})$ defined by

$$(e^{ixP}f)(t) = f(t+x) \quad \text{and} \quad (e^{iyQ}f)(t) = e^{iyt}f(t) \tag{16}$$

for every $f \in L^2(\mathbb{R})$ and $x, y, t \in \mathbb{R}$. Let ξ be a vector of norm 1 (in other words, ξ is a wave function) and set

$$\xi_{xy} = e^{iyQ}e^{-ixP}\xi \quad (x, y \in \mathbb{R}).$$

Then one obtains a probability density

$$\varrho(x, y) = \frac{1}{2\pi}\langle \xi_{xy}, D\xi_{xy}\rangle$$

on the phase space $\mathbb{R}^2$ for any statistical operator D on $\mathcal{H}$. It is an uncertainty relation in the phase space of quantum mechanics that the density ϱ can not be arbitrarily localized. The entropic uncertainty relation states that

$$H(D, \xi) = -\int_{\mathbb{R}^2} \varrho(x, y) \log \varrho(x, y) \, dx \, dy \geq \log 2\pi e. \tag{17}$$

In the most interesting applications the entropies are connected with a limiting procedure. This can be seen already in formula (3), which was the origin of the Boltzmann-Gibbs entropy $H(p_1, p_2, \ldots, p_k)$. In thermodynamics the limiting procedure is the thermodynamical limit, and the emerging entropy density is a macroscopic variable, moreover, it is one of the ingredients of the Gibbs variational principle. Statistical mechanics provides much room for entropic variational expressions. To take an axample, in a d dimensional quantum lattice gas model, a copy $M_n(\mathbb{C})$ of the $n \times n$ matrix algebra is associated to each $i \in \mathbb{Z}^d$. For a finite subset $\Lambda \subset \mathbb{Z}^d$ a mean-field local hamiltonian is given by

$$H_\Lambda = \sum_{i \in \Lambda} h_i + \sum_{i \in \Lambda} x_i \times \frac{1}{|\Lambda| - 1} \sum_{\substack{j \in \Lambda \\ i \neq j}} x_j \tag{18}$$

where h_i and x_i are copies of some self-adjoint matrices at the corresponding lattice site. The thermodynamical limit of the local free energy density

$$F_\Lambda(\beta) = -\frac{1}{|\Lambda|\beta} \log \operatorname{Tr} \exp\left(\beta\, H_\Lambda\right)$$

resembles probabilistic large deviation theory and it is the sample of perturbational limit theorems. There entropies appear in the role of a rate function like in the case of classical theory.

In the course of testing a quantum system the upper bound of the available average information is a characteristic of the evolution of the system. This is the content of dynamical entropy providing a quantum extension of the Kolmogorov-Sinai entropy of classical dynamical systems. The quantum dynamical entropy will be obtained by abelian modelling of the dynamical evolution and become a conjugation invariant quantity.

Entropy shows up very naturally in the central limit theorems: Normal fluctuations manifest themselves in the form of a maximum entropy principle. In the course of the central limit the relative entropy with respect to the limiting state converges to 0.

Notes and Remarks. The quantum logic approach is summarized here in order to show similarities and differences in the stochastic description of classical and quantum behavior. There are several books that treat the quantum logical approach in more detail, we may mention [Emch 1972], [Gudder 1979], [Beltrametti and Cassinelli 1981]. We have to emphasize that in books on quantum mechanics states are often understood as pure states in our sense.

The expression (1) is due to [Boltzmann 1877]. In the derivation of the entropy formula (6) von Neumann benefited from earlier arguments by Einstein and Szilárd. The gedanken experiment is analyzed in more detail in Chapter 5 of [Neumann 1932]. The excellent survey [Wehrl 1978] is a suggested overview on entropy and it contains a number of references on the early developments. Since the publication of that paper several "rather tricky and involved" proofs in quantum entropy theory have become natural and standard.

Part I

Entropies
for Finite Quantum Systems

1 Fundamental Concepts

The notion of entropy was introduced by Clausius in order to discuss the thermal behaviour of physical systems. After the work of Bolztmann, Gibbs and others, von Neumann and Shannon contributed to the theory of entropy. Their starting points were very different, von Neumann was motivated by quantum mechanics and Shannon founded communication theory.

The entropy of a state describing a physical system is a quantity expressing the uncertainty or randomness of the system. Shannon regarded this uncertainty attached to a physical system as the amount of information carried by the system, so that the entropy of a state can be read as the information carried by the state. His idea comes from the following consideration : If a physical system has a large uncertainty and one receives information on the system by some procedure, then the so-obtained information is more valuable than that received from a system having less uncertainty.

The *relative entropy* is an information measure representing the uncertainty of a state with respect to another state. Hence it indicates a kind of distance between the two states. In information theory the relative entropy $S(\nu, \mu)$ of two finite probability distributions $\nu = (\kappa_1, \kappa_2, \ldots, \kappa_n)$ and $\mu = (\lambda_1, \lambda_2, \ldots, \lambda_n)$ on an n-point space is usually defined by

$$S(\nu, \mu) = \begin{cases} \sum_i \kappa_i (\log \kappa_i - \log \lambda_i) & \text{if } \lambda_i = 0 \text{ implies } \kappa_i = 0 \\ +\infty & \text{otherwise.} \end{cases} \tag{1.1}$$

We shall regard relative entropy as the most fundamental concept and take it as a starting point for introducing several other kinds of entropy-like quantities. First we shall discuss the generalisation of the above concept of relative entropy to finite quantum systems. In this deductive treatment the basic properties of entropy will be obtained from those of the relative entropy.

We shall follow an algebraic approach and by a finite quantum system we shall mean a finite dimensional C*-algebra. (This does not mean that in this chapter any knowledge of operator algebra theory would be required.) The algebraic viewpoint does not distinguish between isomorphic algebras. For example, the algebra of $n \times n$ complex diagonal matrices is obviously identified with the algebra of n-tuples of complex numbers. To work in the framework of finite quantum systems is advantageous didactically, because all linear operators are bounded in this case, so that complicated domain problems are avoided. Another advantage is that a number of problems can be treated explicitly in finite dimension and some of these techniques and results can be carried over by an approximation procedure to the infinite dimension, which is more interesting for applications in physics.

One knows that a finite dimensional C*-algebra is a direct sum of full matrix algebras. On such an algebra $\mathcal{A}$ there exists a unique linear functional Tr, which takes the value 1 at each minimal projection. It is "tracial" in the sense that

$$\mathrm{Tr}\,ab = \mathrm{Tr}\,ba \qquad (a, b \in \mathcal{A})$$

and the integer $\mathrm{Tr}\,I$ is called the *tracial dimension* of $\mathcal{A}$. (Readers feeling uneasy with this C*-algebraic language may simply think of elements of algebras as finite square matrices.) To explain the relation of Tr and the matrix trace, assume that $\mathcal{A}$ is a direct sum

$$\oplus_{j=1}^{k} M_{n_j}(\mathbb{C})$$

where $M_{n_j}(\mathbb{C})$ is the full matrix algebra of $n_j \times n_j$ complex matrices. If tr_j is the usual matrix trace on $M_{n_j}(\mathbb{C})$ then

$$\mathrm{Tr}\,(a_1 \oplus a_2 \oplus \ldots \oplus a_k) = \sum_{j=1}^{k} \mathrm{tr}_j\,a_j\,.$$

In particular, $\mathrm{Tr}\,I = \sum_{j=1}^{k} n_j$.

Every functional ω on $\mathcal{A}$ is uniquely represented by a *density operator* $D_\omega \in \mathcal{A}$ in the form

$$\omega(a) = \mathrm{Tr}\,D_\omega\,a \qquad (a \in \mathcal{A})\,. \tag{1.2}$$

If ω is positive then $D_\omega \geq 0$ and the normalisation $\omega(I) = 1$ implies $\mathrm{Tr}\,D_\omega = 1$.

The *relative entropy* of ω with respect to φ is defined by

$$S(\omega, \varphi) = \begin{cases} \mathrm{Tr}\,D_\omega(\log D_\omega - \log D_\varphi) & \text{if } \mathrm{supp}\,D_\varphi \geq \mathrm{supp}\,D_\omega \\ +\infty & \text{otherwise.} \end{cases} \tag{1.3}$$

Here $\mathrm{supp}\,D_\psi$ denotes the smallest projection p such that $\psi(p) = \psi(I)$. In particular, $S(\omega, \varphi)$ is always finite if the density of φ has strictly positive eigenvalues. (Such a φ is called faithful.) When D_φ commutes with D_ω and their eigenvalue lists are $(\lambda_1, \lambda_2, \ldots, \lambda_n)$ and $(\kappa_1, \ldots, \kappa_n)$ respectively, then $S(\,.\,,\,.\,)$ reduces to the classical expression (1.1). Although we mostly speak of the relative entropy of states it is convenient to allow ω and φ in the definition of $S(\omega, \varphi)$ to be arbitrary positive functionals. It is straightforward that the relative entropy of arbitrary positive functionals obeys the transformation rule

$$S(\lambda\omega, \mu\varphi) = \lambda S(\omega, \varphi) - \lambda \log \mu - \eta(\lambda)\omega(I), \qquad (\lambda, \mu \in \mathbb{R}^+) \tag{1.4}$$

which is called *scaling property* in the sequel. The notation $\eta(t)$ for the continuous function $-t \log t$ is standard in information theory.

The entropy $S(\omega)$ of a functional ω is defined by means of its density operator as

$$S(\omega) = \operatorname{Tr} \eta(D_\omega). \tag{1.5}$$

This notion was introduced by von Neumann in 1927 and we shall term it *von Neumann entropy* or shortly entropy. Up to a sign the entropy is the relative entropy with respect to the functional Tr, that is,

$$S(\omega) = -S(\omega, \operatorname{Tr}) = -\sum_i \kappa_i \log \kappa_i = \sum_i \eta(\kappa_i).$$

The unnormalised Tr plays the role of a uniform distribution on $\mathcal{A}$ since it associates the same value to each minimal projection. In terms of the normalised trace $\tau = \operatorname{Tr}/\operatorname{Tr}(I)$ we have $S(\omega) = -S(\omega, \tau) + \omega(I) \log \operatorname{Tr}(I)$. Observe that our definition (1.5) includes the so-called Shannon entropy of a finite probability distribution. If $\mathcal{A} = \mathbb{C}^n$ and $\omega(c_1, c_2, \ldots, c_n) = \sum_i p_i c_i$ with a probability distribution $(p_1, p_2, \ldots, p_n)$ then $S(\omega) = \sum_i \eta(p_i)$. The Shannon entropy $\sum_i \eta(p_i)$ represents the information carried by the random variable which takes the value x_i with probability p_i $(i = 1, 2, \ldots, n)$.

The aim of this chapter is to obtain the most important properties of the von Neumann entropy and of the relative entropy by elementary methods in the frame of finite quantum systems. Later on these properties will be reproved in more general situations but by means of heavier mathematical tools.

Proposition 1.1 If φ and ω are states on a finite quantum system then the inequality

$$S(\omega, \varphi) \geq \tfrac{1}{2}\operatorname{Tr}(D_\omega - D_\varphi)^2$$

holds.

Proof. The inequality is trivial if $\operatorname{supp} D_\varphi$ is strictly smaller than $\operatorname{supp} D_\omega$. Hence we assume that $\operatorname{supp} D_\varphi = I$. Let

$$D_\varphi = \sum_i \lambda_i p_i \qquad \text{and} \qquad D_\omega = \sum_j \kappa_j q_j$$

be the spectral decompositions. Since for some $\xi \in (x, y)$

$$-\eta(x) + \eta(y) + (x - y)\eta'(y) = -\tfrac{1}{2}(x - y)^2 \eta''(\xi) \geq \tfrac{1}{2}(x - y)^2 \tag{1.6}$$

we have

$$\operatorname{Tr} p_i q_j \left(-\eta(D_\omega) + \eta(D_\varphi) + (D_\omega - D_\varphi)\eta'(D_\varphi) - \tfrac{1}{2}(D_\omega - D_\varphi)^2\right)$$
$$= \left(-\eta(\kappa_j) + \eta(\lambda_i) + (\kappa_j - \lambda_i)\eta'(\lambda_i) - \tfrac{1}{2}(\kappa_j - \lambda_i)^2\right)\operatorname{Tr} p_i q_j \geq 0.$$

Summation over i and j gives

$$\text{Tr}\left(-\eta(D_\omega) + \eta(D_\varphi) + (D_\omega - D_\varphi)(I - \log D_\varphi) - \tfrac{1}{2}(D_\omega - D_\varphi)^2\right) \geq 0$$

which was to be proven.   □

It follows that for states ω and φ the quantity $S(\omega, \varphi)$ is always nonnegative and $S(\omega, \varphi) = 0$ occurs only in the case $\omega = \varphi$. To prove convexity, a key property of the relative entropy, we need some preparation.

Lemma 1.2 Let A be a positive operator and V a contraction. Then

$$(V^*AV)^t \geq V^*A^tV$$

holds for every $0 < t < 1$.

Proof. For a projection P and an invertible positive operator T one obtains from the decomposition

$$PT^{-1}P = P(PTP)^{-1}P + PT^{-1}P^\perp(P^\perp T^{-1}P^\perp)^{-1}P^\perp T^{-1}P \tag{1.7}$$

the inequality

$$PT^{-1}P \geq P(PTP)^{-1}P. \tag{1.8}$$

Therefore, we have

$$(\lambda + PSP)^{-1} \leq P(\lambda + S)^{-1}P + \lambda^{-1}P^\perp \tag{1.9}$$

for $\lambda > 0$ and $S \geq 0$. In order to estimate $(V^*AV)^t$ we use the formula

$$T^t = \frac{\sin t\pi}{\pi} \int_0^\infty \lambda^{t-1} - \lambda^t(\lambda + T)^{-1}\, d\lambda. \tag{1.10}$$

The contraction V may be written in the form PUP where U is a unitary in a larger Hilbert space and P is the orthogonal projection onto the domain of V. Hence (1.9) yields

$$\begin{aligned}
(\lambda + V^*AV)^{-1} &= P(\lambda P + PU^*PAPUP)^{-1}P \leq P(\lambda + U^*PAPU)^{-1}P \\
&= PU^*(\lambda + PAP)^{-1}UP \\
&\leq PU^*P(\lambda + PAP)^{-1}PUP + \lambda^{-1}PU^*P^\perp UP \\
&= V^*(\lambda + A)^{-1}V + \lambda^{-1}(I - V^*V).
\end{aligned}$$

Finally, we infer

$$\begin{aligned}
(V^*AV)^t &= \frac{\sin t\pi}{\pi} \int_0^\infty \lambda^{t-1} - \lambda^t(\lambda + V^*AV)^{-1}\, d\lambda \\
&\geq \frac{\sin t\pi}{\pi} V^* \left(\int_0^\infty \lambda^{t-1} - \lambda^t(\lambda + A)^{-1}\, d\lambda \right) V = V^*A^tV,
\end{aligned}$$

and the proof is complete.   □

Lemma 1.3 Let $\mathcal{A}_1$ and $\mathcal{A}_2$ be finite quantum systems, $\alpha : \mathcal{A}_1 \to \mathcal{A}_2$ a linear mapping and $S_1, T_1 \in \mathcal{A}_1$, $S_2, T_2 \in \mathcal{A}_2$ positive operators. Assume that T_1 and T_2 are invertible and the following conditions hold.

(i) $\alpha(a^*a) \geq \alpha(a)^*\alpha(a)$ $(a \in \mathcal{A}_1)$,
(ii) $\operatorname{Tr} S_2 \, \alpha(a) \leq \operatorname{Tr} S_1 \, a$ $(a \in \mathcal{A}_1^+)$,
(iii) $\operatorname{Tr} T_2 \, \alpha(a) \leq \operatorname{Tr} T_1 \, a$ $(a \in \mathcal{A}_1^+)$.

Then

$$\operatorname{Tr} \alpha(x)^* S_2^t \alpha(x) T_2^{1-t} \leq \operatorname{Tr} x^* S_1^t x T_1^{1-t}$$

for every $0 < t < 1$ and $x \in \mathcal{A}_1$.

Proof. $\mathcal{A}_1$ and $\mathcal{A}_2$ become Hilbert spaces with the Hilbert-Schmidt inner product:

$$\langle a, b \rangle = \operatorname{Tr} a^* b.$$

Set a linear operator $V : \mathcal{A}_1 \to \mathcal{A}_2$ by

$$V \, a \, T_1^{1/2} = \alpha(a) \, T_2^{1/2} \qquad (a \in \mathcal{A}_1).$$

The simple calculation

$$\|\alpha(a)T_2^{1/2}\|^2 = \operatorname{Tr} T_2 \, \alpha(a)^* \alpha(a) \leq \operatorname{Tr} T_2 \, \alpha(a^*a) \leq \operatorname{Tr} T_1 \, a^* a = \|a \, T_1^{1/2}\|^2$$

shows that V is a contraction.

We introduce two other operators Δ and D by

$$\Delta(a \, T_1^{1/2}) = S_1 \, a \, T_1^{-1/2} \qquad (a \in \mathcal{A}_1)$$
$$D(b \, T_2^{1/2}) = S_2 \, b \, T_2^{-1/2} \qquad (b \in \mathcal{A}_2)$$

Δ and D are positive and

$$\Delta^t(a \, T_1^{1/2}) = S_1^t \, a \, T_1^{1/2-t} \qquad (a \in \mathcal{A}_1)$$
$$D^t(b \, T_2^{1/2}) = S_2^t \, a \, T_2^{1/2-t} \qquad (b \in \mathcal{A}_2)$$

hold for any $t > 0$.

The chain of inequalities

$$\langle V^* D V a T_1^{1/2}, a T_1^{1/2} \rangle$$
$$= \langle D\alpha(a)T_2^{1/2}, \alpha(a)T_2^{1/2} \rangle = \langle S_2\alpha(a)T_2^{1/2}, \alpha(a)T_2^{-1/2} \rangle$$
$$= \operatorname{Tr} S_2\alpha(a)\alpha(a)^* \leq \operatorname{Tr} S_2\alpha(aa^*) \leq \operatorname{Tr} S_1 \, aa^* = \langle \Delta a T_1^{1/2}, a T_1^{1/2} \rangle$$

tells us that

$$V^* D V \leq \Delta. \tag{1.11}$$

According to Lemma 1.2 we have

$$V^* D^t V \leq (V^* D V)^t$$

and one obtains from (1.10) and (1.11)

$$(V^* D V)^t \leq \Delta^t$$

for $0 < t < 1$. Hence

$$\langle V^* D^t V x T_1^{1/2}, x T_1^{1/2} \rangle \leq \langle \Delta^t x T_1^{1/2}, x T_1^{1/2} \rangle$$

which is identical to the stated inequality.   □

Lemma 1.3 is a version of the celebrated *Lieb concavity* result. The linear mapping $\alpha : \mathcal{A}_1 \to \mathcal{A}_2$ is called *Schwarz mapping* when the inequality

$$\alpha(a^* a) \geq \alpha(a)^* \alpha(a)$$

holds for every $a \in \mathcal{A}_1$. If T and S are densities of the linear functionals ω and φ, their relative entropy can be obtained by differentiation of

$$\mathrm{Tr}\, S^t\, T^{1-t}$$

at $t = 1$.

Let $\omega_1, \omega_2, \omega, \varphi_1, \varphi_2, \varphi$ be states of a finite quantum system $\mathcal{A}$ such that

$$\omega = \lambda \omega_1 + (1 - \lambda)\omega_2 \qquad \varphi = \lambda \varphi_1 + (1 - \lambda)\varphi_2$$

with some $0 < \lambda < 1$. We choose $\mathcal{A}_1 = \mathcal{A}$ and $\mathcal{A}_2 = \mathcal{A} \oplus \mathcal{A}$. Let $\alpha : \mathcal{A}_1 \to \mathcal{A}_2$ be the homomorphism defined by

$$a \mapsto a \oplus a \qquad (a \in \mathcal{A}).$$

Set the states ω_{12} and φ_{12} of $\mathcal{A}_2$ as

$$\omega_{12}(a \oplus b) = \lambda \omega_1(a) + (1 - \lambda)\omega_2(b) \quad (a, b \in \mathcal{A})$$
$$\varphi_{12}(a \oplus b) = \lambda \varphi_1(a) + (1 - \lambda)\varphi_2(b) \quad (a, b \in \mathcal{A}).$$

If S_1, T_1, S_2, T_2 are the densities of $\varphi, \omega, \varphi_{12}, \omega_{12}$ respectively, then the hypothesis of Lemma 1.3 holds. We have

$$(1 - t)^{-1}(1 - \mathrm{Tr}\, S_2^t\, T_2^{1-t}) \geq (1 - t)^{-1}(1 - \mathrm{Tr}\, S_1^t\, T_2^{1-t})$$

and letting $t \nearrow 1$ yields

$$S(\omega, \varphi) \leq S(\omega_{12}, \varphi_{12}).$$

Since

$$S(\omega_{12}, \varphi_{12}) = \lambda S(\omega_1, \varphi_1) + (1 - \lambda)S(\omega_2, \varphi_2),$$

what we have proven is the joint convexity of the relative entropy.

Theorem 1.4 If $\omega_1, \omega_2, \varphi_1, \varphi_2$ are states of a finite quantum system and $0 < \lambda < 1$ then

$$S(\lambda \omega_1 + (1 - \lambda)\omega_2, \lambda \varphi_1 + (1 - \lambda)\varphi_2) \leq \lambda S(\omega_1, \varphi_1) + (1 - \lambda)S(\omega_2, \varphi_2).$$

Proof. Strictly speaking, in the above proof we have used the invertibility of the densities of φ and φ_{12}. If $\mathrm{supp}\,\omega_1 \leq \mathrm{supp}\,\varphi_1$ and $\mathrm{supp}\,\omega_2 \leq \mathrm{supp}\,\varphi_2$ do not hold then the stated inequality is trivial. Therefore one may assume that

$$I = \mathrm{supp}\,\varphi_1 \geq \mathrm{supp}\,\varphi_2 \geq \mathrm{supp}\,\omega_2.$$

The density of φ is then invertible and a small alteration of the above proof will help. (With $p = \mathrm{supp}\,\varphi_2$, take $\mathcal{A}_2 = \mathcal{A} \oplus p\mathcal{A}p$ and $\alpha(a) = a \oplus pap$.)   □

Lemma 1.3 includes also the monotonicity of the relative entropy under Schwarz mappings. The next result is *Uhlmann's monotonicity theorem*.

Theorem 1.5 Let $\alpha : \mathcal{A}_1 \to \mathcal{A}_2$ be a unital Schwarz mapping between the finite quantum systems $\mathcal{A}_1$ and $\mathcal{A}_2$. Then for states φ and ω of $\mathcal{A}_2$ the inequality

$$S(\omega \circ \alpha, \varphi \circ \alpha) \leq S(\omega, \varphi)$$

holds.

Proof. The proof proceeds by differentiation from Lemma 1.3.   □

Let $\mathcal{A}$ be a finite quantum system and $\mathcal{A}_0$ a subalgebra of $\mathcal{A}$. Theorem 1.5 tells us that

$$S(\omega|\mathcal{A}_0, \; \varphi|\mathcal{A}_0) \leq S(\omega, \varphi) \tag{1.12}$$

for arbitrary states ω and φ. Note that equality must hold in (1.12) if there exists a Schwarz mapping $\beta : \mathcal{A} \to \mathcal{A}_0$ that leaves both φ and ω invariant. Later on we shall discuss this possibility in detail. Choosing the trivial subalgebra we can conclude from (1.12) that the relative entropy of states is nonnegative.

Some properties of the entropy can be deduced from those of the relative entropy.

Proposition 1.6 Let $\mathcal{A}$ be a finite quantum system with states φ and ω. Then for every $0 < \lambda < 1$ the following inequality holds

$$\lambda S(\varphi) + (1 - \lambda)S(\omega) \leq S(\lambda\varphi + (1 - \lambda)\omega)$$
$$\leq \lambda S(\varphi) + (1 - \lambda)S(\omega) + H(\lambda, 1 - \lambda)$$

where $H(\lambda, 1 - \lambda) = \eta(\lambda) + \eta(1 - \lambda)$.

Proof. The first inequality is an immediate consequence of the convexity of the relative entropy. In order to obtain the second inequality we benefit from the formula

$$\mathrm{Tr}\, A(\log(A + B) - \log A)$$
$$= \int_0^\infty \mathrm{Tr}\, A(A + t)^{-1} B(A + B + t)^{-1}\, dt \geq 0 \qquad (A, B \geq 0)$$

and infer

$$\mathrm{Tr}\, \lambda D_\varphi \log(\lambda D_\varphi + (1 - \lambda)D_\omega) \geq \mathrm{Tr}\, \lambda D_\varphi \log \lambda D_\varphi$$

and

$$\mathrm{Tr}\, (1 - \lambda)D_\omega \log(\lambda D_\varphi + (1 - \lambda)D_\omega) \geq \mathrm{Tr}\, (1 - \lambda)D_\omega \log(1 - \lambda)D_\omega \,.$$

Adding the latter two inequalities we obtain the second inequality of the proposition.   □

Since η is continuous on $\mathbb{R}^+$, the entropy is a continuous functional on the states. In some cases, a more concrete estimate for the continuity might be useful. We recall that if φ and ω are states with densities D_φ and D_ω, respectively, then

$$\|\varphi - \omega\| = \mathrm{Tr}\,|D_\varphi - D_\omega|\,,$$

where $|A|$ stands for $(A^*A)^{1/2}$. Separating the negative and nonnegative eigenvalues of $D_\varphi - D_\omega$ we may write

$$D_\varphi - D_\omega = x - y \tag{1.13}$$

where x and y are nonnegative operators with disjoint supports. So

$$|D_\varphi - D_\omega| = x + y\,.$$

(Let us note that (1.13) is called Jordan decomposition.)

Lemma 1.7 Let $\lambda_1 \geq \lambda_2 \geq \ldots \geq \lambda_n$ and $\mu_1 \geq \mu_2 \geq \ldots \geq \mu_n$ be the eigenvalues of the self-adjoint matrices A and B, respectively. Then

$$\mathrm{Tr}\,|A - B| \geq \sum_{i=1}^{n} |\lambda_i - \mu_i|\,.$$

Proof. Let $x - y$ be the Jordan decomposition of $A - B$. So $\mathrm{Tr}\,|A - B| = \mathrm{Tr}\,x + \mathrm{Tr}\,y$. Consider $C = B + x = A + y$ with eigenvalues $\nu_1 \geq \nu_2 \geq \ldots \geq \nu_n$. Since $C \geq A, B$ we have

$$\nu_i \geq \lambda_i, \mu_i \qquad (i = 1, 2, \ldots, n)$$

which implies

$$2\nu_i - \lambda_i - \mu_i \geq |\lambda_i - \nu_i|\,.$$

Summing up we obtain

$$\sum_{i=1}^{n} |\lambda_i - \mu_i| \leq \sum_{i=1}^{n} (2\nu_i - \lambda_i - \mu_i) = \mathrm{Tr}\,(2C - A - B) = \mathrm{Tr}\,x + \mathrm{Tr}\,y$$

and the lemma is proven. □

The von Neumann entropy is the trace of a continuous function of the density matrix, hence it is an obviously continuous functional on the states. However, a more precise estimate for the continuity will be required in approximations. Such an estimate is due to *Fannes*.

Proposition 1.8 Let φ and ω be states of a finite quantum system $\mathcal{A}$ with tracial dimension $d = \mathrm{Tr}\,I$. If $\|\varphi - \omega\| < \frac{1}{3}$, the inequality

$$|S(\varphi) - S(\omega)| \leq \|\varphi - \omega\| \log d + \eta(\|\varphi - \omega\|)$$

holds.

Proof. Let $\lambda_1 \geq \ldots \geq \lambda_d$ and $\mu_1 \geq \ldots \geq \mu_d$ be the eigenvalues of the densities D_φ and D_ω. We set $\varepsilon_i = |\lambda_i - \mu_i|$ and $\varepsilon = \sum_i \varepsilon_i$. From the elementary inequality

$$|\eta(s) - \eta(t)| \leq \eta(s-t) \qquad 0 \leq s-t < \tfrac{1}{3}$$

we have

$$|S(\varphi) - S(\omega)| \leq \sum_{i=1}^d |\eta(\lambda_i) - \eta(\mu_i)| \leq \sum_{i=1}^d \eta(\varepsilon_i)$$

and

$$\sum_{i=1}^d \eta(\varepsilon_i) = \varepsilon \sum_{i=1}^d \eta\left(\frac{\varepsilon_i}{\varepsilon}\right) - \varepsilon \log \varepsilon \leq \varepsilon \log d + \eta(\varepsilon).$$

Here the following facts were used. By the previous lemma $\varepsilon \leq \|\varphi - \omega\|$ and η is monotone on the interval $[0, \tfrac{1}{3}]$. $\square$

The *strong subadditivity* of the entropy follows easily from the monotonicity of the relative entropy (Theorem 1.5). The strong subadditivity property is related to the composition of three different systems and it is essential in the proof of the existence of the global entropy density functional of lattice systems. Let $\mathcal{A}_1, \mathcal{A}_2, \mathcal{A}_3$ be finite quantum systems and set $\mathcal{A}_{123} = \mathcal{A}_1 \otimes \mathcal{A}_2 \otimes \mathcal{A}_3$, $\mathcal{A}_{12} = \mathcal{A}_1 \otimes \mathcal{A}_2$ and $\mathcal{A}_{23} = \mathcal{A}_2 \otimes \mathcal{A}_3$. For a state φ_{123} of $\mathcal{A}_{123}$ we denote by $\varphi_{12}, \varphi_{23}, \varphi_2$ its restrictions to $\mathcal{A}_{12}, \mathcal{A}_{23}, \mathcal{A}_2$, respectively. The strong subadditivity asserts the following.

Proposition 1.9 $S(\varphi_{123}) + S(\varphi_2) \leq S(\varphi_{12}) + S(\varphi_{23})$.

When the algebra $\mathcal{A}_2$ is trivial we have

$$S(\varphi_{13}) \leq S(\varphi_1) + S(\varphi_3) \tag{1.14}$$

which is referred to as subadditivity.

Proof. In order to prove the strong subadditivity we start with the identities

$$S(\varphi_{123}, \mathrm{Tr}_{123}) = S(\varphi_{12}, \mathrm{Tr}_{12}) + S(\varphi_{123}, \varphi_{12} \otimes \mathrm{Tr}_3) \tag{1.15}$$

and

$$S(\varphi_2, \mathrm{Tr}_2) + S(\varphi_{23}, \varphi_2 \otimes \mathrm{Tr}_3) = S(\varphi_{23}, \mathrm{Tr}_{23}) \tag{1.16}$$

where Tr with a subscript denotes the trace functional in the corresponding algebra. (Note that Tr_{123} restricted to $\mathcal{A}_i$ differs from Tr_i by a factor of $\mathrm{Tr}_i(I)/\mathrm{Tr}_{123}(I)$.) Due to the monotonicity we have

$$S(\varphi_{123}, \varphi_{12} \otimes \mathrm{Tr}_3) \geq S(\varphi_{23}, \varphi_2 \otimes \mathrm{Tr}_3). \tag{1.17}$$

So that addition of (1.15)–(1.17) yields

$$S(\varphi_{123}, \mathrm{Tr}_{123}) + S(\varphi_2, \mathrm{Tr}_2) \geq S(\varphi_{12}, \mathrm{Tr}_{12}) + S(\varphi_{23}, \mathrm{Tr}_{23})$$

which is the assertion, up to sign.   □

Let $\mathcal{A}$ be a finite quantum system with Hamiltonian $H \in \mathcal{A}^{sa}$. The *free energy functional* at the inverse temperature $\beta = (\kappa_{\mathrm{B}} T)^{-1}$

$$F(\varphi) = \varphi(H) - \frac{1}{\beta} S(\varphi)$$

is defined for the states of the system. (κ_{B} is Boltzmann's constant.) Next we show that there is a unique canonical state φ_c determined by the thermodynamic stability condition, namely, by the condition that φ_c should minimize the free energy functional at temperature T.

Proposition 1.10 If $\Phi(\beta) = -\frac{1}{\beta} \log \mathrm{Tr}\, e^{-\beta H}$, then

$$F(\varphi) \begin{cases} = \Phi(\beta) & \text{for } \varphi = \varphi_c \\ > \Phi(\beta) & \text{otherwise,} \end{cases}$$

where φ_c is the canonical state given by the density $\exp(-\beta(H - \Phi(\beta)))$.

Proof. The proof follows from the observation

$$\frac{1}{\beta} S(\varphi, \varphi_c) = F(\varphi) - \Phi(\beta).$$

Since $S(\varphi, \varphi_c) \geq 0$, with equality if and only if $\varphi_c = \varphi$ (see Proposition 1.1), we have

$$F(\varphi) \geq \Phi(\beta)$$

and equality holds only in the case $\varphi = \varphi_c$.   □

In order to treat a natural generalisation of Proposition 1.10 to relative entropy we introduce *perturbed functionals*. Let φ be a state (or a positive functional) on a finite quantum system and let $h \in \mathcal{A}$ be self-adjoint. Set D_φ for the density and p for the support projection of φ. The the perturbed functional φ^h is given by the density matrix

$$p \exp(\log D_\varphi + php)$$

where $\log D_\varphi$ is understood as $p(\log(D_\varphi + p^\perp))$. Note that the functional φ^h is not normalized and $\operatorname{supp} \varphi^h = \operatorname{supp} \varphi$.

Proposition 1.11 For a state φ of $\mathcal{A}$ with support projection $p \in \mathcal{A}$ and $h = h^* \in \mathcal{A}$ the following variational formula holds

$$\log \varphi^h(I) = \max\{\omega(php) - S(\omega, \varphi) : \omega \text{ is a state of } \mathcal{A}\}.$$

The maximum is attained at the state $\omega = \varphi^h / \varphi^h(I)$.

Proof. The quantity $S(\omega, \varphi) - \omega(php)$ is exactly $S(\omega, \varphi^h)$. According to the monotonicity of the relative entropy

$$S(\omega, \varphi^h) \geq \omega(I)(\log \omega(I) - \log \varphi^h(I)),$$

and equality holds if and only if $\omega = \varphi^h / \varphi^h(I)$. $\square$

The frame of finite dimensional C*-algebras contains two extremes. One extreme is the case when the centre of the algebra $\mathcal{A}$ is $\mathcal{A}$ itself, i.e. $\mathcal{A}$ is commutative and the other extreme is when the centre reduces to scalar multiples of the identity. (The latter algebras, i.e. algebras with trivial centre are called factors.) When $\mathcal{A}$ is written in the form

$$\mathcal{A} = \oplus_{j=1}^k M_{n_j}(\mathbb{C}),$$

then the identity of the summand $M_{n_j}(\mathbb{C})$ is a central projection of $\mathcal{A}$, and $\mathcal{A}$ is a factor if and only if $k = 1$. On the other hand $\mathcal{A}$ is commutative if and only if $n_j = 1$ for every $j = 1, 2, \ldots, k$. Let us consider inclusions $\mathcal{B} \subset \mathcal{A}$ with finite dimensional C*-algebras $\mathcal{A}$ and $\mathcal{B}$. When $\mathcal{A}$ is commutative it is represented by complex functions on a finite space and to each subalgebra $\mathcal{B}$ there corresponds a partition of the space such that $\mathcal{B}$ consists of functions taking a constant value on each set of the partition. If $\mathcal{A}$ and $\mathcal{B}$ are factors, for example if $\mathcal{A} = M_n(\mathbb{C})$ and $\mathcal{B} = M_m(\mathbb{C})$ then, one can see that the inclusion $\mathcal{B} \subset \mathcal{A}$ forces $n = tm$ for an integer t, and up to an isomorphism $\mathcal{A} = \mathcal{B} \otimes M_t(\mathbb{C})$. The integer t is termed the multiplicity of $\mathcal{B}$ in $\mathcal{A}$ and

$$t = \frac{\mathrm{Tr}_\mathcal{A} b}{\mathrm{Tr}_\mathcal{B} b} \qquad (b \in \mathcal{B}).$$

(Note that in case of a factor our steadily used functional Tr coincides with the common matrix trace.) Turning to the general case, we set

$$\mathcal{B} = \oplus_{i=1}^l M_{m_i}(\mathbb{C})$$

and assume $\mathcal{B} \subset \mathcal{A}$. The inclusion representation $\pi : \mathcal{B} \to \mathcal{A}$ restricted to $M_{m_i}(\mathbb{C})$ is decomposed into k summands and $\pi(M_{m_i}(\mathbb{C}))I_{n_j}$ has a certain multiplicity $\mu(i, j)$ in $M_{n_j}(\mathbb{C})$. (Here I_{n_j} is written for the identity of $M_{n_j}(\mathbb{C})$ and it is a central projection of $\mathcal{A}$). The matrix $\mu(i, j)$ with l rows and k columns describes the inclusion $\mathcal{B} \subset \mathcal{A}$, and it is termed inclusion matrix.

Proposition 1.12 Let $\mathcal{B} \subset \mathcal{A}$ be finite quantum systems and $\tau = \mathrm{Tr}_\mathcal{A} / \mathrm{Tr}_\mathcal{A} I$ the normalized trace functional on $\mathcal{A}$. Then there exists a unique linear mapping $E : \mathcal{A} \to \mathcal{B}$ such that

(i)   if $a \in \mathcal{A}^+$ then $E(a) \in \mathcal{B}^+$,
(ii) $E(b) = b$ for every $b \in \mathcal{B}$,
(iii) $E(ab) = E(a)b$ holds for every $a \in \mathcal{A}$ and $b \in \mathcal{B}$,
(iv) $\tau(E(a)) = \tau(a)$ for every $a \in \mathcal{A}$.

Proof. Endow $\mathcal{A}$ with the Hilbert-Schmidt inner product $\langle a_1, a_2 \rangle = \tau(a_2^* a_1)$. One can observe readily that $E : \mathcal{A} \to \mathcal{B}$ shares the properties (i)–(iv) if and only if E is the orthogonal projection onto $\mathcal{B}$ with respect to the Hilbert-Schmidt inner product. This supplies us with both the existence and the uniqueness of E. □

A linear mapping $E : \mathcal{A} \to \mathcal{B}$ fulfilling conditions (i)–(iii) in the above proposition is usually called a *conditional expectation*. This terminology is borrowed from probability theory and motivated by the case of a commutative $\mathcal{A}$. Conditional expectations are frequently used tools.

The constant multiple of the linear mapping E of Proposition 1.12 is sometimes called partial trace. We are speaking here of the mapping

$$F(a) = \frac{\mathrm{Tr}_{\mathcal{A}}\, I}{\mathrm{Tr}_{\mathcal{B}}\, I} E(a) \qquad (a \in \mathcal{A}) \tag{1.18}$$

which possesses the property

$$\mathrm{Tr}_{\mathcal{A}}\, a = \mathrm{Tr}_{\mathcal{B}}\, F(a) \qquad (a \subset \mathcal{A})$$

and sends the density of a state φ of $\mathcal{A}$ to the density of $\varphi|\mathcal{B}$.

When $\mathcal{B} \subset \mathcal{A}$, then for arbitrary states φ and ω on $\mathcal{A}$ the inequality

$$S(\omega, \varphi) - S(\omega|\mathcal{B}, \varphi|\mathcal{B}) \geq 0 \tag{1.19}$$

holds due to the monotonicity of the relative entropy (cf. Theorem 1.5 and (1.17)). We shall see that in certain situations the difference on the left hand side of (1.19) is a relative entropy quantity itself.

Theorem 1.13 Let $\mathcal{B} \subset \mathcal{A}$ be finite quantum systems and φ a state of $\mathcal{A}$ with invertible density. Assume that there exists a conditional expectation $E : \mathcal{A} \to \mathcal{B}$ such that $\varphi \circ E = \varphi$. Then for every state ω of $\mathcal{A}$ we have

$$S(\omega, \varphi) - S(\omega|\mathcal{B}, \varphi|\mathcal{B}) = S(\omega, \omega \circ E) \tag{1.20}$$

Proof. We denote by $D_\varphi^{\mathcal{A}}$, $D_\omega^{\mathcal{A}}$, $D_\varphi^{\mathcal{B}}$, $D_\omega^{\mathcal{B}}$ and D the densities of states φ, ω, $\varphi|\mathcal{B}$, $\omega|\mathcal{B}$ and $\omega \circ E$, respectively. The existence of E implies the strong condition

$$(D_\omega^{\mathcal{B}})^t (D_\varphi^{\mathcal{B}})^{-t} = D^t (D_\varphi^{\mathcal{A}})^{-t} \qquad (t \in \mathbb{R}) \tag{1.21}$$

for these densities. Now we multiply (1.21) by $D_\omega^{\mathcal{A}}$ and take $\mathrm{Tr}_{\mathcal{A}}$. The derivative of

$$\mathrm{Tr}_{\mathcal{A}}\, D_\omega^{\mathcal{A}} (D_\omega^{\mathcal{B}})^t (D_\varphi^{\mathcal{B}})^{-t} = \mathrm{Tr}_{\mathcal{B}} (D_\omega^{\mathcal{B}})^{1+t} (D_\varphi^{\mathcal{B}})^{-t}$$

at $t = 0$ gives $S(\varphi|\mathcal{B}, \omega|\mathcal{B})$. On the other hand, the derivative of

$$\mathrm{Tr}_{\mathcal{A}}\, D_\omega^{\mathcal{A}} D^t (D_\varphi^{\mathcal{A}})^{-t}$$

equals

$$\mathrm{Tr}_{\mathcal{A}}\, D_\omega^{\mathcal{A}} (\log D - \log D_\varphi^{\mathcal{A}}) = S(\omega, \varphi) - S(\omega, \omega \circ E)\,,$$

which is the conditional expectation property (1.20).   □

We shall discuss condition (1.21) in terms of the theory of operator algebras later on in more detail. The conditional expectation property (1.20) will be used in the next chapter as the crucial postulate for the relative entropy.

Let $\mathcal{A}, \mathcal{B}, \varphi, E$ be as in Theorem 1.13 and ω_0 a state on $\mathcal{B}$. It is a consequence of the last theorem that the functional $\omega \mapsto S(\omega, \varphi)$ takes its least value on the convex set $\{\omega : \omega|\mathcal{B} = \omega_0\}$ at $\omega = \omega_0 \circ E$. This minimization problem is important for statistical inference.

In order to improve the lower bound of Proposition 1.1 we need the following lemma.

Lemma 1.14 For all $0 \leq u \leq v \leq 1$ the inequality

$$2(u - v)^2 \leq u \log \frac{u}{v} + (1 - u) \log \frac{1 - u}{1 - v}$$

holds.

Theorem 1.15 For states φ and ω of a finite quantum system $\mathcal{A}$

$$S(\omega, \varphi) \geq \tfrac{1}{2} \|\omega - \varphi\|^2 \,.$$

Proof. Let p be the spectral projection of $D_\varphi - D_\omega$ corresponding to the positive eigenvalues and take $q = I - p$. We may assume that

$$u \equiv \operatorname{Tr} D_\omega p \leq \operatorname{Tr} D_\varphi p \equiv v \,.$$

Set $\mathcal{B}$ for the two dimensional commutative subalgebra of $\mathcal{A}$ generated by the projection p. We have

$$\|\varphi - \omega\| = \operatorname{Tr} (D_\varphi - D_\omega)p - \operatorname{Tr} (D_\varphi - D_\omega)q = 2(v - u)$$

and Lemma 1.14 tells us

$$2(v - u) \leq \sqrt{2S(\omega|\mathcal{B}, \omega|\mathcal{B})} \,.$$

The rest of the proof follows from the monotonicity

$$S(\omega|\mathcal{B}, \varphi|\mathcal{B}) \leq S(\omega, \varphi) \,,$$

established in (1.12).   □

Inspection of the proof of Theorem 1.15 reveals that the lemma contained the result in the (two dimensional) commutative case and the idea was to reduce the general case to the commutative one. To compute the norm $\|\varphi - \omega\|$ it is enough to know these states on commutative subalgebras and

$$\|\varphi - \omega\| = \max\{\|\varphi|\mathcal{A}_0 - \omega|\mathcal{A}_0\| \, : \, \mathcal{A}_0 \subset \mathcal{A} \text{ is commutative}\} \,.$$

In contrast to this property of the norm our next aim is to show that in general

$$S(\omega, \varphi) > \sup\{S(\omega|\mathcal{A}_0, \varphi|\mathcal{A}_0) \, : \, \mathcal{A}_0 \subset \mathcal{A} \text{ is commutative}\} \,. \tag{1.22}$$

Proposition 1.16 Let φ_0 and φ_1 be faithful states of the finite quantum system $\mathcal{A}$ and $\mathcal{B} \subset \mathcal{A}$ a commutative subalgebra. If

$$S(\varphi_1, \varphi_0) = S(\varphi_1|\mathcal{B}, \varphi_0|\mathcal{B})$$

then D_{φ_0} commutes with D_{φ_1} and $D_{\varphi_0} D_{\varphi_1}^{-1} \in \mathcal{B}$.

Proof. For $0 \le t \le 1$ we set

$$\varphi_t = (1 - t)\varphi_0 + t\varphi_1$$

and we write simply D_t for the density of φ_t. With the notation

$$K(t) = S(\varphi_t, \varphi_0) - tS(\varphi_1, \varphi_0)$$
$$L(t) = S(\varphi_t|\mathcal{B}, \varphi_0|\mathcal{B}) - tS(\varphi_1|\mathcal{B}, \varphi_0|\mathcal{B})$$
$$y(t) = K(t) - L(t)$$

we have

$$y(0) = y(1) = 0 \qquad \text{and} \qquad y(t) \ge 0 \qquad (0 \le t \le 1).$$

We are going to show that y is convex. To this end we estimate y''.
   Using the formula (3.16) we infer

$$K'(t) = \text{Tr}_{\mathcal{A}} S \log(tS + D_0) - \text{Tr}_{\mathcal{A}}(D_1 \log D_1 - D_0 \log D_0)$$

where $S = D_1 - D_0$. Now by means of the representation

$$\log x = \int_0^\infty (1 + t)^{-1} - (x + t)^{-1}\, dt$$

we obtain

$$K''(t) = \text{Tr}_{\mathcal{A}} \int_0^\infty S(D_t + u)^{-1} S(D_t + u)^{-1}\, du\,, \tag{1.23}$$

and similarly

$$L''(t) = \text{Tr}_{\mathcal{B}} \int_0^\infty F(S)(F(D_t) + u)^{-1} F(S)(F(D_t) + u)^{-1}\, du\,,$$

where $F : \mathcal{A} \to \mathcal{B}$ is the partial trace. Since $\mathcal{B}$ is supposed to be commutative we have

$$L''(t) = \text{Tr}_{\mathcal{B}}\left(F(S)^2 F(D_t)^{-1}\right). \tag{1.24}$$

If the spectral resolution of D_t is $\sum_i \lambda_i^{(t)} p_i^{(t)}$, then from (1.23)

$$K''(t) = \sum_{i,j} \text{Tr}\, S p_i^{(t)} S p_j^{(t)} / \text{Lm}\,(\lambda_i^{(t)}, \lambda_j^{(t)})\,,$$

where Lm stands for the logarithmic mean: $\text{Lm}\,(\alpha, \beta) = (\alpha - \beta)/(\log \alpha - \log \beta)$ if $\alpha \ne \beta$, and $\text{Lm}\,(\alpha, \alpha) = \alpha$. The elementary inequality $\text{Lm}\,(\alpha, \beta) \le (\alpha + \beta)/2$

holds and if α is different from β then the logarithmic mean is strictly smaller than the arithmetic one. Hence

$$K''(t) \geq \operatorname{Tr} SW(t) \,,$$

where $W(t) = 2\sum_{i,j}(p_i^{(t)} S p_j^{(t)})/(\lambda_i^{(t)} + \lambda_j^{(t)})$ is a self-adjoint element of $\mathcal{A}$. It is easy to check that

$$D_t W(t) + W(t)D_t = 2S \,.$$

Hence

$$\operatorname{Tr} SW(t) = \tfrac{1}{2}\operatorname{Tr}(D_t W(t)^2 + W(t)D_t W(t)) = \operatorname{Tr} D_t W(t)^2 \,.$$

With the choice of $V(t) = F(S)F(D_t)^{-1}$ the identity

$$\operatorname{Tr} SW(t) = \operatorname{Tr} D_t V(t)^2 + \operatorname{Tr} D_t(W(t) - V(t))^2 + 2\operatorname{Tr}(SV(t) - D_t V(t)^2)$$

gives

$$\operatorname{Tr} SW(t) = \operatorname{Tr} D_t V(t)^2 + \operatorname{Tr} D_t(W(t) - V(t))^2 \tag{1.25}$$

and

$$K''(t) \geq \operatorname{Tr} SW(t) \geq \operatorname{Tr} D_t V(t)^2 = \operatorname{Tr}_{\mathcal{B}} F(D_t)V(t)^2 = L''(t) \,.$$

The last inequality yields $y'' \geq 0$, establishing the convexity of y.

Since

$$y'(0) = S(\varphi_1, \varphi_0) - S(\varphi_1|\mathcal{B}, \varphi_0|\mathcal{B}) = 0,$$

y must vanish on the whole interval $[0,1]$ due to our hypothesis. This implies that

$$K''(t) = \operatorname{Tr} SW(t)$$

and for $i \neq j$ the coefficient of $(\operatorname{Lm}(\lambda_i^{(t)}, \lambda_j^{(t)}))^{-1}$, which is

$$\operatorname{Tr} S p_i^{(t)} S p_i^{(t)} \,,$$

vanishes, too. The identity

$$\operatorname{Tr}|SD_t - D_t S|^2 = \sum_{i,j}(\lambda_i^{(t)} - \lambda_j^{(t)})^2 \operatorname{Tr} S p_i^{(t)} S p_j^{(t)}$$

shows that $SD_t = D_t S$ or equivalently, D_0 and D_1 commute. From (1.25) we conclude

$$W(t) = V(t) \,,$$

which for $t = 1$ reads as

$$F(D_0)F(D_1)^{-1} = D_0 D_1^{-1} \,.$$

This condition implies that $D_0 D_1^{-1} \in \mathcal{B}$. $\qquad\square$

By some compactness arguments one can see that the supremum in (1.22) is in fact a maximum. Hence, in the light of Proposition 1.16, if the densities of φ and ω do not commute, then (1.22) holds with strict inequality.

It is understood in probability theory that the notion of (Shannon, or measure theoretic) entropy has successful applications in a variety of subjects because it determines the asymptotic behaviour of certain probabilities in the course of independent trials. Now we will discuss these phenomena for finite quantum systems and to do this a kind of law of large numbers is required.

Let $\mathcal{A}$ be a finite quantum system with a faithful state ω. The n-fold algebraic tensor product $\mathcal{A}_n = \mathcal{A} \otimes \ldots \otimes \mathcal{A}$ is again a finite quantum system, and the product functional $\omega_n = \omega \otimes \ldots \otimes \omega$ is a state of $\mathcal{A}_n$. Using the obvious identifications the inclusion $(\mathcal{A}_n, \omega_n) \subset (\mathcal{A}_m, \omega_m)$ holds for $n \leq m$, and we set

$$(\mathcal{A}_\infty, \omega_\infty) = \cup \{(\mathcal{A}_n, \omega_n) : n \in \mathbb{N}\}.$$

On the $*$-algebra $\mathcal{A}_\infty$ the right shift endomorphism γ is defined for $a_1 \otimes a_2 \otimes \ldots \otimes a_n \in \mathcal{A}_n$ as

$$\gamma(a_1 \otimes a_2 \otimes \ldots \otimes a_n) = I \otimes a_1 \otimes a_2 \otimes \ldots \otimes a_n \in \mathcal{A}_{n+1}$$

and ω_∞ is obviously invariant under γ. Now perform the GNS-construction with the state ω_∞ and arrive at the triplet $(\pi, \mathcal{H}, \Omega)$. We identify $\mathcal{A}_\infty$ through its faithful representation π with a subalgebra of the generated von Neumann algebra $\mathcal{M} = \pi(\mathcal{A}_\infty)'' \subset \mathcal{B}(\mathcal{H})$. The normal state

$$\omega(a) = \langle \Omega, a\Omega \rangle \qquad (a \in \mathcal{M})$$

is an extension of ω_∞, and the endomorphism γ extends to $\mathcal{M}$ so that the relation $\omega \circ \gamma = \omega$ is preserved. (For the sake of simpler notation we do not use a new letter for the extension.)

The following result may be called the weak law of large numbers for independent finite quantum systems.

Proposition 1.17 In the situation described above the following statements hold.

(i)   If $a \in \mathcal{M}$ and $\gamma(a) = a$ then $a \in \mathbb{C}\,I$.

(ii) For every $a \in \mathcal{M}$ the sequence $S_n(a) = n^{-1}(a + \gamma(a) + \ldots + \gamma^{n-1}(a))$ converges to $\omega(a)I$ in the strong operator topology.

(iii) If $a \in \mathcal{M}^{sa}$ and $I \subset \mathbb{R}$ is a closed interval such that $\omega(a) \notin I$, and p_n is the spectral projection of $S_n(a)$ corresponding to the interval I, then $p_n \to 0$ in the strong operator topology.

Proof. We apply the statistical ergodic theorem for the isometry $V : \mathcal{H} \to \mathcal{H}$ defined by the formula

$$Va\Omega = \gamma(a)\Omega \qquad (a \in \mathcal{M}).$$

Set E for the projection onto the fixed point space of V. In order to show condition (i) we prove that E is of rank one. It is easy to see that for $a, b \in \mathcal{A}_\infty$ the relation

$$\omega(aS_n(b)) \to \omega(a)\omega(b) \qquad (1.26)$$

holds. Equivalently,

$$\langle a^*\Omega, \, n^{-1}(I + V + \ldots + V^{n-1})b\Omega \rangle \to \langle \Omega, a\Omega \rangle \langle \Omega, b\Omega \rangle \,.$$

Therefore we have

$$\langle a^*\Omega, \, Eb\Omega \rangle = \langle \Omega, a\Omega \rangle \langle \Omega, b\Omega \rangle$$

for every $a, b \in \mathcal{M}$. The vector Ω being cyclic the equality

$$Eb\Omega = \langle \Omega, b\Omega \rangle \Omega = \omega(b)\Omega$$

must hold and E is really of rank one.

The ergodicity condition (i) implies easily that the state ω is faithful on $\mathcal{M}$. Hence the vector Ω is cyclic for the commutant $\mathcal{M}'$ too. The statistical ergodic theorem tells us that

$$S_n(a)b'\Omega \to \omega(a)b'\Omega$$

for every $b' \in \mathcal{M}'$. Since $(S_n(a))_n$ is norm bounded and $\mathcal{M}'\Omega$ is dense, we are able to conclude that $s_n(a) \to \omega(a)I$ strongly. Statement (ii) is proved and (iii) follows. $\qquad \square$

Let us fix a positive number $\varepsilon < 1$. For a while we shall say that a projection $Q_n \in \mathcal{A}_n$ is rather sure if $\omega_n(Q_n) \geq 1 - \varepsilon$. On the other hand, the size of Q_n, i.e. the cardinality of a maximal pairwise orthogonal family of projections contained in Q_n, is given by $\mathrm{Tr}_n Q_n$. (The subscript n in Tr_n indicates that the algebraic trace functional on $\mathcal{A}_n$ is meant here.) The theorem below says that it is the von Neumann's entropy of ω which governs asymptotically the size of the rather sure projections: A rather sure projection in $\mathcal{A}_n$ contains at least $\exp(nS(\omega))$ pairwise orthogonal minimal projections.

Theorem 1.18 Under the above conditions and with the above notation the limit relation

$$\lim_{n \to \infty} \tfrac{1}{n} \inf\{\log \mathrm{Tr}_n Q_n : Q_n \in \mathcal{A}_n \text{ is a projection, } \omega_n(Q_n) \geq 1 - \varepsilon\} = S(\omega)$$

holds.

Proof. If D_n denotes the density of ω_n then one can see easily that

$$-\log D_n = \sum_{i=0}^{n-1} \gamma^i(-\log D_1)$$

where γ stands for the right shift. The sequence $(\gamma^i(-\log D_1))_i$ behaves as independent identically distributed random variables with respect to the state

ω_∞. More precisely, the previous proposition applies for $a = -\log D_1$ and tells us that

$$-\tfrac{1}{n} \log D_n \to S(\omega)I$$

strongly. Let $P(n, \delta)$ be the spectral projection of the self-adjoint operator $-n^{-1} \log D_n$ corresponding to the interval $(S(\omega) - \delta, \, S(\omega) + \delta)$. According to (iii) of Proposition 1.17 one has

$$P(n, \delta) \to I \qquad\qquad (1.27)$$

strongly for every $\delta > 0$. In particular,

$$\omega(P(n, \delta)) = \langle P(n, \delta)\Omega, \Omega \rangle \to 1$$

as $n \to \infty$, and $P(n, \delta)$ is a rather sure projection if n is large enough. It follows from the definition of $P(n, \delta)$ that

$$D_n \, P(n, \delta) \exp(n\, S(\omega) - n\delta) \le P(n, \delta) \le D_n \, \exp(n\, S(\omega) + n\delta) \qquad (1.28)$$

which gives

$$\tfrac{1}{n} \log \mathrm{Tr}_n \, P(n, \delta) \le S(\omega) + \delta \, .$$

Since $\delta > 0$ was arbitrary we establish

$$\limsup_{n \to \infty} \tfrac{1}{n} \inf\{\log \mathrm{Tr}_n \, Q_n \, : \, Q_n \} \le S(\omega) \, . \qquad\qquad (1.29)$$

To prove that $S(\omega)$ is actually the limit we shall argue by contradiction. Assume that there exist a sequence $n(1) < n(2) < \ldots$ of integers, a number $t > 0$ and projections $Q(n(k)) \in \mathcal{A}_{n(k)}$ $(k = 1, 2, \ldots)$ such that

(i) $\omega_\infty(Q(n(k)) \ge 1 - \varepsilon,$
(ii) $\log \mathrm{Tr}_{n(k)} Q(n(k)) \le n(k)(S(\omega) - t).$

The bounded sequence $(Q(n(k)))_k$ has a weak limit point in the von Neumann algebra $\mathcal{M}$, say $T \in \mathcal{M}$. Instead of selecting a subsequence we suppose that $Q(n(k)) \to T$ weakly. It is straightforward to show that from (1.27) the weak limit

$$Q(n(k))P(n(k), \delta) \to T$$

follows. Consequently,

$$\liminf_{k \to \infty} \omega_\infty(Q(n(k)) \, P(n(k), \delta)) \ge \omega(T) \ge 1 - \varepsilon \, . \qquad (1.30)$$

Using the first part of (1.28) we estimate

$$\begin{aligned}
\mathrm{Tr}\, Q(n(k)) &\ge \mathrm{Tr}\, Q(n(k)) \, P(n(k), \delta) \\
&\ge \mathrm{Tr}\, D_{n(k)} Q(n(k)) \, P(n(k), \delta) \exp(n\, S(\omega) - n\delta) \\
&= \exp(n\, S(\omega) - n\delta)\omega_\infty(Q(n(k)) \, P(n(k), \delta))
\end{aligned}$$

and

$$\liminf_{k\to\infty} \frac{1}{n(k)} \log \mathrm{Tr}_{n(k)} Q(n(k))$$

$$\geq S(\omega) - \delta + \lim_{k\to\infty} \frac{1}{n(k)} \log \omega_\infty(Q(n(k))\, P(n(k),\delta))\,.$$

The limit term on the right hand side vanishes due to (1.30) and we arrive at a contradiction with (ii) if $0 < \delta < t$. This proves the theorem. □

A variant of Theorem 1.18 holds for the relative entropy. Let us keep the setting of Theorem 1.18 but in addition let φ be a faithful state on the finite quantum system $\mathcal{A}$. We write φ_n for the corresponding product states of $\mathcal{A}_n$ ($n \in \mathbb{N}$). Here φ_n will measure the size of the projections in $\mathcal{A}_n$ while ω_n may be interpreted as probability. Set

$$R(n,\varepsilon) = \inf\{\varphi_n(Q_n) : Q_n \in \mathcal{A}_n \text{ is a projection, } \omega_n(Q_n) \geq 1 - \varepsilon\}$$

for positive ε. For a tracial state φ the proof of Theorem 1.18 works and yields

$$\lim_{n\to\infty} \tfrac{1}{n} \log R(n,\varepsilon) = -S(\omega,\varphi)\,. \tag{1.31}$$

This relation holds actually for an arbitrary state φ and the relation is called the *quantum Stein lemma*.

Before closing this chapter we explain the concept of mutual entropy which plays an important role both in communication theory and in the definition of dynamical entropy.

By a decomposition of a state φ of a quantum system we mean a finite convex combination $\varphi = \sum_i \lambda_i \varphi_i$ in the state space. This decomposition is called *extremal* if every φ_i is pure and it is called *orthogonal* if $\mathrm{supp}\,\varphi_i$ is orthogonal to $\mathrm{supp}\,\varphi_j$ whenever $i \neq j$. For a state of an abelian finite dimensional algebra the extremal decomposition is unique and orthogonal. In case of a general finite quantum system the extremal decomposition is neither unique nor necessarily orthogonal. If the density matrix possesses an eigenvalue with multiplicity, then there exists an infinity of orthogonal extremal decompositions.

A positive unital mapping $\alpha : \mathcal{A}_1 \to \mathcal{A}$ between two finite quantum systems $\mathcal{A}_1$ and $\mathcal{A}$ may be interpreted as a communication channel. The mapping α associates to any state φ of $\mathcal{A}$ a state $\varphi \circ \alpha$ of $\mathcal{A}_1$. On the language of communication theory, φ is an input state and $\varphi \circ \alpha$ is the corresponding output state. The notion of *mutual entropy* represents the amount of information correctly transmitted from the input to the output system. Therefore the mutual entropy does depend on the input state and on the transmission channel. Let $\varphi = \sum_i \lambda_i \varphi_i$ be an orthogonal extremal decomposition. With respect to this decomposition the compound state φ_{co} is defined on the product algebra $\mathcal{A} \otimes \mathcal{A}_1$ as

$$\varphi_{\mathrm{co}} = \sum_i \lambda_i(\varphi_i \otimes \varphi_i \circ \alpha)\,. \tag{1.32}$$

The mutual entropy describing the amount of information transmitted through the channel should indicate some relation between elementary components (pure states) of φ and $\varphi \circ \alpha$. This relation is expressed in the compound state. Let us consider the abelian case. Then $\mathcal{A}_1 = \mathbb{C}^n$, $\mathcal{A} = \mathbb{C}^k$ and a channel is given by a stochastic matrix $(c_{j,i})$ with k rows and n coloums. The formula (1.32) gives the compound measure which contains the correlation between the input and the output states.

The mutual entropy of the input state φ and the channel α for the given decomposition is $S(\varphi_{co}, \varphi \otimes (\varphi \circ \alpha))$, and the mutual entropy $I(\varphi; \alpha)$ is defined to be the supremum of all this relative entropy quantities over all possible orthogonal extremal decompositions of φ. One checks readily that

$$S(\varphi_{co}, \varphi \otimes (\varphi \circ \alpha)) = \sum_i \lambda_i S(\varphi_i \circ \alpha, \varphi \circ \alpha), \tag{1.33}$$

hence we may write

$$I(\varphi; \alpha) = \sup\{\textstyle\sum_i \lambda_i S(\varphi_i \circ \alpha, \varphi \circ \alpha) : \varphi = \textstyle\sum_i \lambda_i \varphi_i\}$$

where the least upper bound is taken over all orthogonal extremal decompositions $\varphi = \sum_i \lambda_i \varphi_i$.

Theorem 1.19 $I(\varphi; \alpha) \le \min\{S(\varphi), S(\varphi \circ \alpha)\}$.

Proof. Since

$$\sum_i \lambda_i S(\varphi_i \circ \alpha, \varphi \circ \alpha) = S(\varphi \circ \alpha) - \sum_i \lambda_i S(\varphi_i \circ \alpha)$$

the second majorization is clear without any assumption on the decomposition. To prove the first estimate we need to assume the extremality of the decomposition $\varphi = \sum_i \lambda_i \varphi_i$ and we use the monotonicity of the relative entropy.

$$\sum_i \lambda_i S(\varphi_i \circ \alpha, \varphi \circ \alpha) \le \sum_i \lambda_i S(\varphi_i, \varphi)$$
$$= -\sum_i \lambda_i \mathrm{Tr}\, \eta(D_{\varphi_i}) + \sum_i \mathrm{Tr}\, \lambda_i (D_{\varphi_i} \log D_\varphi)$$

which equals to $S(\varphi)$, due to $\eta(D_{\varphi_i}) = 0$.   □

The proposition tells us that the quantity of information correctly transmitted through the channel can not exceed the information content of the input state. The difference $S(\varphi) - I(\varphi; \alpha)$ is called *entropy defect*. We note that $S(\varphi \circ \alpha) \le S(\varphi)$ is not true in general.

Finally, we show an example. Let $0 < \lambda < 1$ and define a state φ_λ of $M_2(\mathbb{C})$ by its density

$$D_\lambda = \frac{1}{2} \begin{pmatrix} 1 & 1 - 2\lambda \\ 1 - 2\lambda & 1 \end{pmatrix}.$$

For $\lambda \neq 1/2$ the orthogonal decomposition of φ_λ is unique, in fact

$$D_\lambda = \frac{\lambda}{2} \begin{pmatrix} 1 & -1 \\ -1 & 1 \end{pmatrix} + \frac{1-\lambda}{2} \begin{pmatrix} 1 & 1 \\ 1 & 1 \end{pmatrix}$$

and for the diagonal subalgebra $\mathcal{D} \subset M_2(\mathbb{C})$ we easily obtain

$$I(\varphi_\lambda; \mathcal{D}) = 0. \qquad (\lambda \neq 1/2)$$

For $\lambda = 1/2$ the state $\varphi_{1/2}$ is the tracial one and it possesses several orthogonal decompositions. The decomposition

$$D_{1/2} = \frac{1}{2} \begin{pmatrix} 1 & 0 \\ 0 & 0 \end{pmatrix} + \frac{1}{2} \begin{pmatrix} 0 & 0 \\ 0 & 1 \end{pmatrix}$$

yields the lower bound $I(\varphi_{1/2}; \mathcal{D}) \geq \log 2$. Since $I(\varphi_{1/2}; \mathcal{D}) \leq S(\varphi_{1/2}) = \log 2$, we have

$$I(\varphi_{1/2}; \mathcal{D}) = \log 2.$$

This example shows that $I(\varphi; \mathcal{D})$ is not continuous in φ and it may vanish for a mixed state φ.

Notes and Remarks. In mathematical statistics the relative entropy of two measures was introduced in [Kullback and Leibler 1951] under the name information for discrimination. Rényi later proposed the name information gain but several other expressions are used in the literature of statistics, information and communication theory, such as for example, I-divergence, cross-information etc. In the context of operator algebras $S(\omega, \varphi)$ was defined in [Umegaki 1962] and it was called relative information. The *relative entropy* $S(\omega, \varphi)$ as we use it here is rather an information quantity than an entropy quantity. Indeed, it is convex and decreasing under positive mappings. An entropy quantity is expected to be concave and increasing. (Relying upon the slogan "entropy is the lack of information", one obtains an entropy quantity from an information quantity by changing sign and vice versa.) Concerning $S(\omega, \varphi)$ there are different sign and ordering conventions in the literature. For example, those of the book [Bratteli and Robinson 1981] are related to ours by $-S_{BR}(\omega, \varphi) = S(\omega, \varphi)$.

The quantum-mechanical entropy $S(\varphi) = -\operatorname{Tr} D_\varphi \log D_\varphi$ appeared in [von Neumann 1927] where the form of the density of the canonical state was also established. The information theoretical entropy originated from [Shannon 1948]. Proposition 1.1 is from [Streater 1985]. The first concavity results for trace functions were proved in [Lieb 1973a] by a method based on operator-valued analytic functions and it implies the joint convexity of the relative entropy. The present proof for of this property benefits from the operator monotonicity of the function $x \mapsto x^t$ $(0 < t < 1)$, and deduces joint convexity from monotonicity under Schwarz mappings ([Petz 1986b]). (In fact, the basic properties of entropy and relative entropy are rather equivalent,

[Wehrl 1978] is a good source in this respect.) Theorem 1.5 is a special case of Theorem 5.3 which was proven in [Uhlmann 1977]. The use of relative entropy in mathematical physics was initiated in the papers [Lindblad 1973, 1974, 1975], too. The *strong subadditivity* of the entropy attracted a lot of attention at the beginning of the rigorous quantum statistical mechanics. The first proof is due to [Lieb and Ruskai 1973a] (see also the review [Lieb 1975]). The estimate of Proposition 1.8 is from [Fannes 1973b]. Lemma 1.7 is a simple consequence of standard eigenvalue inequalities, see for example II.6.5 in [Kato 1966]. Concerning conditional expectations in operator algebra theory we refer to [Strătilă 1981] or to the short survey [Petz 1988a]. The conditional expectation property (1.20) was obtained in [Hiai, Ohya and Tsukada 1981] under the strong condition that φ, restricted to the subalgebra $\mathcal{B}$, be tracial. The form presented here is from [Petz 1986a]. Lemma 1.14 is from [Csiszár 1967] and the subsequent Theorem 1.15 is due to [Hiai, Ohya and Tsukada 1981]. The paper [Petz 1985a] contains Proposition 1.16. It is worthwhile to note that no positive lower bound is known for the difference of the left and right hand sides of inequality (1.22).

Proposition 1.17 is a very special case of well-known ergodic theorems in operator algebras, our formulation is not more than a functional analytic formulation of the simplest law of large numbers. More sophisticated limit theorems may be found in [Jajte 1985]. Theorem 1.18 is the analogue of the so-called fundamental theorem of information theory. Its generalization (1.31) was discussed in [Hiai and Petz 1991].

In the quantum context the *mutual entropy* and the *compound state* was discussed in [Ohya 1983].

Comments. In the standard formalism of quantum mechanics, a quantum mechanical system is described by a Hilbert space and physical observables correspond to (self-adjoint) operators acting on the Hilbert space. When $A \in B(\mathcal{H})$ is an observable and $\Phi \in \mathcal{H}$ is a state vector, then $\langle \Phi, A\Phi \rangle$ is interpreted as the expectation value of the observable A in the quantum state Φ.

If the system of interest is a *composite system* of two subsystems described by the Hilbert spaces $\mathcal{H}_1$ and $\mathcal{H}_2$, then the composite system is described by the tensor product Hilbert space $\mathcal{H}_1 \otimes \mathcal{H}_2$. Assume that $\{\xi_j\}$ is a basis of $\mathcal{H}_1$ and $\{\eta_i\}$ is a basis of $\mathcal{H}_2$. Given a *state vector* Φ of the composite system, we have an expansion

$$\Phi = \sum_{i,j} w_{ij}\, \xi_j \otimes \eta_i$$

in the product basis. Let A be a linear operator acting on $\mathcal{H}_1$. It is easy to compute that

$$\langle \Phi, (A \otimes I)\Phi \rangle = \operatorname{Tr} AD_1$$

for an operator D_1 which has the properties $D_1 \geq 0$ and $\operatorname{Tr} D_1 = 1$. Therefore, D_1 is a *statistical operator* or *density matrix* if both Hilbert spaces are of finite dimension. (Actually $D_1 = W^*W$, when w_{ij} is the ij-element of the matrix W.) This argument shows how to arrive at a density matrix from a state vector of a composite system. Note that the statistical operator $|\Phi\rangle\langle\Phi|$ describes the original state of the composite system, and D_1 is the *reduced density matrix* on $\mathcal{H}_1$.

If D_j is a density matrix on $\mathcal{H}_j$ ($j = 1, 2$), then $D_1 \otimes D_2$ is a density matrix on $\mathcal{H}_1 \otimes \mathcal{H}_2$. This density matrix is of product-type and the corresponding state is a product state. The convex hull of densities of product-type is an important subset of all densities acting on $\mathcal{H}_1 \otimes \mathcal{H}_2$. The elements of this set are called *disentangled* (or separable) states, and the other states are *entangled*. The phenomenon of *entanglement* is a remarkable feature of the Hilbert space probability theory.

The Hilbert space of a spin-$\frac{1}{2}$ is $\mathbb{C}^2$ which has a canonical basis $\{|\uparrow\rangle, |\downarrow\rangle\}$. The space $\mathbb{C}^2 \otimes \mathbb{C}^2$ describes a system of two spins and has the basis $|\uparrow\rangle \otimes |\uparrow\rangle$, $|\uparrow\rangle \otimes |\downarrow\rangle$, $|\downarrow\rangle \otimes |\uparrow\rangle$ and $|\downarrow\rangle \otimes |\downarrow\rangle$. The state vector

$$\Phi = \frac{1}{\sqrt{2}}(|\uparrow\rangle \otimes |\downarrow\rangle - |\downarrow\rangle \otimes |\uparrow\rangle) \tag{1.34}$$

bears the name "singlet" and plays a leading role in a paradox due to Einstein, Podolsky and Rosen. The reduced density of $|\Phi\rangle\langle\Phi|$ is

$$\tfrac{1}{2}|\uparrow\rangle\langle\uparrow| + \tfrac{1}{2}|\downarrow\rangle\langle\downarrow|.$$

$|\Phi\rangle\langle\Phi|$ is an extreme point in the convex set of density matrices, hence the convex decomposition

$$|\Phi\rangle\langle\Phi| = \sum_i \lambda_i D_1^i \otimes D_2^i$$

implies $|\Phi\rangle\langle\Phi| = D_1 \otimes D_2$. From the equality of the entropies we infer that $S(D_1) = 0$. Therefore, the reduced density matrix of the right hand side has 0 entropy, while the reduced density of $|\Phi\rangle\langle\Phi|$ is the most mixed state (with entropy $\log 2$). The contradiction shows that the vector state $|\Phi\rangle\langle\Phi|$ is entangled. The argument shows that all vector states are entangled except for the product ones.

The proof of Proposition 1.9 gives that equality holds in the subadditivity (1.14) if and only if $\varphi_{13} = \varphi_1 \otimes \varphi_3$. The case of equality in the strong subadditivity is more delicate, see the comments at the end of Chapter 9.

Theorem 1.18 is an abstract form of Shannon's *noiseless coding theorem* in the quantum setting, see Theorem 1.1 in [Csiszár and Körner 1981]. Schumacher completed this result by a coding-encoding procedure and obtained his *compression theorem* for memoryless sources, see [Schumacher 1995], or [Jozsa and Schumacher 1994]. The compression theorem has many extensions, see, for example, [Winter and Massar 2001].

(1.31) is related to the *hypothesis testing* problem. The state ω is the *null hypothesis* and φ is the *alternative hypothesis*. The problem is to decide which hypothesis is true and in the asymptotic situation the n-fold product states

ω_n and φ_n are given. The decision is performed by a two-valued measurement $\{A_n, I - A_n\}$, where $0 \le A_n \le I$ is an observable and A_n corresponds to the acceptance of ω_n and $I - A_n$ corresponds to the acceptance of φ_n. A_n is called a *test*, moreover $\alpha_n := \omega_n(I - A_n)$ is the *error of the first kind* and $\beta_n := \varphi_n(A_n)$ is the *error of the second kind*. Set

$$\beta^*(n, \varepsilon) = \inf\{\varphi_n(A_n) \; : \; A_n \in \mathcal{A}_n, \quad 0 \le A_n \le I, \quad \omega_n(I - A_n) \le \varepsilon\},$$

which is the infimum of the error of the second kind when the error of the first kind is at most ε. Obviously,

$$\beta^*(n, \varepsilon) \le R(n, \varepsilon),$$

since in the later minimization is over all tests A_n which are projections. (Such a test is called sharp.) The asymptotic behaviour of the two quantities is the same and the first analysis was done in [Hiai and Petz 1991]. In this paper

$$\limsup_{n \to \infty} \frac{1}{n} \log R(n, \varepsilon) \le -S(\omega, \varphi) \tag{1.35}$$

was proved in full generality and

$$\liminf_{n \to \infty} \frac{1}{n} \log R(n, \varepsilon) \ge -S(\omega, \varphi) \tag{1.36}$$

was conjectured on the basis of a weaker form. (1.36) was obtained in [Ogawa and Nagaoka 2000] by the method of quasi-entropies and they proved a strong converse of the *quantum Stein lemma* as well.

2 Postulates for Entropy and Relative Entropy

Von Neumann used arguments of phenomenological thermodynamics to deduce certain properties of the entropy of a state described by a statistical operator. His argument (reproduced also in the Introduction) supports the *mixing condition*

$$S(\lambda\varphi + (1 - \lambda)\psi)$$
$$= \lambda S(\varphi) + (1 - \lambda)S(\psi) - \lambda\log\lambda - (1 - \lambda)\log(1 - \lambda) \qquad (2.1)$$

(for $0 < \lambda < 1$ and disjoint states φ and ψ) and the property

$$S(\varphi) = 0 \quad \text{if} \quad \varphi \quad \text{is pure.} \qquad (2.2)$$

The conditions (2.1) and (2.2) determine a unique functional

$$S(\omega) = \operatorname{Tr}\eta(D_\omega),$$

on states of a finite quantum system. (Here $\eta(t) = -t\log t$.)

Creating the foundations of information theory in 1948, Shannon derived from certain desiderata the formula

$$H(p_1, p_2, \ldots, p_n) = -\sum_{k=1}^{n} p_k\log p_k = \sum_{k=1}^{n}\eta(p_k) \qquad (2.3)$$

for the amount of information contained in a single observation of a random variable ξ which takes the different values $x_1, x_2, \ldots, x_n$ with probabilities $p_k = \operatorname{Prob}(\xi = x_k)$ $(k = 1, 2, \ldots, n)$.

In the present chapter Shannon's axiomatic point of view will be adopted and we shall deduce the von Neumann entropy and the relative entropy from certain postulates. For the sake of simplicity we restrict ourselves to states of finite quantum systems (that is, finite dimensional C*-algebras).

The original set of axioms given by Shannon himself was simplified by Hinčin and Faddeev. In this latter form these axioms are as follows: Let Π denote the set of all finite probability distributions $(p_1, p_2, \ldots p_n)$ $(p_i \geq 0, \sum_i p_i = 1)$. Suppose that a function $I : \Pi \to \mathbb{R}$ satisfies the following conditions.

(i) Continuity: $p \mapsto I(p, 1 - p)$ is continuous on $[0, 1]$.
(ii) Normalization: $I(1/2, 1/2) = \log 2$.

(iii) Symmetry: $I(p_1, p_2, \ldots, p_n)$ is a symmetric function of its variables.
(iv) Recursion: For every $0 \le \lambda < 1$ we have $I(p_1, \ldots, p_{n-1}, \lambda p_n, (1 - \lambda)p_n) = I(p_1, \ldots, p_n) + p_n I(\lambda, 1 - \lambda)$.

Then $I(p_1, p_2, \ldots, p_n)$ must be the quantity $H(p_1, p_2, \ldots, p_n)$ defined by (2.3). Observe that axioms (i)–(iii) determine the function $p \mapsto I(p, 1 - p)$ and the recursion property (iv) tells us how to compute I for an $(n + 1)$-point distribution if the function I is already known for n-point distributions.

Assume that the functional S_0 associates to each state of a finite quantum systems a real number and that the invariance

$$S_0(\varphi \circ \alpha) = S_0(\varphi) \qquad (2.4)$$

holds for every $*$-algebra automorphism α. It follows from (2.4) that if φ is a state of a full matrix algebra then $S_0(\varphi)$ depends only on the eigenvalue list of D_φ. Postulating the properties corresponding to (i)–(iv), we can conclude that for the above mentioned φ we have

$$S_0(\varphi) = \operatorname{Tr} \eta(D_\varphi)$$

In other words, S_0 is the von Neumann entropy. (In fact, the symmetry property (iii) is contained in the invariance (2.4).) This axiomatic approach to the von Neumann entropy, which is based on the axiomatization of Shannon's entropy, is not natural from the point of view of physics, because it is difficult to imagine a reasonable physical interpretation of the recursive condition (iv). We therefore replace the recursion condition (iv) by a physically more acceptable one, which nevertheless, is not very different mathematically.

Below we intend to use the following convention which allows a simple notation. Let R be a functional defined on states of a finite quantum systems. For a probability distribution $(\lambda_1, \ldots, \lambda_n)$ let $R(\lambda_1, \ldots, \lambda_n)$ denote the value of R at the corresponding state of $\mathbb{C}^n$.

Theorem 2.1 Let S_0 be a real valued function on states of finite quantum systems satisfying the following conditions.

(i)   The invariance (2.4) holds.
(ii) $\varphi \mapsto S_0(\varphi)$ is continuous on the state space of any given system.
(iii) If $\varphi_1, \varphi_2, \ldots, \varphi_n$ are pairwise disjoint states and $(\lambda_1, \ldots, \lambda_n)$ is a probability distribution, then $S_0(\lambda_1 \varphi_1 + \lambda_2 \varphi_2 + \ldots + \lambda_n \varphi_n) = \sum_{i=1}^{n} \lambda_i S_0(\varphi_i) + S_0(\lambda_1, \ldots, \lambda_n)$.
(iv) If p is a projection such that $\varphi(p) = 1$ and φ is a state of the algebra $\mathcal{A}$, then $S_0(\varphi) = S_0(\varphi|p\mathcal{A}p)$.
(v)   With the above notation $S_0(1/2, 1/2) = \log 2$.

Then S_0 is identical to the von Neumann entropy.

Proof. First we show that conditions (iii) and (iv) imply the recursion property. The equality

$$S_0(1,0) = 1 \cdot S_0(1) + 0 \cdot S_0(1) + S_0(1,0)$$

is an application of (iii)–(iv) and gives $S_0(1) = 0$. Again by (iii) and (iv) we have

$$S_0(p_1, p_2, \ldots, p_{n-1}, \lambda p_n, (1-\lambda)p_n)$$
$$= p_n S_0(\lambda, 1-\lambda) + (1-p_n)S\left(\frac{p_1}{1-p_n}, \ldots, \frac{p_{n-1}}{1-p_n}\right) + S_0(p_n, 1-p_n)$$

and

$$S_0(p_1, p_2, \ldots, p_n)$$
$$= p_n S_0(1) + (1-p_n)S\left(\frac{p_1}{1-p_n}, \ldots, \frac{p_{n-1}}{1-p_n}\right) + S_0(p_n, 1-p_n).$$

Combination of these equations yields

$$S_0(p_1, p_2, \ldots, p_{n-1},\ \lambda p_n, (1-\lambda)p_n)$$
$$= S_0(p_1, p_2, \ldots, p_n) + p_n S_0(\lambda, 1-\lambda)$$

and $I(.) = S_0(.)$ satisfies all the conditions of the Hinčin-Faddeev characterization of Shannon's entropy of probability distributions. We arrive at

$$S_0(.) = H(.)$$

for every probability distribution. Condition (iv) readily provides

$$S_0(\psi) = S_0(1) = 0$$

for any pure state ψ. To finish the poof it is enough now to apply the mixing condition (iii) for the decomposition $\sum \lambda_i \varphi_i$ of a state φ into pairwise orthogonal pure states and obtain

$$S_0(\varphi) = S_0(\lambda_1, \ldots, \lambda_n) = H(\lambda_1, \ldots, \lambda_n)$$

which is von Neumann's entropy formula. □

The mixing condition (iii) in Theorem 2.1 is the same as von Neumann's condition (2.1). Note that equality in the mixing condition becomes inequality if the states are not assumed to be pairwise disjoint (see Proposition 1.6).

Let $\mathcal{A}$ and $\mathcal{B}$ be finite quantum systems. A linear mapping $\beta : \mathcal{A} \to \mathcal{B}$ is termed doubly stochastic if it is positive, unital and $\mathrm{Tr}_\mathcal{B}\alpha(a) = \mathrm{Tr}_\mathcal{A}a$ for every $a \in \mathcal{A}$. (So the existence of a doubly stochastic $\beta : \mathcal{A} \to \mathcal{B}$ implies that the algebraic dimension of $\mathcal{A}$ and $\mathcal{B}$ are the same.) It is known that

$$S(\varphi \circ \beta) \geq S(\varphi) \tag{2.5}$$

if β is doubly stochastic. This is the monotonicity property of the von Neumann entropy under doubly stochastic mappings (see (3.30)).

Theorem 2.2 Let S_0 be a real valued function on states of finite quantum systems such that

(i)   The monotonicity (2.5) holds.
(ii) $\varphi \mapsto S_0(\varphi)$ is continuous on the state space of the algebra $\mathbb{C}^2$.
(iii) If $\mathcal{B} = \mathcal{A}_1 \otimes \mathcal{A}_2$ and φ is a state of $\mathcal{B}$ then $S_0(\varphi) \leq S_0(\varphi|\mathcal{A}_1) + S_0(\varphi|\mathcal{A}_2)$, moreover, equality holds when φ is a product state.
(iv) If φ is a state of $\mathcal{B}$ and p is a projection in $\mathcal{B}$ such that $p \geq \text{supp } \varphi$ then
$S(\varphi) = S(\varphi|p\mathcal{B}\,p)$.

Then there exists a constant c such that $S_0 = c\,S$, i.e. S_0 is a multiple of the von Neumann entropy.

Proof. Let φ be a state on the finite quantum system $\mathcal{A}$ and $D_\varphi = \sum_i \lambda_i q_i$ the spectral decomposition of the density of φ. Consider the commutative subalgebra $\mathcal{A}_0$ generated by the spectral projections q_i. Since the embedding $\mathcal{A}_0 \to \mathcal{A}$ is doubly stochastic monotonicity (i) provides

$$S_0(\varphi|\mathcal{A}_0) \geq S_0(\varphi).$$

On the other hand,

$$\beta(a) = \sum_i (\text{Tr } aq_i)q_i$$

is a doubly stochastic mapping from $\mathcal{A}$ onto $\mathcal{A}_0$ as well. Hence $S_0(\varphi) \leq S_0(\varphi|\mathcal{A}_0)$ and we have

$$S_0(\varphi) = S_0(\varphi|\mathcal{A}_0).$$

By means of (i) we have succeeded in reducing the whole problem to commutative algebras. Now we consider S_0 on probability distributions. It also follows from (i) that $S_0(p_1, \ldots, p_n)$ $(\sum_i p_i = 1, p_i \geq 0)$ is a symmetric function of its arguments.

There exists a highly nontrivial result due to Aczél, Forte and Ng which says that a symmetric, additive and subadditive (condition (iii)) and also expansible (condition (iv)) function on the finite probability distributions has necessarily the form

$$S_0(p_1, p_2, \ldots, p_n) = c_1 H(p_1, p_2, \ldots, p_n) + c_2 \log N(p_1, \ldots, p_n)$$

where c_1 and c_2 are constants, $H(p_1, p_2, \ldots, p_n)$ is the Shannon entropy and $N(p_1, \ldots, p_n)$ is the number of non-zeros in $(p_1, p_2, \ldots, p_n)$. The functional $\log N$ (also called Hartley entropy) is obviously non-continuous. Hence the continuity hypothesis (ii) yields $c_2 = 0$. Up to a constant factor S_0 is the Shannon entropy on commutative algebras and the above reduction step shows that it is the von Neumann entropy in general. $\square$

Certainly the conditions of Theorem 2.2 are very strong and it seems very likely that the von Neumann entropy can be characterized by some weaker form of these postulates. However, this does not seem to be really significant, because all the postulates are physically pregnant properties of the entropy. Monotonicity may be replaced by other conditions forcing $S_0(\varphi)$ to be a symmetric function of the eigenvalues of the statistical operator D_φ.

The first effort to axiomatize quantum relative entropy was made by Donald in 1986. He took a modified version of von Neumann's conditions (2.1) and (2.2), and tried to give a heuristic probabilistic interpretation for them. The variant of (2.1) reads

$$S(\lambda\omega_1 + (1-\lambda)\omega_2, \varphi)$$
$$= \lambda S(\omega_1, \varphi) + (1-\lambda)S(\omega_2, \varphi) - \lambda \log \lambda - (1-\lambda)\log(1-\lambda) \qquad (2.6)$$

for $0 < \lambda < 1$ with disjoint states ω_1 and ω_2. Furthermore, he required that

$$S(\omega_0, \varphi) = -\omega_0(\log D_\varphi) \qquad (2.7)$$

for every pure state ω_0. Compared with (2.6) the interpretation of condition (2.7) is less natural. Nevertheless, one sees immediately that through the orthogonal decomposition of ω the postulates (2.6) and (2.7) give

$$S(\omega, \varphi) = \operatorname{Tr} D_\omega (\log D_\omega - \log D_\varphi).$$

Our crucial postulate for the relative entropy includes the notion of conditional expectation. Let us recall that in the setting of operator algebras conditional expectation (or projection of norm one) is defined as a positive unital idempotent linear mapping onto a subalgebra.

Let us list the properties of the relative entropy functional that will be used in an axiomatic characterization:

(i) *Conditional expectation property*: Assume that $\mathcal{A}$ is a subalgebra of $\mathcal{B}$ and that there exists a projection of norm one E of $\mathcal{B}$ onto $\mathcal{A}$ such that $\varphi \circ E = \varphi$. Then, for every state ω of $\mathcal{B}$, $S(\omega, \varphi) = S(\omega|\mathcal{A}, \varphi|\mathcal{A}) + S(\omega, \omega \circ E)$.

(ii) *Invariance property*: For every automorphism α of $\mathcal{B}$ we have $S(\omega, \varphi) = S(\omega \circ \alpha, \varphi \circ \alpha)$.

(iii) *Direct sum property*: Assume that $\mathcal{B} = \mathcal{B}_1 \oplus \mathcal{B}_2$. Let $\varphi_{12}(a \oplus b) = \lambda\varphi_1(a) + (1-\lambda)\varphi_2(b)$ and $\omega_{12}(a \oplus b) = \lambda\omega_1(a) + (1-\lambda)\omega_2(b)$ for every $a \in \mathcal{B}_1, b \in \mathcal{B}_2$ and some $0 < \lambda < 1$. Then $S(\omega_{12}, \varphi_{12}) = \lambda S(\omega_1, \varphi_1) + (1-\lambda)S(\omega_2, \varphi_2)$.

(iv) *Nilpotence property*: $S(\varphi, \varphi) = 0$.

(v) *Measurability property*: The function $(\omega, \varphi) \mapsto S(\omega, \varphi)$ is measurable on the state space of the finite dimensional C*-algebra $\mathcal{B}$ (when φ is assumed to be faithful).

The properties (i)–(v) are well-known properties of the relative entropy functional. Among them the most crucial is the conditional expectation property. The quantity

$$S_{\mathrm{BS}}(\omega, \varphi) = \operatorname{Tr} D_\omega \, \log(D_\omega^{1/2} D_\varphi^{-1} D_\omega^{1/2}) \qquad (2.8)$$

shares the properties (ii)–(v) and coincides with the relative entropy for commuting densities. (Further information on S_{BS} can be found in Chapter 7).

The relative entropy $S(\omega, \varphi)$ is the "informational divergence" of ω from φ. In this spirit the conditional expectation property has a rather natural interpretation. The informational divergence $S(\omega, \varphi)$ has two components. One component is the divergence of ω from φ on the subalgebra $\mathcal{A}$. The other component comes from the extension procedure of a state on the subalgebra to the whole algebra $\mathcal{B}$. Relative to a state φ (or rather to the conditional expectation E), the natural extension of $\omega|\mathcal{A}$ to $\mathcal{B}$ is obviously $\omega \circ E$. Hence the second component of $S(\omega, \varphi)$ is the informational divergence of ω from $\omega \circ E$. (Note that if the φ-preserving conditional expectation of $\mathcal{B}$ onto $\mathcal{A}$ does not exist then $\omega|\mathcal{A}$ does not have a natural extension to $\mathcal{B}$ relative to φ and in this case our argument does not lead to any conclusion.)

The interpretation of the invariance, direct sum and nilpotence properties is obvious. The measurability axiom is merely technical. The nilpotence property (iv) follows from the conditional expectation property: One has to choose the trivial subalgebra $\mathcal{A} = \mathcal{B}$.

Theorem 2.3 If a real valued functional $R(\omega, \varphi)$ defined for faithful states φ and arbitrary states ω of finite quantum systems shares the properties (i)–(v) then there exists a constant $c \in \mathbb{R}$ such that

$$R(\omega, \varphi) = c \operatorname{Tr} D_\omega (\log D_\omega - \log D_\varphi) .$$

Proof. The proof consists of several steps. We will show that $R(\omega, \varphi) = c\, S(\omega, \varphi)$ holds for increasing classes of states.

Consider the three dimensional commutative algebra $\mathbb{C}^3$. Its states correspond to probability distributions (p_1, p_2, p_3) (i.e. $0 \le p_i$, $p_1 + p_2 + p_3 = 1$).

$\square$

Lemma 2.4 For probability distributions (p_1, p_2, p_3) and (q_1, q_2, q_3) the recursive relation

$$R((p_1, p_2, p_3), (q_1, q_2, q_3)) = R((p_1 + p_2, p_3), (q_1 + q_2, q_3))$$
$$+ (p_1 + p_2) R\left(\left(\frac{p_1}{p_1 + p_2}, \frac{p_2}{p_1 + p_2} \right), \left(\frac{q_1}{q_1 + q_2}, \frac{q_2}{q_1 + q_2} \right) \right) \qquad (2.9)$$

holds.

We benefit from the conditional expectation property in the situation $\mathbb{C}^2 \simeq \{(c_1, c_1, c_2) : c_1, c_2 \in \mathbb{C}\} \subset \mathbb{C}^3$. There exists a conditional expectation $E : \mathbb{C}^3 \to \mathbb{C}^2$ preserving the state $\varphi = \varphi_{(q_1, q_2, q_3)}$ and it given by

$$E : (c_1, c_2, c_3) \mapsto \left(\frac{q_1 c_1 + q_2 c_2}{q_1 + q_2}, \frac{q_1 c_1 + q_2 c_2}{q_1 + q_2}, c_3 \right) .$$

The state $\omega_{(p_1, p_2, p_3)} \circ E$ corresponds to the measure

$$\left(\frac{p_1+p_2}{q_1+q_2}q_1, \frac{p_1+p_2}{q_1+q_2}q_2, p_3\right)$$

and we obtain

$$R((p_1,p_2,p_3),(q_1,q_2,q_3)) = R((p_1+p_2,p_3),(q_1+q_2,q_3))$$
$$+R\left((p_1,p_2,p_3),\left(\frac{p_1+p_2}{q_1+q_2}q_1, \frac{p_1+p_2}{q_1+q_2}q_2, p_3\right)\right).$$

Due to the direct sum condition the last term here equals to the last term of (2.9) and the lemma follows.   □

Interchanging in $\mathbb{C}^3$ the second and third coordinates by means of the invariance condition we conclude the equation

$$R((p_1+p_2,p_3),(q_1+q_2,q_3))$$
$$+(p_1+p_2)R\left(\left(\frac{p_1}{p_1+p_2},\frac{p_2}{p_1+p_2}\right),\left(\frac{q_1}{q_1+q_2},\frac{q_2}{q_1+q_2}\right)\right) =$$
$$R((p_1+p_3,p_2),(q_1+q_3,q_2)) \tag{2.10}$$
$$+(p_1+p_3)R\left(\left(\frac{p_1}{p_1+p_3},\frac{p_3}{p_1+p_3}\right),\left(\frac{q_1}{q_1+q_3},\frac{q_3}{q_1+q_3}\right)\right).$$

With the notation

$$F(x,y) = R((1-x,x),(1-y,y))$$

equation (2.10) is of the following form:

$$F(x,y) + (1-x)F\left(\frac{u}{1-x},\frac{v}{1-y}\right)$$
$$= F(u,v) + (1-u)F\left(\frac{x}{1-u},\frac{y}{1-v}\right). \tag{2.11}$$

A lengthy but elementary analysis yields that the only measurable solution of the functional equation (2.11) is

$$F(x,y) = c\left(x\log\frac{x}{y} + (1-x)\log\frac{1-x}{1-y}\right).$$

(See pp. 204–207 of [Aczél and Daróczy 1975].)

The recursion (2.9) remains true if p_3 and q_3 are replaced by $p_3,p_4,\ldots,p_n$ and $q_3,q_4,\ldots;q_n$ respectively. In this way we obtain that $R(\omega,\varphi) = c\,S(\omega,\varphi)$ whenever φ and ω are states of a commutative finite dimensional C*-algebra.

Now let φ and ω be states of an algebra $\mathcal{B}$ such that the densities D_φ and D_ω commute. Let $\mathcal{A}$ be the maximal abelian subalgebra generated by D_φ and D_ω. If E is the conditional expectation of $\mathcal{B}$ onto $\mathcal{A}$ that preserves Tr then $\varphi\circ E = \varphi$ and $\omega\circ E = \omega$. The conditional expectation property tells us that

$$R(\omega,\varphi) = R(\omega|\mathcal{A},\varphi|\mathcal{A}) + R(\omega,\omega\circ E).$$

By the nilpotence the second term vanishes and we arrive at

$$R(\omega, \varphi) = c\, S(\omega, \varphi) \tag{2.12}$$

for commuting states.   □

The next step is $\mathcal{B} = M_n(\mathbb{C})$. Our aim is to show that (2.12) holds for arbitrary states on $\mathcal{B}$. (As always, φ is supposed to be faithful.)

Lemma 2.5 If $\sigma = \lambda\sigma_1 + (1-\lambda)\sigma_2 \quad (0 < \lambda < 1)$ then

$$\lambda R(\sigma_1, \varphi) + (1-\lambda)R(\sigma_2, \varphi)$$
$$= R(\sigma, \varphi) + \lambda R(\sigma_1, \sigma) + (1-\lambda)R(\sigma_2, \sigma). \tag{2.13}$$

Proof. The proof of (2.13) is quite similar to that of Lemma 2.4. The conditional expectation property should be applied to $\mathcal{B} \oplus \mathcal{B}$ and to its diagonal subalgebra $\{b \oplus b : b \in \mathcal{B}\}$. The mapping

$$E(a \oplus b) = (\lambda a + (1-\lambda)b) \oplus (\lambda a + (1-\lambda)b)$$

is a conditional expectation leaving the state

$$\varphi_{12}(a \oplus b) = \lambda\varphi(a) + (1-\lambda)\varphi(b)$$

invariant. The argument is completed by reference to the invariance and direct sum properties.   □

Now we resume the determination of $S(\omega, \varphi)$ for states of $\mathcal{B} = M_n(\mathbb{C})$. We choose a basis such that the density of φ is diagonal. Then the density of ω is of the form

$$\begin{pmatrix} A_k & 0 \\ 0 & D_{n-k} \end{pmatrix} = D_\omega$$

where $A_k \in M_k(\mathbb{C})$ and $D_{n-k} \in M_{n-k}(\mathbb{C})$ is a diagonal matrix. We are going to prove (2.12) by mathematical induction on k. If $k = 0$, then D_φ and D_ω commute, and (2.12) holds. Now we fix an integer $k > 0$ and carry out the induction step from $k - 1$ to k.

Let U be a diagonal unitary matrix such that

$$U_{ii} = \begin{cases} 1, & \text{if } i \neq k; \\ -1, & \text{if } i = k. \end{cases}$$

Then

$$D_{\sigma_2} = U D_\omega U \equiv \mathrm{Ad}_U D_\omega$$

differs from $D_{\sigma_1} \equiv D_\omega$ only in the sign of the entries in the kth row and in the kth column, except for the diagonal entry. The density

$$D_\sigma = \tfrac{1}{2}(D_{\sigma_1} + D_{\sigma_2})$$

is of the form

$$\begin{pmatrix} A_{k-1} & 0 \\ 0 & D_{n-k+1} \end{pmatrix}$$

where D_{n-k+1} is an $(n-k+1) \times (n-k+1)$ diagonal matrix. From the induction hypothesis we have

$$R(\sigma, \varphi) = c\, S(\sigma, \varphi)\,. \tag{2.14}$$

Write (2.13) with $\tau = \mathrm{Tr}\,/n$ and $\lambda = \frac{1}{2}$. Then

$$\tfrac{1}{2} R(\sigma_1, \tau) + \tfrac{1}{2} R(\sigma_2, \tau) = R(\sigma, \tau) + \tfrac{1}{2} R(\sigma_1, \sigma) + \tfrac{1}{2} R(\sigma_2, \sigma)\,. \tag{2.15}$$

The invariance, more precisely, the set of relations $\tau(U\,.\,U) = \tau(\,.\,)$, $\sigma_1(U\,.\,U) = \sigma_2(\,.\,)$, $\sigma(U\,.\,U) = \sigma(\,.\,)$, ensures

$$R(\sigma_1, \tau) = R(\sigma_2, \tau) \quad \text{and} \quad R(\sigma_1, \sigma) = R(\sigma_2, \sigma)\,.$$

Therefore

$$R(\sigma_1, \tau) = R(\sigma, \tau) + R(\sigma_1, \sigma)\,,$$

and (2.12) yields

$$R(\sigma_1, \sigma) = c\, S(\sigma_1, \sigma)\,. \tag{2.16}$$

In our special case

$$R(\sigma_1, \varphi) = R(\sigma_2, \varphi) \quad \text{and} \quad R(\sigma_1, \sigma) = R(\sigma_2, \sigma)$$

hold due to the invariance under the automorphism Ad_U. Hence (2.13) reads as

$$R(\sigma_1, \varphi) = R(\sigma, \varphi) + R(\sigma_1, \sigma)\,,$$

where both terms on the right hand side have been compared with the relative entropy ((2.14) and (2.16)). So (2.12) holds for all states φ and ω on $M_n(\mathbb{C})$. Since a finite dimensional C*-algebra is a direct sum of full matrix algebras, the direct sum property extends (2.12) to all C*-algebras of finite dimension, and the proof Theorem 2.3 is complete. $\qquad\qquad\square$

By more complicated mathematical tools the concept of relative entropy may be extended to states of infinite dimensional C*-algebras. If we restrict ourselves to C*-algebras with good approximation property (nuclear algebras), then the above characterization may be modified easily. It is sufficient to choose instead of (ii) the monotonicity and instead of measurability the lower semi-continuity property given below.

(ii') *Monotonicity property:* For every completely positive unital mapping α we have $S(\omega, \varphi) \geq S(\omega \circ \alpha, \varphi \circ \alpha)$.

(v') *Lower semicontinuity:* The functional $(\omega, \varphi) \mapsto S(\omega, \varphi)$ is lower semicontinuous with respect to the weak* topology.

Postulates (i), (ii$'$), (iii), (iv) and (v$'$) constitute the definition of the relative entropy functional up to a constant factor.

Notes and Remarks. Von Neumann has showed that any pure state may be converted reversibly and adiabaticaly into any other pure state. Hence the same entropy has to be assigned to all pure states and (2.2) may be regarded as a kind of scaling. The monograph [Aczél and Daróczy 1975] contains various characterizations of the Shannon entropy. Formally these results may be transformed into the context of finite quantum systems. However, from the physical point of view, the two characterization theorems of this chapter seem to be the only natural ones. Theorem 2.1 is a variant of 2.2.4 in [Thirring 1983] and Theorem 2.2 in the quantum setting is due to [Ochs 1975]. In the proof of Theorem 2.2 the name of Hartley appears, the idea of measuring information regardless its content dates back to him (1928). Theorem 2.3 appears in [Petz 1992a] but it is strongly related to the work [Donald 1987a]. Condition (2.13) is one of Donald's axioms and the subsequent mathematical induction comes from him. The conditional expectation property was obtained in [Petz 1986a] and it is proven here as Theorem 1.13 and 5.15. Formula (2.8) for the relative entropy was suggested in [Belavkin and Staszewski 1982], see also Chapter 7. Characterization of the relative entropy of states of infinite dimensional operator algebras is treated in [Petz 1994b].

3 Convex Trace Functions

This chapter is concerned with certain convex functionals $F : \mathcal{A} \to \mathbb{R}$ defined on a finite quantum system $\mathcal{A}$. We recall that F is convex if

$$F(\lambda A + (1 - \lambda)B) \leq \lambda F(A) + (1 - \lambda)F(B)$$

for $0 < \lambda < 1$ and F is termed concave if $-F$ is convex. These functionals involve the trace operation which plays an important role in quantum statistical mechanics, and all the inequalities discussed here have been motivated by considerations in physics. The simplest example for the functionals we have in mind is the functional

$$D \mapsto \operatorname{Tr} \eta(D)$$

defined on the positive part $\mathcal{A}^+$ of a finite quantum system $\mathcal{A}$ and is related to von Neumann's entropy. (Here $\eta(t) = -t \log t$ and if $\sum_i \lambda_i p_i$ is the spectral decomposition of D then $\eta(D)$ is understood as $\sum_i \eta(\lambda_i)p_i$.) Von Neumann recognized that the concavity of the entropy functional is based on the fact that η is a concave function on $\mathbb{R}^+$. More precisely, he observed the following.

Proposition 3.1 Let $\mathcal{A}$ be a finite quantum system and $f : [\alpha, \beta] \to \mathbb{R}$ a convex function. Then the functional

$$F(A) = \operatorname{Tr} f(A)$$

is convex on the set $\{A \in \mathcal{A}^{sa} : \operatorname{Sp}(A) \subset [\alpha, \beta]\}$.

Proof. First we note that for a pairwise orthogonal family (p_i) of minimal projections with $\sum p_i = I$ we have

$$\operatorname{Tr} f(B) \geq \sum_i f(\operatorname{Tr} B \, p_i). \tag{3.1}$$

Indeed, using the convexity of f we deduce (3.1) as follows. Let $\sum_j s_j q_j$ be the spectral decomposition of B. Then

$$\operatorname{Tr} f(B) = \sum_j f(s_j) \operatorname{Tr} q_j = \sum_i \sum_j f(s_j) \operatorname{Tr} q_j p_i$$

$$\geq \sum_i f(\textstyle\sum_j s_j \operatorname{Tr} q_j p_i) = \sum_i f(\operatorname{Tr} B \, p_i).$$

To prove the proposition we write $\sum_i \mu_i p_i$ for the spectral decomposition of $A = \lambda B_1 + (1 - \lambda) B_2$. Applying (3.1) twice we infer

$$\lambda \operatorname{Tr} f(B_1) + (1 - \lambda) \operatorname{Tr} f(B_2)$$
$$\geq \lambda \sum_i f(\operatorname{Tr} B_1 p_i) + (1 - \lambda) \sum_i f(\operatorname{Tr} B_2 p_i)$$
$$\geq \sum_i f(\lambda \operatorname{Tr} B_1 p_i + (1 - \lambda) \operatorname{Tr} B_2 p_i) = \sum_i f(\operatorname{Tr} A p_i) = \operatorname{Tr} f(A),$$

which is the convexity of the functional F.   □

We next turn to Lieb's concavity theorem, which is much deeper than the previous proposition. It originated from the convexity of a skew information quantity introduced by Wigner and Yanase.

Proposition 3.2 On the positive part $\mathcal{A}^+$ of a finite quantum system $\mathcal{A}$ the functional

$$A \mapsto \operatorname{Tr} A^t K A^{1-t} K^*$$

is concave for every $0 < t < 1$ and $K \in \mathcal{A}$.

Proof. Let $0 < \lambda < 1$ and $A = \lambda A_1 + (1 - \lambda) A_2$. We are going to show that

$$\operatorname{Tr} A^t K A^{1-t} K^* \geq \lambda \operatorname{Tr} A_1^t K A_1^{1-t} K^* + (1 - \lambda) \operatorname{Tr} A_2^t K A_2^{1-t} K^*. \tag{3.2}$$

By simple approximation one can assume that A_1, A_2 and K are invertible.

Endowed with the Hilbert-Schmidt inner product $\langle A, B \rangle = \operatorname{Tr} A^* B$ the linear space $\mathcal{A}$ becomes a Hilbert space $\mathcal{H}$. We write $\mathcal{K}$ for $\mathcal{H} \oplus \mathcal{H}$. It is straightforward to verify that with fixed K and A the mapping

$$T : BKA^{1/2} \mapsto \sqrt{\lambda} BKA_1^{1/2} \oplus \sqrt{1 - \lambda} BKA_2^{1/2} \qquad (B \in \mathcal{A})$$

is an isometry $T : \mathcal{H} \to \mathcal{K}$. (Note that $KA^{1/2}$ is invertible and so T is everywhere defined.)

Introduce the operators $\Delta_{12} \in B(\mathcal{K})$ and $\Delta \in B(\mathcal{H})$ by the formulae

$$\Delta_{12}(X \oplus Y) = A_1 X A_1^{-1} \oplus A_2 Y A_2^{-1} \qquad (X \oplus Y \in \mathcal{K}),$$
$$\Delta B = ABA^{-1} \qquad (B \in \mathcal{H}).$$

Both Δ and Δ_{12} are positive operators. They have the following properties.

(i) $\langle \Delta^t K A^{1/2}, K A^{1/2} \rangle = \operatorname{Tr} A^t K A^{1-t} K^*$,
(ii) $\langle \Delta_{12}^t (\sqrt{\lambda} K A_1^{1/2} \oplus \sqrt{1 - \lambda} K A_2^{1/2}), (\sqrt{\lambda} K A_1^{1/2} \oplus \sqrt{1 - \lambda} K A_2^{1/2}) \rangle$
      $= \lambda \operatorname{Tr} A_1^t K A_1^{1-t} K^* + (1 - \lambda) \operatorname{Tr} A_2^t K A_2^{1-t} K^*$,
(iii) $T^* \Delta_{12} T = \Delta$.

It follows from these properties and from the definition of T that our claim (3.2) is equivalent to the inequality

$$\langle \Delta_{12}^t T K A^{1/2}, T K A^{1/2} \rangle \leq \langle \Delta^t K A^{1/2}, K A^{1/2} \rangle. \tag{3.3}$$

According to Lemma 1.2 we have

$$T^* \Delta^t_{12} T \le (T^* \Delta_{12} T)^t = \Delta^t$$

which clearly implies (3.3).   □

Corollary 3.3 For every $k \in \mathcal{A}$ and $0 < t < 1$ the functional

$$(a, b) \mapsto \mathrm{Tr}\, a^t k b^{1-t} k^*$$

is concave on $\mathcal{A}^+ \times \mathcal{A}^+$.

Proof. Let $\mathcal{B} = \mathcal{A} \otimes M_2(\mathbb{C})$ be the *-algebra of 2×2 matrices with entries in $\mathcal{A}$. If we apply Proposition 3.2 for

$$A = \begin{pmatrix} a & 0 \\ 0 & b \end{pmatrix} \quad \text{and} \quad K = \begin{pmatrix} 0 & 0 \\ k & 0 \end{pmatrix}$$

then the corollary is obtained.   □

Since

$$\frac{d}{dt} \mathrm{Tr}\, A^t B^{1-t} \Big|_{t=0} = -\mathrm{Tr}\, B(\log B - \log A)$$

Lieb's concavity theorem yields the joint convexity of the relative entropy, Theorem 1.4. Wigner, Yanase and Dyson suggested the concept of "skew information" of a state φ relative to a self-adjoint observable K in the form

$$I_p(\varphi, K) = \mathrm{Tr}\, D_\varphi K^2 - \mathrm{Tr}\, D_\varphi^{1-p} K D_\varphi^p K \qquad (3.4)$$

where $0 < p < 1$. The convexity of $I_p(\varphi, K)$ as a function of φ is a necessary property for $I_p(\varphi, K)$ for being regarded a reasonable notion of information. The first term in (3.4) is linear and the second term is convex due to Proposition 3.2. So convexity in φ is ensured.

Let us write $\mathcal{A}^{++}$ for the positive invertible elements of $\mathcal{A}$. For $A \in \mathcal{A}^{++}$ define a linear transformation of $\mathcal{H}$ by

$$T_A(K) = \int_0^\infty (t + A)^{-1} K (t + A)^{-1} \, dt \,. \qquad (3.5)$$

Since

$$Q(A, K) \equiv \langle T_A(K), K \rangle = \int_0^\infty \mathrm{Tr}\, (t + A)^{-1} K (t + A)^{-1} K^* \, dt \qquad (3.6)$$

is nonnegative, the operator T_A is positive (definite). In a basis in which A is diagonal one can compute T_A explicitly. Namely,

$$(T_A(K))_{ij} = K_{ij} / \mathrm{Lm}\, (a_i, a_j)$$

where the a_i's are the eigenvalues of A and $\mathrm{Lm}\, (x, y)$ stands for the so-called logarithmic mean

$$\mathrm{Lm}\,(x,y) = \begin{cases} (x-y)/(\log x - \log y) & x \neq y \\ x & x = y \end{cases}$$

(One knows that $\sqrt{xy} \leq \mathrm{Lm}\,(x,y) \leq \frac{1}{2}(x+y)$.) The inverse transformation is

$$(T_A^{-1}(K))_{ij} = K_{ij}\mathrm{Lm}\,(a_i, a_j)\,,$$

but this is the same as

$$T_A^{-1}(K) = \int_0^1 A^t K A^{1-t}\,dt\,, \tag{3.7}$$

as we can see by calculating the integral.

Proposition 3.4 For a finite quantum system $\mathcal{A}$ the functional

$$(A, K) \mapsto Q(A, K)$$

(defined in (3.6)) is jointly convex on the domain $\mathcal{A}^{++} \times \mathcal{A}$.

Proof. Keeping the notation $\mathcal{K} = \mathcal{H} \oplus \mathcal{H}$ from the proof of Proposition 3.2 we set $A = \lambda A_1 + (1-\lambda)A_2$ and define two quadratic forms on $\mathcal{K}$ as follows.

$$D(K_1 \oplus K_2) = \lambda Q(A_1, K_1) + (1-\lambda)Q(A_2, K_2)$$
$$N(K_1 \oplus K_2) = Q(A, \lambda K_1 + (1-\lambda)K_2).$$

Note that both forms are non-degenerate. In terms of D and N the majorization $N \leq D$ is to be shown.

Let d and n be the corresponding sesquilinear forms on $\mathcal{K}$ that is

$$D(\xi) = d(\xi, \xi), \quad N(\xi) = n(\xi, \xi) \quad (\xi \in \mathcal{K})\,.$$

There exists an operator X on $\mathcal{K}$ such that

$$d(\xi, \eta) = n(X\xi, \eta) \quad (\xi, \eta \in \mathcal{K})$$

and our aim is to show that its eigenvalues are not less than 1. If $X(K \oplus L) = \gamma(K \oplus L)$, we have

$$d(K \oplus L, K' \oplus L') = \gamma n(K \oplus L, K' \oplus L')$$

for every K', $L' \in \mathcal{H}$. This is rewritten in terms of the Hilbert-Schmidt inner product as

$$\lambda\langle T_{A_1}(K), K'\rangle + (1-\lambda)\langle T_{A_2}(L), L'\rangle$$
$$= \gamma\langle T_A(\lambda K + (1-\lambda)L), \lambda K' + (1-\lambda)L'\rangle\,,$$

which is equivalent to the equations

$$T_{A_1}(K) = \gamma T_A(\lambda K + (1-\lambda)L)$$

and

$$T_{A_2}(L) = \gamma T_A(\lambda K + (1-\lambda)L)\,.$$

We infer

$$T_A^{-1}(M) = \lambda T_{A_1}^{-1}(\gamma M) + (1 - \lambda)T_{A_2}^{-1}(\gamma M)$$

with the new notation $M \equiv T_A(\lambda K + (1 - \lambda)L)$. In another form

$$\int_0^1 A^t M A^{1-t} \, dt = \gamma\lambda \int_0^1 A_1^t M A_1^{1-t} \, dt + \gamma(1 - \lambda) \int_0^1 A_2^t M A_2^{1-t} \, dt.$$

To finish the proof we multiply by M^*, take the trace and use Proposition 3.2:

$$\gamma\lambda \int_0^1 \text{Tr}\, A_1^t M A_1^{1-t} M^* \, dt + \gamma(1 - \lambda) \int_0^1 \text{Tr}\, A_2^t M A_2^{1-t} M^* \, dt$$

$$= \int_0^1 \text{Tr}\, A^t M A^{1-t} M^* \, dt$$

$$\geq \int_0^1 (\lambda \text{Tr}\, A_1^t M A_1^{1-t} M^* + (1 - \lambda)\text{Tr}\, A_2^t M A_2^{1-t} M^* \, dt.$$

Now, clearly, $\gamma \geq 1$ must hold.   □

For $A \in \mathcal{A}^{++}$ and $K \in \mathcal{A}^{sa}$ the operator valued function

$$x \mapsto \log(A + xK)$$

is real analytic in a neighbourhood of $0 \in \mathbb{R}$. From the integral representation

$$\log B = \int_0^\infty (1 + t)^{-1} - (t + B)^{-1} \, dt$$

and the formula

$$\frac{d}{dx}(A + xK)^{-1}\Big|_{x=0} = -A^{-1}KA^{-1}$$

one can deduce

$$\frac{d}{dx} \log(A + xK)\Big|_{x=0} = \int_0^\infty (A + t)^{-1}K(A + t)^{-1} \, dt = T_A(K). \qquad (3.8)$$

One computes similarly that

$$\frac{d^2}{dx^2} \log(A + xK)\Big|_{x=0}$$

$$= -2 \int_0^\infty (A + t)^{-1}K(A + t)^{-1}K(A + t)^{-1} \, dt \equiv -R_A(K). \qquad (3.9)$$

The following lemma is rather straightforward but it will be useful in the sequel.

Lemma 3.5 Let $\mathcal{C}$ be a convex cone in a vector space and $F : \mathcal{C} \to \mathbb{R}$ a convex function such that

(i) $\lim_{x \to +0} x^{-1}(F(A + xB) - F(A)) \equiv G(A, B)$ exists for all $A, B \in \mathcal{C}$
(ii) $F(\lambda A) = \lambda F(A)$ for every $\lambda > 0$ and $A \in \mathcal{C}$.

Then $G(A, B) \leq F(B)$.

Proof. The homogeneity and convexity of F yield

$$F(A + xB) = (1 + x)F((1 + x)^{-1}A + x(1 + x)^{-1}B) \leq F(A) + xF(B).$$

Now subtract $F(A)$ from both sides of the inequality, divide it by $x > 0$ and take the limit $x \to +0$. □

Lemma 3.6 Let $\mathcal{A}$ be a finite quantum system. For $A, B \in \mathcal{A}^{++}$ and $K, M \in \mathcal{A}^{sa}$ the inequality

$$-\operatorname{Tr} BR_A(K) + 2\operatorname{Tr} MT_A(K) \leq \operatorname{Tr} MT_B(M)$$

holds.

Proof. Let us consider the functional

$$(A, K) \mapsto Q(A, K) = \operatorname{Tr} T_A(K)K^*$$

which is convex due to Proposition 3.4. It is also homogeneous of order 1:

$$Q(\lambda A, \lambda K) = \int_0^\infty \operatorname{Tr} (A + t/\lambda)^{-1}K(A + t/\lambda)^{-1}K^* \, dt = \lambda Q(A, K).$$

Application of the previous lemma yields

$$\lim_{x \to +0} x^{-1}Q(A + xB, K + xM) - Q(A, K) \leq Q(B, M) \tag{3.10}$$

provided that the limit exists. We compute

$$\lim_{x \to +0} x^{-1}(Q(A + xB, K + xM) - Q(A, K + xM))$$
$$= \lim_{x \to +0} x^{-1}(Q(A + xB, K) - Q(A, K)) = -\operatorname{Tr} BR_A(K) \tag{3.11}$$

and one obtains from

$$Q(A, K + xM) - Q(A, K)$$
$$= x\operatorname{Tr} T_A(K)M + x\operatorname{Tr} T_A(M)K + x^2\operatorname{Tr} T_A(M)M$$

the relation

$$\lim_{x \to +0} x^{-1}(Q(A, K + xM) - Q(A, K)) = 2\operatorname{Tr} MT_A(K). \tag{3.12}$$

(3.11) and (3.12) supply the left hand side of (3.10) and provide the proof the lemma. □

Now we are in a position to prove a deep result due to Lieb.

Theorem 3.7 Let $\mathcal{A}$ be a finite quantum system and $L \in \mathcal{A}^{sa}$ a fixed observable. Then the functional

$$A \mapsto F_L(A) = \operatorname{Tr} \exp(L + \log A)$$

defined on $\mathcal{A}^{++}$ is concave.

Proof. The function

$$f(x) = \text{Tr} \exp(L + \log(A + xK))$$

of the real variable x is differentiable in a neighbourhood of 0. The theorem is equivalent to the statement

$$\frac{d^2 f}{dx^2}\bigg|_{x=0} \leq 0$$

for all choices of $A \in \mathcal{A}^{++}$ and $K, L \in \mathcal{A}^{sa}$. Indeed, if f is concave then putting $A = A_2$ and $K = A_1 - A_2$ we get

$$\text{Tr} \exp(L + \log(A_2 + \lambda(A_1 - A_2)))$$
$$\geq \lambda \text{Tr} \exp(L + \log(A_2 + (A_1 - A_2))) + (1 - \lambda)\text{Tr} \exp(L + \log A_2).$$

To compute the second derivative of f we may benefit from the norm convergent expansions

$$\log(A + xK) = \log A + x\, T_A(K) - \tfrac{1}{2}x^2 R_A(K) + \ldots$$

and

$$e^{a+b} = e^a + \int_0^1 e^{(1-s)a} b\, e^{sa}\, ds + \int_0^1 \int_0^s e^{(1-s)a} b\, e^{(s-u)a} b\, e^{ua}\, du\, ds + \ldots\,.$$

With the simplification $B \equiv \exp(L + \log A)$ we have

$$\exp(L + \log(A + xK)) = B + x \int_0^1 B^{1-s} T_A(K) B^s\, ds$$

$$+ x^2 \left(-\frac{1}{2}\int_0^1 B^{1-s} R_A(K) B^s\, ds\right.$$

$$\left. + \int_0^1 \int_0^s B^{1-s} T_A(K) B^{s-u} T_A(K) B^u du\, ds\right) + \ldots$$

and

$$\frac{d^2 f}{dx^2}\bigg|_{x=0} = -\text{Tr}\, B R_A(K) + \text{Tr} \int_0^1 T_A(K) B^y T_A(K) B^{1-y}\, dy$$

$$= -\text{Tr}\, B R_A(K) + \text{Tr}\, T_A(K) T_B^{-1}(T_A(K)).$$

To complete the proof we use Lemma 3.6 which gives

$$\frac{d^2 f}{dx^2}\bigg|_{x=0} \leq \text{Tr}\, M T_B(M) - \text{Tr}\, M T_A(K) = 0$$

where M stands for $T_B^{-1}(T_A(K))$. $\square$

It may be noted that the value $\varphi^h(I)$ at I of the unnormalized perturbed functional of a state φ on a finite quantum system has the form $F_L(A)$ with $A = D_\varphi$, $L = h$. Hence $\varphi^h(I)$ and $\log \varphi^h(I)$ are concave functionals on the state space.

In statistical mechanics Golden has proved that if A and B are hermitian and nonnegative definite matrices then the inequality

$$\operatorname{Tr} e^A e^B \geq \operatorname{Tr} e^{A+B} \tag{3.13}$$

holds. He observed that this inequality may be used to obtain lower bounds for the Helmholtz free-energy function by partitioning of the Hamiltonian. Independently, Thompson proved (3.13) for hermitian A and B without the requirement of definiteness and applied the inequality to obtain an upper bound for the partition function of an anti-ferromagnetic chain. Nowadays (3.13) is termed *Golden-Thompson inequality* and it is a basic tool in quantum statistical mechanics. It is really an inequality except for the trivial case $[A, B] = 0$ and one may deduce it from the monotonicity of the relative entropy (see Chapter 7).

An extension of the Golden-Thompson inequality (due to Lieb) may be derived from Theorem 3.7.

Theorem 3.8 Let $\mathcal{A}$ be a finite quantum system and $A, B, C \in \mathcal{A}^{sa}$. Then the inequality

$$\operatorname{Tr} e^C \, T_{\exp(-A)} \, (e^B) \geq \operatorname{Tr} e^{A+B+C}$$

holds, and, for commuting A and B, it reads

$$\operatorname{Tr} e^C e^A e^B \geq \operatorname{Tr} e^{A+B+C} \, .$$

Proof. We recall that $T_X(Y)$ was defined in (3.5). We choose $L = A + C$, $\alpha = e^{-A}$, $\beta = e^B$ and conclude from Theorem 3.7 that the functional

$$\alpha \mapsto -\operatorname{Tr} e^{L+\log \alpha}$$

is convex on the cone $\mathcal{A}^{++}$. It is also homogeneous of order 1 and the hypothesis of Lemma 3.5 is fulfilled. So

$$-\operatorname{Tr} e^{A+B+C} = -\operatorname{Tr} e^{L+\log \beta} \geq -\frac{d}{dx} \operatorname{Tr} e^{L+\log(\alpha+x\beta)}\Big|_{x=0}. \tag{3.14}$$

The power series of $e^{L+\log(\alpha+x\beta)}$ starts with the terms

$$e^C + x \int_0^1 e^{(1-s)C} T_\alpha(\beta) e^{sC} \, ds$$

$$+ x^2 \int_0^1 \int_0^s e^{(1-s)C} T_\alpha(\beta) e^{(s-u)C} T_\alpha(\beta) e^{uC} \, du \, ds \, .$$

Hence the derivative in (3.14) is $-\operatorname{Tr} e^C T_\alpha(\beta)$ and the first inequality is shown.

If A commutes with B then simply

$$T_\alpha(\beta) = \int_0^\infty (e^{-A} + t)^{-1} e^B (e^{-A} + t)^{-1} \, dt$$

$$= e^B \int_0^\infty (e^{-A} + t)^{-2} \, dt = e^B e^A$$

and this gives the second statement of the theorem.   □

The obvious generalization

$$\mathrm{Tr}\, e^{A+B+C} \leq \mathrm{Tr}\, e^A e^B e^C$$

of the Golden-Thompson inequality is false. (However, if two of the three matrices commute then it is true as it was stated in the last theorem.) Because of its importance we present another proof of the inequality, which, on the other hand, shows a different generalization. We need the following two matrix results, which we use here without proof.

Lemma 3.9 If $\alpha_1, \alpha_2, \ldots, \alpha_n$ and $\kappa_1, \kappa_2, \ldots, \kappa_n$ are the eigenvalues of the $n \times n$ matrices x and xx^* respectively, arranged so that $|\alpha_1| \geq |\alpha_2| \geq \ldots \geq |\alpha_n| \geq 0$ and $\kappa_1 \geq \kappa_2 \geq \ldots \kappa_n \geq 0$, then for any increasing function $f(x)$ of the positive argument x such that $f(e^t)$ is a convex function of t,

$$\sum_{i=1}^{k} f(\kappa_i) \geq \sum_{i=1}^{k} f(|\alpha_i|^2) \qquad k = 1, 2, \ldots, n\,.$$

Lemma 3.10 For any complex $n \times n$ matrices A and B,

$$\lim_{s \to \infty} (e^{A/s} e^{B/s})^s = e^{A+B}\,.$$

With the choice $f(x) = x^s$ and $k = n$ Lemma 3.9 yields

$$\mathrm{Tr}\,(XX^*)^s \geq |\mathrm{Tr}\, X^{2s}|\,. \tag{3.15}$$

Theorem 3.11 If $\mathcal{A}$ is a finite quantum system then

$$\mathrm{Tr}\, e^{(A+A^*)/2} e^{(B+B^*)/2} \geq |\mathrm{Tr}\, e^{A+B}|$$

for every $A, B \in \mathcal{A}$.

Proof. Since $\mathcal{A}$ may be represented by matrices the exponential product formula in Lemma 3.10 and (3.15) are applicable. Substituting $X = AB$ into (3.15) we have

$$\mathrm{Tr}\,((ABB^*A^*)^s) \geq |\mathrm{Tr}\,((AB)^{2s})|\,,$$

where the left hand side is nothing else but $\mathrm{Tr}\,(BB^*A^*A)^s$. Setting $s = 2^{k-1}$ with a positive integer k and using (3.15) gives

$$|\mathrm{Tr}\,(AB)^{2^k}| \leq \mathrm{Tr}\,(BB^*A^*A)^{2^{k-1}} \leq |\mathrm{Tr}\,((BB^*A^*A)^2)^{2^{k-2}}|$$
$$\leq \mathrm{Tr}\,((BB^*A^*A)(BB^*A^*A)^*)^{2^{k-2}}$$
$$= \mathrm{Tr}\,((A^*A)^2(BB^*)^2)^{2^{k-2}}\,.$$

By a repeated application of this argument we easily infer

$$\mathrm{Tr}\,((A^*A)^{2^{k-1}}(BB^*)^{2^{k-1}}) \geq |\mathrm{Tr}\,((AB)^{2^k})|\,.$$

Now replace A by $\exp(2^{-k}A)$ and B by $\exp(2^{-k}B)$:

$$\operatorname{Tr}\left((e^{2^{-k}A^*}e^{2^{-k}A})^{2^k/2}(e^{2^{-k}B}e^{2^{-k}B^*})^{2^k/2}\right) \geq \left|\operatorname{Tr}\left((e^{2^{-k}A}e^{2^{-k}B})^{2^k}\right)\right|.$$

The obvious continuity of Tr together with the exponential product formula allows us to conclude the theorem. $\qquad\square$

Corollary 3.12 If A and B are self-adjoint then

$$|\operatorname{Tr} e^{A+iB}| \leq \operatorname{Tr} e^A.$$

The Golden-Thompson upper bound for $\operatorname{Tr} e^{A+B}$ is complemented by the *Peierls-Bogoliubov inequality*. Before turning to it one we show another convex trace function.

Proposition 3.13 For a finite quantum system $\mathcal{A}$ the functional

$$A \mapsto \log \operatorname{Tr} e^A$$

is convex on $\mathcal{A}^{sa}$.

Proof. As we saw above, the statement is equivalent to the convexity of the function

$$f(t) = \log \operatorname{Tr}(e^{A+tB}) \qquad (t \in \mathbb{R}) \tag{3.16}$$

for every $A, B \in \mathcal{A}^{sa}$. To show this we prove that $f''(0) \geq 0$. First observe that

$$\frac{d}{dt}\operatorname{Tr} F(A+tB) = \operatorname{Tr}(BF'(A+tB)), \tag{3.17}$$

which is checked for a polynomial F by an easy direct computation and it can be extended to a more general F by means of polynomial approximation. (3.17) tells us that

$$f'(t) = \frac{\operatorname{Tr} e^{A+tB}B}{\operatorname{Tr} e^{A+tB}}.$$

In the computation of the second derivative we use the identity

$$e^{A+tB} = e^A + t\int_0^1 e^{uA}Be^{(1-u)(A+tB)}\,du. \tag{3.18}$$

In order to write $f''(0)$ in a convenient form we introduce the inner product

$$(X, Y) = \int_0^1 \operatorname{Tr} e^{tA}X^*e^{(1-t)A}Y\,dt = \langle T_{e^A}^{-1}(X), Y\rangle \tag{3.19}$$

(see (3.7) for the definition of the operator T_D^{-1}). Since T_D and T_D^{-1} are positive definite for $D \in \mathcal{A}^{++}$, the sesquilinear form $(.\,,.)$ is indeed an inner product. (It is frequently termed Bogoliubov inner product.) Now

$$f''(0) = \frac{(I,I)(B,B) - (I,B)^2}{(\operatorname{Tr} e^A)^2} \geq 0$$

due to the Schwarz inequality. $\qquad\square$

Corollary 3.14 If $A, B \in \mathcal{A}^{sa}$ then

$$\log \operatorname{Tr} e^{A+B} \geq \log \operatorname{Tr} e^A + \frac{\operatorname{Tr} B e^A}{\operatorname{Tr} e^A} \tag{3.20}$$

and

$$|\log \operatorname{Tr} e^{A+B} - \log \operatorname{Tr} e^A| \leq \|B\|. \tag{3.21}$$

Proof. The first inequality is of the form $f(1) \geq f(0) + f'(0)$ which follows immediately from the convexity of f. The second one is a consequence of the estimate $|f'(t)| \leq \|B\|$. □

The above proof allows us to analyze the case of equality in (3.20). The equality is equivalent to $f''(0) = 0$ and this happens in the only case $B = \lambda I$ for some $\lambda \in \mathbb{R}$. Hence in all nontrivial situation the Peierls-Bogoliubov inequality is a strict inequality. We note also that in terms of state perturbation, the Golden-Thompson and Peierls-Bogoliubov bounds read as

$$e^{\varphi(h)} \leq \varphi^h(I) \leq \varphi(e^h) \tag{3.22}$$

for a state on a finite quantum system $\mathcal{A}$ and $h \in \mathcal{A}^{sa}$. Here the density of φ is not required to be invertible. In fact (3.22) is true on a much higher level of generality as it will be seen later on (in Chapter 12).

In course of the proof of Theorem 3.11 the inequality

$$\operatorname{Tr} (X^{1/2} Y X^{1/2})^q \leq \operatorname{Tr} X^{q/2} Y^q X^{q/2}$$

was obtained for $q = 2^k$ and for positive matrices X and Y. In fact,

$$\operatorname{Tr} (X^{1/2} Y X^{1/2})^{rp} \leq \operatorname{Tr} (X^{r/2} Y^r X^{r/2})^p \tag{3.23}$$

holds for every $r \geq 1$ and $p > 0$. The inequality (3.23) implies that the function

$$p \mapsto \operatorname{Tr} (e^{pB/2} e^{pA} e^{pB/2})^{1/p} \tag{3.24}$$

is increasing for $p > 0$. Its limit at $p = 0$ is $\operatorname{Tr} e^{A+B}$. Hence for every $p > 0$

$$\operatorname{Tr} e^{A+B} \leq \operatorname{Tr} (e^{pB/2} e^{pA} e^{pB/2})^{1/p} \tag{3.25}$$

which extends the *Golden-Thompson inequality* (3.13).

We shall see that (3.25) gives a lower bound for the relative entropy of two states of a finite quantum system $\mathcal{A}$. For $X \in \mathcal{A}^{++}$ with $\operatorname{Tr} X = 1$ and for $B \in \mathcal{A}^{sa}$

$$\operatorname{Tr} X(\log X - B) \geq \operatorname{Tr} XA - \log \operatorname{Tr} e^{A+B}$$

holds due to Proposition 1.11. Replacing A with $\frac{1}{p} \log e^{-pB/2} X^p e^{-pB/2}$ and using (3.25) we arrive at

$$\operatorname{Tr} X(\log X - B) \geq \frac{1}{p} \operatorname{Tr} X \log e^{-pB/2} X^p e^{-pB/2}$$

which reads in terms of the relative entropy as

$$S(\omega, \varphi) \geq \frac{1}{p} \operatorname{Tr} D_\omega \log D_\varphi^{-p/2} D_\omega^p D_\varphi^{-p/2} \tag{3.26}$$

for every $p > 0$. (The limit of the right hand side as $p \to 0$ is just the left hand side.)

Some particular cases of the following simple and useful observation are sometimes called *Klein inequalities*.

Proposition 3.15 If f_k and g_k are functions $[\alpha, \beta] \to \mathbb{R}$ such that for some $c_k \in \mathbb{R}$

$$\sum_k c_k f_k(x) g_k(y) \geq 0$$

for every $x, y \in [\alpha, \beta]$, then

$$\sum_k c_k \operatorname{Tr} f_k(A) g_k(B) \geq 0$$

whenever A, B are self-adjoint elements of a finite quantum system with $\operatorname{Sp}(A), \operatorname{Sp}(B) \subset [\alpha, \beta]$.

Proof. Let $A = \sum \lambda_i p_i$ and $B = \sum \mu_j q_j$ be the spectral decompositions. Then

$$\sum_k c_k \operatorname{Tr} f_k(A) g_k(B) = \sum_k \sum_{i,j} c_k \operatorname{Tr} p_i f_k(A) g_k(B) q_j$$

$$= \sum_{i,j} \operatorname{Tr} p_i q_j \sum_k c_k f_k(\lambda_i) g_k(\mu_j) \geq 0$$

due to the hypothesis. □

In particular, if f is convex then

$$f(x) - f(y) - (x - y) f'(y) \geq 0$$

and

$$\operatorname{Tr} f(A) \geq \operatorname{Tr} f(B) + \operatorname{Tr} (A - B) f'(B). \tag{3.27}$$

Replacing f by $\eta(t) = -t \log t$ we see that the relative entropy of two states is nonnegative. This fact is a traditional application of the Klein inequality. In the proof of Proposition 1.1 the stronger estimate

$$-\eta(x) + \eta(y) + (x - y) \eta'(y) \geq \tfrac{1}{2}(x - y)^2$$

made possible a hidden use of the Klein inequality.

From the inequality $1 + \log x \leq x$ $(x > 0)$ one obtains

$$\gamma^{-1}(a - a^{1-\gamma} b^\gamma) \leq a(\log a - \log b) \leq \gamma^{-1}(a^{1+\gamma} b^{-\gamma} - a)$$

for $a, b, \gamma > 0$. If T and S are nonnegative invertible operators of a finite quantum system then Proposition 3.15 gives

$$\gamma^{-1} \text{Tr} \, (S - S^{1-\gamma} T^\gamma) \leq \text{Tr} \, S(\log S - \log T)$$
$$\leq \gamma^{-1} \text{Tr} \, (S^{1+\gamma} T^{-\gamma} - S) \tag{3.28}$$

which provides a lower as well as an upper estimate for the relative entropy.

Proposition 3.16 Let $\mathcal{A}$ and $\mathcal{B}$ be finite quantum systems, $\alpha : \mathcal{A} \to \mathcal{B}$ a positive unital mapping and $f : \mathbb{R} \to \mathbb{R}$ a convex function. Then

$$\text{Tr} \, f(\alpha(A)) \leq \text{Tr} \, \alpha(f(A))$$

for every $A \in \mathcal{A}^{sa}$.

Proof. By means of the spectral decompositions $A = \sum_j \nu_j q_j$ and $\alpha(A) = \sum_i \mu_i p_i$ we have

$$\mu_i = \text{Tr} \, (\alpha(A) p_i) / \text{Tr} \, p_i = \sum_j \nu_j \text{Tr} \, (\alpha(q_j) p_i) / \text{Tr} \, p_i$$

whereas the convexity of f yields

$$f(\mu_i) \leq \sum_j f(\nu_j) \text{Tr} \, (\alpha(q_j) p_i) / \text{Tr} \, p_i \, .$$

Therefore,

$$\text{Tr} \, f(\alpha(A)) = \sum_i f(\mu_i) \text{Tr} \, p_i \leq \sum_{i,j} f(\nu_j) \text{Tr} \, (\alpha(q_j) p_i) = \text{Tr} \, \alpha(f(A)) \, ,$$

which was to be proven.   □

The first proposition of this chapter becomes a consequence of Proposition 3.16 if we define $\alpha : \mathcal{A} \oplus \mathcal{A} \to \mathcal{A}$ by $\alpha(A \oplus B) = \lambda A + (1 - \lambda) B$ for $0 < \lambda < 1$.

A positive unital mapping $\alpha : \mathcal{A} \to \mathcal{A}$ is called doubly stochastic if $\text{Tr} \circ \alpha = \text{Tr}$. It is easy to see that a linear mapping $\alpha : \mathcal{A} \to \mathcal{A}$ is doubly stochastic if and only if such is its dual α^* (with respect to the Hilbert-Schmidt inner product). A state φ is called *more mixed* than the state ψ (written $\varphi \succ \psi$) if there exists a doubly stochastic map β such that $\varphi = \psi \circ \beta$. It follows readily from Proposition 3.16 that

$$\varphi \succ \psi \quad \text{implies} \quad \text{Tr} \, f(D_\varphi) \leq \text{Tr} \, f(D_\psi) \tag{3.29}$$

for every convex function $f : \mathbb{R}^+ \to \mathbb{R}$. In particular,

$$\varphi \succ \psi \quad \text{implies} \quad S(\varphi) \geq S(\psi). \tag{3.30}$$

The more mixed the state, the bigger is its entropy.

A typical application of the inequality (3.30) concerns the coarse-grained density matrix. Let us write the $n \times n$ matrices in the form of 2×2 matrices with matrix entries. Then the transformation

$$\begin{pmatrix} A & B \\ C & D \end{pmatrix} \mapsto \begin{pmatrix} A & 0 \\ 0 & D \end{pmatrix}$$

is double stochastic and we have

$$S\left(\begin{pmatrix} A & B \\ B^* & D \end{pmatrix}\right) \leq S\left(\begin{pmatrix} A & 0 \\ 0 & D \end{pmatrix}\right)$$

for a density matrix. Another application of (3.30) is the inequality

$$S(\varphi) \leq \sum_i \eta(a_{ii})$$

where (a_{ii}) is the diagonal of the density matrix of φ.

The most mixed state is the normalized trace functional, $\mathrm{Tr}\,/\mathrm{Tr}\,I$. This is clear from the following eigenvalue characterization of the partial order $\succ$. Let

$$\lambda_1^{(i)} \geq \lambda_2^{(i)} \geq \ldots \geq \lambda_n^{(i)}$$

be the decreasingly ordered eigenvalue list of the density matrix of the state φ_i $(i = 1, 2)$. Then $\varphi_1 \succ \varphi_2$ if and only if

$$\sum_{i=1}^{k} \lambda_i^{(1)} \leq \sum_{i=1}^{k} \lambda_i^{(2)} \qquad (k = 1, 2, \ldots, n). \tag{3.31}$$

Proposition 3.17 Let φ be a state on a finite quantum system $\mathcal{A}$ and $f : \mathbb{R}^+ \to \mathbb{R}^+$ a convex function with $f(0) = 0$. If the state ψ is determined by the density

$$D_\psi = \frac{f(D_\varphi)}{\mathrm{Tr}\, f(D_\varphi)}$$

then $\varphi \succ \psi$ holds.

Proof. Set $\lambda_1, \lambda_2, \ldots, \lambda_n$ for the decreasingly ordered eigenvalue list of D_φ. Under the hypothesis on f the inequality $f(x)y \leq f(y)x$ holds for $0 \leq x \leq y$. Hence for $i \leq j$ we have $\lambda_j f(\lambda_i) \geq \lambda_i f(\lambda_j)$ and

$$\begin{aligned} &(f(\lambda_1) + \ldots + f(\lambda_k))(\lambda_{k+1} + \ldots + \lambda_n) \\ &\qquad \geq (\lambda_1 + \ldots + \lambda_k)(f(\lambda_{k+1}) + \ldots + f(\lambda_n)). \end{aligned}$$

Adding to both sides the term $(f(\lambda_1) + \ldots + f(\lambda_k))(\lambda_1 + \ldots + \lambda_k)$ we arrive at

$$(f(\lambda_1) + \ldots + f(\lambda_k)) \sum_{i=1}^{n} \lambda_i \geq (\lambda_1 + \ldots + \lambda_k) \sum_{i=1}^{n} f(\lambda_i).$$

This shows that the sum of the k largest eigenvalues of $f(D_\varphi)/\mathrm{Tr}\, f(D_\varphi)$ must exceed that of D_φ, $\lambda_1 + \ldots + \lambda_k$. The criterium (3.31) ensures now $\varphi \succ \psi$. $\square$

The canonical (Gibbs) state φ_c^β at inverse temperature $\beta = (kT)^{-1}$ possesses the density $e^{-\beta H}/\mathrm{Tr}\, e^{-\beta H}$ (see Proposition 1.10). Choosing $f(x) = x^{\beta'/\beta}$ with $\beta' > \beta$ the Proposition 3.17 tells us that

$$\varphi_c^{\beta'} \succ \varphi_c^\beta \,, \tag{3.32}$$

that is, at higher temperature the canonical state is more mixed. (The most mixed tracial state is canonical at infinite temperature.) According to (3.30) the entropy $S(\varphi_c^\beta)$ is an increasing function of the temperature.

Now we turn to an important special case of Proposition 3.16. Let $\mathcal{A}_i$ be a finite quantum system ($i = 1, 2, 3$). We fix a tracial state τ_i on $\mathcal{A}_i$ ($i = 1, 3$) and consider the usual trace functional Tr_2 on $\mathcal{A}_2$. Set $\tau = \tau_1 \otimes \mathrm{Tr}_2 \otimes \tau_3$ and write F_3, F_{12}, F_{23} for the partial trace mappings of $\mathcal{A}$ onto $\mathcal{A}_3, \mathcal{A}_1 \otimes \mathcal{A}_2, \mathcal{A}_2 \otimes \mathcal{A}_3$, respectively. Assume that $p_{23} \in \mathcal{A}_2 \otimes \mathcal{A}_3$ is an operator of norm 1. We define

$$\alpha(a_{12}) = F_{13}((a_{12} \otimes I_3)(I_1 \otimes p_{23}))$$

for $a_{12} \in \mathcal{A}_1 \otimes \mathcal{A}_2$. In this way we obtain a linear mapping $\alpha : \mathcal{A}_1 \otimes \mathcal{A}_2 \to \mathcal{A}_1 \otimes \mathcal{A}_3$. Since

$$\alpha(I_{12}) = F_{13}(I_1 \otimes P_{23}) = I_1 \otimes F_3(P_{23}),$$

α is unital if $F_3(P_{23}) = I$. Obviously, $\|\alpha\| \leq 1$ thanks to the assumption $\|p_{23}\| = 1$, and we obtain that α is a positive unital mapping. (Recall that for a unital mapping β the positivity of β is equivalent to $\|\beta\| \leq 1$.) We introduce another mapping $\gamma : \mathcal{A}_1 \otimes \mathcal{A}_3 \to \mathcal{A}_1 \otimes \mathcal{A}_2$ by the formula

$$\gamma(a_{13}) = F_{12}(a_{13} P_{23})\,.$$

The following result is an abstract form of the *Berezin-Lieb inequality*.

Proposition 3.18 Using the above notation we assume that $P_{23} \in \mathcal{A}_2 \otimes \mathcal{A}_3$ is such that $F_3(P_{23}) = F_2(P_{23}) = I$. If $h : \mathbb{R} \to \mathbb{R}$ is a convex function, $a_{12} = a_{12}^* \in \mathcal{A}_1 \otimes \mathcal{A}_2$ and $b_{13} = b_{13}^* \in \mathcal{A}_1 \otimes \mathcal{A}_3$ then the inequalities

(i) $\tau_1 \otimes \tau_3(h(\alpha(a_{12})) \leq \tau_1 \otimes \mathrm{Tr}_2(h(a_{12}))$,
(ii) $\tau_1 \otimes \tau_3(h(\gamma(b_{13})) \leq \tau_1 \otimes \mathrm{Tr}_2(h(b_{13}))$

hold.

Proof. We have seen that under the above conditions on P_{23} the mappings α and γ are positive and unital. Taking into account $(\tau_1 \otimes \tau_3) \circ \alpha = \tau_1 \otimes \mathrm{Tr}_2$ and its analogue with γ, we deduce (i) and (ii) from Proposition 3.16. □

That $\mathcal{A}_1$ is finite dimensional is not necessary in the last proposition. We can recapture the Berezin-Lieb inequality by choosing $\mathcal{A}_3$ to be a commutative von Neumann algebra L^∞. Let $(X, \mathcal{S}, \mu)$ be a probability space, $\mathcal{H}$ a Hilbert space and $P : X \to \mathcal{B}(\mathcal{H})$ a measurable projection valued function. One may consider P as a projection in the algebra $\mathcal{B}(\mathcal{H}) \otimes L^\infty(X, \mathcal{S}, \mu)$. Let

$\mathcal{K}$ be another Hilbert space and A a selfadjoint operator on $\mathcal{K} \otimes \mathcal{H}$. On the commutative algebra $L^{\infty}(\mu)$ let us consider the integration with respect to μ as a tracial state. Then the lower symbol of A is a function $A^l : X \to \mathcal{B}(\mathcal{K})$ defined as

$$A^l(\xi) = F_2(A(I \otimes P(\xi))) \tag{3.33}$$

We have the three-fold tensor product $\mathcal{B}(\mathcal{K}) \otimes \mathcal{B}(\mathcal{H}) \otimes L^{\infty}(X, \mathcal{S}, \mu)$ of the last proposition. $P \equiv P_{23}$ is a projection in $\mathcal{B}(\mathcal{H}) \otimes L^{\infty}(\mu)$ and if $P : \Omega \to \mathcal{B}(\mathcal{H})$ is such that

$$\int P(\xi)\, d\mu(\xi) = I \tag{3.34}$$

and

$$\mathrm{Tr}_2\, P(\xi) = 1 \qquad (\xi \in X),$$

then the above conditions on P_{23} hold. If $\mathcal{K}$ is one dimensional then the lower symbol is a common function and part (i) of the proposition takes the form

$$\int h(A^l(\xi))\, d\mu(\xi) \leq \mathrm{Tr}\, h(A). \tag{3.35}$$

This is the Berezin-Lieb inequality, which gives a lower estimate for $\mathrm{Tr}\, h(A)$ by means of the lower symbol. If there exists a function $A^u \in L^{\infty}(\mu)$, called upper symbol, such that

$$A = \int A^u(\xi)P(\xi)\, d\mu(\xi), \tag{3.36}$$

then (ii) provides an upper estimate

$$\mathrm{Tr}\, h(A) \leq \int h(A^u(\xi))\, d\mu(\xi). \tag{3.37}$$

In a typical application of these inequalities $h = \exp$ and the projection-valued function P is given by coherent vectors.

Notes and Remarks. For the sake of simplicity all inequalities of the chapter are stated and proved for finite quantum systems. Most of the proofs work without alteration if Tr is replaced by a tracial state of a C*-algebra. More care has to be taken however, if we want to allow the semi-finite trace of $\mathcal{B}(\mathcal{H})$ with an infinite dimensional $\mathcal{H}$ to take the place of the tracial state.

Theorem 3.7 was obtained in [Lieb 1973a], this paper gave the title of this chapter. Theorem 3.8 is from that paper, too. The *Golden-Thompson inequality* appeared in [Golden 1965] and [Thompson 1965]. The key to the proof is (3.15). It was recognized in [Thompson 1971] that if $F(X)$ is a certain function of the eigenvalues of the complex matrix X such that (3.15) holds when F is written instead of Tr then $F(e^{A+B}) \leq F(e^A e^B)$ for hermitian A and B. For example, if $I_s(X)$ denotes the sth coefficient in the characteristic polynomial of X then $I_s(e^{A+B}) \leq I_s(e^A e^B)$ ([Lenard 1971]). Extension of the Golden-Thompson inequality to operators of an infinite dimensional Hilbert space is contained in [Ruskai 1972] and [Breitenbecker and Grümm 1972]. Theorem 3.11 is from [Cohen, Friedland, Kato and Kelly 1982]. The inequality $\mathrm{Tr}\, e^A e^{A^*} \leq \mathrm{Tr}\, e^{A+A^*}$ with a rather surprising direction of the inequality sign was found in [Bernstein 1988] for a (non-symmetric) real matrix A. (The equality holds only for normal A, see [So 1992].) The interested reader may find lots of useful inequalities in [Simon 1979]. The trace inequality (3.23) was proved in [Araki 1990], for Hilbert space operators, too. See [Kosaki 1992] for a generalization. It is proven in [Friedland and So 1994] that if the function (3.24) is constant in an interval then A and B commute. (3.26) appeared in [Hiai and Petz 1991] and a similar lower bound was found in [Hiai and Petz 1993]. The bounds (3.28) for the relative entropy are from [Ruskai and Stillinger 1990]. Proposition 3.16 is generalized in [Petz 1987] to semi-finite traces of von Neumann algebras.

The *more mixed* relation $\succ$ was introduced into mathematical physics by Uhlmann. Essentially the whole monograph [Alberti and Uhlmann 1981] is devoted to the analysis of this partial order. The reader is warned that in the mathematical literature essentially the same partial order is denoted by the opposite direction of the symbol (and it is sometimes termed majorization). See [Marshall and Olkin 1979] for the majorization of probability distributions and the lecture notes [Ando 1982] for matrices. [Fack and Kosaki 1986] makes a generalization to von Neumann algebras.

Proposition 3.17 is due to [Wehrl 1974]. The *Berezin-Lieb inequalities* appeared in [Berezin 1972] and [Lieb 1973b] and they were used to compute the classical limit of quantum partition functions also in [Simon 1980]. Concerning coherent states and symbols of operators the monograph [Perelomov 1986] is a good source.

Part II

Entropies
for General Quantum Systems

4 Modular Theory and Auxiliaries

This chapter is aimed to provide a very concise overview of some important results of the modular theory of operator algebras. The modular theory is technically related to quantum entropy. On infinite dimensional operator algebras the definition of the relative entropy requires the spatial derivative operator of a positive normal functional ψ on a von Neumann algebra $\mathcal{M}$ acting on a Hilbert space $\mathcal{H}$ with respect to a positive normal functional φ' on the commutant $\mathcal{M}' \subset B(\mathcal{H})$. The spatial derivative is a generalization of the modular operator and its measure theoretic forefather is the Radon-Nikodym derivative of two measures. Besides spatial derivatives and relative modular operators we shall mention some results on generalized conditional expectations and on the resolvent convergence of unbounded operators.

The *lineal* of ψ is the set of all so-called ψ-bounded vectors. Formally,

$$D(\mathcal{H}, \psi) = \{\xi \in \mathcal{H} : \|a\xi\| \leq C_\xi \psi(aa^*) \text{ for all } a \in \mathcal{M}\}. \tag{4.1}$$

The lineal is a linear manifold with closure supp ψ. (Recall that supp ψ is the smallest projection p in $\mathcal{M}$ such that $\psi(p) = \psi(I)$. As usual, we identify closed subspaces of a Hilbert space with the corresponding orthogonal projections.) When ψ is a vector state of the form $\psi(a) = \langle \Psi, a\Psi \rangle$ $(a \in \mathcal{M})$ then $D(\mathcal{H}, \psi) = \mathcal{M}'\Psi$ and, obviously

$$D(\mathcal{H}, \psi)^- = [\mathcal{M}'\psi] = \text{supp } \psi. \tag{4.2}$$

Set $(\Psi, \pi_\psi, \mathcal{H}_\psi)$ for the GNS-triplet corresponding to the state ψ. It is possible to define for $\xi \in D(\mathcal{H}, \psi)$ a bounded operator

$$R^\psi(\xi) : \mathcal{H}_\psi \to \mathcal{H}$$

such that

$$R^\psi(\xi)\pi_\psi(a)\Psi = a\xi \qquad (a \in \mathcal{M}). \tag{4.3}$$

It is easy to check that $a\,R^\psi(\xi) = R^\psi(\xi)$ $(a \in \mathcal{M})$ and this implies

$$R^\psi(\xi_1)R^\psi(\xi_2)^* \in \mathcal{M}' \qquad (\xi_1, \xi_2^* \in D(\mathcal{H}, \psi)).$$

We introduce the notation

$$\Theta^\psi(\xi) = R^\psi(\xi)R^\psi(\xi)^*.$$

It is noteworthy that in the case $\psi(a) = \langle \Psi, a\Psi \rangle$ we have

$$\Theta^\psi(a'\Psi) = a'[\mathcal{M}\Psi]a'^*.$$

The proof of the following assertions is straightforward.

Lemma 4.1 If $\psi \leq \lambda\omega$ then $D(\mathcal{H}, \psi) \subset D(\mathcal{H}, \omega)$ and $\Theta^\omega(\xi) \leq \lambda^2\Theta^\psi(\xi)$ for any $\xi \in D(\mathcal{H}, \psi)$.

Lemma 4.2 Let $\mathcal{M}_0$ be a von Neumann subalgebra of $\mathcal{M}$ and let ω stand for $\psi|\mathcal{M}_0$. Then $D(\mathcal{H}, \psi) \subset D(\mathcal{H}, \omega)$ and $\Theta^\omega(\xi) \leq \Theta^\psi(\xi)$ for $\xi \in D(\mathcal{H}, \psi)$.

For $\xi \in D(\mathcal{H}, \psi)$ and $\eta \in D(\mathcal{H}, \psi)^\perp$ we define a functional q as follows.

$$q(\xi + \eta) = \varphi'(\Theta^\psi(\xi))$$

where φ' is a fixed normal state on $\mathcal{M}'$. This q turns out to be a densely defined quadratic form that is lower semi-continuous and hence closable as well. The form representation theorem tells us that there exists a positive self-adjoint operator $\Delta(\varphi'/\psi)$ such that

(i) $\|\Delta(\varphi'/\psi)^{1/2}\zeta\|^2 = q(\zeta)$ $(\zeta \in \mathcal{D}(q))$,
(ii) $\mathcal{D}(q)$ is a core for $\Delta(\varphi'/\psi)^{1/2}$.

The *spatial derivative* operator $\Delta(\varphi'/\psi)$ was introduced by *Connes* as a spatial generalization of the relative modular operator (treated below). We note that the definition of the spatial derivative operator $\Delta(\varphi'/\psi)$ makes sense if the linear functional φ' is replaced by a so-called weight. Let $\mathcal{N}$ be a von Neumann algebra. A mapping $w : \mathcal{N}^+ \to \mathbb{R}^+ \cup \{+\infty\}$ is called *normal weight* if the following conditions hold.

(i) $w(a + b) = w(a) + w(b)$ $(a, b \in \mathcal{N}^+)$
(ii) $w(\lambda a) = \lambda w(a)$ $(\lambda \in \mathbb{R}^+, a \in \mathcal{N}^+, 0 \times \infty = 0)$
(iii) w is weak* lower semi-continuous.
(iv) For every $0 \neq a \in \mathcal{N}^+$ there exists $0 \neq b \leq a$ such that $w(b)$ is finite.

The most common example of a normal weight is the canonical trace on $B(\mathcal{H}) \equiv \mathcal{N}$. Let $(\xi_i)_i$ be an orthogonal basis of the Hilbert space $\mathcal{H}$. For $A \in B(\mathcal{H})^+$ the definition

$$\text{Tr}\, A = \sum_i \langle A\xi_i, \xi_i \rangle$$

gives the canonical trace which is independent of the basis.

We shall give some properties of $\Delta(\varphi'/\psi)$ without proof.

$$\text{supp}\, \Delta(\varphi'/\psi) = \text{supp}\, \varphi' \,\text{supp}\, \psi \tag{4.4}$$
$$\Delta(\varphi'_1 + \varphi'_2/\psi) = \Delta(\varphi'_1/\psi) + \Delta(\varphi'_2/\psi). \tag{4.5}$$

Here, on the right hand side, the + sign stands for the form sum of the positive operators. If $\text{supp}\, \varphi'_1 \perp \text{supp}\, \varphi'_2$, we have an orthogonal sum due to (4.4). For a non-faithful state φ' it is a frequently used trick to add an orthogonal state ω' with $\operatorname{supp}\varphi' \oplus \operatorname{supp}\omega' = I$. Then $\varphi' + \omega'$ is a faithful functional on $\mathcal{M}'$ and we have

$$\Delta(\varphi'/\psi) = \Delta(\varphi' + \omega'/\psi)\operatorname{supp}\varphi'.\tag{4.6}$$

This relation allows us to deduce some statements concerning a non-faithful state from the faithful case.

In the definition of $\Delta(\varphi'/\psi)$ the algebra and its commutant play symmetric roles. Hence the reciprocal spatial derivative $\Delta(\psi/\varphi')$ is defined as well. For a positive self-adjoint operator A and $z \in \mathbb{C}$ the power A^z denotes the sum of 0 on $(\operatorname{supp}A)^\perp$ and the usual power A^z on $\operatorname{supp}A$. With this convention we have

$$\Delta(\varphi'/\psi)^z = \Delta(\psi/\varphi')^{-z} \qquad (z \in \mathbb{C}).\tag{4.7}$$

If φ' is faithful then the operator

$$\Delta(\psi/\varphi')^z \Delta(\omega/\varphi')^{-z} \qquad (z \in \mathbb{C})$$

is independent of φ'. The bounded operator

$$[D\psi, D\omega]_t \equiv \Delta(\psi/\varphi')^{it}\Delta(\omega/\varphi')^{-it} \qquad (t \in \mathbb{R})\tag{4.8}$$

is affiliated with the von Neumann algebra $\mathcal{M}$. This one-parameter family $[D\psi, D\omega]_t$ is called the *Connes (or Radon-Nikodym) cocycle* of the functionals ψ and ω. If $\operatorname{supp}\psi$ and $\operatorname{supp}\omega$ are commuting projections then $[D\psi, D\omega]_t$ is a partial isometry with initial and final projections $\operatorname{supp}\psi\operatorname{supp}\omega$. In particular, for faithful states it is a family of unitaries. Under some support conditions the chain rule

$$[D\psi, D\omega]_t = [D\psi, D\sigma]_t[D\sigma, D\omega]_t \qquad (t \in \mathbb{R})\tag{4.9}$$

is now obvious.

Still assuming that φ' is faithful the mapping

$$\sigma_t^\varphi(a) = \Delta(\varphi/\varphi')^{it}a\Delta(\varphi/\varphi')^{-it} \qquad (t \in \mathbb{R},\ a \in \mathcal{M})\tag{4.10}$$

is independent of φ' and maps into the reduced von Neumann algebra $p\mathcal{M}p$ with $p = \operatorname{supp}\varphi$. The group σ^φ is an automorphism group on $p\mathcal{M}p$ and it is termed the modular group of the normal state φ. It follows from (4.8) that the Connes cocycle intertwines between the modular groups.

The Connes cocycle is an operator-valued function on the real line. The exponent in formula (4.8) shows that it is more convenient to consider the function

$$it \mapsto [D\psi, D\omega]_t,\tag{4.11}$$

that is, we regard the cocycle as a function defined of the imaginary line of the complex plain. If a majorization relation $\psi \le \lambda\omega$ holds, then (4.11) admits an analytic continuation to the strip $\{z \in \mathbb{C} : 0 \le \operatorname{Re}z \le 1/2\}$ as a bounded-operator-valued function. The method of analytic continuation is an efficient tool in the modular theory.

Spatial derivatives are the most convenient to use and are the most powerful when the algebra is in standard form. Let $\mathcal{M}$ be a von Neumann algebra. Its *standard form* is a quadruple $(\mathcal{H}, \pi, \mathcal{P}, J)$ where $\mathcal{H}$ is a Hilbert space, π is a faithful normal representation of $\mathcal{M}$ on $\mathcal{H}$, J is an anti-unitary operator with $J^2 = I$ (called *modular conjugation*) and $\mathcal{P}$ is a closed cone (called natural positive cone) in the Hilbert space $\mathcal{H}$. These objects are connected to each other through several useful relations. Below we describe these relations and doing this we shall identify $\mathcal{M}$ with $\pi(\mathcal{M})$ through π.

Every positive normal functional φ of $\mathcal{M}$ has a unique vector representative $\xi_\varphi \in \mathcal{P}$, i.e.

$$\varphi(a) = \langle \xi_\varphi, a\xi_\varphi \rangle \qquad (a \in \mathcal{M}). \tag{4.12}$$

The mapping $\varphi \mapsto \xi_\varphi$ is norm continuous. More precisely, the estimates

$$\|\xi_\varphi - \xi_\omega\|^2 \le \|\varphi - \omega\| \le \|\xi_\varphi - \xi_\omega\| \, \|\xi_\varphi + \xi_\omega\| \tag{4.13}$$

hold. A vector $\xi \in \mathcal{P}$ is cyclic if and only if it is separating. The (conjugate linear) *-isomorphism

$$j(a) = JaJ \qquad (a \in \mathcal{M})$$

maps $\mathcal{M}$ onto its commutant $\mathcal{M}'$. The modular conjugation J leaves $\mathcal{P}$ fixed and

$$aj(a)\mathcal{P} \subset \mathcal{P} \qquad (a \in \mathcal{M}).$$

Whenever φ is a faithful positive normal functional on $\mathcal{M}$ the set

$$\{aj(a)\xi_\varphi : a \in \mathcal{M}\}$$

is dense in $\mathcal{P}$. The natural positive cone is self-dual in the following sense

$$\xi \in \mathcal{P} \iff \langle \xi, \eta \rangle \ge 0 \text{ for every } \eta \in \mathcal{P}. \tag{4.14}$$

Let φ and ω be normal states on $\mathcal{M}$. On the domain $\mathcal{M}\xi_\omega + [\mathcal{M}\xi_\omega]^\perp$ one can define a closable operator $S^0_{(\varphi,\omega)}$ by the formula

$$S^0_{(\varphi,\omega)}(a\xi_\omega + \eta) = (\operatorname{supp}\varphi)a^*\xi_\varphi \qquad (a \in \mathcal{M}, \, \eta \perp \mathcal{M}\xi_\omega).$$

The closure of $S^0_{(\varphi,\omega)}$ admits a polar decomposition

$$S_{(\varphi,\omega)} = J\Delta(\varphi,\omega)^{1/2}$$

where J is the modular conjugation and $\Delta(\varphi,\omega)$ is the spatial derivative operator of the state φ of $\mathcal{M}$ with respect to the vector state ω induced by $\xi_\omega \in \mathcal{P}$ on the commutant $\mathcal{M}'$. $\Delta(\varphi,\omega)$ is termed *relative modular operator*. In connection with (4.4) and (4.7) we have

$$\operatorname{supp}\Delta(\varphi,\omega) = (\operatorname{supp}\varphi)[\mathcal{M}\xi_\omega], \quad \Delta(\varphi,\omega) = J\Delta(\omega,\varphi)^{-1}J. \tag{4.15}$$

The relative modular operator was defined and investigated in detail by *Araki* as the relativization of the modular operator $\Delta(\varphi, \varphi)$.

Let us interrupt the review of the relative modular operator theory and turn to generalized conditional expectations, because this concept may be treated quite naturally in connection with the standard representation. Let $\mathcal{M}$ be a von Neumann algebra with normal state φ and with a von Neumann subalgebra $\mathcal{M}_0$. We write φ_0 for $\varphi | \mathcal{M}_0$ and p_0 for $\operatorname{supp} \varphi_0$.

Let $(\mathcal{H}, \pi, \mathcal{P}, J)$ and $(\mathcal{H}_0, \pi_0, \mathcal{P}_0, J_0))$ be the standard representations of the von Neumann algebra $\mathcal{M}$ and $\mathcal{M}_0$) respectively, and let ξ_φ and ξ_{φ_0} the vector representatives of φ and φ_0 in the natural positive cones. We set a partial isometry $V : \mathcal{H}_0 \to \mathcal{H}$ with initial projection $[\pi_0(\mathcal{M}_0)\xi_{\varphi_0}]$ by the formula

$$V \pi_0(a)\xi_{\varphi_0} = \pi(a)\xi_\varphi \qquad (a \in \mathcal{M}).$$

One can verify that

$$V^* \pi(\mathcal{M})' V \subset (\pi_0(\mathcal{M}_0))'.$$

Hence there exists a unique element $x \in p_0 \mathcal{M}_0 p_0$ such that

$$\pi_0(x) = J_0 V^* J \pi(a) J V J_0 \tag{4.16}$$

for a fixed $a \in \mathcal{M}$. The *generalized conditional expectation* $E_\varphi : \mathcal{M} \to p_0 \mathcal{M}_0 p_0$ at the operator a is defined as this element x. The following proposition is straightforward from the construction.

Proposition 4.3 $E_\varphi : \mathcal{M} \to p_0 \mathcal{M}_0 p_0$ is a completely positive mapping and $\operatorname{supp} E_\varphi = \operatorname{supp} \varphi$, $\varphi \circ E_\varphi = \varphi$ and $E_\varphi(\operatorname{supp} \varphi) = p_0$.

Let $\mathcal{M}$ be a von Neumann algebra and $\mathcal{M}_0$ be its subalgebra. A conditional expectation E of $\mathcal{M}$ onto $\mathcal{M}_0$ is defined to be a linear mapping such that

(i)   If $a \geq 0$ then $E(a) \geq 0$.
(ii) $E(a) = a$ for every $a \in \mathcal{M}_0$.
(iii) $E(ab) = E(a)b$ for every $a \in \mathcal{M}$ and $b \in \mathcal{M}_0$.

In fact, (iii) follows from (i) and (ii), but this nontrivial fact is not required here. If there is a faithful normal state φ on $\mathcal{M}$ such that $\varphi \circ E = \varphi$ then the conditional expectation E is faithful and normal. We are dealing with such conditional expectations in most cases. Yet, a conditional expectation onto a subalgebra that preserves a given state does not always exists. For example, if $\mathcal{M} = M_2(\mathbb{C}) \otimes M_2(\mathbb{C})$ and $\mathcal{M}_0 = M_2(\mathbb{C}) \otimes \mathbb{C} \subset \mathcal{M}$, then for a state φ of $\mathcal{M}$ a conditional expectation $E : \mathcal{M} \to \mathcal{M}_0$ (preserving φ) exists if and only if φ is a product state. The *Takesaki theorem* below clarifies the existence of the conditional expectation by means of the modular group of the state.

Theorem 4.4 Let $\mathcal{M}_0$ be a von Neumann subalgebra of the von Neumann algebra $\mathcal{M}$ and let φ be a faithful normal state of $\mathcal{M}$. Then the φ-preserving conditional expectation of $\mathcal{M}$ onto $\mathcal{M}_0$ exists if and only if $\mathcal{M}_0$ is stable under the modular group of φ. When the conditional expectation exists, then it is unique.

The generalized conditional expectation constructed above is not surjective and does not possess the module property (iii) for every $b \in \mathcal{M}_0$. Nevertheless, a reminiscent of these features is present.

Theorem 4.5 Under the conditions of the previous theorem let E_φ stand for the generalized conditional expectation of $\mathcal{M}$ into $\mathcal{M}_0$. For a fixed $b \in \mathcal{M}_0$ the condition

$$E_\varphi(ab) = E_\varphi(a)b \qquad (a \in \mathcal{M})$$

holds if and only if $\sigma_t^\varphi(b) \in \mathcal{M}_0$ for every $t \in \mathbb{R}$.

Theorem 4.6 Let $(\mathcal{M}_i)_i$ be an increasing net of von Neumann a subalgebras of the von Neumann algebra $\mathcal{M}$ and let φ be a faithful normal state of $\mathcal{M}$. Assume that $\cup_i \mathcal{M}_i$ is strongly dense in $\mathcal{M}$ and stand E_i for the φ-preserving generalized conditional expectation into $\mathcal{M}_i$. Then

$$\lim_i E_i(a) = a \quad \text{strongly for every } a \in \mathcal{M}.$$

The existence of the φ-preserving conditional expectation means a maximal coincidence of the modular structure of $\mathcal{M}$ and $\mathcal{M}_0$. For example, the condition $\sigma_t^\varphi(b) \in \mathcal{M}_0$ $(t \in \mathbb{R})$ is equivalent to $\sigma_t^\varphi(b) = \sigma_t^{\varphi_0}(b)$ $(t \in \mathbb{R})$. If the algebra $\mathcal{M}$ acts in a standard way on the Hilbert space $\mathcal{H}$ (carrying a positive cone $\mathcal{P}$ and a modular conjugation J) then the action of $\mathcal{M}_0$ restricted to $\mathcal{H}_0 = [\mathcal{M}_0 \xi_\varphi]$ supplies a standard representation of $\mathcal{M}_0$ with cone $\mathcal{P}_0 = \mathcal{H}_0 \cap \mathcal{P}$ and conjugation $J_0 = J|\mathcal{H}_0$. If $\xi_{\omega_0} \in \mathcal{P}_0$ is the vector representative of the normal state ω_0 on $\mathcal{M}_0$ then $\xi_{\omega_0} \in \mathcal{P}$ represents the state $\omega_0 \circ E$ of $\mathcal{M}$. Therefore,

$$\Delta(\omega_0 \circ E, \varphi)|\mathcal{H}_0 = \Delta(\omega_0, \varphi_0) \tag{4.17}$$

and

$$[D(\omega_0 \circ E), D(\psi_0 \circ E)]_t = [D\omega_0, D\psi_0]_t \qquad (t \in \mathbb{R}) \tag{4.18}$$

for normal states ω_0 and ψ_0 of $\mathcal{M}_0$.

For defining the continuity of the relative modular operator as a function of two states the appropriate notion of convergence is the strong resolvent one. Let (A_n) be a sequence of self-adjoint operators acting in a Hilbert space $\mathcal{H}$. We say that (A_n) converges to A in the sense of strong resolvent convergence if for some $z \in \mathbb{C}$ (bounded away from the spectra of A_n and A) $(A_n - z)^{-1} \to (A - z)^{-1}$ strongly. In case of a norm bounded sequence the strong resolvent convergence is equivalent to the simple strong convergence. This fact follows also from the following two propositions.

Proposition 4.7 Let (A_n) be a sequence of self-adjoint operators and A be another self-adjoint operator. Assume that there is a core $\mathcal{D}$ for A such that $\mathcal{D} \subset \mathcal{D}(A_n)$ and $A_n \xi \to A\xi$ for every $\xi \in \mathcal{D}$. Then $A_n \to A$ in the sense of strong resolvent convergence.

Let (h_n) be a bounded sequence of self-adjoint operators and suppose that $h_n \to h$ strongly. It is a consequence of the previous proposition that $H + h_n \to H + h$ strongly in the resolvent sense for any self-adjoint operator H. (The domain of H may play the role of the core $\mathcal{D}$ in the previous proposition.)

Proposition 4.8 Assume that the sequence (A_n) of self-adjoint operators converges to the self-adjoint operator A in the sense of strong resolvent convergence. Let $f : \mathbb{R} \to \mathbb{C}$ be a bounded Borel measurable function which is continuous except on a closed set of spectral measure zero with respect to the operator A. Then $f(A_n) \to f(A)$ strongly.

Let the von Neumann algebra $\mathcal{M}$ be in standard form and let (φ_n) and (ψ_n) be sequences of positive normal functionals of $\mathcal{M}$ converging in norm to the positive normal functionals φ and ψ, respectively. The estimate (4.12) shows that for the representing vectors from the natural positive cone the relations

$$\xi_{\varphi_n} \to \xi_\varphi \quad \text{and} \quad \xi_{\psi_n} \to \xi_\psi \tag{4.19}$$

hold. It follows that

$$[\mathcal{M}'\xi_{\psi_n}][\mathcal{M}'\xi_\psi] \to [\mathcal{M}'\xi_\psi]$$

strongly. For $x \in \mathcal{M}[\mathcal{M}'\xi_\psi]$ we have

$$\|\Delta(\varphi_n, \psi_n)^{1/2} x\xi_{\psi_n} - \Delta(\varphi, \psi)^{1/2} x\xi_\psi\|$$
$$= \|[\mathcal{M}'\xi_{\psi_n}]x^*\xi_{\psi_n} - [\mathcal{M}'\xi_\psi]x^*\xi_\psi\|$$
$$\leq \|([\mathcal{M}'\xi_{\psi_n}][\mathcal{M}'\xi_\psi] - [\mathcal{M}'\xi_\psi])x^*\xi_\psi\| + \|[\mathcal{M}'\xi_{\psi_n}](x^*\xi_{\psi_n} - x^*\xi_\psi)\|$$

which tends to 0. The vectors

$$\mathcal{D} = \{(1 + \Delta(\varphi, \psi)^{1/2})x\xi_\psi \ : \ x \in [\mathcal{M}'\xi_\psi]\}$$

are dense in $[\mathcal{M}\xi]$ and for $\eta \in \mathcal{D}$

$$[1 + \Delta(\varphi_n\psi_n)^{1/2}]^{-1}\eta \to [1 + \Delta(\varphi, \psi)^{1/2}]^{-1}\eta . \tag{4.20}$$

Indeed,

$$\|([1 + \Delta(\varphi_n, \psi_n)^{1/2}]^{-1} - [1 + \Delta(\varphi, \psi)^{1/2}]^{-1})[1 + \Delta(\varphi, \psi)^{1/2}]x\xi_\psi\|$$
$$\leq \|[1 + \Delta(\varphi_n, \psi_n)^{1/2}]^{-1}([1 + \Delta(\varphi, \psi)^{1/2}]x\xi_\psi - [1 + \Delta(\varphi_n\psi_n)^{1/2}]\xi_{\psi_n})$$
$$+ (x\xi_{\psi_n} - x\xi_\psi)\|$$
$$\leq 2\|x(\xi_{\psi_n} - \xi_\psi)\| + \|\Delta(\varphi, \psi)^{1/2}x\xi_\psi - \Delta(\varphi_n, \psi_n)^{1/2}x\xi_{\psi_n}\| .$$

In this way we have shown the following result due to *Araki*.

Proposition 4.9 Let $\varphi_n, \varphi, \psi_n, \psi \in \mathcal{M}_*^+$ ($n \in \mathbb{N}$) and assume $\varphi_n \to \varphi$, $\psi_n \to \psi$ in norm . Then

$$[1 + \Delta(\varphi_n, \psi_n)^{1/2}]^{-1}[\mathcal{M}\xi_\psi] \to [1 + \Delta(\varphi, \psi)^{1/2}]^{-1}[\mathcal{M}\xi_\psi]$$

strongly in the standard representation.

If ψ is faithful then the proposition expresses the strong resolvent convergence $\Delta(\varphi_n, \psi_n)^{1/2} \to \Delta(\varphi, \psi)^{1/2}$. Since we can not restrict ourselves to faithful ψ always, a modification of the strong resolvent convergence is needed in order to handle the projection $Q' \equiv [\mathcal{M}\xi_\psi] \in \mathcal{M}'$.

Lemma 4.10 Let (H_n) be a sequence of positive self-adjoint operators on a Hilbert space $\mathcal{H}$. Let H be a positive self-adjoint operator and P a projection commuting with H. Assume that

$$\lim_n (1 + H_n)^{-1}P = (1 + H)^{-1}P \quad \text{strongly.}$$

Then for every continuous bounded function $f : \mathbb{R}^+ \to \mathbb{C}$ with $\lim_{x\to\infty} f(x) = 0$ we have

$$\lim_n f(H_n)P = f(H)P \quad \text{strongly.}$$

Proof. By means of mathematical induction one proves easily that

$$\lim_n (1 + H_n)^{-k}P = (1 + H)^{-k}P \quad \text{strongly}$$

for every $k \in \mathbb{N}$. The Weierstrass approximation theorem tells us that

$$\lim_n g((1 + H_n)^{-1})P = g((1 + H)^{-1})P \quad \text{strongly}$$

for every continuous function $g : [0, 1] \to \mathbb{C}$ and an appropriate choice of g gives the statement.   □

Lemma 4.11 Let (H_n) and (k_n) be sequences of self-adjoint operators and let P be a projection. Assume that (k_n) is bounded, $k_n \to k$ strongly and there exists a self-adjoint operator H such that

$$\lim_n (i + H_n)^{-1}P = (i + H)^{-1}P \quad \text{strongly.}$$

If P commutes with k and H then

$$\lim_n (i + H_n + k_n)^{-1}P = (i + H + k)^{-1}P \quad \text{strongly.}$$

Proof. Note that $H_n + k_n$ and $H + k$ are self-adjoint operators with domains $\mathcal{D}(H_n)$ and $\mathcal{D}(H)$, respectively. Therefore, if $P\eta = \eta$ then $\xi = (i + H + k)^{-1}\eta$ is in the domain of H and we may set $\xi_n \equiv (i + H_n)^{-1}(i + H)\xi$. Due to the assumptions we have $P\xi = \xi$ and

$$(i + H_n + k_n)\xi_n \to (i + H)\xi + k\xi = \eta, \quad \xi_n \to \xi.$$

Writing

$$[(i + H_n + k_n)^{-1} - (i + H + k)^{-1}]\eta$$
$$= (i + H_n + k_n)^{-1}[\eta - (i + H_n + k_n)\xi_n] + (\xi_n - \xi)$$

we observe that both terms on the right hand side tend to 0.   □

This lemma will be used in connection with the perturbation of states. Besides it is noteworthy that it contains the additivity of the strong resolvent convergence when one of the two sequences is norm-bounded.

Notes and Remarks. The first expose of the modular theory was given in [Takesaki 1970] and a good outline may be found in [Bratteli and Robinson 1979]. For a thorough treatment of the subject we refer to [Strătilă and Zsidó 1979] and [Strătilă 1981].

The *spatial derivative operator* was introduced in [Connes 1980]. The notion is mostly discussed in the case when one of the two states (or weights) is faithful. Concerning the *standard form* of a von Neumann algebra the original source is [Haagerup 1975]. From several points of view the Hilbert space of the standard representation is the analogue of an L^2 function space and $\mathcal{P}$ is something like the cone of positive functions. The highly technical paper [Araki and Masuda 1982] treats this subject and it contains the best account on properties of the *relative modular operator* and on the *Connes cocycle* of states (even weights) without any assumption of faithfulness. The analytic continuation of the Connes cocycle is due to [Connes 1973], see also 3.13 in [Strătilă 1981]. Further examples of analytic continuation of products of spatial derivatives appear in [Cecchini and Petz 1989b].

The *generalized conditional expectation* with respect to a faithful normal state φ appeared in [Accardi and Cecchini 1982] under the name φ-conditional expectation and Theorem 4.5 was obtained there, too. Theorem 4.4 is due to [Takesaki 1972], and a review of the topic is in [Petz 1988a]. It seems that generalized conditional expectation does not fit the notion of relative entropy. (An extension of Theorem 5.15 with generalized conditional expectation does not work.)

Theorem 4.6 is a typical martingale type convergence result of noncommutative probability. It appeared independently in [Hiai and Tsukada 1984] and [Petz 1984]. The former paper treats the decreasing case as well. (The decreasing sequence of generalized conditional expectations does not converge always but that of conditional expectations does.)

The strong resolvent convergence is standard in the theory of unbounded operators, see [Kato 1966] and [Reed and Simon 1972]. For a quick introduction to self-adjoint operators the few pages of [Faris 1975] are very suitable.

Proposition 4.9 was achieved in [Araki 1977] and the subsequent convergence lemmas are tailored for relative modular operator convergence after [Donald 1990].

5 Relative Entropy
of States of Operator Algebras

Let $\mathcal{M}$ be a von Neumann algebra with normal states φ and ω. In this chapter the relative entropy $S(\omega, \varphi)$ will be defined by means of the spatial derivative operator. Of course, for finite quantum systems the new definition reduces to the previous one, which was based upon density matrices (see Chapter 1). We note that $S(\omega, \varphi)$ should be regarded as the entropy of the state ω with respect to φ. In order to define the relative entropy of states of a C*-algebra the enveloping von Neumann algebra will be used.

Let $\mathcal{M}$ act on a Hilbert space $\mathcal{H}$ and $\omega = \omega_\xi$ be a vector state given by a vector $\xi \in \mathcal{H}$. The vector $\xi \in \mathcal{H}$ induces a vector state ω'_ξ on the commutant $\mathcal{M}'$ of $\mathcal{M}$. The spatial derivative $\Delta(\varphi/\omega'_\xi)$ is at our disposal. This is a positive self-adjoint operator with support $[\mathcal{M}\xi]$ supp φ. (Here supp φ, the support projection of φ, belongs to $\mathcal{M}$ and $[\mathcal{M}\xi]$ stands for the orthogonal projection onto the closure of the linear manifold $\mathcal{M}\xi$. The latter projection is an element of $\mathcal{M}'$.) *Araki* defined the *relative entropy* as follows.

$$S(\omega, \varphi) = \begin{cases} +\infty & \text{if } \xi \notin \text{supp } \varphi \\ -\langle \log \Delta(\varphi/\omega'_\xi)\xi, \xi \rangle & \text{otherwise.} \end{cases} \tag{5.1}$$

Note that $\xi \in \text{supp } \varphi$ is equivalent to supp $\omega \le$ supp φ. Let

$$\Delta(\varphi/\omega'_\xi) = \int_0^\infty \lambda \, dE_\lambda$$

be the spectral decomposition. More precisely, (5.1) is meant as

$$-\langle \log \Delta(\varphi/\omega'_\xi)\xi, \xi \rangle = -\int_0^1 \log \lambda \, d\langle E_\lambda, \xi, \xi \rangle - \int_1^\infty \log \lambda \, d\langle E_\lambda \xi, \xi \rangle. \tag{5.2}$$

Since $\log \lambda \le \lambda$ and

$$\int_0^\infty \lambda \, d\langle E_\lambda \xi, \xi \rangle = \|\Delta(\varphi/\omega'_\xi)^{1/2}\xi\|^2 = \varphi(I)$$

we see that the second term on the right hand side of (5.2) is always finite. So $S(\omega, \varphi)$ is finite or $+\infty$ depending on the integral $\int_0^1 \log \lambda \, d\langle E_\lambda \xi, \xi \rangle$. Up to this point it is not clear whether this definition is independent of the auxiliary vector ξ.

It is desirable to get rid of the domain problem in (5.1) caused by the logarithmic function. The following equivalent definition is essentially due to *Uhlmann* who embedded it into a quadratic interpolation machinery.

$$S(\omega, \varphi) = - \lim_{t \to +0} t^{-1} (\|\Delta(\varphi/\omega'_\xi)^{t/2}\xi\|^2 - \|\xi\|^2). \tag{5.3}$$

Since for a given $\lambda > 1$ (or $0 < \lambda < 1$)

$$\lim_{t \to +0} t^{-1}(\lambda^t - 1) = \log \lambda \tag{5.4}$$

increasingly (or decreasingly), the monotone integral convergence theorem ensures that (5.1) and (5.3) are equivalent.

Proposition 5.1 Let φ and ω be normal positive functionals of the von Neumann algebra $\mathcal{M}$ and let $\lambda, \mu > 0$. Then

(i) $S(\mu\omega, \lambda\varphi) = \mu S(\omega, \varphi) - \mu\omega(I)(\log \lambda - \log \mu),$
(ii) $S(\omega, \varphi) \geq \omega(I)(\log \omega(I) - \log \varphi(I)).$

Proof. Property (i) is straightforward from any of the two definitions of the relative entropy. To prove (ii) we use the inequality

$$\|\Delta^{t/2}\xi\| \leq \|\xi\|^{1-t}\|\Delta^{1/2}\xi\|^t$$

and formula (5.3). We have

$$t^{-1}(\|\Delta^{t/2}\xi\|^2 - \|\xi\|^2) \leq t^{-1}(\|\xi\|^{2(1-t)}\|\Delta^{1/2}\xi\|^{2t} - \|\xi\|^2)$$

with $\Delta = \Delta(\varphi/\omega'_\xi)$ and the limit $t \to +0$ yields (ii).   □

Some basic properties of the relative entropy are related to the theory of *operator monotone* and *operator convex functions*. Let us recall that a real function f is operator monotone if

$$A \geq B \quad \text{implies} \quad f(A) \geq f(B) \tag{5.5}$$

whenever A and B are bounded self-adjoint operators with spectrum in the domain of f and acting on the same Hilbert space. Similarly, g is operator convex if

$$g(\lambda X + (1 - \lambda)Y) \leq \lambda g(X) + (1 - \lambda)g(Y) \tag{5.6}$$

for $0 < \lambda < 1$ and operators X and Y satisfying the conditions above. Operator monotonicity of the function $\lambda \mapsto \lambda^t$ ($0 < t < 1$) was used implicitly in Chapter 1. In fact, Lemma 1.2 is strong enough to provide both operator monotonicity and operator concavity of the function $\lambda \mapsto \lambda^t$. For monotonicity one writes Lemma 1.2 with $V = A^{-1/2}B^{1/2}$; to prove concavity one chooses

$$A = \begin{pmatrix} X & 0 \\ 0 & Y \end{pmatrix} \quad \text{and} \quad V = \begin{pmatrix} \sqrt{\lambda} & 0 \\ \sqrt{1-\lambda} & 0 \end{pmatrix}.$$

In this way inequalities (5.5) and (5.6) may be concluded. The limit (5.4) yields that the logarithm is operator monotone and operator concave.

The following lemma still bears the marks of interpolation theory.

Lemma 5.2 Let H_i be positive self-adjoint operators on Hilbert spaces $\mathcal{H}_i$ $(i = 1, 2)$. If $T : \mathcal{H}_1 \to \mathcal{H}_2$ is a bounded operator such that

(i) $T(\mathcal{D}(H_1)) \subset \mathcal{D}(H_2)$,
(ii) $\|H_2 T\xi\| \leq \|T\| \, \|H_1\xi\|$ $(\xi \in \mathcal{D}(H_1))$,

then for every $0 < t < 1$ and $\xi \in \mathcal{D}(H_1)$ the estimate

$$\|H_2^t T\xi\| \leq \|T\| \, \|H_1^t\xi\|$$

holds.

We omit the detailed proof. The interested reader may compare 1.18.10 in [Triebel 1978] with the Calderon-Lions interpolation theorem (IX. 20 in [Reed and Simon 1975]). Another – more direct – proof might go as follows. Assume that $\|T\| = 1$. Then

$$T^* H_2^2 T \leq H_1^2 \qquad \text{and} \qquad T^* H_2^{2t} T \leq H_1^{2t}$$

are the hypothesis and claim, respectively. From the hypothesis

$$(T^* H_2^2 T)^t \leq H_1^{2t} \tag{5.7}$$

by the operator monotonicity of the function λ^t. So it is sufficient to combine (5.7) with the inequality

$$T^* H_2^{2t} T \leq (T^* H_2^2 T)^t \tag{5.8}$$

(which is an extension of Lemma 1.2 to unbounded operators). The justification of these steps can be done.

Lemma 5.2 leads to the monotonicity theorem for relative entropy. A unital mapping α of a C*-algebra $\mathcal{A}_1$ into $\mathcal{A}_2$ is called Schwarz mapping if the Schwarz inequality

$$\alpha(a^* a) \geq \alpha(a)^* \alpha(a) \qquad (a \in \mathcal{A}_1)$$

holds. We recall that the physically relevant completely positive maps do satisfy the Schwarz inequality. Now we prove the *Uhlmann monotonicity theorem* for Schwarz mappings.

Theorem 5.3 Let $\mathcal{M}_1$ and $\mathcal{M}_2$ be von Neumann algebras with positive normal functionals φ_1, ω_1 and φ_2, ω_2, respectively. Let $\alpha : \mathcal{M}_1 \to \mathcal{M}_2$ be a Schwarz mapping such that $\varphi_2 \circ \alpha \leq \varphi_1$ and $\omega_2 \circ \alpha \leq \omega_1$. Then

$$S(\omega_1, \varphi_1) \leq S(\omega_2, \varphi_2).$$

Proof. We assume that $\mathcal{M}_i$ acts on a Hilbert space $\mathcal{H}_i$ and ω_i is a vector functional induced by $\xi_i \in \mathcal{H}_i$ $(i = 1, 2)$. Since

$$\|\alpha(a)\xi_2\|^2 = \omega_2(\alpha(a)^*\alpha(a)) \leq \omega_2(\alpha(a^* a)) \leq \omega_1(a^* a) = \|a\xi_1\|^2,$$

the correspondence $T : a\xi_1 \to \alpha(a)\xi_2$ is a linear contraction defined on $\mathcal{M}_1$ and extends to a contraction of $\mathcal{H}_1$ in an obvious way. We write Δ_i for the spatial derivative operator $\Delta(\varphi_i/\omega'_{\xi_i})$. Remember that $\mathcal{M}_i\xi_i + (\mathcal{M}_i\xi_i)^\perp$ is a core for $\Delta_i^{1/2}$. We show that $T\mathcal{D}(\Delta_1^{1/2}) \subset \mathcal{D}(\Delta_2^{1/2})$. Let $\eta \in \mathcal{D}(\Delta_1^{1/2})$. Then there exists a sequence $\zeta_n = a_n\xi_1 + \eta_n$ such that $a_n \in \mathcal{M}_1, \eta_n \in [\mathcal{M}_1\xi_1]^\perp, \zeta_n \to \eta$ and $\Delta_1^{1/2}\eta_n \to \Delta_1^{1/2}\eta$. The estimate

$$\|\Delta_2^{1/2}T\zeta_n - \Delta_2^{1/2}T\zeta_m\|^2 = \varphi_2(\alpha(a_n - a_m)^*\alpha(a_n - a_m))$$
$$\leq \|\Delta_1^{1/2}(a_n - a_m)\xi_1\|^2 = \|\Delta_1^{1/2}\zeta_n - \Delta_1^{1/2}\zeta_m\|^2$$

yields that $(\Delta_2^{1/2}T\zeta_n)_n$ forms a Cauchy sequence. Therefore $T\eta$ must be in the domain of $\Delta_2^{1/2}$ and $\|\Delta_2^{1/2}T\zeta\| \leq \|\Delta_1^{1/2}\eta\|$. The conditions of Lemma 5.2 have been verified for $H_i = \Delta_i^{1/2}$. The inequality

$$t^{-1}(\|\Delta_2^{t/2}\xi_2\|^2 - \|\xi_2\|^2) \leq t^{-1}(\|\Delta_1^{t/2}\xi_1\|^2 - \|\xi_1\|^2)$$

is at our disposal and the limit $t \to +0$ completes the proof. $\square$

One of the consequences of the rather general Uhlmann monotonicity theorem is that the quantity $S(\omega, \varphi)$ is in fact independent of the auxiliary vector representative ξ of ω used in the definitions (5.1) and (5.3). It follows also that restricting the states to a subalgebra we obtain smaller relative entropy. In particular, property (ii) of Proposition 5.1 is recaptured and the relative entropy of two states is again nonnegative. Monotonicity of the relative entropy under completely positive unital mappings is plausible on a physical ground as well.

Theorem 5.4 The relative entropy functional is jointly convex, that is,

$$S(\lambda\omega_1 + (1-\lambda)\omega_2, \lambda\varphi_1 + (1-\lambda)\varphi_2) \leq \lambda S(\omega_1, \varphi_1) + (1-\lambda)S(\omega_2, \varphi_2),$$

if $0 < \lambda < 1$ and $\varphi_1, \varphi_2, \omega_1, \omega_2$ are normal positive functionals of a von Neumann algebra.

Proof. Joint convexity may be reduced to monotonicity by duplication of the algebra (see the proof of Theorem 1.4). On $\mathcal{M} \oplus \mathcal{M}$ we consider the functionals φ_{12} and ω_{12} defined by the expressions

$$\varphi_{12}(a \oplus b) = \lambda\varphi_1(a) + (1-\lambda)\varphi_2(b) \qquad (a, b \in \mathcal{M}),$$
$$\varphi_{12}(a \oplus b) = \lambda\omega_1(a) + (1-\lambda)\omega_2(b) \qquad (a, b \in \mathcal{M}).$$

Assume that ω_1 and ω_2 are vector induced functionals. Then so is ω_{12}. We write $\omega'_1, \omega'_2, \omega'_{12}$ for the corresponding functionals on the commutants. We have

$$\Delta(\varphi_{12}/\omega'_{12}) = \Delta(\lambda\varphi_1/\lambda\omega'_1) \oplus \Delta((1-\lambda)\varphi_2/(1-\lambda)\omega'_2)$$
$$= \Delta(\varphi_1/\omega'_1) \oplus \Delta(\varphi_2/\omega'_2),$$

and the definition of relative entropy directly gives

$$S(\omega_{12}, \varphi_{12}) = \lambda S(\omega_1, \varphi_1) + (1-\lambda)S(\omega_2, \varphi_2).$$

On the diagonal algebra $\{a \oplus a : a \in \mathcal{M}\} \subset \mathcal{M} \oplus \mathcal{M}$ the states φ_{12} and ω_{12} recover the convex combinations $\lambda\varphi_1 + (1-\lambda)\varphi_2$ and $\lambda\omega_1 + (1-\lambda)\omega_2$, respectively. Now monotonicity of the relative entropy under restriction to a subalgebra makes the proof complete. $\qquad\qquad\qquad\qquad\qquad\qquad\qquad\square$

It is worthwhile to note that by virtue of the scaling identity (i) of Proposition 5.1 the joint convexity of the relative entropy is equivalent to the following inequality, which concerns finite sums of positive normal functionals.

$$S(\textstyle\sum_i \omega_i, \sum_i \varphi_i) \le \sum_i S(\omega_i, \varphi_i). \tag{5.9}$$

Let $\mathcal{M}$ be a finite quantum system. Its standard representation is obtained by viewing $\mathcal{M}$ as a Hilbert space with the Hilbert-Schmidt inner product $\langle a, b \rangle = \operatorname{Tr} a^* b$, and representing $a \in \mathcal{M}$ by the left multiplication operator $\pi(a)b = ab$. Let φ and ω be states on $\mathcal{M}$ with densities D_φ and D_ω, respectively. They may be regarded as states on $\pi(\mathcal{M})$. Then $\xi = D_\omega^{1/2}$ becomes the representing vector for ω. It follows from direct calculation that

$$\Delta(\varphi/\omega'_\xi) = \pi(D_\varphi)\pi'(D_\omega^{-1})$$

where π' is the anti-representation of $\mathcal{M}$ by right multiplications. Hence assuming $\operatorname{supp} \omega \le \operatorname{supp} \varphi$ (i.e., $\operatorname{supp} D_\omega \le \operatorname{supp} D_\varphi$) we have

$$S(\omega, \varphi) = -\langle \log \pi(D_\varphi)\pi'(D_\omega^{-1})D_\omega^{1/2}, D_\omega^{1/2} \rangle = \operatorname{Tr} D_\omega(\log D_\omega - \log D_\varphi).$$

This is the usual relative entropy formula for finite quantum systems.

Let $(X, \mathcal{S})$ be a measurable space and μ, ν be two probability measures on the σ-algebra $\mathcal{S}$. Assume that ν is absolutely continuous with respect to μ and write $d\nu/d\mu \equiv F \in L^1(X, \mathcal{S}, \mu)$. The measures are naturally regarded as normal states of the abelian von Neumann algebra $\mathcal{M} = L^\infty(X, \mathcal{S}, \mu)$ acting on the Hilbert space $\mathcal{H} \equiv L^2(X, \mathcal{S}, \mu)$ by multiplication. The vector representative of the state ν is $F^{1/2} \in L^2(X, \mathcal{S}, \mu)$. Since $\mathcal{M}$ is maximal abelian we have $\mathcal{M} = \mathcal{M}'$ and $\nu = \nu'$. The spatial derivative is easily seen to be M_F^{-1} where M_F is the multiplication by F. Now,

$$S(\nu, \mu) = -\langle \log M_F^{-1} F^{1/2}, F^{1/2} \rangle = \int F \log F \, d\mu.$$

This is the *Kullback-Leibler informational divergence* of the measures μ and ν.

Theorem 5.5 $\|\varphi - \omega\|^2 \le 2S(\omega, \varphi)$ for normal states φ and ω of a von Neumann algebra.

Proof. Consider the Jordan decomposition $\psi_+ - \psi_-$ of the normal hermitian functional $\varphi - \omega$. Then $\|\varphi - \omega\| = \|\psi_+\| + \|\psi_-\|$ and ψ_+, ψ_- are positive normal functionals with orthogonal supports e and f. Consider the two dimensional subalgebra generated by the projection $e \in \mathcal{M}$. The monotonicity of the relative entropy tells us that

$$S(\omega, \varphi) \geq \omega(e) \log \frac{\omega(e)}{\varphi(e)} + \omega(e^{\perp}) \log \frac{\omega(e^{\perp})}{\varphi(e^{\perp})} \,.$$

It follows from the Jordan decomposition that

$$\|\varphi - \omega\| = (\varphi - \omega)(e) - (\varphi - \omega)(e^{\perp}) = 2(\varphi(e) - \omega(e)) \,.$$

Finally, invocation of Lemma 1.14 completes the proof.   □

Corollary 5.6 If $S(\omega, \varphi) = 0$ for normal states φ and ω of a von Neumann algebra then $\varphi = \omega$.

The corollary shows that the relative entropy $S(\omega, \varphi)$ is a suitable quantity to measure how much ω differs from φ. Let us emphasise that the relative entropy is not a metric. This negative statement remains true if $S(\omega, \varphi)$ is replaced by its symmetric version $S(\omega, \varphi) + S(\varphi, \omega)$ or by any reasonable function of $S(\omega, \varphi)$ and $S(\varphi, \omega)$. In spite of this discouraging fact, it will turn out that there are certain analogues where the relative entropy seems to play the role of squared Euclidean distance.

The spatial derivative operator involved in the definition of the relative entropy is a representation dependent self-adjoint operator. It would be useful to have at our disposal a formula expressing the relative entropy of two states by means of operator algebraic terms in a representation independent way. We recall that the Connes cocycle is a one-parameter family of contractions. For normal states φ and ω of the von Neumann algebra $\mathcal{M}$ acting on the Hilbert space $\mathcal{H}$ the definition of the Connes cocycle is

$$[D\varphi, D\omega]_t = \Delta(\varphi/\psi')^{\mathrm{it}} \, \Delta(\omega/\psi')^{-\mathrm{it}} \qquad (t \in \mathbb{R}) \,,$$

where ψ' is a faithful normal functional on the commutant $\mathcal{M}'$.

Theorem 5.7 If φ and ω are normal state on a von Neumann algebra such that $S(\omega, \varphi)$ is finite, then

$$S(\omega, \varphi) = \mathrm{i} \lim_{t \to +0} t^{-1}(\omega([D\varphi, D\omega]_t) - 1) \,.$$

Proof. Due to our hypothesis $p \equiv \mathrm{supp}\, \varphi \geq \mathrm{supp}\, \omega$. If the von Neumann algebra $\mathcal{M}$ is replaced by $p\mathcal{M}p$, both sides of the requested equality remain unchanged. Hence we may assume that φ is faithful. Let $\mathcal{M}$ act on a Hilbert space and let $\omega \equiv \omega_\xi$ be a vector state. We may choose a faithful functional $\psi' = \omega'_\xi + \nu'$ on $\mathcal{M}'$ so that the support of the vector state ω'_ξ on $\mathcal{M}'$ and that of ν' are orthogonal. Then

$$[D\varphi, D\omega]_t \xi = \Delta(\varphi, \psi')^{\mathrm{it}} \Delta(\omega, \psi')^{\mathrm{it}} \xi = \Delta(\varphi, \omega'_\xi)^{\mathrm{it}} \xi \,.$$

Let $\int_0^\infty \lambda \, dE_\lambda$ be the spectral decomposition of the self-adjoint operator $\Delta(\varphi, \omega'_\xi)$ and write $d\mu(\lambda)$ for $d\langle E_\lambda \xi, \xi \rangle$. Then

$$-\mathrm{it}^{-1}(\omega([D\varphi, D\omega]_t) - 1) = \int_0^\infty \frac{\lambda^{\mathrm{it}} - 1}{\mathrm{it}} \, d\mu(\lambda).$$

By means of the estimate

$$\left| \log \lambda - \frac{\lambda^{it} - 1}{it} \right| \leq 2|\log \lambda|$$

we conclude from the dominated convergence theorem to the limit relation

$$\lim_{t \to +0} \int_0^\infty \frac{\lambda^{it} - 1}{it} d\mu(\lambda) = \int_0^\infty \log \lambda \, d\mu(\lambda),$$

which is equivalent to the theorem.   □

In principle Theorem 5.7 gives the possibility to compute relative entropy. Next we discuss another formula due to Kosaki, which is too complicated for computational purposes but which, on the other hand, shows all the main properties of the relative entropy.

Lemma 5.8 Let H be a positive self-adjoint operator on a Hilbert space $\mathcal{H}$ and D be a core for $H^{1/2}$. Then we have

$$\langle H(t+H)^{-1}\xi, \xi \rangle = \inf\{\|H^{1/2}h\|^2/t + \|g\|^2 : \xi = h + g \text{ and } h, g \in D\}$$

for every $t > 0$ and $\xi \in D$.

Proof. Since $(H/t)(1 + (H/t))^{-1} = H(t+H)^{-1}$ we may assume that $t = 1$. Using the spectral theorem, one may check easily that

$$\|(1+H)^{1/2}\eta\| \leq \|(2+H)^{1/2}\eta\| \leq \sqrt{2}\|(1+H)^{1/2}\eta\|.$$

Thus the graph norm of $H^{1/2}$ and $(1+H)^{1/2}$ are equivalent and the two operators have the same cores. In particular, D is a core for $(1+H)^{1/2}$. The identity

$$\|H^{1/2}h\|^2 + \langle g, g \rangle$$
$$= \langle H(1+H)^{-1}\xi, \xi \rangle + \|(1+H)^{1/2}(H(1+H)^{-1}\xi - g)\|^2 \qquad (5.10)$$

$(\xi = h + g)$ shows that

$$\langle H(1+H)^{-1}\xi, \xi \rangle \leq \|H^{1/2}h\|^2 + \|g\|^2.$$

Now we examine the last term in (5.10). The vector $H(1+H)^{-1}\xi$ belongs to $\mathcal{D}(H^{1/2}) = \mathcal{D}((1+H)^{1/2})$ and by an appropriate choice of $g \in D$ the last term of (5.10) can be made arbitrarily small.   □

Next we describe the situation in which the previous lemma will be used. Let $\mathcal{M}$ be a von Neumann algebra acting on a Hilbert space $\mathcal{H}$ in the standard way. For normal states φ and ω of $\mathcal{M}$ we have their vector representatives Φ and Ω from the natural positive cone. Write Δ for the spatial derivative $\Delta(\varphi/\omega')$, where ω' is the vector state of $\mathcal{M}'$ implemented by $\Omega \in \mathcal{H}$. Δ is a positive self-adjoint operator and $\mathcal{M}\Omega \oplus (\mathcal{M}\Omega)^\perp$ is a core for $\Delta^{1/2}$. Assume that N is a linear subspace of the von Neumann algebra $\mathcal{M}$, $I \in N$ and N is dense in $\mathcal{M}$ with respect to the strong* operator topology. It follows easily from the definition of Δ that under these conditions $N\Omega$ is dense in $\mathcal{M}\Omega$ with respect to the graph norm of $\Delta^{1/2}$. Consequently, $D = N\Omega \oplus (\mathcal{M}\Omega)^\perp$ is a core for $\Delta^{1/2}$.

Lemma 5.9 For each $t > 0$ we have

$$\langle \Delta(t + \Delta)^{-1}\Omega, \Omega \rangle = \inf\{\omega(x^*x) + t^{-1}\varphi(yy^*) : x, y \in N, \ x + y = I\}.$$

Proof. According to the previous lemma and to the above discussion the left hand side is equal to

$$\inf\{\|x\Omega + \eta\|^2 + t^{-1}\|\Delta^{1/2}(y\Omega - \eta)\|^2 \\ : x, y \in \mathcal{M}, \ \eta \in (\mathcal{M}\Omega)^\perp, \ x\Omega + y\Omega = \Omega\}. \tag{5.11}$$

Notice that

$$\|x\Omega + \eta\|^2 = \|x\Omega\|^2 + \|\eta\|^2, \quad \|\Delta^{1/2}(y\Omega - \eta)\|^2 = \varphi(y\,p_\omega\,y^*)$$

and

$$x\Omega + y\Omega = \Omega \quad \Longleftrightarrow \quad (x + y)p_\omega = p_\omega$$

where p_ω is the support projection of ω. Therefore, (5.11) is equal to

$$\inf\{\omega(x^*x) + t^{-1}\varphi(y\,p_\omega\,y^*) : x, y \in \mathcal{M}, \ (x + y)p_\omega = p_\omega\},$$

which is obviously majorized by

$$\inf\{\omega(x^*x) + t^{-1}\varphi(yy^*) : x, y \in \mathcal{M}, \ x + y = I\}, \tag{5.12}$$

since $\varphi(y\,p_\omega\,y^*) \le \varphi(yy^*)$. It is easy to see that the reverse majorization is also valid. If $x, y \in \mathcal{M}$ satisfy $(x + y)p_\omega = p_\omega$ then $\bar{x} = x\,p_\omega + (I - p_\omega)$ and $\bar{y} = y\,p_\omega$ satisfy $\bar{x} + \bar{y} = I$, and

$$\omega(\bar{x}^*\bar{x}) + t^{-1}\varphi(\bar{y}\,\bar{y}^*) = \omega(x^*x) + t^{-1}\varphi(y\,p_\omega\,y^*).$$

Finally, the infimum in (5.12) remains the same if the condition $x, y \in \mathcal{M}$ is replaced by $x, y \in N$, because $\omega(x^*x) + t^{-1}\varphi(yy^*)$ is continuous with respect to the strong* topology. □

Consider for a while functions $g : \mathbb{R}^+ \to \mathbb{R}$ admitting the integral decomposition

$$g(\lambda) = \alpha + \beta\lambda + \int_0^\infty \lambda(t + \lambda)^{-1}(1 + t)\,d\mu(t), \tag{5.13}$$

where α, β are real numbers, $\beta \ge 0$ and μ is a finite Radon measure on $(0, \infty)$. (One knows that decomposition (5.13) is characteristic for the operator monotone functions on $\mathbb{R}^+$, but we do not need this result.)

Let $\int \lambda\,dE_\lambda$ be the spectral decomposition of Δ. The expression

$$\langle g(\Delta)\Omega, \Omega \rangle \equiv \int g(\lambda)\,d\langle E_\lambda\Omega, \Omega \rangle \tag{5.14}$$

is always finite if g is of the form (5.13). Indeed, $g(\lambda) \le t\lambda + s$ for some constants $t, s \ge 0$ and Ω is in the domain of $\Delta^{1/2}$. In order to compute (5.14) we note first that

$$\int \alpha \, d\langle E_\lambda \Omega, \Omega \rangle = \alpha \langle (\text{supp } \Delta)\Omega, \Omega \rangle = \alpha \omega(p_\varphi)$$

and

$$\int \beta \lambda \, d\langle E_\lambda \Omega, \Omega \rangle = \beta \|\Delta^{1/2}\Omega\|^2 = \beta \varphi(p_\omega) \,.$$

Using the Fubini theorem we infer

$$\int g(\lambda) \, d\langle E_\lambda \Omega, \Omega \rangle$$

$$= \alpha \omega(p_\varphi) + \beta \varphi(p_\omega) + \int \left(\int \lambda(t+\lambda)^{-1}(1+t) \, d\mu(t) \right) d\langle E_\lambda \Omega, \Omega \rangle$$

$$= \alpha \omega(p_\varphi) + \beta \varphi(p_\omega) + \int \left(\int \lambda(t+\lambda)^{-1} d\langle E_\lambda \Omega, \Omega \rangle \right) (1+t) \, d\mu(t)$$

$$= \alpha \omega(p_\varphi) + \beta \varphi(p_\omega) + \int \langle \Delta(t+\Delta)^{-1}\Omega, \Omega \rangle (1+t) \, d\mu(t) \,.$$

Lemma 5.9 tells us that $\langle \Delta(t+\Delta)^{-1}\Omega, \Omega \rangle$ may be expressed as a certain infimum. What we are going to show now is that in this formula for $\langle g(\Delta)\Omega, \Omega \rangle$ the order of infimum and integration may be interchanged.

Proposition 5.10 Let N be a linear subspace, dense in $\mathcal{M}$ in the strong* operator topology, of the von Neumann algebra $\mathcal{M}$ and assume that $I \in N$. For any $\varphi, \omega \in \mathcal{M}_*^+$ and any $g : \mathbb{R}^+ \to \mathbb{R}$ of the form (5.13) we have

$$\langle g(\Delta)\Omega, \Omega \rangle = \alpha \omega(p_\varphi) + \beta \varphi(p_\omega)$$

$$+ \inf \left\{ \int_0^\infty [\omega(y(t)^*y(t)) + t^{-1}\varphi(x(t)x(t)^*](1+t) \, d\mu(t) \right\},$$

where the infimum is taken over all step functions $x(\cdot) : \mathbb{R}^+ \to N$ such that the range of x is finite, $x(t) = 0$ for sufficiently small t and $x(t) = I$ for t large enough. Moreover, $y(t)$ stands for $I - x(t)$.

Proof. Due to Lemma 5.9 and to the above computation it is sufficient to construct a step function $x : \mathbb{R}^+ \to N$ such that

$$\left| \int_0^\infty (\omega(y(t)^*y(t)) + t^{-1}\varphi(x(t)x(t)^*))(1+t) \, d\mu(t) \right.$$

$$\left. - \int_0^\infty \langle \Delta(t+\Delta)^{-1}\Omega, \Omega \rangle (1+t) \, d\mu(t) \right| < \varepsilon$$

for a given $\varepsilon > 0$. For $t > K$ let $x(t) = I$. For K sufficiently large

$$\left| \int_K^\infty t^{-1}\varphi(I)(1+t) \, d\mu(t) - \int_K^\infty \langle \Delta(t+\Delta)^{-1}\Omega, \Omega \rangle (1+t) \, d\mu(t) \right| < \frac{\varepsilon}{3} \,.$$

Similarly, for $\delta > 0$ sufficiently small

$$\left| \int_0^\delta \omega(I)(1+t) \, d\mu(t) - \int_0^\delta \langle \Delta(t+\Delta)^{-1}\Omega, \Omega \rangle (1+t) \, d\mu(t) \right| < \frac{\varepsilon}{3}$$

and we may choose $x(t) = 0$ for $t < \delta$. Hence we have to approximate on the compact interval $[\delta, K]$ with accuracy $\varepsilon/3$. For the sake of simplicity we restrict ourselves to the case $d\mu(t) = f(t)dt$ with a piecewise continuous function. Then, by means of a standard continuity argument and Lemma 5.9, the approximating step function is easily produced.   □

Proposition 5.10 does not apply directly to the relative entropy, a further approximation procedure is needed. We set

$$g_n(\lambda) = \log(\lambda + n^{-1}) \qquad (\lambda \geq 0) \tag{5.15}$$

for $n = 1, 2, \ldots$ and notice that

$$g_n(\lambda) = -\log n + \int_{1/n}^{\infty} \lambda(t + \lambda)^{-1} t^{-1}\, dt. \tag{5.16}$$

This function is of the form (5.13) if we choose $\alpha = -\log n$, $\beta = 0$ and for the measures $(1+t)d\mu(t) = \chi_{(1/n,\infty)}(t)t^{-1}dt$. (Note that the Radon-Nikodym derivative $d\mu(t)/dt$ is piecewise continuous in this case and the restricted proof of Proposition 5.10 works for g_n.)

Theorem 5.11 Assume that N is a linear subspace of a von Neumann algebra $\mathcal{M}$ and that I is contained in N. If N is dense in the strong*-operator topology, then, for any $\varphi, \omega \in \mathcal{M}_*^+$, we obtain

$$S(\omega, \varphi) = \sup \sup \left\{ \omega(I) \log n - \int_{1/n}^{\infty} \left(\omega(y(t)^* y(t)) + t^{-1}\varphi(x(t)x(t)^*) \right) \frac{dt}{t} \right\},$$

where the first sup is taken over all natural numbers n, the second one is over all step functions $x : (1/n, \infty) \to N$ with finite range and where $y(t) = I - x(t)$.

Proof. First we treat the case supp $\varphi \equiv p_\varphi \geq p_\omega \equiv$ supp ω. Proposition 5.10 tells us that

$$-\langle g_n(\Delta)\Omega, \Omega \rangle$$
$$= \sup \left\{ \omega(p_\varphi) \log n - \int_{1/n}^{\infty} \left(\omega(y(t)^* y(t)) + t^{-1}\varphi(x(t)x(t)^*) \right) \frac{dt}{t} \right\}.$$

As $n \to \infty$ we have $-g_n(\lambda) \nearrow -\log \lambda$ and using the monotone convergence theorem of integrals on $(0,1)$ and the dominated convergence theorem on $[1, \infty)$, we infer

$$-\langle g_n(\Delta)\Omega, \Omega \rangle \nearrow S(\omega, \varphi).$$

This completes the proof in the case supp $\varphi \geq$ supp ω. Else the double supremum should be $+\infty$. Choosing $x(t) = I - p_\varphi$ for $0 < t < 1$ and $x(t) = I$ for $t > 1$ we have

$$\omega(I - p_\varphi) \log n - \varphi(I)$$

inside the double sup sign. If $\omega(I - p_\varphi) \neq 0$, then

$$\sup \{\omega(I - p_\varphi) \log n - \varphi(I) : n \in \mathbb{N}\} = +\infty \,,$$

and the variational formula holds in the degenerate case, too. $\qquad\square$

We shall refer to the expression in Theorem 5.11 as *Kosaki's formula*. It yields properties of the relative entropy based upon the operator monotonicity of the logarithm function.

Corollary 5.12 The relative entropy of positive normal functionals of von Neumann algebras possesses the following properties.

(i) The map $(\omega, \varphi) \mapsto S(\omega, \varphi) \in \mathbb{R} \cup \{+\infty\}$ defined on $\mathcal{M}_*^+ \times \mathcal{M}_*^+$ is convex and lower semi-continuous with respect to the $\sigma(\mathcal{M}_*, \mathcal{M})$-topology (i.e., topology of pointwise convergence on $\mathcal{M}_*$.)

(ii) $\varphi_1 \leq \varphi_2$ implies $S(\omega, \varphi_1) \geq S(\omega, \varphi_2)$.

(iii) Let $\mathcal{M}_0$ be another von Neumann algebra, and $\alpha : \mathcal{M}_0 \to \mathcal{M}$ be a unital normal Schwarz mapping. Then for every $\varphi, \psi \in \mathcal{M}_*^+$ we have $S(\omega \circ \alpha, \varphi \circ \alpha) \leq S(\omega, \varphi)$.

(iv) Let $(\mathcal{M}_i)_i$ be an increasing net of von Neumann subalgebras of $\mathcal{M}$ with the property $(\cup_i \mathcal{M}_i)'' = \mathcal{M}$. Then the increasing net $(S(\omega|\mathcal{M}_i, \varphi|\mathcal{M}_i))_i$ converges to $S(\omega, \varphi)$.

Proof. According to Kosaki's formula $S(\omega, \varphi)$ is the supremum of continuous affine functionals. This provides (i). Property (ii) is immediate.

To prove (iii) it suffices to note that if $x_0 : \mathbb{R}^+ \to \mathcal{M}_0$ is a step function, then $x = \alpha(x_0)$ is also a step function $\mathbb{R}^+ \to \mathcal{M}$ with $\varphi(x(t)x(t)^*) \leq \varphi \circ \alpha(x_0(t)x_0(t)^*)$ and similarly for φ and y.

In the proof of (iv) (called martingale property) we choose $N = \cup_i \mathcal{M}_i$ which is strong* dense in $\mathcal{M}$. Let $x : [1/n, \infty) \to N$ be a step function with finite range. If we choose i big enough then the range of x lies in $\mathcal{M}_i$ and

$$\omega(I) \log n - \int_{1/n}^{\infty} \left(\omega(y(t)^*y(t)) + t^{-1}\varphi(x(t)x(t)^*) \right) \frac{dt}{t} \leq S(\varphi|\mathcal{M}_i, \omega|\mathcal{M}_i) \,.$$

Taking the limit in i on the right hand side as well as the double supremum on the left hand side we obtain

$$S(\omega, \varphi) \leq \lim_i S(\omega|\mathcal{M}_i, \varphi|\mathcal{M}_i) \,.$$

This yields property (iv) because according to (iii) we have $S(\omega|\mathcal{M}_i, \varphi|\mathcal{M}_i) \leq S(\omega, \varphi)$. $\qquad\square$

The *martingale property*
martingale property (iv) of Corollary 5.12 allows a simplification when one works with a hyper-finite von Neumann algebra. By definition such an algebra contains an increasing sequence of finite dimensional subalgebras with weakly dense union. On finite dimensional algebras the density operator technique suffices to obtain results on relative entropy and in some cases properties go through the increasing limit. For example, the lower semi-continuity of the relative entropy is inherited in this way.

Corollary 5.13 Let (p_n) be a sequence of projections in the von Neumann algebra $\mathcal{M}$ such that $p_n \to I$ strongly. If $\mathcal{M}_n = p_n\mathcal{M}p_n + \mathbb{C}(I - p_n)$, then

$$S(\omega|\mathcal{M}_n, \varphi|\mathcal{M}_n) \to S(\omega, \varphi)$$

as $n \to \infty$ for every normal state φ and ω.

Proof. Due to the monotonicity we have $S(\omega|\mathcal{M}_n, \varphi|\mathcal{M}_n) \le S(\omega, \varphi)$. Using Kosaki's formula we assume that

$$\log m - \int_{1/m}^{\infty} \left(t^{-1}\omega(y(t)^*y(t)) + t^{-2}\varphi(x(t)x(t)^*)\right) dt$$

approximates $S(\omega, \varphi)$ for an appropriate step function $x : [1/m, \infty) \to \mathcal{M}$ with $x(t) = I$ for t large enough. Set $x_n(t) = p_n x(t)p_n + (I - p_n)$ and $y(t) = I - x_n(t)$. Then

$$S(\omega, \varphi) \ge S(\omega|\mathcal{M}_n, \varphi|\mathcal{M}_n)$$

$$\ge \log m - \int_{1/m}^{\infty} \left(t^{-1}\omega(y_n(t)^*y_n(t)) + t^{-2}\varphi(x_n(t)x_n(t)^*)\right) dt\,,$$

and since

$$\log m - \int_{1/m}^{\infty} \left(t^{-1}\omega(y_n(t)^*y_n(t)) + t^{-2}\varphi(x_n(t)x_n(t)^*)\right) dt$$

$$\to \log m - \int_{1/m}^{\infty} \left(t^{-1}\omega(y(t)^*y(t)) + t^{-2}\varphi(x(t)x(t)^*)\right) dt$$

(as $n \to \infty$) we can conclude the theorem.   □

The von Neumann entropy is monotone under doubly stochastic mappings, that is

$$S(\omega \circ \alpha, \mathrm{Tr}) \le S(\omega, \mathrm{Tr})\,,$$

whenever α is a positive unital mapping preserving Tr (cf. (3.30)). One might wonder why the condition

$$\alpha(a)^*\alpha(a) \le \alpha(a^*a) \tag{5.17}$$

is not needed in the case of the von Neumann entropy contrary to property (iii) of Corollary 5.12. A quick look at Kosaki's formula sheds light on this point. Assume that φ is tracial which means that $\varphi(a^*a) = \varphi(aa^*)$ $(a \in \mathcal{M})$. Then

$$|\varphi(y)| \le \varphi(|y|) \qquad (y \in \mathcal{M}), \tag{5.18}$$

and we have

$$\varphi((I - y)(I - y)^*) = \varphi(I + yy^*) - \varphi(y + y^*) \ge \varphi((1 - |y|)^*(1 - |y|))\,.$$

We can therefore restrict the second supremum in Kosaki's formula to step functions with self-adjoint values. For a self-adjoint (or more generally, for a normal) $a \in \mathcal{M}$ and for an arbitrary unital positive mapping α the inequality (5.17) always holds (see 3.2.1 in [Bratteli and Robinson 1979]). In this way we arrived at

Corollary 5.14 If φ is tracial and α is a positive unital normal mapping of a von Neumann algebra $\mathcal{M}_0$ into $\mathcal{M}$, then $S(\omega \circ \alpha, \varphi \circ \alpha) \leq S(\omega, \varphi)$.

By means of the representation

$$- \log \lambda = \int_0^\infty \left((t+1)^{-1} - \lambda (t+\lambda)^{-1} \right) \frac{dt}{t} \qquad (\lambda > 0)$$

one can avoid the double sup sign in Kosaki's formula and obtain the relationship

$$S(\omega, \varphi) = \sup \left\{ \int_0^\infty \left((t+1)^{-1} \omega(I) - \omega(y(t)^* y(t)) - t^{-1} \varphi(x(t)x(t)^*) \right) \frac{dt}{t} \right\},$$

where the supremum is taken over all bounded step functions with a countable range in N. This expression is less powerful than Kosaki's formula. For example, due to the presence of step functions with infinite range it does not show the lower semi-continuity of the relative entropy.

Kosaki's formula reveals important, nevertheless not fully characteristic, properties of the relative entropy. We say this because the formula is based upon operator convexity of the function $x \mapsto -\log x$ yet several other entropy-like quantities also share similar properties.

Now we turn to the *conditional expectation property* (used already in Chapter 2 for the axiomatic determination of the relative entropy).

Theorem 5.15 Let $\mathcal{M}$ be a von Neumann algebra and $\mathcal{M}_1$ a von Neumann subalgebra of $\mathcal{M}$. Assume that there exists a faithful normal conditional expectation E of $\mathcal{M}$ onto $\mathcal{M}_1$. If φ_1 and ω are states of $\mathcal{M}_1$ and $\mathcal{M}$, respectively, then

$$S(\omega, \varphi_1 \circ E) = S(\omega | \mathcal{M}_1, \varphi_1) + S(\omega, \omega \circ E). \qquad (5.19)$$

The proof of (5.19) proceeds in several steps.

Lemma 5.16 Equality (5.19) holds if φ_1 is faithful and there are positive constants λ and μ such that $\varphi_1 \leq \lambda \omega_1 \leq \mu \varphi_1$ (writing ω_1 for $\omega | \mathcal{M}_1$).

Proof. Equality (5.19) is obvious if $S(\omega, \varphi_1 \circ E) = \infty$ and $S(\omega, \omega \circ E) = \infty$. Therefore we may assume that one of them is finite. Let us see that in all the three quantities in (5.19) are finite this case. Indeed,

$$S(\omega, \varphi_1 \circ E) \leq S(\omega, \lambda \mu^{-1} \omega \circ E) = \log \mu - \log \lambda + S(\omega, \omega_1 \circ E),$$
$$S(\omega_1, \varphi_1) \leq S(\omega_1, \lambda \mu^{-1} \omega_1) = \log \mu - \log \lambda,$$
$$S(\omega, \omega \circ E) \leq S(\omega, \lambda^{-1} \varphi_1 \circ E) = \log \lambda + S(\omega, \varphi_1 \circ E).$$

We use the differentiation formula of Theorem 5.7 involving the Connes cocycle of states:

$$S(\psi_2, \psi_1) = i \lim_{t \to +0} \frac{\psi_1([D\psi_1, D\psi_2]_t - I)}{t}$$

whenever ψ_1 and ψ_2 are normal states and $S(\psi_1, \psi_2)$ is finite. The chain rule (4.9) of the Connes cocycles says that

$$[D(\varphi_1 \circ E), D\omega]_t = [D(\varphi_1 \circ E), D(\omega \circ E)]_t [D(\omega \circ E), D\omega]_t.$$

Note that under our assumptions $\omega \circ E$ is faithful and

$$[D(\varphi_1 \circ E), D(\omega \circ E)]_t = [D\varphi_1, D\omega_1]_t$$

for every real t (cf. (4.11)). It is convenient to use a short hand notation, say, u_t, v_t, w_t, for the three cocycles, respectively. We have

$$\frac{\omega(u_t) - 1}{t} = \frac{\omega(v_t) - 1}{t} + \frac{\omega(w_t) - 1}{t} + t^{-1} \langle (w_t - I)\Omega, (v_t^* - I)\Omega \rangle$$

and let us show that the third term of the right side tends to 0 as $t \to 0$. Here $(w_t - I)\Omega \to 0$ since the cocycle w_t is so-continuous. On the other hand, $v_t = [D\varphi_1, D\omega_1]_t$ and the function

$$it \mapsto v_t \Omega$$

admits an analytical extension to the strip $\{z \in \mathbb{C} : -1/2 < \mathrm{Re}\, z < 1/2\}$ (see (4.11)). Therefore, $t^{-1}(v_t - I)\Omega$ is bounded in a neighbourhood of 0.   □

Lemma 5.17 Equality (5.19) holds if there are positive constants λ and μ such that $\varphi_1 \leq \lambda\omega_1 \leq \mu\varphi_1$.

Proof. Let $p_1 = \mathrm{supp}\, \varphi_1$. Consider the von Neumann algebras $\mathcal{N} = p_1 \mathcal{M} p_1$ and $\mathcal{N}_1 = p_1 \mathcal{M}_1 p_1$. The mapping $F = E|\mathcal{N}$ is a faithful normal conditional expectation of $\mathcal{N}$ onto $\mathcal{N}_1$. Using the fact that in case $\mathrm{supp}\, \psi_1, \mathrm{supp}\, \psi_2 \leq p_1$

$$S(\psi_1, \psi_2) = S(\psi_1 | p_1 \mathcal{M} p_1, \psi_2 | p_1 \mathcal{M} p_1),$$

we can reduce the statement to the faithful case contained in Lemma 5.16. □

Lemma 5.18 (5.19) holds if $\varphi_1 \leq \lambda\omega_1$ with a positive constant λ.

Proof. For $\varepsilon > 0$, let $\nu_1^\varepsilon = (1 - \varepsilon)\varphi_1 + \varepsilon\omega_1$. Then $\nu_1^\varepsilon \leq ((1 - \varepsilon)\lambda + \varepsilon)\omega_1$ and $\omega_1 \leq \varepsilon^{-1}\nu_1^\varepsilon$. We can benefit from Lemma 5.17 to get

$$S(\omega, \nu_1^\varepsilon \circ E) = S(\omega_1, \nu_1^\varepsilon) + S(\omega, \omega \circ E).$$

Next the limit $\varepsilon \to +0$. We may assume that $S(\omega, \omega_1 \circ E)$ is finite. By convexity

$$S(\omega, \nu_1^\varepsilon \circ E) \leq (1 - \varepsilon)S(\omega, \varphi_1 \circ E) + \varepsilon S(\omega, \omega_1 \circ E),$$

and we infer

$$\limsup_{\varepsilon \to 0} S(\omega, \nu_1^\varepsilon \circ E) \leq S(\omega, \varphi_1 \circ E).$$

A combination with lower semi-continuity yields

$$\lim_{\varepsilon \to 0} S(\omega, \nu_1^\varepsilon \circ E) = S(\omega, \varphi_1 \circ E).$$

By a similar argument

$$\lim_{\varepsilon \to 0} S(\omega_1, \nu_1^\varepsilon) = S(\omega_1, \varphi_1)$$

and the proof of the lemma is complete.   □

Now we show (5.19) without any additional hypothesis. For $1 > \varepsilon > 0$ let $\psi^\varepsilon = (1 - \varepsilon)\omega + \varepsilon\varphi_1 \circ E$. Then $\varphi_1 \le \varepsilon^{-1}(\psi^\varepsilon|\mathcal{M}_1)$ and by Lemma 5.18 we have

$$S(\psi^\varepsilon, \varphi_1 \circ E) = S(\psi^\varepsilon|\mathcal{M}_1, \varphi_1) + S(\psi^\varepsilon, \psi^\varepsilon \circ E).$$

Applying convexity and lower semi-continuity as in the proof of Lemma 5.18 and letting $\varepsilon \to +0$ we conclude (5.19).   □

Let E be as in Theorem 5.15 and φ be a normal state such that $\varphi \circ E = \varphi$. For a fixed normal state ω_1 of $\mathcal{M}_1$ write $\mathcal{K}$ for the set of all normal extensions of ω_1 to $\mathcal{M}$. Then $\mathcal{K}$ is a convex set and we can ask for an $\omega \in \mathcal{K}$ which minimizes $S(\omega, \varphi)$. If there exists a $\psi \in \mathcal{K}$ such that $S(\psi, \varphi)$ is finite, then the minimizer is unique, and it is $\omega_1 \circ E$. This is evident from (5.19), but we also have the following norm estimate from Theorem 5.5.

Corollary 5.19 With the above notation we have

$$\tfrac{1}{2}\|\omega - \omega_1 \circ E\|^2 + S(\omega_1, \varphi|\mathcal{M}_1) \le S(\omega, \varphi), \qquad (\omega \in \mathcal{K}).$$

Theorem 5.15 may be applied to deduce some identities and inequalities concerning relative entropy.

Corollary 5.20 Let $\varphi_1 \otimes \varphi_2$ and ω_{12} be normal states of the tensor product von Neumann algebra $\mathcal{M}_1 \otimes \mathcal{M}_2$ and let $\omega_i = \omega_{12}|\mathcal{M}_i$ $(i = 1, 2)$. Then

$$S(\omega_{12}, \varphi_1 \otimes \varphi_2) = S(\omega_1, \varphi_1) + S(\omega_{12}, \omega_1 \otimes \varphi_2). \tag{5.20}$$

Proof. The equality holds trivially if the support of ω_{12} is strictly larger then that of $\varphi_1 \otimes \varphi_2$. Hence we may assume that $\varphi_1 \otimes \varphi_2$ is faithful. The mapping $a \otimes b \mapsto \varphi_2(b)a$ extends to a faithful normal conditional expectation E preserving the state $\varphi_1 \otimes \varphi_2$. Hence Theorem 5.15 applies and $\omega_{12} \circ E = \omega_1 \otimes \varphi_2$ yields the corollary.   □

According to the previous corollary

$$S(\omega_{12}, \omega_1 \otimes \varphi_2) = S(\omega_2, \varphi_2) + S(\omega_{12}, \omega_1 \otimes \omega_2) \tag{5.21}$$

and a combination of (5.20) and (5.21) gives the identity

$$S(\omega_{12}, \varphi_1 \otimes \varphi_2) = S(\omega_1, \varphi_1) + S(\omega_2, \varphi_2) + S(\omega_{12}, \omega_1 \otimes \omega_2). \tag{5.22}$$

□

Corollary 5.21 With the above notation we have

$$S(\omega_{12}, \varphi_1 \otimes \varphi_2) \geq S(\omega_1, \varphi_1) + S(\omega_2, \varphi_2).$$

Moreover, if the equality holds and the relative entropies are finite, then $\omega = \omega_1 \otimes \omega_2$.

Proof. The proof is nothing else but a look at (5.22) and Theorem 5.5. □

Proposition 5.22 Let φ and $\omega_1, \omega_2, \ldots, \omega_n$ be arbitrary positive normal functionals on the von Neumann algebra $\mathcal{M}$. Then the equality

$$S(\omega, \varphi) + \sum_{i=1}^{n} S(\omega_i, \omega) = \sum_{i=1}^{n} S(\omega_i, \varphi)$$

holds with $\omega = \sum_{i=1}^{n} \omega_i$.

Proof. We carry out the proof for $n = 2$. Upon rescaling it can be assumed that φ and ω are states. Define $\mathcal{N} = \mathcal{M} \oplus \mathcal{M}$ and set the functionals $\varphi_{12}, \omega_{12}, \psi_{12}$ by the following formulas :

$$\varphi_{12}(a \oplus b) = \tfrac{1}{2}(\varphi(a) + \varphi(b))$$
$$\omega_{12}(a \oplus b) = \omega_1(a) + \omega_2(b)$$
$$\psi_{12}(a \oplus b) = \tfrac{1}{2}(\omega_1(a) + \omega_1(b) + \omega_2(a) + \omega_2(b))$$

for $a, b \in \mathcal{M}$. The mapping $E : a \oplus b \mapsto \tfrac{1}{2}(a+b) \oplus \tfrac{1}{2}(a+b)$ is a φ_{12} preserving faithful normal conditional expectation onto the diagonal subalgebra $\mathcal{N}_0 = \{a \oplus a : a \in \mathcal{M}\}$. Since $\psi_{12} = \omega_{12} \circ E$, Theorem 5.15 gives

$$S(\omega_1, \varphi/2) + S(\omega_2, \varphi/2)$$
$$= S(\omega_1 + \omega_2, \varphi) + S\left(\omega_1, \frac{\omega_1 + \omega_2}{2}\right) + S\left(\omega_2, \frac{\omega_1 + \omega_2}{2}\right)$$

which is equivalent to the statement. □

The equality in Proposition 5.22 will be called *Donald's identity*.

The simplest definition of relative entropy has been given by statistical operators (Chapter 1),but statistical operators are not available in general due to the lack of trace functional and it is the spatial derivative operator which becomes the crucial object in the definition. For positive functionals of an arbitrary C*-algebra $\mathcal{A}$ the relative entropy may be defined via the enveloping von Neumann algebra $\mathcal{A}^{**}$. Let us recall that the second dual $\mathcal{A}^{**}$ of $\mathcal{A}$ is the double commutant of $\mathcal{A}$ in its universal representation. Every state ψ of $\mathcal{A}$ admits a unique normal extension $\tilde{\psi}$ to $\mathcal{A}^{**}$. We set

$$S(\psi_1, \psi_2) = S(\tilde{\psi}_1, \tilde{\psi}_2),$$

since the right hand side is defined for the normal states $\tilde{\psi}_1$ and $\tilde{\psi}_2$. Kosaki's formula supplies us with an equivalent definition.

$$S(\nu, \psi) = \sup\sup\left\{\nu(I)\log n - \int_{1/n}^{\infty}\left(\nu(y(t)^*y(t)) - t^{-1}\psi(x(t)x(t)^*)\right)\frac{dt}{t}\right\}.$$

Here the double supremum is taken as in Theorem 5.11. The GNS-const-ruction may be used to have a third equivalent. Let $(\mathcal{H}, \Phi, \pi)$ stand for the GNS-triplet for the unital C*-algebra $\mathcal{A}$ with the positive functional φ of $\mathcal{A}$. Let ω be a positive functional on $\mathcal{A}$. We write $\bar{\omega}$ for the normal state of $\pi(\mathcal{A})''$ such that

$$\bar{\omega}(\pi(a)) = \omega(a) \qquad (a \in \mathcal{A}),$$

provided a normal functional with this property exists. (When $\bar{\omega}$ exists it is said that ω is quasi-contained in φ.) We check that

$$S(\omega, \varphi) = \begin{cases} S(\bar{\omega}, \bar{\varphi}) & \text{if } \bar{\omega} \text{ exists,} \\ +\infty & \text{otherwise.} \end{cases}$$

When $\bar{\omega}$ exists, the equality $S(\omega, \varphi) = S(\bar{\omega}, \bar{\varphi})$ follows obviously from Kosaki's formula. Now assume that ω is not quasi-contained in φ. One knows that in this case the central support z_ω of $\tilde{\omega}$ is not smaller than the central support z_φ of $\tilde{\varphi}$. Therefore, there exists a projection p in $\mathcal{A}^{**}$ such that $\tilde{\varphi}(p) = 0$ and $\tilde{\omega}(p) \neq 0$. This yields $S(\tilde{\omega}, \tilde{\varphi}) = +\infty$. The properties of the relative entropy functional of states of a C*-algebra are easily deduced from the von Neumann algebra case. We summarize the most important ones.

Proposition 5.23 The relative entropy of positive functionals of a C*- algebra possesses the following properties.

(i) $(\omega, \varphi) \mapsto S(\omega, \varphi)$ is convex and weakly lower semi-continuous.
(ii) $\|\varphi - \omega\|^2 \leq 2S(\omega, \varphi)$ if $\varphi(I) = \omega(I) = 1$.
(iii) $S(\omega, \varphi_1) \geq S(\omega, \varphi_2)$ if $\varphi_1 \leq \varphi_2$.
(iv)   For a unital Schwarz map $\alpha : \mathcal{A}_0 \to \mathcal{A}_1$ the relation $S(\omega \circ \alpha, \varphi \circ \alpha) \leq S(\omega, \varphi)$ holds.
(v)   For $\omega = \sum_i^n \omega_i$ we have $S(\omega, \varphi) + \sum_{i=1}^n S(\omega_i, \omega) = \sum_{i=1}^n S(\omega_i, \varphi)$.
(vi)   Let $(\mathcal{A}_i)_i$ be an increasing net of C*-subalgebras of the C*-algebra $\mathcal{A}$ so that $\cup_i \mathcal{A}_i$ is dense in $\mathcal{A}$. Then the increasing net $(S(\omega|\mathcal{A}_i, \varphi|\mathcal{A}_i))_i$ converges to $S(\omega, \varphi)$ for states ω and φ of $\mathcal{A}$.

It is worth noting that one can infer the strict convexity of the relative entropy in the second variable. Let $\varphi, \omega_1, \omega_2$ be states and assume that $S(\varphi, \omega_1)$ and $S(\varphi, \omega_2)$ are finite. The Donald identity (v) gives

$$\lambda S(\omega_1, \varphi) + (1 - \lambda)S(\omega_2, \varphi) - S(\lambda\omega_1 + (1 - \lambda)\omega_2, \varphi)$$
$$= \lambda S(\omega_1, \lambda\omega_1 + (1 - \lambda)\omega_2) + (1 - \lambda)S(\omega_2, \lambda\omega_1 + (1 - \lambda)\omega_2).$$

If for $0 < \lambda < 1$ the left hand side vanishes, then $\omega_1 = \omega_2$ according to property (ii). The strict convexity is also clear from the following estimate.

Proposition 5.24 Let φ and ω_i be states on a C*-algebra $(i = 1, 2, \ldots, n)$. Then

$$0 \geq S\left(\sum_{i=1}^{n} \lambda_i \omega_i, \varphi\right) - \sum_{i=1}^{n} \lambda_i S(\omega_i, \varphi) \geq \sum_{i=1}^{n} \lambda_i \log \lambda_i$$

provided that $\lambda_i \geq 0$ and $\sum_{i=1}^{n} \lambda_i = 1$.

Proof. We concentrate on the second inequality. By the Donald identity

$$S\left(\sum_{i=1}^{n} \lambda_i \omega_i, \varphi\right) = \sum_{i=1}^{n} \lambda_i S(\omega_i, \varphi) - \sum_{j=1}^{n} \lambda_j S\left(\omega_j, \sum_{i=1}^{n} \lambda_i \omega_i\right).$$

Since $S\left(\omega_j, \sum_{i=1}^{n} \lambda_i \omega_i\right) \leq S(\omega_j, \lambda_j \omega_j) = -\log \lambda_j$ the statement is proved. $\square$

Proposition 5.25 Let $(\varphi_{j,k})$ be a finite family of positive functionals on a C*-algebra. If $\varphi = \sum_{j,k} \varphi_{j,k}$, $\varphi_j^{(1)} = \sum_k \varphi_{j,k}$ and $\varphi_k^{(2)} = \sum_j \varphi_{j,k}$, then

$$\sum_{j,k} S(\varphi_{j,k}, \varphi) \geq \sum_j S(\varphi_j^{(1)}, \varphi) + \sum_k S(\varphi_k^{(2)}, \varphi).$$

Proof. The ingredients of the proof are the joint convexity (cf. (5.9)) and Donald's identity (v) in the previous proposition. Due to Donald's identity we have

$$S(\varphi_j^{(1)}, \varphi) + \sum_n S(\varphi_{j,n}, \varphi_j^{(1)}) = \sum_n S(\varphi_{j,n}, \varphi)$$

for every index j. It follows that

$$\sum_j S(\varphi_j^{(1)}, \varphi) + \sum_k S(\varphi_k^{(2)}, \varphi) - \sum_{j,k} S(\varphi_{j,k}, \varphi)$$

$$= \sum_k S(\varphi_k^{(2)}, \varphi) - \sum_{j,k} S(\varphi_{j,k}, \varphi_j^{(1)}). \tag{5.23}$$

From the joint convexity

$$S(\varphi_k^{(2)}, \varphi) = S\left(\sum_j \varphi_{j,k}, \sum_j \varphi_j^{(1)}\right) \leq \sum_j S(\varphi_{j,k}, \varphi_j^{(1)}).$$

Summing up for the subscript k and comparing with (5.23) we arrive at the stated inequality. $\square$

If we visualize the decomposition $(\varphi_{j,k})$ in a matrix arrangement so that $(\varphi_j^{(1)})_j$ and $(\varphi_k^{(2)})_k$ are the vertical and horizontal marginals, respectively. This motivates the term *marginal inequality*

Theorem 5.26 Let $\mathcal{K}$ be a closed convex set in the state space of the C*-algebra $\mathcal{A}$ and let φ be a state of $\mathcal{A}$. If $F : \mathcal{K} \to \mathbb{R} \cup \{+\infty\}$ is a lower semi-continuous convex functional and

$$d = \inf\{F(\omega) + S(\omega, \varphi) : \omega \in \mathcal{K}\}$$

is finite, then there exists a unique $\psi \in \mathcal{K}$ such that $F(\psi) + S(\varphi, \psi) = d$.

Proof. We choose a sequence $(\omega_n) \subset \mathcal{K}$ such that $\lim_{n\to\infty}\{F(\omega_n)+S(\omega_n,\varphi)\}$ $= d$. From the convexity of F and from identity (v) of Proposition 5.23 we infer

$$F(\omega_n) + S(\omega_n,\varphi) + F(\omega_m) + S(\omega_m,\varphi)$$
$$\geq 2F\left(\frac{\omega_n + \omega_m}{2}\right) + 2S\left(\frac{\omega_n + \omega_m}{2},\varphi\right)$$
$$+ S\left(\omega_m, \frac{\omega_n + \omega_m}{2}\right) + S\left(\omega_n, \frac{\omega_n + \omega_m}{2}\right)$$
$$\geq 2d + S\left(\omega_n, \frac{\omega_n + \omega_m}{2}\right) + S\left(\omega_m, \frac{\omega_n + \omega_m}{2}\right).$$

Hence

$$S\left(\omega_n, \frac{\omega_n + \omega_m}{2}\right) \leq \varepsilon$$

if n and m are sufficiently large. Inequality (ii) of Proposition 5.22 ensures that (ω_n) is a Cauchy sequence that converges in norm to a state $\psi \in \mathcal{K}$. From lower semi-continuity

$$F(\psi) + S(\psi,\varphi) \leq \liminf_n\{F(\omega_n) + S(\omega_n,\varphi)\} \leq d$$

and ψ is a minimizer. Since any minimizing sequence is convergent, the minimizer must be unique. □

Proposition 5.27 Let φ be a positive normal functional of the von Neumann algebra $\mathcal{M}$ and let $t \in \mathbb{R}$. Then the set

$$\mathcal{K}(\varphi,t) = \{\omega \in \mathcal{M}_+^* : S(\omega,\varphi) \leq t\}$$

consists of normal functionals and it is a convex compact set with respect to the $\sigma(\mathcal{M}_*, \mathcal{M})$ topology.

Proof. If $S(\omega,\varphi)$ is finite, then ω is quasi-contained in φ, and so ω is normal. Therefore the $\sigma(\mathcal{M}_*, \mathcal{M})$ and the $\sigma(\mathcal{M}^*, \mathcal{M})$ topologies coincide on $\mathcal{K}(\varphi,t)$. To show compactness it suffices to note that $\mathcal{K}(\varphi,t)$ is closed and bounded. Boundedness follows from

$$\omega(I)\,(\log\omega(I) - \log\varphi(I)) \leq S(\omega,\varphi)$$

and $\mathcal{K}(\varphi,t)$ is closed and convex as a consequence of lower semi-continuity and convexity of the relative entropy. □

Suppose that $\mathcal{A}$ and $\mathcal{B}$ are C*-algebras, and $\mathcal{A} \odot \mathcal{B}$ is their algebraic tensor product. Forming the completion of $\mathcal{A} \odot \mathcal{B}$ with respect to a C*-norm we obtain a C*-algebra which can reasonably be regarded as a "C*-algebra tensor product" of $\mathcal{A}$ and $\mathcal{B}$. If there are several C*-algebraic norms on $\mathcal{A} \odot \mathcal{B}$ then the C*-algebra tensor product is not unique. Among the possible C*-algebraic norms of $\mathcal{A} \odot \mathcal{B}$ there is a smallest norm and completion with respect to this minimal norm yields the *injective tensor product*. It is also called the "spatial" tensor product because it can be obtained in representations as well. On $\mathcal{A} \odot \mathcal{B}$ there is also a maximal norm and completion with respect to this is the *projective tensor product*. Product states may be formed in any tensor product. If $\varphi_{12} = \varphi_1 \otimes \varphi_2$ is a product state on the projective tensor product $\mathcal{A}_1 \otimes \mathcal{A}_2$ then its GNS-representation is the tensor product of the representations corresponding to φ_1 and φ_2, respectively.

Proposition 5.28 Let $\varphi_1 \otimes \varphi_2$ and ω_{12} be states of the projective tensor product C*-algebra $\mathcal{A}_1 \otimes \mathcal{A}_2$ let $\omega_i = \omega_{12}|\mathcal{A}_i$ $(i = 1, 2)$. Then

$$S(\omega_{12}, \varphi_1 \otimes \varphi_2) \geq S(\omega_1, \varphi_1) + S(\omega_2, \varphi_2).$$

Proof. If ω_{12} is not quasi-contained in $\varphi_1 \otimes \varphi_2$, then the stated inequality holds trivially with $S(\omega_{12}, \varphi_1 \otimes \varphi_2) = +\infty$. If ω_{12} is quasi-contained in $\varphi_1 \otimes \varphi_2$ then we may pass to the GNS-representation and may refer to the von Neumann algebraic version of Corollary 5.21.   □

In some cases the C* tensor product is unique. A C*-algebra $\mathcal{A}$ is said to be *nuclear* if, for every C*-algebra $\mathcal{B}$, there is only one C*-norm on $\mathcal{A} \odot \mathcal{B}$. Finite dimensional and abelian C*-algebras are nuclear. A C*-algebra $\mathcal{A}$ is called *AF-algebra* if it contains an increasing sequence of finite dimensional subalgebras such that their union is norm dense in $\mathcal{A}$. It can be proved that the inductive limit of nuclear C*-algebras is itself nuclear. In particular, every AF-algebra is nuclear. Let $(\alpha_i : \mathcal{A}_i \to \mathcal{A})_i$ be a net of unital completely positive mapping defined on finite dimensional algebras. We shall call $(\alpha_i)_i$ a norm approximating net if for each i there exists a unital completely positive mapping $\beta_i : \mathcal{A} \to \mathcal{A}_i$ such that $||\alpha_i \circ \beta_i(a) - a|| \to 0$ $(a \in \mathcal{A})$. (The net $\alpha_i \circ \beta_i$ approximates the identity of $\mathcal{A}$ in the topology of pointwise norm convergence.) The class of nuclear C*-algebras is characterized by the existence of a *norm approximating net*.

Theorem 5.29 Let $\mathcal{A}$ be a nuclear C*-algebra with states φ and ω. If $(\alpha_i : \mathcal{A}_i \to \mathcal{A})_i$ is a norm approximating net then

$$S(\omega, \varphi) = \lim_i S(\omega \circ \alpha_i, \varphi \circ \alpha_i).$$

Consequently, $S(\omega, \varphi)$ is the lowest upper bound of the quantities $S(\omega \circ \alpha, \varphi \circ \alpha)$ where α ranges over all completely positive unital mappings from a finite dimentional algebra into $\mathcal{A}$.

Proof. Since $S(\omega \circ \alpha, \varphi \circ \alpha) \leq S(\omega, \varphi)$ holds for any completely positive mapping α, we show that, given an approximating net $(\alpha_i : \mathcal{A}_i \to \mathcal{A})_i$, and numbers $0 \leq u < S(\omega, \varphi)$, $0 < \varepsilon$,

$$S(\omega \circ \alpha_i, \varphi \circ \alpha_i) \geq u - \varepsilon \tag{5.24}$$

holds for i sufficiently large.

The main ingredient of the proof will be Kosaki's formula. There exists an $n \in \mathbb{N}$ and a step function $x : [1/n, \infty) \to \mathcal{A}$ such that it has finite range, $x(t) = I$ for large t and

$$\log n - \int_{1/n}^{\infty} \left(t^{-1} \omega(y(t)^* y(t)) + t^{-2} \varphi(x(t) x(t)^*) \right) dt \geq u.$$

For large i we have

$$\log n - \int_{1/n}^{\infty} \left(t^{-1} \omega(\alpha_i \circ \beta_i(y(t)^* y(t))) + t^{-2} \varphi(\alpha_i \circ \beta_i(x(t) x(t)^*)) \right) dt \geq u - \varepsilon,$$

where $\|\alpha_i \circ \beta_i(a) - a\| \to 0$ for every $a \in \mathcal{A}$. So, writing $x_i(t)$ and $y_i(t)$ for $\beta_i(x_i(t))$ and $\beta_i(y_i(t))$, respectively, we obtain from the Schwarz inequality

$$\log n - \int_{1/n}^{\infty} \left(t^{-1} (\omega \circ \alpha_i)(y_i(t)^* y_i(t)) + t^{-2} (\varphi \circ \alpha_i)(x_i(t) x_i(t)^*) \right) dt \geq u - \varepsilon,$$

and Kosaki's formula yields (5.24) for i sufficiently large.   □

Injective von Neumann algebras admit a weak* approximating net for the identity, and with arguments similar to those of the preceding proof one obtains the following.

Theorem 5.30 Let $\mathcal{M}$ be an injective von Neumann algebra with normal states φ and ω. Then the relative entropy $S(\omega, \varphi)$ is the supremum of all the quantities $S(\omega \circ \alpha, \varphi \circ \alpha)$, where α runs over all completely positive unital mappings from a finite dimentional algebra into $\mathcal{M}$.

Completely positive unital mappings from a finite dimentional algebra will be called testing channels in Part III and they will play a crucial role in the definition of dynamical entropy.

Notes and Remarks. Umegaki's relative entropy defined by means of statistical operators was extended to faithful normal states of arbitrary von Neumann algebras in [Araki 1976]. Allowing nonfaithful states Araki established the main properties of the relative entropy in [Araki 1977] by means of a rather complicated unbounded operator technique. For example, in this paper the monotonicity under taking a subalgebra was proved only in specific situations and lower norm semicontinuity was obtained. Inequality (5.7) follows from Lemma D of [Araki and Masuda 1982] and (5.8) is proved in the appendix of [Petz 1985b]. The important Theorem 5.3 appeared in [Uhlmann 1977]. Our proof is the translation of Uhlmann's abstract interpolation method to operator theory. In fact, Uhlmann defined a relative entropy for states of a *-algebra by interpolating between the seminorms $\|a\|_0 = \varphi(a^* a)^{1/2}$ and $\|a\|_1 = \omega(aa^*)^{1/2}$.

The norm estimate of Theorem 5.5 is due to [Hiai, Ohya and Tsukada 1981]. *Kosaki's formula* was based upon some ideas of [Pusz and Woronowicz 1978] and our presentation follows the original article [Kosaki 1986a]. A similar expression was used in [Donald 1985] in order to establish the weak lower semicontinuity of the relative entropy functional. Lemma 5.8 is already in [Kosaki 1982]. Concerning operator monotone functions the monograph [Donogue 1974] is the standard reference but [Hansen and Pedersen 1982] may also be suggested. Corollary 5.13 seems to be new. It was proved by Gardner that the tracial property for a state is equivalent to (5.18). The reference and characterizations of tracial states may be found in [Petz and Zemánek 1988]. Theorem 5.15 is due to [Petz 1990b]. Its weaker form and Theorem 5.7 are in [Petz 1986a]. It must be noted that (5.19) was proved also in [Hiai, Ohya and Tsukada 1981] under the restrictive condition that the subalgebra $\mathcal{M}_1$ is in the centralizer of the faithful normal state φ given on a von Neumann algebra.

Equality (5.22) was communicated to the authors by M. Donald. Corollary 5.21 is the superadditivity of the relative entropy. Donald's identity appeared in [Donald 1987a] for injective von Neumann algebras. The presented proof is from [Petz 1991a]. The *marginal inequality* was first used in [Connes 1985] in connection with the definition of the dynamical entropy. Our proof follows the advice of [Araki 1987a]. The strict convexity of relative entropy in the second variable was emphasized first in [Donald 1987a] but an independent and quite different proof is found in [Petz 1986c].

Concerning tensor products of C*-algebras we refer to the books [Takesaki 1979] and [Kadison and Ringrose 1986]. A C*-algebra possessing a norm approximating net is usually called *semidiscrete*. The equivalence of nuclearity and semidiscreteness was proved in [Choi and Effros 1979]. The works [Effros 1978] and [Lance 1982] review the relation of injectivity, semidiscreteness and nuclearity. Most physically important operator algebras are nuclear or injective. The algebra of the canonical commutation relation is a nonseparable nuclear C*-algebra.

Comments. The relative entropy can be used to define a measure for entanglement, [Vedral at al 1998a]. Let $\mathcal{A}$ be a tensorproduct $M_n(\mathbb{C}) \otimes M_k(\mathbb{C})$ and be $\mathcal{K}$ be the closed convex hull of product densities. (Elements of $\mathcal{K}$ are called *separable or disentangled states*.) Given a density matrix $D \in \mathcal{A}$, there is a unique minimizer D_0 of the relative entropy $S(D, \rho)$ when ρ runs over $\mathcal{K}$,

$$E(D) := \inf\{S(D, \rho) : \rho \in \mathcal{K}\}.$$

For a pure state the entanglement measure is the von Neumann entropy of the reduced density matrix [Vedral at al 1998b].

6 From Relative Entropy to Entropy

In contrast with the commutative case, the state space of a quantum system is not a *Choquet simplex* in the sense that its states admit several *extremal decompositions*. For example, in $\mathcal{A} = M_2(\mathbb{C})$ the general form of a density matrix is

$$D = \frac{1}{2} \begin{pmatrix} 1+a & b+\mathrm{i}c \\ b-\mathrm{i}c & 1-a \end{pmatrix} \tag{6.1}$$

where a, b, c are real numbers and $a^2 + b^2 + c^2 \leq 1$. Thanks to the affine correspondence $D \longleftrightarrow (a,b,c)$ we can visualize the state space as a ball of radius 1 whose surface points correspond to pure states.

Let φ be a state of a finite quantum system and $\varphi = \sum_i \lambda_i \psi_i$ be an extremal decomposition (that is, every ψ_i is pure). Approaching from information theory one might think that the entropy of φ is $-\sum_i \lambda_i \log \lambda_i$. This, however, would not be satisfactory because the λ_i's are not in general the probabilities of mutually exclusive events. In fact,

$$S(\varphi) \leq -\sum_i \lambda_i \log \lambda_i, \tag{6.2}$$

and equality holds if and only if the extremal decomposition $\sum_i \lambda_i \psi_i$ is orthogonal. This can be deduced from the following proposition. Inequality (6.2) may be interpreted by saying that in the sense of information content, the most economical extremal decomposition is the orthogonal one.

In this chapter it will be observed that the von Neumann entropy can be obtained from the decompositions of the given state by means of relative entropy. This will allow us the extension of the entropy of states to arbitrary C*-algebras. Although the generalization can be made, the new entropy is infinite in the type II and III cases. In order to have a finite quantity one may restrict the class of allowed decompositions. What we get in this way is a concept of entropy of a state with respect to a convex subset of the state space.

Proposition 6.1 Let φ be a state of a finite quantum system $\mathcal{A}$. If φ is a convex combination $\sum_j \mu_j \varphi_j$ of states then

$$S(\varphi) \geq \sum_j \mu_j S(\varphi_j, \varphi).$$

When all the φ_j's are pure, the equality sign applies.

Proof. It suffices to prove the equality because inequality follows by the convexity of the relative entropy. By simple computation we have

$$\sum_j \mu_j S(\varphi_j, \varphi) = \sum_j \mu_j \operatorname{Tr} \eta(D_{\varphi_j}) + \operatorname{Tr} \eta(D_\varphi)$$

and the first term on the right hand side vanishes when all the D_{φ_j}'s are projections.   □

Now let $\varphi = \sum_i \lambda_i \psi_i$ be an extremal decomposition. Combining Proposition 6.1 with the monotonicity of the relative entropy in the first variable we infer (6.2) as follows:

$$S(\varphi) = \sum_i \lambda_i S(\psi_i, \varphi) \le \sum_i \lambda_i S(\lambda_i \psi_i, \psi_i) = -\sum_i \lambda_i \log \lambda_i .$$

Now we turn to the *von Neumann entropy* of normal states of type I factors (or direct sum of type I factors). On these algebras the functional Tr should be understood as the faithful normal semifinite trace which takes the value 1 at each minimal projection. If $\mathcal{M} = B(\mathcal{H})$, the density D_φ of any normal state φ with respect to Tr is a compact contraction and this is the reason why the von Neumann entropy

$$S(\varphi) = \operatorname{Tr} \eta(D_\varphi) \tag{6.3}$$

is easy to handle (compared with the Segal entropy, subject of the next chapter). Let φ be a normal state on $B(\mathcal{H})$ and let $(\mathcal{H}_n)$ be an increasing sequence of subspaces of $\mathcal{H}$ such that $\cup_n \mathcal{H}_n$ is dense in the support of φ. Then

$$S(\varphi) = \lim_{n \to \infty} S(\varphi | B(\mathcal{H}_n)) \tag{6.4}$$

follows from the definition. This simple observation makes room for an easy approximation argument. The concavity and the *strong subadditivity* may be obtained in this way from the corresponding finite dimensional results (Proposition 1.6 and 1.9).

Proposition 6.2 The von Neumann entropy is lower semi-continuous and possesses the following concavity property

$$\lambda S(\varphi_1) + (1 - \lambda)S(\varphi_2) \le S(\lambda \varphi_1 + (1 - \lambda)\varphi_2)$$
$$\le \lambda S(\varphi_1) + (1 - \lambda)S(\varphi_2) + \eta(\lambda) + \eta(1 - \lambda) .$$

Proposition 6.3 Let φ_{123} be a normal state of $B(\mathcal{H}_1) \otimes B(\mathcal{H}_2) \otimes B(\mathcal{H}_3)$ with partials φ_1, φ_2, φ_3, φ_{12} and φ_{23}. Then the strong subadditivity

$$S(\varphi_{123}) + S(\varphi_2) \le S(\varphi_{12}) + S(\varphi_{23})$$

holds.

The next lemma is an easy piece of functional analysis.

Lemma 6.4 Let φ_{12} be a pure state on $B(\mathcal{H}_1) \otimes B(\mathcal{H}_2)$ and let D_1 and D_2 be the density of the partials $\varphi_1 \equiv \varphi_{12}|B(\mathcal{H}_1)$ and $\varphi_2 \equiv \varphi_{12}|B(\mathcal{H}_1)$, respectively. Then the nonzero eigenvalues of D_1 are the same as those of D_2, multiplicities included. In particular, $S(\varphi_1) = S(\varphi_2)$.

Proof. Let (ξ_i) and (η_j) be bases in $\mathcal{H}_1$ and $\mathcal{H}_2$, respectively. φ being pure, it is induced by a vector $\sum_{i,j} c_{ij}\xi_i \otimes \eta_j$. One computes

$$D_1 = \sum_{i,i',k} c_{ik}\bar{c}_{i'k}|\xi_i\rangle\langle\xi_{i'}|$$

and

$$D_2 = \sum_{j,j',k} c_{kj}\bar{c}_{kj'}|\xi_j\rangle\langle\xi_{j'}| \, .$$

This shows that the nonzero eigenvalues of D_1 are those of the matrix $(c_{ij})(c_{ij})^*$ while the nonzero eigenvalues of D_2 are those of $(c_{ij})^*(c_{ij})$. It is known that the spectrum of ab and ba are the same except for the point 0, so we arrive at the statement on the spectrum of D_1 and D_2. $S(\varphi_1) = S(\varphi_2)$ is a trivial consequence. $\qquad\square$

The previous lemma, with the help of a purification argument will yield the *triangle inequality* for the von Neumann entropy, Proposition 6.5. If $D_\varphi = \sum_i \lambda_i E_{ii}$ is the density operator of the normal state φ of $B(\mathcal{H})$ expressed by means of matrix units $(E_{ij})_{ij}$, then the operator

$$\sum_{k,l} \sqrt{\lambda_k \lambda_l} E_{kl} E_{kl} \tag{6.5}$$

is the density of a pure state on $B(\mathcal{H}) \otimes B(\mathcal{H}) = B(\mathcal{H} \otimes \mathcal{H})$ which gives the initial state φ as partials. This trick will be the key to the proof of the triangle inequality. The state with density (6.5) is a vector state induced by the vector

$$\sum_i \sqrt{\lambda_i}\xi_i \otimes \xi_i \tag{6.6}$$

if the projection E_{ii} corresponds to the vector ξ_i. This *purification* procedure is not strictly unique, instead of the vector (6.6) one could take

$$\sum_i \sqrt{\lambda_i}\xi_i \otimes \eta_i \tag{6.7}$$

in $\mathcal{H} \otimes \mathcal{H}'$ where $(\eta_i)_i$ is a set of pairwise orthogonal unit vectors in $\mathcal{H}'$. All purifications of the given normal state φ of $B(\mathcal{H})$ are of this form. In quantum theory not all the states of a composite system are written as convex compositions of product states. A non-product state like (6.7) can not be a convex combination. Such states are of deeply quantum theoretical nature and have caused so many problems for philosophers.

Proposition 6.5 Let ω_{12} be a normal state of $B(\mathcal{H}_1) \otimes B(\mathcal{H}_2) = B(\mathcal{H}_1 \otimes \mathcal{H}_2)$ with partials ω_1 and ω_2. Then

$$|S(\omega_1) - S(\omega_2)| \leq S(\omega_{12}).$$

Proof. On a three-fold-product $B(\mathcal{H}_1) \otimes B(\mathcal{H}_2) \otimes B(\mathcal{H}_3)$ we can have a pure state ω_{123} which supplies the partial ω_{12} on the first two factors. According to Lemma 6.4 we have $S(\omega_1) = S(\omega_{23})$. So the strong subadditivity

$$S(\omega_{123}) + S(\omega_2) \leq S(\omega_{12}) + S(\omega_{23})$$

yields $S(\omega_2) - S(\omega_1) \leq S(\omega_{12})$. Exchanging the subscripts 1 and 2 the triangle inequality follows.   □

The next result is the continuity of the entropy at bounded energy.

Proposition 6.6 Let $0 \leq H \in B(\mathcal{H})$ be a self-adjoint operator such that $\mathrm{Tr}\, e^{-\beta H} < +\infty$ for every $\beta > 0$. Then for every $t > 0$ the entropy is continuous on the set

$$\{\varphi \in \mathfrak{S}_*(B(\mathcal{H})) : \quad \varphi(H) \leq t\}.$$

Proof. Due to lower semi-continuity it is sufficient to prove that

$$S(\varphi) \geq \limsup_{n \to \infty} S(\varphi_n) \tag{6.8}$$

if $\varphi_n \to \varphi$ and φ_n, φ are in the given set. Let us denote the positive functional with density $e^{-\beta H}$ by ω_β. Then

$$S(\varphi) = -S(\varphi, \omega_\beta) + \beta\varphi(H),$$

and similarly with φ_i in place of φ. From the lower semi-continuity of the relative entropy we have

$$-S(\varphi, \omega_\beta) + \beta\varphi(H) \geq \limsup_{n \to \infty} -S(\varphi_n, \omega_\beta) + \beta\varphi(H)$$

$$\geq \limsup_{n \to \infty} (-S(\varphi_n, \omega_\beta) + \beta\varphi(H)) + \limsup_{n \to \infty} (\beta\varphi(H) - \beta\varphi_n(H))$$

$$\geq \limsup_{n \to \infty} S(\varphi_n) - 2\beta C.$$

Since $\beta > 0$ can be arbitrary small we arrive at (6.8).   □

Proposition 6.1 and its consequence (6.2) remain valid if $\mathcal{A}$ is a von Neumann algebra that is a direct sum of type I factors and φ is an arbitrary normal state.

Proposition 6.1 allows the following definition of the entropy of states of arbitrary C*-algebras.

$$S(\varphi) = \sup\left\{\sum_i \lambda_i S(\varphi_i, \varphi) : \sum_i \lambda_i \varphi_i = \varphi\right\}. \tag{6.9}$$

Here the supremum is over all decompositions of φ into finite (or equivalently countable) convex combinations of other states. Apparently the background uniform distribution provided by the trace functional in finite quantum systems is not present in this definition.

Some properties of $S(\varphi)$ are immediate from those of the relative entropy. The quantity $S(\varphi)$ is nonnegative and vanishes if and only if, φ is a pure state. Moreover, the entropy is lower semi-continuous because it is the supremum of lower semi-continuous relative entropy functionals (see (i) in Proposition 5.23). The invariance of $S(\varphi)$ under automorphisms is obvious as well.

Proposition 6.7 Let $\mathcal{A}_0 \subset \mathcal{A}$ be C*-algebras and assume that there exists a conditional expectation E of $\mathcal{A}$ onto $\mathcal{A}_0$ preserving a given state φ of $\mathcal{A}$. Then

$$S(\varphi) \geq S(\varphi|\mathcal{A}_0). \tag{6.10}$$

Proof. Indeed, if $\sum \lambda_i \psi_i = \varphi|\mathcal{A}_0$ for some states ψ_i of $\mathcal{A}_0$ then $\sum \lambda_i \psi_i \circ E$ is a decomposition of φ. The rest follows from $S(\psi_i, \varphi|\mathcal{A}_0) \leq S(\psi_i \circ E, \varphi)$. $\square$

The existence of the conditional expectation preserving the given state is an essential premise in Proposition 6.7. Without it the monotonicity property (6.10) does not hold. The simplest counterexamples comes from the fact that in the quantum case a pure state of the algebra can have a restriction which is a mixed state on the subalgebra.

Proposition 6.8 Let $\mathcal{A} = \mathcal{A}_1 \oplus \mathcal{A}_2$ be C*-algebras and $\varphi = \lambda\varphi_1 \oplus (1-\lambda)\varphi_2$ a state of $\mathcal{A}$ $0 < \lambda < 1$. Then

$$S(\varphi) = \lambda S(\varphi_1) + (1-\lambda)S(\varphi_2) - \lambda \log \lambda - (1-\lambda)\log(1-\lambda).$$

Now let φ be a normal state on a von Neumann algebra $\mathcal{M}$. Then a decomposition $\varphi = \sum \lambda_i \varphi_i$ is necessarily built from normal states φ_i. Hence, we may restrict ourselves to normal states φ_i in the definition (6.9). If p is the support projection of φ then $S(\psi, \varphi) = S(\psi|p\mathcal{M}p, \varphi|p\mathcal{M}p)$, whenever $\psi(p) = 1$ for the state ψ. Consequently

$$S(\varphi) = S(\varphi|p\mathcal{M}p). \tag{6.11}$$

One can easily see that in any von Neumann algebra $\mathcal{M}$ the identity decomposes uniquely into the sum $z_d + z_c$ of two central projections in such a way that z_c does not include minimal central projections and z_d is the orthogonal sum of minimal central projections.

Lemma 6.9 Let φ be a faithful normal state on a von Neumann algebra $\mathcal{M}$ and assume $S(\varphi) < +\infty$. Then $\mathcal{M}$ is a countable direct sum of factors.

Proof. Let $z_d + z_c$ be the decomposition of the identity into discrete and continuous central projections described above. Arguing by contradictions we will prove that $z_c = 0$.

With the notation $\mathcal{M}_d = z_d\mathcal{M}$, $\mathcal{M}_c = z_c\mathcal{M}$ and $\lambda = \varphi(z_c)$ we have

$$\varphi = \lambda\varphi_c \oplus (1 - \lambda)\varphi_d$$

where φ_c and φ_d are some normal states on $\mathcal{M}_c$ and $\mathcal{M}_d$, respectively. The assumptions $S(\varphi) < +\infty$ and $\lambda \neq 0$ imply $S(\varphi_c) < +\infty$ according to Proposition 6.3. For any given $n \in \mathbb{N}$ we can find pairwise orthogonal central projections $e_1, e_2 \ldots, e_n$ in $\mathcal{M}_c$ such that $\varphi_c(e_i) = n^{-1}$ $(i = 1, 2, \ldots, n)$. Let $\mathcal{A}_0$ be the commutative subalgebra generated by those projections. Since there exists a conditional expectation of $\mathcal{M}_c$ onto $\mathcal{A}_0$ we infer

$$S(\varphi_c) \geq S(\varphi_c|\mathcal{A}_0) = \log n$$

from Proposition 6.2. Here n may be arbitrary large and the contradiction shows that $z_c = 0$. So $\mathcal{M} = \mathcal{M}_d$ is a direct sum of factors. This direct sum must be countable because $\mathcal{M}$ admits a faithful normal state. $\qquad\square$

Lemma 6.10 Let φ be a faithful normal state on a von Neumann algebra $\mathcal{M}$. If $\mathcal{M}$ is a type II or type III factor, then $S(\varphi) = +\infty$.

Proof. We consider the centralizer subalgebra

$$\mathcal{M}_\varphi = \{a \in \mathcal{M} \ : \ \varphi(ab) = \varphi(ba) \text{ for every } b \in \mathcal{M}\} \qquad (6.12)$$

of $\mathcal{M}$. It follows from (6.12) and from Theorem 4.4 that for any subalgebra $\mathcal{A}_0$ of $\mathcal{M}_\varphi$ there exists a conditional expectation $\mathcal{M}_\varphi \to \mathcal{A}$ preserving φ. According to Proposition 6.7 we have $S(\varphi) \geq S(\varphi|\mathcal{A}_0)$ for any subalgebra $\mathcal{A}_0$ of $\mathcal{M}_\varphi$. If $\mathcal{A}_0$ is a maximal abelian subalgebra of $\mathcal{M}$ then $\mathcal{M}$ being continuous $\mathcal{A}$ can not contain a minimal projection. (A minimal projection in $\mathcal{A}_0$ would be minimal in $\mathcal{M}$, too.) If $\mathcal{M}_\varphi$ contains a maximal abelian subalgebra $\mathcal{A}_0$ of $\mathcal{M}$, we can apply the argument in the proof of the previous lemma to obtain $S(\varphi|\mathcal{A}_0) = +\infty$.

Now we shall look for a maximal abelian subalgebra $\mathcal{A}_0$ of $\mathcal{M}$ which is contained in $\mathcal{M}_\varphi$. Assume that $\mathcal{M}$ is of type II. Then the modular group of φ is inner, in other words, there exists a one-parameter group of unitaries $u_t \in \mathcal{M}$ such that

$$\sigma_t^\varphi(a) = u_t\, a\, u_t^* \qquad (t \in \mathbb{R})\,.$$

Let $\mathcal{A}_0$ be a maximal abelian subalgebra of $\mathcal{M}_\varphi$ containing the commuting family $\{u_t \ : \ t \in \mathbb{R}\}$. From the characterization

$$\mathcal{M}_\varphi = \{a \in \mathcal{M} \ : \ \sigma_t^\varphi(a) = a \text{ for every } t \in \mathbb{R}\} \qquad (6.13)$$

of the *centralizer* one can see that $\mathcal{A}_0$ is maximal abelian in $\mathcal{M}$.

If $\mathcal{M}$ is a type III_λ factor $0 \leq \lambda < 1$, then it is known that $\mathcal{M}_\varphi$ includes a maximal abelian subalgebra $\mathcal{A}_0$ of $\mathcal{M}$ (see 29.9 in [Strătilă 1981]).

It remains to handle the type III_1 factors. In this case we benefit from the homogeneity of the state space (see [Connes and Størmer 1978]). This property means that any two faithful normal states are approximately unitarily equivalent. We take projections $e_1, e_2, \ldots, e_n$ in $\mathcal{M}$ such that $\varphi(e_1) = \varphi(e_2) = \ldots = \varphi(e_n) = 1/n$ and define a faithful normal state ψ by

$$\psi(a) = \sum_{i=1}^{n} \varphi(e_i \, a \, e_i) \qquad (a \in \mathcal{M}) \,.$$

Obviously,

$$S(\psi) \geq \log n \,. \tag{6.14}$$

Furthermore, due to the homogeneity of the state space there exists a sequence (u_n) of unitaries of $\mathcal{M}$ such that $\|\psi(\,.\,) - \varphi(u_n \,.\, u_n^*)\| \to 0$. Hence

$$S(\psi) \leq \liminf_{n \to \infty} S(\varphi(u_n \,.\, u_n^*)) = S(\varphi) \tag{6.15}$$

by the lower semi-continuity and unitary invariance of the entropy. From (6.14) and (6.15) we conclude that $S(\varphi) = +\infty$. □

The next theorem will describe the the entropy functional on normal states of von Neumann algebras. Observe that (6.9) is a somewhat trivial extension of the von Neumann entropy in the sense that for factors of type I or type II the entropy is always infinite. It is understood in measure theoretic information theory that the entropy of a non-atomic measure must be infinite. Due to the lack of minimal projections the type II and type III von Neumann algebras resemble non-atomic measure spaces.

Theorem 6.11 Let φ be normal state of a von Neumann algebra $\mathcal{M}$ and let p be the support projection of φ. If $p\mathcal{M}p$ is a countable direct sum of type I factors then $S(\varphi) = \operatorname{Tr}\eta(D_\varphi)$ where D_φ is the density of φ with respect to the canonical semifinite normal trace Tr on $p\mathcal{M}p$. Otherwise, $S(\varphi) = \infty$.

Proof. Due to the expansive property (6.11) we may assume that φ is faithful. It follows from the previous lemmas that it suffices to consider the case when $\mathcal{M}$ is a countable direct sum of type I factors. We have to show that $S(\varphi)$ defined by (6.9) coincides with the von Neumann entropy. Let $\varphi = \sum_i \lambda_i \varphi_i$ be a convex decomposition of φ and let D_i be the density of φ_i. Then

$$\sum_i \lambda_i S(\varphi_i, \varphi) = -\sum_i \lambda_i \operatorname{Tr}\eta(D_i) + \operatorname{Tr}\eta(D_\varphi) \leq \operatorname{Tr}\eta(D_\varphi)$$

and we have $S(\varphi) \leq \operatorname{Tr}\eta(D_\varphi)$. The converse inequality comes from the orthogonal extremal decomposition of φ. □

Now we turn to the *entropy of states* on C*-algebras. Let ψ be a state of a C*-algebra $\mathcal{A}$. We write $\bar{\psi}$ for the vector state induced by the cyclic vector Ψ on the von Neumann algebra $\pi_\psi(\mathcal{A})''$ when (H_ψ, Ψ, π_ψ) is the GNS-triplet corresponding to ψ.

Lemma 6.12 With the above notation we have $S(\psi) = S(\bar{\psi})$.

Proof. The key to the proof is the fact that a finite relative entropy may be computed in the GNS representation of the reference state. This gives $S(\psi) \geq S(\bar{\psi})$. On the other hand, if $\psi = \sum_i \lambda_i \psi_i$ is a convex decomposition in the state space of $\mathcal{A}$, then we can find (normal) states $\bar{\psi}_i$ of $\pi_\psi(\mathcal{A})''$ such that $\psi_i = \bar{\psi}_i \circ \pi_\psi$ and $\bar{\psi} = \sum_i \lambda_i \bar{\psi}_i$. Since $S(\psi_i, \psi) = S(\bar{\psi}_i, \bar{\psi})$, the converse inequality $S(\psi) \leq S(\bar{\psi})$ follows. $\square$

Theorem 6.13 Let ψ be a state of a C*-algebra $\mathcal{A}$. Then

$$S(\psi) = \inf \left\{ \sum_i \eta(\lambda_i) \right\},$$

where the infimum is taken over all possible decompositions $\psi = \sum_i \lambda_i \psi$ into pure states. If ψ is not a countable convex combination of pure states, then $S(\psi) = \infty$.

Proof. Lemma 6.12 allows us to reduce the theorem to the von Neumann algebra version. We have to consider two cases. If ψ is not a countable convex combination of pure states, then the von Neumann algebra $p\pi_\psi(\mathcal{A})''p$ is not a countable direct sum of type I factors. (p is the support projection of $\bar{\psi}$.) Theorem 6.11 tells us that $S(\bar{\psi}) = \infty$ in this case. If ψ is a countable convex combination of pure states then $p\pi_\psi(\mathcal{A})''p$ is a countable direct sum of type I factors and we may refer to Theorem 6.11 again. $\square$

Theorem 6.14 Let φ and ω be states of a C*-algebra $\mathcal{A}$ and let $0 < \lambda < 1$. Then

$$\lambda S(\varphi) + (1 - \lambda)S(\omega) \leq S(\lambda\varphi + (1 - \lambda)\omega)$$
$$\leq \lambda S(\varphi) + (1 - \lambda)S(\omega) + H(\lambda, 1 - \lambda),$$

where $H(\lambda, 1 - \lambda) = \eta(\lambda) + \eta(1 - \lambda)$.

Proof. The second inequality is almost obvious from the previous theorem. If $\varphi = \sum_i \lambda_i \varphi_i$ and $\omega = \sum_j \mu_j \omega_j$ then $\lambda\varphi + (1 - \lambda)\omega$ is written as $\sum_i \lambda\lambda_i\varphi_i + \sum_j(1 - \lambda)\mu_j\omega_j$, and this gives the second inequality by taking the infimum.

The only case when the first inequality is not trivial is $S(\lambda\varphi + (1-\lambda)\omega) < \infty$. Then, according to Lemma 6.12, $S(\lambda\varphi + (1 - \lambda)\omega)$ is computed on the von Neumann algebra $p\mathcal{M}p$ of the GNS-representation. From the concavity of the von Neumann entropy (see Proposition 6.2) we have

$$\lambda S(\bar{\varphi}) + (1 - \lambda)S(\bar{\omega}) \leq S(\lambda\varphi + (1 - \lambda)\omega)$$

where $\bar{\varphi}$ and $\bar{\omega}$ are respectively the normal extensions of φ and ω to the von Neumann algebra $p\mathcal{M}p$. Since $S(\varphi) \leq S(\bar{\varphi})$ and $S(\omega) \leq S(\bar{\omega})$, we conclude the concavity of the entropy on C*-algebras. $\square$

Theorem 6.15 Let $\varphi_1 \otimes \varphi_2$ be a product state on the projective tensor product C*-algebra $\mathcal{A}_1 \otimes \mathcal{A}_2$. Then

$$S(\varphi_1 \otimes \varphi_2) = S(\varphi_1) + S(\varphi_2).$$

Proof. The key to the proof is the fact that the GNS-representation for respect to $\varphi_1 \otimes \varphi_2$ is the tensor product of the GNS-representations with φ_1 and φ_2. The theorem reduces to the additivity of the von Neumann entropy under tensor product if φ_1 and φ_2 are type I states. Otherwise we have an equality in the trivial form $\infty = \infty$.   □

Similarly to the previous proof one can get the subadditivity of the entropy under tensor product from the type I situation.

Theorem 6.16 Let φ_{12} be a state on the projective tensor product C*-algebra $\mathcal{A}_1 \otimes \mathcal{A}_2$. Then

$$S(\varphi_{12}) \leq S(\varphi_1) + S(\varphi_2)$$

where φ_1 and φ_2 are the marginal states.

As an extension of the definition (6.9), the entropy of a subalgebra can be defined as follows.

$$H_\varphi(\mathcal{A}) = \sup\left\{\sum_i \lambda_i S(\varphi_i|\mathcal{A}, \varphi|\mathcal{A}) : \sum_i \lambda_i \varphi_i = \varphi\right\} \tag{6.16}$$

for $\varphi \in \mathfrak{S}(\mathcal{B})$ and $\mathcal{A} \subset \mathcal{B}$. So we have

$$S_\mathcal{A}(\varphi) \leq S(\varphi|\mathcal{A}) \tag{6.17}$$

and equality holds when there exists a conditional expectation $E : \mathcal{B} \to \mathcal{A}$ leaving φ invariant. Property (6.17) has an obvious intuitive meaning. The partial information carried by the restriction of a state can not exceed the information carried by the state itself.

The proof of Theorem 6.15 shows an unpleasant side of the definition (6.7). To conclude the additivity under tensor product – which must be an essential property of such a concept – we used pretty much from the structure theory of von Neumann algebras. If $\varphi_1 \otimes \varphi_2 \in \mathfrak{S}(\mathcal{B}_1 \otimes \mathcal{B}_2)$ and $\mathcal{A}_1 \otimes \mathcal{A}_2 \subset \mathcal{B}_1 \otimes \mathcal{B}_2$ then

$$H_{\varphi_1 \otimes \varphi_2}(\mathcal{A}_1 \otimes \mathcal{A}_2) \geq H_{\varphi_1}(\mathcal{A}_1) + H_{\varphi_2}(\mathcal{A}_2) \tag{6.18}$$

obviously holds, but it seems to be difficult to show equality. In contrast to we have this problem concerning additivity, the monotonicity is built into the definition (due to the Uhlmann theorem):

$$H_\varphi(\mathcal{A}_1) \leq H_\varphi(\mathcal{A}_2) \text{ whenever } \mathcal{A}_1 \subset \mathcal{A}_2 \subset \mathcal{B}. \tag{6.19}$$

Now we generalize further the von Neumann entropy. In this extension not only a state of an arbitrary C*-algebra appears but a convex set in the state space is used as a reference system. The *mixture entropy* we are going to define expresses the degree of mixedness with respect to the given reference system. Let $\mathcal{A}$ be a C*-algebra, let $\mathcal{S}$ be a weak* compact and convex subset of the set $\mathfrak{S}$ of all states of $\mathcal{A}$ and write $\mathcal{E}(\mathcal{S})$ for the set of all extreme points of $\mathcal{S}$. From the Krein-Milman theorem, $\mathcal{S}$ is equal to the weak* closure of the convex hull of $\mathcal{E}(\mathcal{S})$. Every state $\varphi \in \mathcal{S}$ has a maximal measure μ pseudo-supported on $\mathcal{E}(\mathcal{S})$ such that

$$\varphi = \int_{\mathcal{S}} \omega \, d\mu. \tag{6.20}$$

The measure μ giving the above decomposition is typically not unique and we denote the set of all such measures by $M_\varphi(\mathcal{S})$. The fact that the set $M_\varphi(\mathcal{S})$ is not empty is a result of Choquet theory and $\mathcal{S}$ is called *Choquet simplex* if $M_\varphi(\mathcal{S})$ is a singleton for every $\varphi \in \mathcal{S}$. Take

$$H^{\mathcal{S}}(\mu) = \sup\{-\textstyle\sum_k \mu(A_k) \log \mu(A_k) : A_k \in P(\mathcal{S})\}, \tag{6.21}$$

where $P(\mathcal{S})$ is the set of all finite measurable partitions of $\mathcal{S}$. Then the mixture entropy of a state $\varphi \in \mathcal{S}$ with respect to $\mathcal{S}$ is defined as

$$S^{\mathcal{S}}(\varphi) = \inf\{H^{\mathcal{S}}(\mu) : \mu \in M_\varphi(\mathcal{S})\}. \tag{6.22}$$

This entropy does depend on the set $\mathcal{S}$ chosen, so that it represents the uncertainty of the state measured from the reference system $\mathcal{S}$. This fact opens a window to the cases where the usual von Neumann entropy is infinite and does not provide the appropriate information. Note that (6.21) gives infinite for non-atomic measures hence in (6.22) we may restrict ourselves to atomic measures. $S^{\mathcal{S}}(\varphi) = 0$ if and only if φ is an extreme point of $\mathcal{S}$. When the reference system $\mathcal{S}$ is the full state space we recover the extended von Neumann entropy $S(\varphi)$ discussed in detail in the first part of the chapter. Recall that concavity and lower semi-continuity of the extended von Neumann entropy required a lot of effort. Namely, all these properties are obvious only from the form (6.9). Concerning mixture entropy, we mention the following general result.

Theorem 6.17 Let $\mathcal{S}$ be a metrizable compact convex subset of the state space of a C*-algebra and assume that the set of all extremal points $\mathcal{E}(\mathcal{S})$ of $\mathcal{S}$ is weak* closed. Then the mixture entropy $S^{\mathcal{S}}$ is weak* lower semi-continuous on $\mathcal{S}$.

Note that if $\mathcal{S}$ is a Choquet simplex then the definition of $S^{\mathcal{S}}(\varphi)$ simplifies as follows.

$$S^{\mathcal{S}}(\varphi) = \sup\{-\textstyle\sum_k \mu(A_k) \log \mu(A_k) : A_k \in P(\mathcal{S})\}, \tag{6.23}$$

where μ is the measure from the barycentric decomposition (6.20) of φ. In this case $S^{\mathcal{S}}(\varphi)$ coincides with the classical entropy of the measure μ which is well-known to be lower semi-continuous.

To show further interesting examples for the mixture entropy, we recall the definition of a C*-dynamical system. Let G be a group and $\sigma : G \to \mathbf{Aut}(\mathcal{A})$ be a homomorphism of G into the group of all automorphisms of the C*-algebra $\mathcal{A}$. We call $(\mathcal{A}, \sigma)$ a *C*-dynamical system*. (If the group G is topological then the appropriate continuity hypothesis is made.) The set

$I(\sigma)$ of all σ-invariant states is a closed convex subset of the state space and the mixture entropy with respect to the reference system $I(\sigma)$ may be considered. Assume that σ is a continuous action of $\mathbb{R}$. Denote then by $K(\sigma)$ the set of all states which satisfies the KMS condition with respect to σ. (See Chapter 15 for the definition of the KMS condition.) Then $K(\sigma) \subset I(\sigma)$ and $K(\sigma)$ is a Choquet simplex.

Proposition 6.18 Let σ be an automorphic action of a group G on a von Neumann algebra $\mathcal{M}$ and assume the existence of a σ-invariant faithful normal state φ of $\mathcal{M}$. Then

$$S^{I(\sigma)}(\varphi) = S(\varphi|\mathcal{N}) = H_\varphi(\mathcal{N})$$

for the fixed point algebra $\mathcal{N}$.

Proof. First of all, since the normal part $I_n(\sigma)$ of $I(\sigma)$ is a face in $I(\sigma)$ and closed in the norm topology, a countable extremal decomposition of φ in $I_n(\sigma)$ is equivalent to that in $I(\sigma)$. What we need now is the Kovács-Szűcs ergodic theorem. This gives the existence of a normal conditional expectation of $\mathcal{M}$ onto $\mathcal{N}$ and makes and affine isomorphism between $I_n(\sigma)$ and the normal state space of $\mathcal{N}$. The von Neumann entropy is the mixture entropy for the full state space as well as for the normal state space. Hence the first equality is proven and the second one is the consequence of the conditional expectation again.   □

When $\sigma = \sigma^\varphi$ is the modular automorphism group associated with a faithful normal state φ, we immediately get the following corollary.

$$S^{I(\sigma^\varphi)}(\varphi) = S(\varphi|\mathcal{Z}_\varphi) \tag{6.24}$$

for the centralizer $\mathcal{Z}_\varphi$ of φ.

Proposition 6.19 For φ be a faithful normal state of the von Neumann algebra $\mathcal{M}$ which satisfies the KMS condition with respect to a group σ of automophisms. Then $S^{K_n(\sigma)}(\varphi) = H_{\mathcal{Z}}(\varphi)$, where $\mathcal{Z}$ stands for the centre of $\mathcal{M}$.

Proof. Since φ is faithful, there exists a normal conditional expectation E from $\mathcal{M}$ onto $\mathcal{Z}$ preserving all $\psi \in K_n(\sigma)$. By the same reasons mentioned in the proof of the previous proposition, we get the desired result.   □

Notes and Remarks. Inequality (6.2) has been known for a long time, see [Jaynes 1956]. The *strong subadditivity* was proved in [Lieb and Ruskai 1973a] after a conjecture in [Landford and Robinson 1968]. The latter paper contained the proof of the subadditivity which is equivalent to the concavity of the entropy. The *triangle inequality* of Proposition 6.5 is from [Araki and Lieb 1970] and a survey on related entropy inequalities is in [Lieb 1975]. Proposition 6.6 is from [Wehrl 1978]. The definitions (6.7) and (6.16) were introduced in the paper [Narnhofer and Thirring 1985], from where the title to this chapter was borrowed. Theorem 6.11 and its lemmas were proved in [Hiai 1991a]. In [Manuceau, Naudts and Verbeure 1972] the expression $S(\psi) = \inf\{\sum_i \eta(\lambda_i)\}$ was taken as a definition and the characterization of this entropy was given in terms of the commutant. The equivalence of the two definitions, Theorem 6.13 in our text, was explicitly pointed out in [Muraki, Ohya and Petz 1992]. Concerning tensor products of algebras we refer to [Takesaki 1979] and [Kadison and Ringrose 1986].

[Bratteli and Robinson 1979] treats in detail the *barycentric decomposition* in locally convex spaces as well as the application to the state space of a C*-algebra. The mixture entropy S^S was introduced in [Ohya 1984]. Several related results are contained in [Ohya 1986a], [Muraki 1991] and [Muraki, Ohya and Petz 1992]. [Ohya 1989] tells about physical applications of the mixture entropy.

Part 5.3 of [Bratteli and Robinson 1981] is devoted to *KMS-states*, see also [Sakai 1991]. The *Kovács-Szűcs ergodic theorem* was published in [Kovács and Szűcs 1966].

7 Functionals of Entropy Type

The subject of this chapter is a family entropy-like functionals including Segal's generalization of the von Neumann entropy as well as variations of the concept of relative entropy. The chapter is closed with conditional entropy.

When von Neumann introduced the entropy

$$S(\varphi) = \operatorname{Tr} \eta(D_\varphi) \tag{7.1}$$

of a state φ, it turned out immediately that some important properties of the entropy functional are due to the concavity of the function $\eta(t) = -t \log t$. The quasi-entropy functional

$$S_f(\varphi) = \operatorname{Tr} f(D_\varphi) \tag{7.2}$$

is worthy of study for some functions $f : \mathbb{R}^+ \to \mathbb{R}$. (Certain results in Chapter 3 may be formulated in terms of quasi-entropies.)

The relative entropy of the normal states ω and φ of the von Neumann algebra $\mathcal{M}$ was defined by means of the logarithm of the relative modular operator $\Delta(\varphi, \omega)$. In principle, the logarithmic function might have been replaced by an arbitrary function $f : \mathbb{R}^+ \to \mathbb{R}$. For $k \in \mathcal{M}$ we call the quantity

$$S_f^k(\omega, \varphi) = \langle k\Omega, \, f(\Delta(\varphi, \omega)) k\Omega \rangle \tag{7.3}$$

quasi-entropy. (Recall that Ω is the representative of ω from the natural positive cone, while $\mathcal{M}$ is in the standard representation.) When speaking of a quasi-entropy we always assume that f is continuous. Physically, most of the functions f are irrelevant but some are useful. Here are a few examples in the context of finite quantum systems.

1. $f(t) = -\log t$

    $$S_f^k(\omega, \varphi) = \operatorname{Tr}\left(D_\omega^{1/2} k^* k D_\omega^{1/2} \log D_\omega - k D_\omega k^* \log D_\varphi\right)$$

2. $f(t) = t^\alpha$

    $$S_f^k(\omega, \varphi) = \operatorname{Tr}\left(D_\omega^{1-\alpha} k^* D_\varphi^\alpha k\right)$$

3. $f(t) = t \log t$

    $$S_f^k(\omega, \varphi) = \operatorname{Tr}\left(D_\varphi \log D_\varphi k k^* - D_\varphi k \log D_\omega k^*\right)$$

4. $f(t) = ut + v$

$$S_f^k(\omega, \varphi) = u\varphi(kk^*) + v\omega(k^*k)$$

The *skew information*

$$I_p(\varphi, k) = \operatorname{Tr} D_\varphi k^* k - \operatorname{Tr} D_\varphi^{1-p} k^* D_\varphi^p k$$

of a state φ relative to an operator k is a quasi-entropy corresponding to the function $f(t) = t - t^p$ (cf. (3.4)).

The relationship

$$S(\omega, \varphi) = -S_\eta^I(\varphi, \omega) \tag{7.4}$$

suggested by the finite system case holds for arbitrary von Neumann algebras as well.

The following results show that the quasi-entropy $S_f^k(\omega, \varphi)$ behaves well if f is an *operator monotone function*. If $k = I$ then instead of $S_f^I(\omega, \varphi)$ we write $S_f(\omega, \varphi)$.

Proposition 7.1 Let $f : [0, \infty) \to \mathbb{R}$ be an operator monotone function with $f(0) \geq 0$. Assume that $\mathcal{M}_0$ and $\mathcal{M}$ are von Neumann algebras with positive normal functionals φ_0, ω_0 and φ, ω, respectively. If $\alpha : \mathcal{M}_0 \to \mathcal{M}$ is a unit preserving Schwarz mapping such that

$$\omega \circ \alpha \leq \omega_0 \quad \text{and} \quad \varphi \circ \alpha \leq \varphi_0$$

then for every $k \in \mathcal{M}_0$ we have

$$S_f^k(\omega_0, \varphi_0) \geq S_f^{\alpha(k)}(\omega, \varphi).$$

Proposition 7.2 Assume that $f : [0, \infty) \to \mathbb{R}$ is operator monotone and $f(0) = 0$. Let $\varphi_1, \varphi_2, \varphi, \omega_1, \omega_2, \omega$ be positive normal functionals on the von Neumann algebra $\mathcal{M}$ such that

$$\lambda\varphi_1 + \mu\varphi_2 \leq \varphi \quad \text{and} \quad \lambda\omega_1 + \mu\omega_2 \leq \omega.$$

If $k \in \mathcal{M}$ and $\lambda, \mu > 0$ then

$$\lambda S_f^k(\omega_1, \varphi_1) + \mu S_f^k(\omega_2, \varphi_2) \leq S_f^k(\omega, \varphi).$$

The proof of these propositions will be omitted. Proposition 7.1 implies that

$$S_f(\omega, \varphi) \leq f(1) \tag{7.5}$$

holds for states ω and φ under the stated conditions. One can prove that in (7.5) equality holds for a non-affine f if and only if $\varphi = \omega$. Proposition 7.2 yields the joint concavity of $S_f^k(\omega, \varphi)$.

Quasi-entropies have implicitly appeared in the deduction of Kosaki's formula. Proposition 5.10 contains a variational formula concerning $S_g(\omega, \varphi)$. The quantity

$$P_A(\omega, \varphi) = \langle \Omega, \Delta(\varphi, \omega)^{1/2} \Omega \rangle \tag{7.6}$$

is sometimes interpreted as a *transition probability* between the states ω and φ. It lies always between 0 and 1. Furthermore, $P_A(\omega, \varphi) = 0$ if and only if ω and φ are orthogonal and $P_A(\omega, \varphi) = 1$ if and only if $\omega \equiv \varphi$. Since $\Delta(\varphi, \omega)^{1/2} \Omega$ is the representing vector of φ from the natural positive cone, $P_A(\omega, \varphi)$ is a symmetric function of the two arguments. The transition probability $P_A(\omega, \varphi)$ extends the notion of affinity of measures to the operator algebraic context. Observe that Propositions 7.1 and 7.2 may be applied to $P_A(\omega, \varphi)$.

For $0 \leq \alpha \leq 1$ one can define

$$S_\alpha(\omega, \varphi) = \frac{1}{1 - \alpha} \log \langle \Phi, \Delta(\omega, \varphi)^\alpha \Phi \rangle, \tag{7.7}$$

where Φ is the vector representative of the state φ from the natural positive cone of the standard representation. On the basis of the similarity with the *relative Rényi entropy*

$$I_\alpha((q_i), (p_i)) = \frac{1}{1 - \alpha} \log \sum_i q_i^\alpha p_i^{1-\alpha}$$

of the probability distributions (q_i) and (p_i) we shall term S_α relative Rényi entropy of order α. In term of densities we have

$$S_\alpha(\omega, \varphi) = \frac{1}{1 - \alpha} \log \operatorname{Tr} D_\omega^\alpha D_\varphi^{1-\alpha}. \tag{7.8}$$

We note that a simple matrix trick allows to compute the relative Rényi entropy from the skew information. In fact,

$$I_p \left(\begin{pmatrix} D_\varphi & 0 \\ 0 & D_\omega \end{pmatrix}, \begin{pmatrix} 0 & 0 \\ I & 0 \end{pmatrix} \right) = 1 - \operatorname{Tr} D_\varphi^{1-p} D_\omega^p.$$

Since

$$\langle \Phi, \Delta(\omega, \varphi)^\alpha \Phi \rangle = \langle \Omega, \Delta(\varphi, \omega)^{1-\alpha} \Omega \rangle,$$

one can deduce easily by means of (5.3) that

$$\lim_{\alpha \to 1} S_\alpha(\omega, \varphi) = S(\omega, \varphi). \tag{7.9}$$

Therefore the relative Rényi entropy may be used for the approximation of the relative entropy.

One can try to extend the notion of relative entropy to the case, where the reference functional is a normal weight. Let the von Neumann algebra $\mathcal{M}$ act on a Hilbert space $\mathcal{H}$ so that $\omega = \omega_\xi$ is a vector state given by a vector $\xi \in \mathcal{H}$. As before, we write ω'_ξ for the state of the commutant $\mathcal{M}'$ induced $\xi \in \mathcal{H}$. Let ψ be a normal weight on $\mathcal{M}$ and let

$$\Delta(\psi / \omega'_\xi) = \int_0^\infty \lambda \, dE_\lambda$$

be the spectral decomposition of the spatial derivative operator. If $\operatorname{supp}\omega \le$ $\operatorname{supp}\psi$, then we set

$$S(\omega, \psi) = -\int_{-\infty}^{1} \log \lambda \, d\langle \xi, E_\lambda \xi \rangle - \int_{1}^{\infty} \log \lambda \, d\langle \xi, E_\lambda \xi \rangle, \qquad (7.10)$$

provided that the right hand side is not of the form $\infty - \infty$. In principle, the methods used to establish properties of the relative entropy of states are effective enough to apply them to $S(\omega, \psi)$. Since $S(\omega, \psi)$ has very restricted practical use, we shall not develop a general theory for it. We content ourself with some considerations of the case of a tracial normal weight. Recall that $w\psi$ is called tracial if $\psi(a^*a) = \psi(aa^*)$ for every $a \in \mathcal{M}$. To emphasise the tracial property we write τ for a faithful tracial normal weight. (In a slightly different terminology τ is called a faithful normal semi-finite trace; the von Neumann algebra should be semi-finite to admit the existence of such a τ.) To a normal state φ there corresponds a positive operator D_φ such that

$$\varphi(a) = \tau(a^{1/2} D_\varphi a^{1/2}) \qquad (a \in \mathcal{M}^+).$$

Since D_φ is not bounded, the way of writing $\tau(a^{1/2} D_\varphi a^{1/2})$ is somewhat formal but it can be made precise. Any way D_φ is a positive operator such that the projection-valued measure of its spectral decomposition $D_\varphi = \int \lambda \, E_\lambda$ has its values in the von Neumann algebra. The quantity

$$S_\tau(\varphi) = \int_{0}^{1} \eta(\lambda) \, d\tau(E_\lambda) + \int_{1}^{\infty} \eta(\lambda) \, d\tau(E_\lambda) \qquad (7.11)$$

is called *Segal entropy*. (Note that $d\tau(E_\lambda)$ is a measure on the real line and $S_\tau(\varphi)$ is always well-defined if the density D_φ is bounded.) Choosing $\mathcal{M} = B(\mathcal{H})$ and $\tau = \mathrm{Tr}$ one observes that the von Neumann entropy is included in the concept of Segal entropy. Unlike to the Segal entropy the von Neumann entropy always takes a definite nonnegative value (but it can be $+\infty$). Indeed, if $\mathcal{M} = B(\mathcal{H})$ and $\tau = \mathrm{Tr}$, then a statistical operator D_φ is a contraction, and

$$S(\varphi) = \int_{0}^{1} \eta(\lambda) \, d\tau(E_\lambda).$$

Let $\mathcal{M}_1$ and $\mathcal{M}_2$ be von Neumann algebras with faithful normal tracial weights τ_1 and τ_2, respectively. If $\beta : \mathcal{M}_1 \to \mathcal{M}_2$ is a positive unital ultra-weakly continuous mapping then $\varphi_2 \circ \beta$ is a normal state on $\mathcal{M}_1$ whenever φ_2 is a normal state of $\mathcal{M}_2$. Assume that $\tau_2 \circ \beta = \tau_1$. Under these assumptions we have the following

Proposition 7.3 Suppose that the density of φ_2 with respect to τ_2 is bounded. Then

$$S_{\tau_1}(\varphi_2 \circ \beta) \ge S_{\tau_2}(\varphi_2).$$

Proof. We consider the transpose map $\alpha : \mathcal{M}_2 \to \mathcal{M}_1$ defined by

$$\tau_1(b_1\alpha(a_2)) = \tau_2(\beta(b_2)a_2) \qquad (b_1 \in \mathcal{M}_1^+, \; a_2 \in \mathcal{M}_2^+).$$

The map α sends density D_2 of φ_2 into the density of $\varphi_2 \circ \beta$. Since α is a unital positive contraction, we can benefit from the following non-commutative version of the *Jensen inequality*.

$$\tau_1(\eta(\alpha(D_2))) \geq \tau_1(\alpha(\eta(D_2))). \tag{7.12}$$

Taking into account $\tau_1 \circ \alpha = \tau_2$ we are able to conclude the proposition. $\square$

Let (P_i) be a family of pairwise orthogonal projections on a Hilbert space $\mathcal{H}$ such that $\sum_i P_i = I$. The mapping $\beta : B(\mathcal{H}) \to B(\mathcal{H})$ defined by

$$\beta(a) = \sum_i P_i a P_i$$

is a unital positive mapping and $\mathrm{Tr} \circ \beta = \mathrm{Tr}$ holds. In this situation Proposition 7.3 may be applied to a normal state φ yielding

$$S(\varphi) \leq \mathrm{Tr}\, \eta(\textstyle\sum_i P_i D_\varphi P_i). \tag{7.13}$$

The density matrix $\sum_i P_i D_\varphi P_i$ is obtained from D_φ by deleting some off-diagonal matrix entries and $\sum_i P_i D_\varphi P_i$ is sometimes called a *coarse-grained density matrix*. The coarse-graining increases the degree of mixedness and leads to a larger entropy. Inequality 7.13 is called also the *Klein-von Neumann theorem*.

The quantum-mechanical Rényi entropies, often called α-*entropies*, are defined for $0 < \alpha \neq 1$ by the formula

$$S_\alpha(\varphi) = \frac{1}{1-\alpha} \log \mathrm{Tr}\, D_\varphi^\alpha. \tag{7.14}$$

If α tends to 1 decreasingly, one obtains the von Neumann entropy as a limiting case. The case $\alpha = 2$ is comparatively simple computationally, because there is no need to diagonalize the density matrix (or something equally difficult). $S_2(\varphi) = 2 \log \sum_i \|D_\varphi \xi_i\|^2$ for any orthonormal basis (ξ_i).

Corollary 7.4 Let τ be a faithful normal tracial weight on the von Neumann algebra $\mathcal{M}$ and let $\mathcal{M}_0$ be a von Neumann subalgebra of $\mathcal{M}$. If the state φ of $\mathcal{M}$ possesses a bounded density with respect to τ and $\tau_0 \equiv \tau|\mathcal{M}_0$ is normal, then

$$S_{\tau_0}(\varphi|\mathcal{M}_0) \geq S_\tau(\varphi).$$

Proof. This follows from Proposition 7.3 by choosing β to be the embedding of $\mathcal{M}_0$ into $\mathcal{M}$. $\square$

Corollary 7.5 The Segal entropy is a concave function on the states with bounded density.

Proof. Let φ_1 and φ_2 be states with bounded density and $0 < \lambda < 1$. Set $\beta : \mathcal{M} \to \mathcal{M} \oplus \mathcal{M}$ by $\beta(a) = a \oplus a$. If

$$\varphi_{12}(a \oplus b) = \lambda\varphi_1(a) + (1 - \lambda)\varphi_2(b),$$

then φ_{12} is a normal state of $\mathcal{M} \oplus \mathcal{M}$ and it possesses a bounded density with respect to the tracial weight

$$\tau_{12}(a \oplus b) = \lambda\tau(a) + (1 - \lambda)\tau(b).$$

Since $\varphi_{12} \circ \beta = \lambda\varphi_1 + (1 - \lambda)\varphi_2$, Proposition 7.3 reads

$$S_\tau(\lambda\varphi_1 + (1 - \lambda)\varphi_2) \geq S_{\tau_{12}}(\varphi_{12}) = \lambda S_\tau(\varphi_1) + (1 - \lambda)S_\tau(\varphi_2)$$

proving the concavity. $\qquad\qquad\square$

The rigorous handling of the Segal entropy requires a lot of technicalities of non-commutative measure theory. We shall not give an account on these methods but we shall show how eigenvalue arguments can be transformed into the continuous case by means of the concept of spectral scale. Let $\mathcal{M}$ be a von Neumann algebra with a fixed tracial normal weight τ. Suppose that $a \in \mathcal{M}^+$ has spectral decomposition

$$\int_0^\infty \lambda \, dP^a(\lambda),$$

where P^a is a projection-valued measure supported on $[0, \|a\|]$. If $\tau(a) < +\infty$, we define the spectral scale of a as

$$\lambda_t(a) = \min\{S : \tau(P^a(S, \infty)) \leq t\} \qquad (t \in \mathbb{R}^+) \qquad (7.15)$$

$t \to \lambda_t(a)$ is a non-increasing function which is continuous from the right. The spectral scale imitates the decreasing list of eigenvalues of a positive compact operator. One can see that

$$\tau(a) = \int_0^\infty \lambda_t(a) \, dt, \qquad (7.16)$$

or, more generally

$$\tau(f(a)) = \int_0^\infty f(\lambda_t(a)) \, dt. \qquad (7.17)$$

Proposition 7.6 The Segal entropy is norm lower semi-continuous on the set $\{\varphi \in \mathfrak{S}_*(\mathcal{M}) : \|D_\varphi\| \leq C\}$ for every $C > 0$.

Proof. If $\|D_\varphi\| \leq C$, then the function $t \mapsto \eta(\lambda_t(D_\varphi))$ is bounded. Since the limit

$$\lim_{n \to \infty} \int_0^n \eta(\lambda_t(D_\varphi)) \, dt$$

is increasing (for large n) and it gives $S_\tau(\varphi)$, we shall show that

$$\varphi \mapsto \int_0^n \eta(\lambda_t(D_\varphi)) \, dt \tag{7.18}$$

is norm continuous. The continuous version

$$\|a - b\|_1 \geq \int_0^\infty |\lambda_t(a) - \lambda_t(b)| \, dt \tag{7.19}$$

of Lemma 1.7 could be used now. (In (7.19) $\| \ \|_1$ is the trace norm and the proof of (7.19) is exactly the same as that of Lemma 1.7.) From (7.19) we observe that $\|\varphi_i - \varphi\| \to 0$ implies

$$\lambda_t(D_{\varphi_i}) \to \lambda_t(D_\varphi)$$

for almost all $t \in \mathbb{R}$. Hence the dominated integral convergence theorem tells us that

$$\int_0^n \eta(\lambda_t(D_{\varphi_i})) \, dt \to \int_0^n \eta(\lambda_t(D_\varphi)) \, dt \, ,$$

and the proof is complete. $\qquad\qquad\qquad\qquad\qquad\qquad\qquad\qquad\qquad\square$

Subadditivity, that is, an appropriate extension of (1.9) holds for the Segal entropy. This will not be proved here. The Segal entropy may be obtained as a greatest lower bound, provided it is finite. If $S_\tau(\varphi)$ is defined and finite, then

$$S_\tau(\varphi) = \inf\{\textstyle\sum_i \eta(\lambda_i)\tau(p_i)\} \, , \tag{7.20}$$

where $\lambda_i = \varphi(p_i)/\tau(p_i)$, the infimum is over all partitions of unity $p_1, p_2, \ldots, p_n$ such that $\tau(p_i) < \infty$, and the p_i's are pairwise orthogonal projections.

The L^∞ space corresponding to a localizable measure μ may be considered as a von Neumann algebra. The integration with respect to μ

$$\tau(f) = \int f \, d\mu \qquad (f \in L^\infty(\mu))$$

yields a normal tracial weight. Any normal state φ is of the form

$$\varphi(f) = \int fg \, d\mu \qquad (f \in L^\infty(\mu)) \, ,$$

where g is a not necessarily a bounded nonnegative measurable function, it is a Radon-Nikodym derivative. Then

$$S_\tau(\varphi) = \int \eta(g) \, d\mu \, . \tag{7.21}$$

This measure theoretic quantity is called the (Boltzmann-Gibbs) differential entropy and it has motivated Segal to introduce the operator algebraic generalization. Let $\mu \equiv \lambda$ be the Lebesgue measure on the real line and g be a probability density there. Setting

$$\lambda_n = \frac{\lambda}{2n}\Big|[-n, n] \quad \text{and} \quad \nu_n(H) = \int_H g \, d\lambda \Big/ \int_{-n}^n g \, d\lambda$$

we can consider the relative entropy of the probability measures λ_n and ν_n. The limit of $S(\nu_n, \lambda_n)$ is never finite. Hence, as a kind of renormalization one takes

$$\lim_{n \to \infty} \left(\log 2n - S(\nu_n, \lambda_n) \right), \tag{7.22}$$

which is the *differential entropy*. When g is a probability density on the real line and the reference measure is understood to be the Lebesgue measure then for $\int \eta(g)\, d\lambda$ we use the notation $H(g)$. If g is bounded, then the above proven properties of the Segal entropy hold for the entropy $H(g)$.

We next discuss a concept of relative entropy of two states of $B(\mathcal{H})$ when a subset $\mathcal{A}$ of operators is distinguished. Let σ and ρ be (not necessarily normal) states on $B(\mathcal{H})$, and let $\mathcal{A} \subset B(\mathcal{H})$. We may think that the set $\mathcal{A}$ is selected for experimental observation or something like that. It is not assumed that $\mathcal{A}$ is an algebra but $I \in \mathcal{A}$ will be required. Following *Donald* we define the quantity $\mathrm{ent}_{\mathcal{A}}(\sigma, \rho)$ by the following postulates.

(i) $\mathrm{ent}_{B(\mathcal{H})}(\sigma, \rho) = S(\sigma, \rho)$ when σ and ρ are normal.
(ii) $\mathrm{ent}_{B(\mathcal{H})}(\sigma, \rho) = \sup\{F(\sigma, \rho) : F$ is w^* lower semi-continuous, convex, and coincides with $S(\sigma, \rho)$ when σ and ρ are normal$\}$.
(iii) $\mathrm{ent}_{\mathcal{A}}(\sigma, \rho) = \inf\{\mathrm{ent}_{B(\mathcal{H})}(\sigma', \rho') : \sigma'|\mathcal{A} = \sigma$ and $\rho'|\mathcal{A} = \rho\}$.

It can be proved that convexity, lower semi-continuity and monotonicity hold for $\mathrm{ent}_{\mathcal{A}}$ when these properties are stated in the appropriate form. It is a reasonable conjecture that for a wide class of operator algebras and states on them the quantity $\mathrm{ent}_{\mathcal{A}}$ reduces to the usual relative entropy. The following result is in this direction. Before stating it let us recall that a von Neumann algebra $\mathcal{M} \subset B(\mathcal{H})$ is called *injective* if there exists a conditional expectation E of $B(\mathcal{H})$ onto $\mathcal{M}$. (We emphasise that the normality of E is not a requirement.) The injective property is independent of the representation: If the stated conditional expectation exists in a representation, then it exists in any other one.

Proposition 7.7 Let $\mathcal{A} \subset B(\mathcal{H})$ be an injective von Neumann algebra and let φ, ω be normal states on $\mathcal{A}$. Then $S(\omega, \varphi) = \mathrm{ent}_{\mathcal{A}}(\omega, \varphi)$.

Now let $\mathcal{A}$ be a C*-algebra with state space $\mathfrak{S}(\mathcal{A})$. Endowed with the weak topology $\mathfrak{S}(\mathcal{A})$ is a compact space. Let $\mathcal{S}$ be a closed convex subset of $\mathfrak{S}(\mathcal{A})$. According to the Krein-Milman theorem $\mathcal{S}$ is the closure of the convex hull of its extreme boundary $\mathcal{E}(\mathcal{S})$. For $\varphi \in \mathcal{S}$ an entropy $S^{\mathcal{S}}(\varphi)$ can be introduced which represents the uncertainty of the given state φ measured from the reference system $\mathcal{S}$.

Every state $\varphi \in \mathcal{S}$ has a maximal measure μ pseudo-supported on $\mathcal{E}(\mathcal{S})$ such that

$$\varphi = \int_{\mathcal{S}} \omega \, d\mu(\omega). \tag{7.23}$$

The measure μ giving this decomposition is not unique in general. When μ is fully atomic, then (7.23) reduces to the convex combination

$$\varphi = \sum_i \lambda_i \varphi_i, \tag{7.24}$$

where $\lambda_i > 0$, $\sum_i \lambda_i = 1$, $\varphi_i \in \mathcal{E}(\mathcal{S})$ and i sums over a countable index set. Ohya defined $S^{\mathcal{S}}(\varphi)$ to be infinite if a decomposition of the form (7.24) does not exist, otherwise $S^{\varphi}(\mathcal{S})$ is the infimum of $\sum_i \eta(\lambda_i)$ over the possible decompositions (7.24). The quantity $S^{\mathcal{S}}$ will be called the *mixture entropy* with respect to $\mathcal{S}$.

It is obvious that $S^{\mathcal{S}}(\varphi) \geq 0$ and $S^{\mathcal{S}}(\varphi) = 0$ if and only if $\varphi \in \mathcal{E}(\mathcal{S})$. It is hard to say anything about $S^{\mathcal{S}}$ in full generality. Its properties depend on the geometry of the compact convex set $\mathcal{S}$. If $\mathcal{S}$ is a Choquet simplex, for example, the set of all KMS-states for a given dynamics, then the standard methods of measure theoretic entropy apply to $S^{\mathcal{S}}$. When $\mathcal{S}$ is the full state space $\mathfrak{S}(\mathcal{A})$ then $S^{\mathcal{S}}$ reduces to the entropy (6.7). This subject was discussed in detail in the previous chapter. The mixture entropy with respect to the whole state space is typically infinite. By the choice of a proper reference system $\mathcal{S} \subset \mathfrak{S}(\mathcal{A})$ the mixture entropy $S^{\mathcal{S}}$ can be made finite and this is the idea which motivates the definition.

Now we define a new kind of entropy for states of C*-algebras which will coincide with the measure theoretic entropy for a measure when it is considered as an integral functional on functions and its value will be twice that of the von Neumann entropy on finite factors.

Let $\mathcal{A}$ be a C*-algebra. An *operational partition of unity* is a k-tuple $\mathcal{V} = (V_1, V_2, \ldots, V_k)$ of elements of $\mathcal{A}$ such that $\sum_{i=1}^{k} V_i^* V_i = I$. To each operational partition of unity a completely positive unital mapping

$$\gamma_{\mathcal{V}}(a) = \sum_{i=1}^{k} V_i^* a V_i \qquad (a \in \mathcal{A}) \tag{7.25}$$

is associated and $\mathcal{V}$ is called φ-invariant for a state φ of $\mathcal{A}$ if $\varphi \circ \gamma_{\mathcal{V}} = \varphi$. The operator matrix

$$\left(V_i^* V_j \right)_{i,j=1}^{k} \in \mathcal{A} \otimes M_k(\mathbb{C})$$

is positive so that

$$\left(\varphi(V_i^* V_j) \right)_{i,j=1}^{k} \tag{7.26}$$

is a density matrix in $M_k(\mathbb{C})$ whenever φ is a state and $\mathcal{V}$ is an operational partition of unity. Following *Lindblad* we define the entropy of a state by means of operational partitions of unity and the related density matrix (7.26). For convenience, let $S(\varphi, \mathcal{V})$ stand for the von Neumann entropy of the density (7.26).

$$S^L(\varphi) = \sup\{S(\varphi, \mathcal{V})\},$$

where the supremum is over all operational partitions of unity which leave φ invariant. First we show that S^L shares the mixing property with other types of entropy-like quantities.

Proposition 7.8 Assume that the C*-algebra $\mathcal{A}$ is a direct sum $\mathcal{A}_1 \oplus \mathcal{A}_2$ and $\varphi_{12}(a \oplus b) = \lambda\varphi_1(a) + (1 - \lambda)\varphi_2(b)$ for every $a \in \mathcal{A}_1, b \in \mathcal{A}_2$ and some $0 < \lambda < 1$. Then

$$S^L(\varphi_{12}) = \lambda S^L(\varphi_1) + (1 - \lambda)S^L(\varphi_2) + H(\lambda, 1 - \lambda).$$

Proof. First we prove $\geq$. Let $\mathcal{V}_1 = (V_1, V_2, \ldots, V_k)$ and $\mathcal{V}_2 = (V_{k+1}, \ldots, V_n)$ be operational partitions of unity invariant with respect to φ_1 and φ_2, respectively. Set

$$W_i = \begin{cases} V_i \oplus 0 & \text{if } 1 \leq i \leq k, \\ 0 \oplus V_i & \text{if } k + 1 \leq i \leq n. \end{cases}$$

One checks that $\mathcal{W} = (W_1, W_2, \ldots, W_n)$ is an operational partition of unity leaving φ_{12} invariant. Since the density

$$\left(\varphi_{12}(W_i^* W_j)\right)_{i,j=1}^n$$

is the direct sum of the densities

$$\lambda\left(\varphi_1(V_i^* V_j)\right)_{i,j=1}^k \quad \text{and} \quad (1 - \lambda)\left(\varphi_2(V_i^* V_j)\right)_{i,j=k+1}^n,$$

a reference to the mixing property (2.1) of the von Neumann entropy yields the equality

$$S(\varphi_{12}, \mathcal{W}) = \lambda S(\varphi_1, \mathcal{V}_1) + (1 - \lambda)S(\varphi_2, \mathcal{V}_2) + H(\lambda, 1 - \lambda).$$

This ensures the inequality $\geq$ in the proposition.

To prove the converse inequality we assume that $\mathcal{W} = (W_1, W_2, \ldots, W_n)$ is an operational partition of unity leaving φ_{12} invariant. The decomposition

$$W_i = V_i \oplus U_i \qquad (1 \leq i \leq n)$$

gives us the invariant operational partitions of unity $\mathcal{V} = (V_1, V_2, \ldots, V_n)$ and $\mathcal{U} = (U_1, U_2, \ldots, U_n)$. We have

$$\varphi_{12}(W_i^* W_j) = \lambda\varphi_1(V_i^* V_j) + (1 - \lambda)\varphi_1(U_i^* U_j),$$

and according to the special convexity property of the entropy (Proposition 1.6) we have

$$S(\varphi_{12}, \mathcal{W}) \leq \lambda S(\varphi_1, \mathcal{V}) + (1 - \lambda)S(\varphi_2, \mathcal{U}) + H(\lambda, 1 - \lambda).$$

The proof is completed by taking the supremum.   □

Proposition 7.9 If φ is a state of the algebra $M_n(\mathbb{C})$ then $S^L(\varphi) = 2S(\varphi)$.

Proof. Let D_φ be the density of φ. We may assume that it is a diagonal matrix expressed as $\sum_i \lambda_i E_{ii}$ by means of the matrix units (E_{ij}). It is readily seen that

$$\mathcal{W} = \{\sqrt{\lambda_{kl}} E_{kl} : 1 \leq k, l \leq n\}$$

is an operational partition of unity which leaves φ invariant. This implies that $S^L(\varphi) \geq S(\varphi, \mathcal{W})$. The density matrix

$$\left(\varphi\left(\left(\sqrt{\lambda_{kl}} E_{kl}\right)^* \left(\sqrt{\lambda_{k'l'}} E_{k'l'}\right)\right)\right)_{(kl)(k'l')}$$

has eigenvalues $\{\lambda_k \lambda_l : 1 \leq k, l \leq n\}$. Hence its von Neumann entropy is $2S(\varphi)$. So $S^L(\varphi) \geq 2S(\varphi)$.

Let us take an arbitrary operational partition of unity $\mathcal{V} = (V_1, V_2, \ldots, V_k)$ which leaves φ invariant. Performing the GNS construction with φ we obtain a Hilbert space $\mathcal{H}$ with cyclic vector Φ. Consider $\mathcal{H} \otimes \mathcal{H}_k$, where $\mathcal{H}_k$ is a k-dimensional Hilbert space with orthonormal basis $\xi_1, \xi_2, \ldots, \xi_k$. The unit vector

$$\Omega = \sum_{l=1}^{k} V_l \Phi \otimes \xi_l \in \mathcal{H} \otimes \mathcal{H}_k$$

induces a state ω_{12} on $B(\mathcal{H}) \otimes B(\mathcal{H}_k) = B(\mathcal{H} \otimes \mathcal{H}_k)$. We claim that $\omega_1 \equiv \omega_{12}|B(\mathcal{H}_k)$ has entropy $\leq 2S(\varphi)$ and $\omega_2 \equiv \omega_{12}|B(\mathcal{H}_k)$ has density

$$\left(\varphi(V_i^* V_j)\right)_{i,j=1}^{k}.$$

The second statement is easily verified by computation, so we can concentrate on the first one and benefit from the subadditivity. Let us denote by π the GNS-representation and by π' the right multiplication anti-representation of $M_n(\mathbb{C})$. Since

$$B(\mathcal{H}) = \pi(M_n(\mathbb{C})) \otimes \pi'(M_n(\mathbb{C})),$$

we refer to the subadditivity of the entropy and have

$$S(\omega_1) \leq S(\omega_1|\pi(M_n(\mathbb{C})) + S(\omega_1|\pi'(M_n(\mathbb{C})). \tag{7.27}$$

Compute:

$$\langle \Omega, (\pi(a) \otimes I)\Omega \rangle = \sum_{l=1}^{k} \langle V_l \Phi, a V_l \Phi \rangle = \varphi(\textstyle\sum_l V_l^* a V_l) = \varphi(a)$$

due to the invariance, and

$$\langle \Omega, (\pi'(a) \otimes I)\Omega \rangle = \sum_{l=1}^{k} \langle V_l \Phi, V_l a \Phi \rangle = \varphi(\textstyle\sum_l V_l^* V_l a) = \varphi(a).$$

Hence both restricted states on the right hand side of (7.27) have entropy $S(\varphi)$.

According to the *triangle inequality* (or to Lemma 6.4) we have $S(\omega_1) = S(\omega_2)$ the state $S(\omega_{12})$ being pure. From (7.27) we know that $S(\omega_1) \leq 2S(\varphi)$. Therefore $2S(\varphi)$ is a bound for arbitrary $S(\varphi, \mathcal{V})$ as well as for $S^L(\varphi)$.   □

Now we are in a position to produce a formula for the Lindblad entropy of a state φ of a finite quantum system $\mathcal{A}$. Let $z_1, z_2, \ldots, z_k$ be the collection of minimal central projections in $\mathcal{A}$. Writing λ_i for $\varphi(z_i)$ we have

$$S^L(\varphi) = \sum_i \lambda_i S^L(\varphi|z_i\mathcal{A}) + H(\lambda_1, \lambda_2, \ldots, \lambda_k)$$

$$= 2 \sum_i \lambda_i S(\varphi|z_i\mathcal{A}) + H(\lambda_1, \lambda_2, \ldots, \lambda_k)$$

$$= 2 \sum_i S(\varphi) - H(\lambda_1, \lambda_2, \ldots, \lambda_k),$$

where the mixing property was used for both the Lindblad and the von Neumann, entropy and $S^L = 2S$ on $z_i\mathcal{A}$ by virtue of Proposition 7.9. Since $H(\lambda_1, \lambda_2, \ldots, \lambda_k)$ is the entropy of φ restricted to the centre $\mathcal{Z}$, we conclude

$$S^L(\varphi) = 2S(\varphi) - S(\varphi|\mathcal{Z}). \tag{7.28}$$

From this form one can not yet see the concavity of the Lindblad entropy. (By definition it is the supremum of concave functionals.) Let n be the tracial dimension of $\mathcal{A}$ and write τ for the normalized canonical trace Tr. Express the von Neumann entropies in (7.28) by means of relative entropies.

$$S(\varphi) = -S(\varphi, \mathrm{Tr}) = -S(\varphi, \tau) + \log n,$$
$$S(\varphi|\mathcal{Z}) = -S(\varphi|\mathcal{Z}, \tau|\mathcal{Z}) + \log k.$$

So the conditional expectation property yields

$$S^L(\varphi) = -2S(\varphi, \tau) + S(\varphi|\mathcal{Z}) + \log \frac{n^2}{k}$$

$$= -S(\varphi, \tau) - S(\varphi, \varphi \circ E) + \log \frac{n^2}{k},$$

where E is the τ-preserving conditional expectation onto $\mathcal{Z}$. Now it is clear that

(i) $\varphi \mapsto S^L(\varphi)$ is concave,
(ii) $S^L(\varphi) \leq \log(n^2/k)$ with equality if and only if $\varphi = \tau$.

It is interesting to consider the *Lindblad entropy* for tracial states. Let Tr denote the tracial functional such that $\mathrm{Tr}\, z_i = (\mathrm{Tr}\, z_i)^2$. This functional is characterized by the property that at a projection p it takes the value of the linear dimension of the algebra $p\mathcal{A}p$. For a tracial state τ of the algebra $\mathcal{A}$ we have

$$S^L(\tau) = \mathrm{Tr}\, \eta(D), \tag{7.29}$$

where D is the density of τ with respect to Tr, that is, $\tau(a) = \mathrm{Tr}\, Da \ (a \in \mathcal{A})$.

Turning back to relative entropy and we discuss some variations of the definitions. For states ω and φ of a C*-algebra $\mathcal{A}$ we set

$$S_{\text{co}}(\omega,\varphi) = \sup\{\textstyle\sum_i \omega(p_i)(\log\omega(p_i) - \log\varphi(p_i))$$
$$: p_i \in \mathcal{A} \quad \text{is projection}, \quad \textstyle\sum_i p_i = I\} \tag{7.30}$$

and

$$S_{\text{cp}}(\omega,\varphi) = \sup\{\textstyle\sum_i \omega(a_i)(\log\omega(a_i) - \log\varphi(a_i))$$
$$: a_i \in \mathcal{A}^+, \quad \textstyle\sum_i a_i = I\}. \tag{7.31}$$

Obviously,

$$S_{\text{co}}(\omega,\varphi) \leq S_{\text{cp}}(\omega,\varphi) \leq S(\omega,\varphi). \tag{7.32}$$

The definition (7.31) may be troublesome if $\mathcal{A}$ does not contain projections at all. To avoid this problem and other degeneracies we restrict ourselves to finite dimensional $\mathcal{A}$ and faithful φ.

The subscripts "co" and "cp" need some explanation. Clearly,

$$S_{\text{co}}(\omega,\varphi) = \sup\{S(\omega|\mathcal{C}, \varphi|\mathcal{C}) : \mathcal{C} \subset \mathcal{A} \text{ is a commutative subalgebra}\} \tag{7.33}$$

and

$$S_{\text{cp}}(\omega,\varphi) = \sup\{S(\omega \circ \alpha, \varphi \circ \alpha) : \alpha : \mathcal{C} \to \mathcal{A} \text{ is a unital}$$
$$\text{completely positive mapping from a commutative algebra}\}.$$

Another reasonable definition for relative entropy is

$$S_{\text{BS}}(\omega,\varphi) = \text{Tr}\, D_\omega \log D_\omega^{1/2} D_\varphi^{-1} D_\omega^{1/2} \tag{7.34}$$

(proposed by Belavkin and Staszewski). For commuting densities $S_{\text{BS}}(\omega,\varphi) = S(\omega,\varphi)$ and from the relation $Xf(X^*X) = f(XX^*)X$ one can see that

$$S_{\text{BS}}(\omega,\varphi) = -\text{Tr}\, D_\varphi \eta(D_\varphi^{-1/2} D_\omega D_\varphi^{-1/2}). \tag{7.35}$$

Some important properties of S_{BS} are contained in the following proposition.

Proposition 7.10 The relative entropy quantity S_{BS} is monotone and additive, that is,

(i) $S_{\text{BS}}(\omega|\mathcal{A}_0, \varphi|\mathcal{A}_0) \leq S_{\text{BS}}(\omega,\varphi)$ if $\mathcal{A}_0 \subset \mathcal{A}$,
(ii) $S_{\text{BS}}(\omega_1 \otimes \omega_2,\ \varphi_1 \otimes \varphi_2) = S_{\text{BS}}(\omega_1,\varphi_1) + S_{\text{BS}}(\omega_2,\varphi_2)$.

Proof. Let τ be a faithful tracial state on $\mathcal{A}$ and let D_ω^τ and D_φ^τ be the densities of ω and φ with respect to τ. Since $\tau(a) = \text{Tr}\, ca$ $(a \in \mathcal{A})$ for an operator c in the centre of $\mathcal{A}$, we have

$$S_{\text{BS}}(\omega,\varphi) = -\tau(D_\varphi^\tau \eta((D_\varphi^\tau)^{-1/2} D_\omega^\tau (D_\omega^\tau)^{-1/2})). \tag{7.36}$$

If $E : \mathcal{A} \to \mathcal{A}_0$ is the τ-preserving conditional expectation then $\kappa_\varphi \equiv E(D_\varphi^\tau)$ and $\kappa_\omega \equiv E(D_\omega^\tau)$ are the densities of $\varphi|\mathcal{A}_0$ and $\omega|\mathcal{A}_0$, respectively. We define a positive unital mapping $\alpha : \mathcal{A} \to \mathcal{A}_0$ by

$$\alpha(X) = \kappa_\varphi^{-1/2} E((D_\varphi^\tau)^{-1/2} X (D_\varphi^\tau)^{-1/2}) \kappa_\varphi^{-1/2} \tag{7.37}$$

and use the inequality

$$\alpha(\eta(X)) \leq \eta(\alpha(X)) \tag{7.38}$$

with $X = (D_\varphi^\tau)^{-1/2} D_\omega^\tau (D_\varphi^\tau)^{-1/2}$. In this way we arrive at the inequality

$$E((D_\varphi^\tau)^{1/2} \eta((D_\varphi^\tau)^{-1/2} D_\omega^\tau (D_\varphi^\tau)^{-1/2})(D_\varphi^\tau)^{1/2})$$
$$\leq \kappa_\varphi^{1/2} \eta(\kappa_\varphi^{-1/2} \kappa_\omega \kappa_\varphi^{-1/2}) \kappa_\varphi^{1/2} \, . \tag{7.39}$$

Applying the functional τ to both sides of the inequality one obtains the monotonicity (i) by virtue of (7.36).

(ii) is obvious from the definition. □

Proposition 7.11 Let φ and ω be faithful states on a finite quantum system $\mathcal{A}$. Then

$$S(\omega, \varphi) \leq S_{\mathrm{BS}}(\omega, \varphi) \, .$$

Proof. First we note that the monotonicity of S_{BS} implies that $S_{\mathrm{co}} \leq S_{\mathrm{BS}}$. We write $\omega_n, \varphi_n, \mathcal{A}_n$ for the n-fold tensor products $\omega \otimes \omega \otimes \ldots \otimes \omega$, $\varphi \otimes \varphi \otimes \ldots \otimes \varphi$ and $\mathcal{A} \otimes \mathcal{A} \otimes \ldots \otimes \mathcal{A}$, respectively. We consider the subalgebra $\mathcal{C}_n$ of $\mathcal{A}_n$ which appears in (7.43). We have by Lemma 7.12

$$S(\omega_n, \varphi_n) - S(\omega_n|\mathcal{C}_n, \ \varphi_n|\mathcal{C}_n) \leq n\varepsilon \tag{7.40}$$

if n is large enough. Since the states $\omega_n|\mathcal{C}_n$ and $\varphi_n|\mathcal{C}_n$ commute, we have

$$S(\omega_n|\mathcal{C}_n, \ \varphi_n|\mathcal{C}_n) = S_{\mathrm{co}}(\omega_n|\mathcal{C}_n, \ \varphi_n|\mathcal{C}_n) \leq S_{\mathrm{BS}}(\omega_n|\mathcal{C}_n, \ \varphi_n|\mathcal{C}_n) \, . \tag{7.41}$$

From these relations we obtain

$$nS(\omega, \varphi) \leq n\varepsilon + S_{\mathrm{BS}}(\omega_n, \varphi_n) = n\varepsilon + nS_{\mathrm{BS}}(\omega, \varphi) \, .$$

Since $\varepsilon > 0$ can be arbitrary small the proposition is proven. □

It is quite remarkable that the previous proposition is equivalent to the inequality

$$\mathrm{Tr}\, A \log AB \geq \mathrm{Tr}\, A(\log A + \log B) \tag{7.42}$$

for nonnegative matrices A and B if B is invertible.

We continue to use the notation of the proof in the previous proposition. We write D_n^ω and D_n^φ for the densities of ω_n and φ_n, respectively. Let $\sum_{i=1}^k \lambda_i e_i$ be the orthogonal decomposition of the density of φ. The projections e_i are assumed to be of rank one. If $\sum_{i=1}^k m_i = n$, we write

$$I(m_1, m_2, \ldots, m_k) = \{(i_1, i_2, \ldots, i_n) : \#\{l : i_l = j\} = m_j, \quad 1 \leq j \leq k\},$$

that is, $I(m_1, m_2, \ldots, m_k)$ consists of all n-tuples of $\{1, 2, \ldots, k\}$ such that the multiplicity of j is m_j ($1 \le j \le k$). Let

$$p(m_1, m_2, \ldots, m_k) = \sum \{e_{i_1} \otimes e_{i_2} \otimes \ldots \otimes e_{i_n}$$
$$: (i_1, i_2, \ldots, i_n) \in I(m_1, m_2, \ldots, m_k)\}.$$

The mapping

$$E_n : a \mapsto \sum \left\{ p(m_1, m_2, \ldots, m_k) a\, p(m_1, m_2, \ldots, m_k) : \sum_{i=1}^{k} n_i = n \right\} \quad (7.43)$$

is a conditional expectation of $\mathcal{A}_n$ onto a subalgebra $\mathcal{C}_n$ of $\mathcal{A}_n$. The spectral projections of D_n^{φ} are sums of certain projections $p(m_1, m_2, \ldots, m_k)$. Therefore,

$$[D_n^{\varphi}, \mathcal{C}_n] = 0. \quad (7.44)$$

If ψ is a state of a finite quantum system, the commutant of the density of ψ is called the centralizer of ψ. It is quite obvious that ψ restricted to its centralizer is tracial. Relation (7.44) yields that $\mathcal{C}_n$ is contained in the centralizer of φ_n.

Lemma 7.12 Let ψ be an arbitrary state on $\mathcal{A}_n$. Then for the above described subalgebra $\mathcal{C}_n$ of $\mathcal{A}_n$,

$$S(\psi, \varphi_n) - S(\psi|\mathcal{C}_n, \varphi_n|\mathcal{C}_n) \le C \log(n + 1)$$

holds.

Proof. We shall show that the constant C may be the tracial dimension of $\mathcal{A}$. Since the conditional expectation defined by (7.43) leaves φ_n invariant, we can apply the conditional expectation property and so $S(\psi, \psi \circ E_n)$ is to be estimated. It follows from the convexity of $S(\psi, \psi \circ E_n)$ that we may assume ψ to be pure. Then its density is a projection and

$$S(\psi, \psi \circ E_n) = S(\psi \circ E_n).$$

Each operator

$$p(m_1, m_2, \ldots, m_k) D_\psi p(m_1, m_2, \ldots, m_k)$$

has rank 0 or 1. Hence the rank of the density of $\psi \circ E_n$ is at most the cardinality

$$\#\{(m_1, m_2, \ldots, m_k) : \textstyle\sum_{i=1}^{k} m_i = n\},$$

which can be majorized by $(n + 1)^k$. Consequently,

$$S(\psi \circ E_n) \le \log(n + 1)^k,$$

which completes the proof. $\qquad\square$

Proposition 7.13 Let φ and ω be states on a finite quantum system $\mathcal{A}$ and assume that φ is faithful. Then

$$S_{co}(\omega, \varphi) = \sup\{\omega(h) - \log\varphi(e^h) : h = h^* \in \mathcal{A}\}.$$

Proof. A reformulation Proposition 1.11 is the following.

$$S(\omega, \varphi) = \sup\{\omega(h) - \log\varphi^h(I) : h = h^* \in \mathcal{A}\}$$

This implies

$$S_{co}(\omega, \varphi) = \sup\{\omega(h) - \log(\varphi|\mathcal{C})^h(I) : h = h^* \in \mathcal{C} \text{ and }$$
$$\mathcal{C} \subset \mathcal{A} \text{ is a commutative sublagebra}\}.$$

Now the proposition is a consequence of the relation $(\varphi|\mathcal{C})^h(I) = \varphi(e^h)$. □

It is remarkable that from Proposition 7.13 one can easily derive the *Golden-Thompson inequality* (3.13). Let A and B be self-adjoint matrices. Assume that $\operatorname{Tr} e^A = 1$ and write X for $e^{A+B}/\operatorname{Tr} e^{A+B}$. So X and e^A are densities and we have

$$\log \operatorname{Tr} e^{A+B} = \operatorname{Tr} XA - S(X, e^B)$$
$$\leq \operatorname{Tr} XA - S_{cp}(X, e^B) \leq \log \operatorname{Tr} e^A e^B$$

which is just (3.13). This proof of the Golden-Thompson inequality is based on the monotonicity of the relative entropy and gives the necessary and sufficient condition for the equality.

$$\operatorname{Tr} e^{A+B} = \operatorname{Tr} e^A e^B \tag{7.45}$$

holds if and only if A commutes with B. This is a consequence of Proposition 1.16.

To start a short discussion on *conditional entropy*, we first look at the concept of conditioning in information theory. Let $(P_i)_{i=1}^n$ and $(Q_j)_{j=1}^m$ be partitions of a probability space (X, μ). For each Q_j a conditional distribution

$$\frac{\mu(P_i \cap Q_j)}{\mu(Q_j)} \qquad (i = 1, 2, \ldots, n)$$

arises which has entropy

$$H(P|Q_j) = -\sum_{i=1}^n \frac{\mu(P_i \cap Q_j)}{\mu(Q_j)} \log \frac{\mu(P_i \cap Q_j)}{\mu(Q_j)}.$$

The conditional entropy $H(P|Q)$ is the expectation of the random variable $j \mapsto H(P|Q_j)$ which takes this value with probability $\mu(Q_j)$. Hence

$$H(P|Q) = -\sum_{j=1}^m \sum_{i=1}^n \mu(P_i \cap Q_j) \log \frac{\mu(P_i \cap Q_j)}{\mu(Q_j)}$$
$$= H(P \vee Q) - H(Q), \tag{7.46}$$

where $\vee$ denotes the join of the two partitions. This expression is in accordance with our intuition that the conditional entropy $H(P|Q)$ measures the additional amount of information provided by P if Q is already known.

Let now $\mathcal{C}$ be a C*-algebra with a state φ. The completely positive unital mappings $\alpha : \mathcal{A} \to \mathcal{C}$ and $\beta : \mathcal{B} \to \mathcal{C}$ will replace the partitions P and Q. Set

$$H_\varphi(\alpha|\beta) = \sup \left\{ \sum_i \lambda_i \big(S(\varphi_i \circ \alpha, \varphi \circ \alpha) \right.$$
$$\left. - S(\varphi_i \circ \beta, \varphi \circ \beta) \big) : \sum_i \lambda_i \varphi_i = \varphi \right\}. \qquad (7.47)$$

Due to the scaling property of the relative entropy this is equivalently written as

$$H_\varphi(\alpha|\beta) = \sup \left\{ \sum_i \big(S(\psi_i \circ \alpha, \varphi \circ \alpha) \right.$$
$$\left. - S(\psi_i \circ \beta, \varphi \circ \beta) \big) : \sum_i \psi_i = \varphi \right\}. \qquad (7.48)$$

Proposition 7.14 The conditional entropy $H_\varphi(\alpha|\beta)$ possesses the following properties.

  (i) $0 \leq H_\varphi(\alpha|\beta) \leq H_\varphi(\alpha)$.
  (ii) $H_\varphi(\alpha \circ \alpha'|\beta) \leq H_\varphi(\alpha|\beta)$.
  (iii) $H_\varphi(\alpha|\beta \circ \beta') \geq H_\varphi(\alpha|\beta)$.
  (iv) $H_\varphi(\alpha|\beta) \leq H_\varphi(\alpha|\gamma) + H_\varphi(\gamma|\beta)$.

Any of these properties is the immediate consequence of the definition and the monotonicity of the relative entropy. (i) tells us that the conditional entropy can not exceed the unconditional one. Furthermore, (ii) and (iii) are monotonicity properties while (iv) is a triangle inequality. Below we use $H_\varphi(\alpha|\beta)$ mostly in the case, where α is the identity of the C*-algebra $\mathcal{A}$ and β is the embedding of a subalgebra $\mathcal{B} \subset \mathcal{A}$. Then we write $H_\varphi(\mathcal{A}|\mathcal{B})$.

Proposition 7.15 Let $\mathcal{M}$ be a von Neumann algebra with a faithful normal state φ and let $\mathcal{N} \subset \mathcal{M}$ be a von Neumann subalgebra. If $H_\varphi(\mathcal{M}|\mathcal{N}) = 0$ then $\mathcal{N} = \mathcal{M}$.

Proof. The hypothesis implies that $S(\omega, \varphi) = S(\omega|\mathcal{N}, \varphi|\mathcal{N})$ for any normal state ω such that $\omega \leq \lambda\varphi$ for some $\lambda > 0$. In the terminology of Chapter 9, this means that $\mathcal{N}$ is sufficient for $\{\varphi, \omega\}$ (see the discussion after Theorem 9.3). Therefore, $[D\omega, D\varphi]_t \in \mathcal{N}$ for every $t \in \mathbb{R}$ and for every $\omega \leq \lambda\varphi$. Since all these Connes cocycles span the algebra $\mathcal{M}$, we arrive at $\mathcal{M} \equiv \mathcal{N}$. $\qquad \square$

Proposition 7.16 Let $\mathcal{M}$ be a von Neumann algebra with a faithful normal state φ and let $\mathcal{N} \subset \mathcal{M}$ be a von Neumann subalgebra such that there exists a conditional expectation E onto $\mathcal{N}$ which preserves φ. Then

$$H_\varphi(\mathcal{M}|\mathcal{N}) = \sup \left\{ \sum_i S(\psi_i, \psi_i \circ E) : \sum_i \psi_i = \varphi \right\},$$

where the supremum is over all decompositions $\sum_i \psi_i = \varphi$ into positive normal functionals.

Proof. A combination of the scaling and the conditional expectation property (see (5.19)) gives

$$S(\psi_i, \varphi) - S(\psi_i | \mathcal{N}, \varphi | \mathcal{N}) = S(\psi_i, \psi_i \circ E)$$

which is sufficient to conclude the proposition.   □

This proposition applies to any von Neumann subalgebra $\mathcal{N}$ if τ is a faithful normal tracial state of a von Neumann algebra $\mathcal{M}$. Then a decomposition $\sum_i \psi_i = \tau$ is equivalent to a partition of unity $(x_1, x_2, \ldots, x_n)$ (that is, $x_i \in \mathcal{M}_+$ and $\sum_i x_i = I$). We have

$$H_\tau(\mathcal{M} | \mathcal{N}) = \sup \left\{ \sum_i \left(\tau(\eta(E_\mathcal{N}(x_i))) - \tau(\eta(x_i)) \right) \right\}, \tag{7.49}$$

where the supremum is over all partition of unities, $E_\mathcal{N}$ is the τ-preserving conditional expectation onto $\mathcal{N}$ and $\eta(t) = -t \log t$. The conditional entropy (7.49) is rarely available in an explicit form. Now we are going to explain the Pimsner-Popa theorem which gives $H_\tau(\mathcal{M} | \mathcal{N})$ in case of a finite dimensional $\mathcal{M}$.

Let $\mathcal{M}$ be a finite dimensional algebra and denote by $z_1^\mathcal{M}, z_2^\mathcal{M}, \ldots, z_m^\mathcal{M}$ its minimal central projections. The canonical trace $\mathrm{Tr}_\mathcal{M}$ takes the value 1 at each minimal projection. So

$$\mathcal{M} z_j^\mathcal{M} = M_{r(j)}(\mathbb{C}) \quad \text{where} \quad r(j) = \mathrm{Tr}_\mathcal{M} z_j^\mathcal{M}.$$

Any tracial state τ is determined by the vector $(t_1, t_2, \ldots, t_m)$, where t_j is the value of τ at a minimal projection in $z_j^\mathcal{M}$. Of course,

$$\sum_{j=1}^m t_j \mathrm{Tr}\, z_j^\mathcal{M} = 1$$

must hold for the trace vector $\mathbf{t} = (t_1, t_2, \ldots, t_m)$ and the dimension vector $\mathbf{r} = (\mathrm{Tr}_\mathcal{M} z_1^\mathcal{M}, \mathrm{Tr}_\mathcal{M} z_2^\mathcal{M}, \ldots, \mathrm{Tr}_\mathcal{M} z_m^\mathcal{M})$. Let now $\mathcal{N}$ be a subalgebra of $\mathcal{M}$ and we write its central decomposition in the form

$$\mathcal{N} = \oplus_{i=1}^n \mathcal{N} z_i^\mathcal{N}.$$

The matrix of the inclusion of $\mathcal{N}$ in $\mathcal{M}$ is defined as follows.

$$\left(\Lambda_\mathcal{N}^\mathcal{M} \right)_{ij} = \frac{\mathrm{Tr}_\mathcal{M} z_i^\mathcal{N} z_j^\mathcal{M}}{\mathrm{Tr}_\mathcal{N} z_i^\mathcal{N}} \quad (1 \leq i \leq n, \ 1 \leq j \leq m). \tag{7.50}$$

Note that the number of rows n is the number of direct summand in the central decomposition of $\mathcal{N}$. If $\mathbf{s}$ is the trace vector of $\tau | \mathcal{N}$ and $\mathbf{p}$ is the dimension vector of $\mathcal{N}$ then we have

$$\begin{aligned}
\Lambda_\mathcal{N}^\mathcal{M} \mathbf{t} &= \mathbf{s}, & \mathbf{t}^t \mathbf{m} &= 1, \\
\left(\Lambda_\mathcal{N}^\mathcal{M} \right)^t \mathbf{p} &= \mathbf{r}, & \mathbf{s}^t \mathbf{n} &= 1,
\end{aligned} \tag{7.51}$$

where $\mathbf{X}^t$ denotes the transpose of $\mathbf{X}$ and all vectors are regarded as column ones. Now we can state the *Pimsner-Popa theorem* which gives a formula for conditional entropy.

Theorem 7.17 Let $\mathcal{N} \subset \mathcal{M}$ be an inclusion of finite dimensional algebras. Using the above notation we have

$$H_\tau(\mathcal{M}|\mathcal{N}) = \sum_{j=1}^{m} r_j \eta(t_j) - \sum_{j=1}^{m} t_j \eta(r_j) - \sum_{i=1}^{n} p_i \eta(s_i) + \sum_{i=1}^{n} s_i \eta(p_i)$$

$$- \sum_{i=1}^{n} \sum_{j=1}^{m} p_i (\Lambda_{\mathcal{N}}^{\mathcal{M}})_{ij} \, t_j \big| \log \left((\Lambda_{\mathcal{N}}^{\mathcal{M}})_{ij}/p_i \right) \big|_+ ,$$

where $|\,.\,|_+$ denotes positive part.

If $(\Lambda_{\mathcal{N}}^{\mathcal{M}})_{ij} \le p_i$ for every i, j then the last term vanishes and the formula for the conditional entropy gets simplified. Another interesting observation is that the other terms are expressed by means of the Lindblad entropy. So

$$H_\tau(\mathcal{M}|\mathcal{N}) = S^L(\tau) - S^L(\tau|\mathcal{N})$$

$$- \sum_{i=1}^{n} \sum_{j=1}^{m} p_i (\Lambda_{\mathcal{N}}^{\mathcal{M}})_{ij} \, t_j \big| \log \left((\Lambda_{\mathcal{N}}^{\mathcal{M}})_{ij}/p_i \right) \big|_+ \qquad (7.52)$$

We discuss *McMillan's type convergence theorem* as the last topics of this chapter. McMillan's theorem is an ergodic type convergence theorem and plays an essential role in information theory. Its full non-commutative extension is still missing but here we give some partial results concerning tracial states. However, we first review McMillan's theorem in its probabilistic form.

Let $(X, \mathcal{F}, \mu)$ be a probability space and T be a measure preserving transformation from X to X. For each finite partition $P = (X_1, X_2, \ldots, X_n)$ of X the entropy function $I(P)$ is defined by

$$I(P) = - \sum_{i=1}^{n} \log \mu(P_i) \chi^{P_i} , \qquad (7.53)$$

where χ^{P_i} stands for the characteristic function of P_i. The entropy of the partition is just the expectation value of the entropy function. The McMillan theorem asserts that

$$\frac{1}{n} I(\vee_{i=0}^{n-1} T^{-i}(P))$$

converges to a T-invariant function h μ-almost everywhere. (Here $\vee_{i=0}^{n-1} T^{-i}(P)$ denotes the join of the partitions $P, T^{-1}P, \ldots, T^{-n+1}P$.) The non-commutative generalization has several obstacles. If instead of partition we consider finite dimensional subalgebra, then we have the problem that two finite dimensional subalgebras may fail to generate a finite dimensional subalgebra. To avoid this difficulty we restrict ourselves to the shift automorphism of an infinite product algebra.

Let $\mathcal{M}$ be a finite dimensional algebra with a tracial state τ and denote by $z_1, z_2, \ldots, z_n$ its minimal central projections. Recall that the pair $(\mathcal{M}, \tau)$

is determined by the trace vector $(t_1, t_2, \ldots, t_m)$ and the dimension vector $(r_1, r_2, \ldots, r_n)$. (They satisfy the relation $\sum_{j=1}^{n} t_j r_j = 1$.) As an extension of (7.52) set

$$I_\tau(\mathcal{M}) = -\sum_{i=1}^{n} \log t_i \, z_i \,, \tag{7.54}$$

for the entropy operator which belongs to the centre. Evidently, we have

$$S(\tau) = -\sum_{i=1}^{n} r_i t_i \log t_i = \tau(I_\tau(\mathcal{M})) \,.$$

Let $\mathcal{M}_i$ be a copy of the finite dimensional C*-algebra $\mathcal{M}$ and consider the infinite C*-algebraic tensor product

$$\mathcal{A} = \otimes_{i=-\infty}^{\infty} \mathcal{M}_i$$

with the right shift automorphism γ and with a γ-invariant tracial state τ. By means of the GNS-represebtation we embed $\mathcal{A}$ densely into a von Neumann algebra $\mathcal{N}$ and we do not use a new letter for the natural extension of τ and γ to $\mathcal{N}$. For a sequence $(A_n)_n$ in $\mathcal{N}$ and for an operator A affiliated with $\mathcal{N}$ we say that $A_n \to A$ τ-almost uniformly if for any $\varepsilon > 0$ there exists a projection $Q \in \mathcal{N}$ such that $\tau(Q) \geq 1 - \varepsilon$ and $\|(A_n - A)Q\| \to 0$. Now we are in a position to state a simple convergence theorem of McMillan's type.

Proposition 7.18 Under the above conditions the operators

$$\frac{1}{n} I_\tau(\vee_{i=0}^{n-1} \gamma^i(\mathcal{M}_0)) = \frac{1}{n} I_\tau(\otimes_{i=0}^{n-1} \mathcal{M}_i)$$

converges τ-almost uniformly to a γ-invariant operator affiliated with $\mathcal{N}$.

Proof. One can observe that the bounded operators $I_\tau(\otimes_{i=0}^{n-1} \mathcal{M}_i)$ belong to the centre of $\mathcal{N}$. In a function representation of the centre we are able to use the commutative McMillan theorem. Since the almost uniform convergence is just an alternative formulation of the almost sure one when the algebra is commutative, the proof is complete in this way. □

Notes and Remarks. The quantity S_f with an arbitrary convex (or concave) function has often appeared in the literature, [Wehrl 1978] contains some references. The analogous use of a function f in the case of the relative entropy was suggested in [Petz 1986b] and the *quasi-entropies* in the context of Neumann algebras were studied in [Petz 1985b]. Proposition 7.1 and 7.2 are from this paper. In classical information theory the term f-*divergence* is used instead of the term quasi-entropy and the function f should be convex rather than operator convex (cf. [Csiszár [1967]]). The transition probability (7.6) was introduced in [Raggio 1982]. Applications and its relation to the other concepts of transition probability were treated in [Raggio 1984].

The *skew information* $I_p(\varphi, k)$ was introduced in [Wigner and Yanase 1963]. The case $I_2(\varphi, k)$ was extended to arbitrary von Neumann algebras in [Connes and Størmer 1976] as a measure of non-commutativity.

Concerning *Rényi entropies* in classical information theory we refer to the works [Rényi 1962], [Rényi 1965] and [Aczél and Daróczi 1975]. The concept of the *Segal entropy* originated from the historically interesting paper [Segal 1960], where essentially the case of bounded density was considered and concavity proved. The characterization (7.20) is from [Padmanabhan 1979]. The non-commutative Jensen inequality in the form (7.12) is a consequence of (7.38); see, for example [Choi 1974]. It is interesting to remark that the possibility of a notion of relative entropy with respect to a non tracial state appears in [Segal 1960], but the author was rather pessimistic concerning this extension. Apparently unaware of the earlier work [Segal 1960], the Segal entropy was redefined in [Ruskai 1973] and its concavity as well as subadditivity were proved in full generality.

The *spectral scale* as a continuous version of the list of eigenvalues (or characteristic values) was introduced in [Ovchinnikov 1970] and in [Fack 1982]. A simple treatment in the case of a finite trace is in [Petz 1985c]. The full power of the spectral scale technique is demonstrated in [Fack and Kosaki 1986] by proving several trace inequalities for unbounded operators. Propositions 7.3 and 7.6 might be new.

The relative entropy quantity ent_A with a different sign convention is discussed in [Donald 1985]. The *mixture entropy* S^S in the case $S = \mathfrak{S}(A)$ was introduced in [Manucea, Naudts and Verbeure 1972], where its concavity was established, too. The use of a general compact convex set S was proposed in [Ohya 1984]. The entropy S_f is the infimum of $\sum_i f(\lambda_i)$ over the decompositions (7.24), provided that $f(0) = f(1) = 0$, see [Uhlmann 1970].

The *Lindblad entropy* was discussed in [Lindblad 1988].

The quantities S_{co} and S_{cp} were mentioned in [Donald 1985]. On the other hand, S_{BS} originated from [Belavkin and Staszewski 1982]. and it is closely related to an operator entropy [Fujii and Kamei 1989], Proposition 7.10 follows from this paper. Proposition 7.11 is from [Hiai and Petz 1991], where it is proved that S_{co}, S_{cp} and S_{BS} show the same asymptotic behaviour as the Umegaki relative entropy. Inequality (7.42) is generalized in [Hiai and Petz 1993].

The Golden-Thompson inequality as a consequence of the monotonicity of the relative entropy and (7.45) are from [Petz 1988c].

The *conditional entropy* was introduced in [Connes and Størmer 1975]. Although its is an algebraic extension of the probabilistic conditional entropy, they used the term "relative entropy". Theorem 7.17 is from [Pimsner and Popa 1986]. A systematic study for a non tracial state was carried out in [Hiai 1991b].

The McMillan theorem may be found on p. 97 of [Martin and England 1981], for example. The entropy operator $I_\tau(\mathcal{M})$ was discussed in [Emch 1974] and in [Ohya 1985]. Proposition 7.18 is from [Ohya, Tsukada and Umegaki 1987]. The monograph [Jajte 1985] contains information on almost uniform convergence and limit theorems.

Comments. A special f-entropy is named after *Tsallis*, see [Tsallis 1988].

$$S_q = \frac{k\left(1 - \sum_i \lambda_i^q\right)}{q - 1} \qquad \left(f(t) = \frac{k}{q-1}(t - t^q)\right).$$

This is a transform of the Rényi entropy and has appeared in many papers recently. Concerning an axiomatization, see [Suyari 2002].

In matrix algebras, there is a correspondence between operational partitions of unity $\mathcal{V} = (V_1, V_2, \ldots, V_k)$ and completely positive unital mappings

$$\gamma_\mathcal{V}(a) = \sum_{i=1}^k V_i^* a V_i. \tag{7.55}$$

(Each completely positive unital mapping has this form but the representation is not unique.) The dual of $\gamma_\mathcal{V}$ is the state transformation

$$\mathcal{E}_\mathcal{V}(D) = \sum_{i=1}^k V_i D V_i^*. \tag{7.56}$$

The von Neumann entropy $S(\varphi, \mathcal{V})$ of (7.26) depends actually only on $\mathcal{E}_\mathcal{V}$, that is, if $\mathcal{E}_\mathcal{V} = \mathcal{E}_{\mathcal{V}'}$, then $S(\varphi, \mathcal{V}) = S(\varphi, \mathcal{V}')$. Hence $S(\varphi, \mathcal{V})$ can be called the *entropy exchange* of the pair $(\varphi, \mathcal{E}_\mathcal{V})$.

Part III

Channeling Transformation
and Coarse Graining

8 Channels and Their Transpose

The transition of a state of a system into another state is always at the centre of interest. Particularly in communication theory the study of state change is indispensable and this state change is described by a *channeling transformation*. Here we consider channels given by a completely positive unital mapping $\alpha : \mathcal{B} \to \mathcal{A}$. The C*-algebras $\mathcal{A}$ and $\mathcal{B}$ represent the *input and output systems*, respectively. To an initial or input state φ of $\mathcal{A}$ the channel associates the output state $\varphi \circ \alpha$.

Composition and *C*-convex combination* are operations on the class of channels.

(i)   If $\beta : \mathcal{A}_1 \to \mathcal{A}_2$ and $\alpha : \mathcal{A}_2 \to \mathcal{A}_3$ are channels then so is $\alpha \circ \beta$.

(ii) If $\alpha : \mathcal{A}_1 \to \mathcal{A}_2$ and $\beta : \mathcal{A}_1 \to \mathcal{A}_2$ are channels and $a, b \in \mathcal{A}_2$ are such that $a^*a + b^*b = I$ then $x \mapsto a^*\alpha(x)a + b^*\beta(x)b$ defines a channel.

(iii) Every conditional expectation is a channel.

Channels arise from different contexts and sometimes they are called by different names. The dynamical evolution of a physical system is described by a semigroup (β_t) of completely positive mappings ($t \in \mathbb{R}$ or $t \in \mathbb{R}^+$) and in this case $\varphi \circ \beta_t$ is not called an output state but rather the time evolution of the initial state φ (Schrödinger picture.)

In measurement theory the observation channel is given by a positive operator-valued measure. Let $(X, \mathcal{S})$ be a measurable space and for every $H \in \mathcal{S}$ let $E(H)$ be a positive operator in a C*-algebra $\mathcal{A}$ such that

(i) $\sum_i E(H_i) = E(\cup_i H_i)$ whenever (H_i) is a pairwise disjoint family in $\mathcal{S}$,

(ii) $E(X) = I$.

For example, if the measurable space is $\mathbb{R}^n$, then for a Borel set H and for a state φ of $\mathcal{A}$ the number $\varphi(E(H))$ is interpreted as the probability that in the given state the observed quantity takes its value in the set H. One can define a mapping

$$\alpha : f \mapsto \int_X f \, dE \tag{8.1}$$

which maps a C*-algebra of bounded functions into algebra $\mathcal{A}$. The mapping α is a channel in the above sense. Motivated by this example a channel

$\alpha : \mathcal{C} \to \mathcal{A}$ will be called *observation channel* if the C*-algebra $\mathcal{C}$ is commutative. The structure of an observation channel $\alpha : \mathcal{C} \to \mathcal{A}$ is particularly simple if $\mathcal{C}$ is of finite dimension. Then α is determined by an n-tuple $(a_1, a_2, \ldots, a_n)$ of nonnegative operators in $\mathcal{A}$ so that $\sum_i a_i = I$. Such a family $(a_1, a_2, \ldots, a_n)$ is called a *partition of unity*. A partition of unity corresponds to a measurement with finitely many outcomes. An important class of measurements is given by orthogonal partitions of unity. Such measurements are often called *simple*. In the case of a simple measurement the corresponding operator-valued measure is projection-valued and equivalently the induced completely positive mapping is multiplicative. A channel with finite dimensional output algebra will be called *finite*.

In quantum theory the observation channels have great importance because the information on a system is obtained through measurements. Let $\mathcal{A}$ be a C*-algebra which is supposed to describe a system and let $\alpha_1 : \mathcal{C}_1 \to \mathcal{A}$ and $\alpha_2 : \mathcal{C}_2 \to \mathcal{A}$ be observations. We say that α_1 is a refinement of α_2 if there exists a channel $\beta : \mathcal{C}_2 \to \mathcal{C}_1$ such that $\alpha_1 \circ \beta = \alpha_2$. This means that performing observation α_1 on any state yields more information than performing α_2. To make the idea clearer, suppose that α_1 corresponds to a partition of unity $(a_1, a_2, \ldots, a_n)$ and α_2 is given by $(b_1, b_2, \ldots, b_m)$. In a state φ the observation α_1 leads to the probability vector

$$p = \big(\varphi(a_1), \varphi(a_2), \ldots, \varphi(a_n)\big)$$

and similarly α_2 leads to

$$q = \big(\varphi(b_1), \varphi(b_2), \ldots, \varphi(b_m)\big).$$

α_1 is a refinement of α_2 if in any state φ the vector q is related to p through a stochastic matrix. Viewing from the other side, α_2 is termed the *coarse graining* of α_1. This terminology will be used below for general channels as well.

In the quantum probabilistic interpretation of a channel $\alpha : \mathcal{B} \to \mathcal{A}$, the C*-algebra $\mathcal{A}$ with a distinguished state φ is a quantum probability space, α is a random variable ("with state space $\mathcal{B}$") and $\varphi \circ \alpha$ is the distribution of the random variable. Sometimes a family $(\alpha_i : \mathcal{B}_i \to \mathcal{A})_i$ of channels with the same input algebra occurs. This will be called a *multiple-channel* (or process, in the probabilistic interpretation).

Let us interrupt the discussion of general channels in order to show some concrete ones which are related to *second quantization*. Let $\mathcal{H}$ be a complex inner product space and write

$$\sigma(f, g) = \text{Im} \langle f, g \rangle \qquad (f, g \in \mathcal{H}). \tag{8.2}$$

We consider a C*-algebra $\text{CCR}(\mathcal{H})$ which is generated by unitary elements $\{W(f) : f \in \mathcal{H}\}$ so that

(i) $W(f)^* = W(-f), \qquad (f \in \mathcal{H})$
(ii) $W(f)W(g) = W(f + g) \exp(-\tfrac{1}{2}\sigma(f, g)) \qquad (f, g \in \mathcal{H})$.

Condition (ii) tells us that $W(f)W(0) = W(0)W(f) = W(f)$. Hence $W(0)$ is the unit of the algebra $\mathrm{CCR}(\mathcal{H})$ and it follows that $W(f)$ is a unitary for every $f \in \mathcal{H}$. Since (ii) is essentially the Weyl form of the canonical commutation relation, the unitaries $W(f)$ are called *Weyl operators*. It can be proven that the above definition determines the C*-algebra $\mathrm{CCR}(\mathcal{H})$ up to isomorphism. In other words, if $\mathcal{A}$ is generated by some unitaries $\{W'(f) : f \in \mathcal{H}\}$ satisfying (i) and (ii), then there exists an isomorphism of $\mathcal{A}$ onto $\mathrm{CCR}(\mathcal{H})$ that sends $W'(f)$ into $W(f)$. We call $\mathrm{CCR}(\mathcal{H})$ the algebra of the canonical commutation relation over the test-function space $\mathcal{H}$. (Weyl algebra is also a common terminology.) As a consequence of the above mentioned uniqueness theorem, every real linear mapping $T : \mathcal{H} \to \mathcal{H}$ such that

$$\sigma(Tf, Tg) = \sigma(f, g) \qquad (f, g \in \mathcal{H}) \tag{8.3}$$

gives rise to an automorphism α_T of $\mathrm{CCR}(\mathcal{H})$ according to the the formula

$$\alpha_T(W(f)) = W(Tf) \qquad (f \in \mathcal{H}). \tag{8.4}$$

For every (complex) linear contraction $A : \mathcal{H} \to \mathcal{H}'$ there exists a completely positive map $\alpha_A : \mathrm{CCR}(\mathcal{H}) \to \mathrm{CCR}(\mathcal{H}')$ such that

$$\alpha_A(W(f)) = W(Af) \exp(\tfrac{1}{4}\|Af\|^2 - \tfrac{1}{4}\|f\|^2) \qquad (f \in \mathcal{H}). \tag{8.5}$$

This α_A is called a *quasi-free map*, and it is a kind of lifting of the contraction A to the algebra of canonical commutation relation. The correspondence $A \mapsto \alpha_A$ is functorial in the sense that

$$\alpha_{AB} = \alpha_A \circ \alpha_B \tag{8.6}$$

whenever $B : \mathcal{H}_1 \to \mathcal{H}_2$ and $A : \mathcal{H}_2 \to \mathcal{H}_3$ are contractions.

After the example of quasi-free completely positive maps we return to general channels and define their mutual entropy with respect to an input state. Let $\alpha : \mathcal{A}_1 \to \mathcal{A}$ be a channel with an input state φ. The *mutual entropy* $I(\varphi; \alpha)$ is defined as follows.

$$I(\varphi; \alpha) = \sup\{\textstyle\sum_i \lambda_i S(\varphi_i \circ \alpha, \varphi \circ \alpha) : \textstyle\sum_i \lambda_i \varphi_i = \varphi\} \tag{8.7}$$

where the least upper bound is taken over all finite *orthogonal decompositions* of φ. Due to the Uhlmann monotonicity theorem we have

$$I(\varphi; \alpha) \geq I(\varphi; \beta \circ \alpha) \quad \text{and} \quad I(\varphi; \alpha) \geq I(\varphi; \alpha \circ \delta) \tag{8.8}$$

for any channels $\delta : \mathcal{A}_2 \to \mathcal{A}_1$ and $\beta : \mathcal{A} \to \mathcal{A}$ such that $\varphi \circ \beta = \varphi$. In particular,

$$I(\varphi; \alpha) = I(\varphi; \gamma \circ \alpha) \tag{8.9}$$

if γ is an automorphism of $\mathcal{A}$ leaving φ invariant. The first inequality of (8.8) tells us that mutual information is decreasing under coarse graining.

Since mutual entropy will play a crucial role in the next chapter we discuss it in detail in the case of a finite dimensional abelian input algebra. Let $\mathcal{C}$ be such an algebra of dimension n and let $\alpha : \mathcal{A}_1 \to \mathcal{C}$ be a channel. Then α is of the form

$$\alpha(a) = \sum_{i=1}^{n} \omega_i(a)p_i \qquad (a \in \mathcal{A})$$

where the p_i's are minimal projections in $\mathcal{C}$ and $\omega_1, \omega_2, \ldots, \omega_n$ are certain states of $\mathcal{A}_1$. For an input state $\mu = (\mu_1, \mu_2, \ldots, \mu_n)$ of $\mathcal{C}$ we have $\mu \circ \alpha = \sum_i \mu_i \omega_i$ and the mutual entropy becomes

$$I(\mu; \alpha) = \sum_i \mu_i S(\omega_i, \mu \circ \alpha). \tag{8.10}$$

Moreover, if $\mathcal{A}_1$ is finite dimensional as well, then

$$I(\mu; \alpha) = S(\mu \circ \alpha) - \sum_i \mu_i S(\omega_i). \tag{8.11}$$

Here the first term is concave in μ (according to the concavity of the entropy) and the second term is affine. Therefore, $\mu \mapsto I(\mu; \alpha)$ is concave under the above restriction. On the other hand, $\alpha \mapsto I(\mu; \alpha)$ is convex functional for arbitrary input and output algebras.

We shall need a simple observation on the *entropy defect*

$$s_\mu(\alpha) \equiv S(\mu) - I(\mu; \alpha) = -\sum_i S(\mu_i \circ \omega_i, \mu \circ \alpha). \tag{8.12}$$

Lemma 8.1 Let $\mathcal{C}_1$ and $\mathcal{C}_2$ be subalgebras of $\mathcal{C}$ so that $\mathcal{C}_1 \vee \mathcal{C}_2 = \mathcal{C}$. Write E_i for the μ-preserving conditional expectation of $\mathcal{C}$ onto $\mathcal{C}_i$ and set $\mu_i \equiv \mu|\mathcal{C}_i$ $(i = 1, 2)$. Then

$$s_{\mu_1}(E_1 \circ \alpha) + s_{\mu_2}(E_2 \circ \alpha) \geq s_\mu(\alpha).$$

Proof. Let $p_1, p_2, \ldots, p_k$ and $q_1, q_2, \ldots, q_l$ be minimal projections of $\mathcal{C}_1$ and $\mathcal{C}_2$, respectively. One can find states ω_{ij} on $\mathcal{A}_1$ such that

$$s_\mu(\alpha) = -\sum_{i,j} S(\mu(p_i q_j)\omega_{ij}, \mu \circ \alpha).$$

Then

$$s_{\mu_1}(E_1 \circ \alpha) = -\sum_i S(\textstyle\sum_j \mu(p_i q_j)\omega_{ij}, \mu \circ \alpha)$$

and

$$s_{\mu_2}(E_2 \circ \alpha) = -\sum_j S(\textstyle\sum_i \mu(p_i q_j)\omega_{ij}, \mu \circ \alpha).$$

In this formulation one can recognize the equivalence of the stated inequality to the *marginal inequality* of Proposition 5.25.   □

Lemma 8.1 contains the subadditivity property of the entropy of subalgebras of a given commutative algebra. With the choice of the injection $\mathbb{C} \to \mathcal{C}$ in place of α the lemma yields

$$S(\mu|\mathcal{C}_1) + S(\mu|\mathcal{C}_2) \geq S(\mu|\mathcal{C}_1 \vee \mathcal{C}_2). \tag{8.13}$$

Proposition 8.2 Let α and α' be channels with finite dimensional input and output algebras $\mathcal{C}$ and $\mathcal{A}$, respectively. Assume that $\mathcal{C}$ is abelian and $\|\alpha - \alpha'\| \leq \varepsilon \leq 1/3$. Then

$$|I(\mu; \alpha) - I(\mu; \alpha')| \leq 2\big(\varepsilon \log d + \eta(\varepsilon)\big)$$

for every input state μ if d stands for the tracial dimension of $\mathcal{A}$.

Proof. We recall that d equals $\mathrm{Tr}_{\mathcal{A}} I_{\mathcal{A}}$. Write $\alpha = (\omega_1, \omega_2, \ldots, \omega_n)$ and $\alpha' = (\omega'_1, \omega'_2, \ldots, \omega'_n)$ in terms of some states ω_i and ω'_i of $\mathcal{A}$ ($1 \leq i \leq n$). From (8.11) we have

$$I(\mu; \alpha) - I(\mu; \alpha') = \big(S(\mu \circ \alpha) - S(\mu \circ \alpha')\big) - \sum_i \mu_i (S(\omega_i) - S(\omega'_i)).$$

Since $\|\omega_i - \omega'_i\| \leq \|\alpha - \alpha'\| \leq \varepsilon \leq 1/3$ by reference to Proposition 1.8 we can conclude the required estimate.   □

Note that the same upper bound holds also for $|s_\mu(\alpha) - s_\mu(\alpha')|$.

Generalizing the entropy (6.9) of a subalgebra we can define

$$H_\varphi(\alpha) = \sup\{\textstyle\sum_i \lambda_i S(\varphi_i \circ \alpha, \varphi \circ \alpha) : \sum_i \lambda_i \varphi_i = \varphi\} \tag{8.14}$$

as the *entropy of the channel* $\alpha : \mathcal{A}_1 \to \mathcal{A}$ with input state φ. (The sup is over all finite convex decompositions.) The relations

$$H_\varphi(\beta \circ \alpha) \leq H_\varphi(\alpha) \quad \text{and} \quad I(\varphi; \alpha) \leq H_\varphi(\alpha) \tag{8.15}$$

are immediate from the definitions and the monotonicity of the relative entropy. The relation between the entropy of a channel and its mutual entropy is intimate and important. First of all if the input algebra is finite dimensional and commutative, then $H_\varphi(\alpha) = I(\varphi; \alpha)$. A decomposition $\sum_i \lambda_i \varphi_i$ of a state φ of the algebra $\mathcal{A}$ is the same as a channel $\beta : \mathcal{A} \to \mathcal{C}$ into a finite dimensional abelian algebra $\mathcal{C}$ on which a state μ is defined so that $\mu \circ \beta = \varphi$. (The state μ corresponds to the probability distribution $(\lambda_i)_i$.) According to (8.10) we have

$$H_\varphi(\alpha) = \sup\{I(\mu; \beta \circ \alpha) : \varphi = \mu \circ \beta\}, \tag{8.16}$$

where β runs over all channels $\mathcal{A} \to \mathcal{C}$ such that $\mathcal{C}$ is finite dimensional and commutative. Moreover, the existence of a state μ on $\mathcal{C}$ is assumed so that $\varphi = \mu \circ \beta$. In Chapter 10 such a pair (β, μ) will be called abelian model of the system (α, φ) and the entropy of multiple channels will be defined in this fashion. Let $\alpha : \mathcal{A} \to \mathcal{C}$ be a channel and assume that $\mathcal{C} \equiv \mathbb{C}^n$ is commutative. Then $\alpha = (\varphi_1, \varphi_2, \ldots, \varphi_n)$, where the φ_i are states of $\mathcal{A}$ ($i = 1, 2, \ldots, n$). Given an input state $\mu = (\mu_1, \mu_2, \ldots, \mu_n)$ we can define a channel $\beta : \mathcal{C} \to \mathcal{A}$ in a natural way. For the sake of simplicity let us suppose that $\mathcal{A}$ is of finite dimension and let φ_i, $\varphi \equiv \sum_i \mu_i \varphi_i$ possess densities D_i and D, respectively ($i = 1, 2, \ldots, n$). The algebra $\mathcal{C}$ being abelian the channel β is determined by a partition of unity in $\mathcal{A}$. The n-tuple

$$(\mu_1 D^{-1/2} D_1 D^{-1/2}, \mu_2 D^{-1/2} D_2 D^{-1/2}, \ldots, \mu_n D^{-1/2} D_n D^{-1/2})$$

is a partition of unity since $\sum_i \mu_i D_i = D$. The so-constructed channel β : $\mathcal{C} \to \mathcal{A}$ has the property $\mu \circ \alpha \circ \beta = \mu$. We shall call β the μ-*transpose* of α, in notation, $\beta = \alpha^{\star,\mu}$. The definition of the transpose is based on the form

$$\langle\!\langle a, b \rangle\!\rangle_\varphi = \operatorname{Tr} a^* D_\varphi^{1/2} b D_\varphi^{1/2} . \tag{8.17}$$

Since $a \geq 0$ if and only if $\langle\!\langle a, b \rangle\!\rangle_\varphi \geq 0$ for every b, $\langle\!\langle \,.\,,\,.\, \rangle\!\rangle_\varphi$ is called a self-polar form. One can easily see that in the above example

$$\langle\!\langle a, \beta(x) \rangle\!\rangle_\varphi = \langle\!\langle \alpha(a), x \rangle\!\rangle_\mu \qquad (a \in \mathcal{A}, \, x \in \mathcal{C}) \tag{8.18}$$

and the dualization (8.18) will be the definition of the transpose in the general case. It remains to find an appropriate extension of the *self-polar form* that does not include the functional Tr.

Let $\mathcal{M}$ be a von Neumann algebra and φ a normal state of $\mathcal{M}$. Assume that $\mathcal{M}$ acts standardly on a Hilbert space $\mathcal{H}$ and ξ_φ is the vector representative of φ from the natural positive cone. Then

$$(a, b) \mapsto \langle\!\langle a, b \rangle\!\rangle_\varphi \equiv \langle a\xi_\varphi, Jb^*\xi_\varphi \rangle \tag{8.19}$$

is the the self-polar form associated with φ. An equivalent definition comes from analytic continuation. The function

$$t \mapsto \varphi(a^* \sigma_t^\varphi(b)) = \langle a\xi_\varphi, \Delta_\varphi^{it} b\xi_\varphi \rangle \qquad (t \in \mathbb{R}) \tag{8.20}$$

admits an analytic extension to the strip $\{z \in \mathbb{C} : -1 \leq \operatorname{Im} z \leq 0\}$, and it follows from

$$\Delta_\varphi^{1/2} b\xi_\varphi = Jb^*\xi_\varphi$$

that the value of the extension at $-i/2$ is exactly $\langle\!\langle a, b \rangle\!\rangle_\varphi$. (It is an advantage of the definition by analytic extension that it also makes sense for C*-algebras if φ is a KMS-state.)

Proposition 8.3 Let $\alpha : \mathcal{M}_1 \to \mathcal{M}_2$ be a channel between the von Neumann algebras $\mathcal{M}_1$ and $\mathcal{M}_2$. Assume that the input state φ is normal and the corresponding output state $\psi \equiv \varphi \circ \alpha$ is faithful and normal. Then there exists a unique channel $\alpha^{\star,\varphi} : \mathcal{M}_2 \to \mathcal{M}_1$ characterized by the relation

$$\langle\!\langle a_2, \alpha(a_1) \rangle\!\rangle_\varphi = \langle\!\langle \alpha^{\star,\varphi}(a_2), a_1 \rangle\!\rangle_\psi \qquad (a_1 \in \mathcal{M}_1, \, a_2 \in \mathcal{M}_2).$$

Proof. Let us fix $a_2 \in \mathcal{M}_2^+$ and use the standard representations of the algebras. The functional $a_1 \mapsto \langle\!\langle a_2, \alpha(a_1) \rangle\!\rangle_\varphi$ is majorized by ψ in the sense that

$$\langle\!\langle a_2, \alpha(a_1) \rangle\!\rangle_\varphi \leq \|a_2\| \psi(a_1) \qquad (a_1 \in \mathcal{M}_1^+)$$

and so

$$\langle\!\langle a_2, \alpha(a_1) \rangle\!\rangle_\varphi = \langle T\xi_\psi, a_1\xi_\psi \rangle \qquad (a_1 \in \mathcal{M}_1^+)$$

with certain positive T in the commutant of $\mathcal{M}_1$. From the Tomita theorem we know that

$$T\xi_\psi = J_1 \alpha^{\star,\varphi}(a_2)\xi_\psi$$

with some positive element $\alpha^{\star,\varphi}(a_2)$ of $\mathcal{M}_1$. In this way we arrive at the existence and uniqueness of a positive mapping $\alpha^{\star,\varphi} : \mathcal{M}_2 \to \mathcal{M}_1$ which satisfies the above duality relation.

We need to show the complete positivity of $\alpha^{\star,\varphi}$, which is equivalent to the inequality

$$\sum_{i=1}^{k}\sum_{j=1}^{k}\langle \alpha^{\star,\varphi}(a_i^* a_j)\eta_i, \eta_j\rangle \geq 0, \tag{8.21}$$

where $a_1, a_2, \ldots, a_k \in \mathcal{M}_2$ and $\eta_1, \eta_2, \ldots, \eta_k$ are vectors in the standard representation Hilbert space of $\mathcal{M}_1$. We may assume that η_i are of the form $J_1 x_i \xi_\psi$ with some $x_i \in \mathcal{M}_1$. Then

$$\sum_{i,j}\langle \alpha^{\star,\varphi}(a_i^* a_j)\eta_i, \eta_j\rangle = \sum_{i,j}\langle \alpha^{\star,\varphi}(a_i^* a_j)(J_1 x_i J_1)\xi_\psi, J_1 x_j \xi_\psi\rangle$$

$$= \sum_{i,j}\langle \alpha^{\star,\varphi}(a_i^* a_j)\xi_\psi, J_1 x_i^* x_j \xi_\psi\rangle = \sum_{i,j}\langle\!\langle \alpha^{\star,\varphi}(a_i^* a_j), x_j^* x_i\rangle\!\rangle_\psi$$

$$= \sum_{i,j}\langle\!\langle a_i^* a_j, \alpha(x_j^* x_i)\rangle\!\rangle_\varphi = \sum_{i,j}\langle a_i^* a_j \xi_\varphi, J_2 \alpha(x_i^* x_j)\xi_\varphi\rangle$$

$$= \sum_{i,j}\langle \alpha(x_i^* x_j) J_2 a_i \xi_\varphi, J_2 a_j \xi_\varphi\rangle,$$

and this is nonnegative due to the complete positivity of α. $\qquad\square$

The trivial example for the transpose is supplied by an automorphism leaving a state invariant. Its transpose is simply the inverse map.

Let $\mathcal{M}_0 \subset \mathcal{M}$ be von Neumann algebras and let φ be a faithful normal state of $\mathcal{M}$. If φ is given by a cyclic and separating vector Φ in a representation, then the generalized conditional expectation $E_\varphi : \mathcal{M} \to \mathcal{M}_0$ is of the form

$$E_\varphi(a) = J_0 P J a J P J_0 \qquad (a \in \mathcal{M})$$

where P is the projection onto the subspace generated by $\mathcal{M}_0\Phi$. One easily verifies that

$$\langle a\Phi, J b_0^* \Phi\rangle = \langle E_\varphi(a)\Phi, J_0 b_0^* \Phi\rangle,$$

that is, the generalized conditional expectation is the transpose of the embedding.

Proposition 8.4 Let $\alpha : \mathcal{M}_1 \to \mathcal{M}_2$ be a channel between the von Neumann algebras $\mathcal{M}_1$ and $\mathcal{M}_2$. Assume that the input state φ and the corresponding output state $\psi \equiv \varphi \circ \alpha$ are faithful and normal. Then the following conditions are equivalent for $a \in \mathcal{M}_1$.

(i) $\alpha(a^*a) = \alpha(a)^*\alpha(a)$ and $\sigma_t^\varphi(\alpha(a)) = \alpha(\sigma_t^\psi(a))$ for every $t \in \mathbb{R}$.
(ii) $\alpha^{\star,\varphi} \circ \alpha(a) = a$.

Furthermore, α restricted to the subalgebra $\mathcal{N}_1 \equiv \{a \in \mathcal{M}_1 : \alpha^{\star,\varphi} \circ \alpha(a) = a\}$ is an isomorphism onto $\mathcal{N}_2 \equiv \{b \in \mathcal{M}_2 : \alpha \circ \alpha^{\star,\varphi}(a) = a\}$.

Proof. It is most convenient to work in the standard representation. Φ and Ψ are the vector representatives of φ and ψ in the positive cones, respectively. Due to the Schwarz inequality the mapping

$$V : a\Psi \mapsto \alpha(a)\Phi \qquad (8.22)$$

defines a contraction. So (ii) is equivalent to the condition

$$\langle Va\Psi, J_2Vx\Psi \rangle = \langle a\Psi, J_1x\Psi \rangle \qquad (x \in \mathcal{M}_1), \qquad (8.23)$$

which may be written as

$$J_1a\Psi = V^*J_2Va\Psi. \qquad (8.24)$$

Now assume (i) and rewrite its second condition in terms of the modular operators $\Delta_1 \equiv \Delta(\psi, \psi)$ and $\Delta_2 \equiv \Delta(\varphi, \varphi)$ as

$$V J_1\Delta_1^{it+1/2}a\Psi = J_2\Delta_2^{it+1/2}Va\Psi \qquad (t \in \mathbb{R}, a \in \mathcal{M}_1). \qquad (8.25)$$

Through analytic continuation this yields at i/2 that

$$V J_1a\Psi = J_2Va\Phi \qquad (a \in \mathcal{M}_1). \qquad (8.26)$$

Now we make a short computation

$$\begin{aligned}
\|V J_1a\Psi\|^2 &= \langle J_2Va\Psi, J_2Va\Psi \rangle \\
&= \varphi(\alpha(a)^*\alpha(a)) = \varphi(\alpha(a^*a)) = \|J_1a\Psi\|^2,
\end{aligned}$$

using first (8.26) followed by the first condition of (i). Since for a contraction A the equality $\|A\xi\| = \|\xi\|$ implies $A^*A\xi = \xi$, we obtain

$$V^*V J_1a\Psi = J_1a\Psi. \qquad (8.27)$$

From (8.26) and (8.27) the condition (8.24) follows and so the implication (i)$\Rightarrow$(ii) is proved.

To prove the converse note first that the mean ergodic theorem implies that $\mathcal{N}_1$ and $\mathcal{N}_2$ are fixed point von Neumann algebras of completely positive mappings with an invariant state. We have seen that $a \in \mathcal{N}_1$ is equivalent to (8.24). So if $a \in \mathcal{N}_1$ then

$$\|a\Psi\| = \|V^*J_2Va\Psi\| \le \|Va\Psi\| \le \|a\Psi\|$$

and $\|Va\Psi\| = \|a\Psi\|$. In other words, $\varphi(\alpha(a^*a)) = \varphi(\alpha(a)^*\alpha(a))$. Since $\alpha(a)^*\alpha(a) \le \alpha(a^*a)$ and φ is faithful, we infer $\alpha(a^*a) = \alpha(a)^*\alpha(a)$. It is well-known (see, for example, 9.2 in [Strătilă 1981]) that the latter condition for a Schwarz mapping α implies

$$\alpha(xa) = \alpha(x)\alpha(a) \quad \text{and} \quad \alpha(a^*x) = \alpha(a)^*\alpha(x) \tag{8.28}$$

for every $x \in \mathcal{M}_1$. In particular, $\alpha|\mathcal{N}_1$ is an isomorphism, and evidently $\alpha(\mathcal{N}_1) = \mathcal{N}_2$. Let $\bar{\psi} = \psi|\mathcal{N}_1$ and $\bar{\varphi} = \varphi|\mathcal{N}_2$. Due to the mean ergodic theorem (of Kovács and Szűcs) there exists a conditional expectation from $\mathcal{M}_1$ onto $\mathcal{N}_1$. Hence σ^φ restricted to $\mathcal{N}_1$ is the modular group of $\bar{\varphi}$. By a similar reasoning we obtain that σ^ψ restricted to $\mathcal{N}_2$ is the modular group of $\bar{\psi}$. Finally, $\alpha|\mathcal{N}_1$ being isomorphism, it intertwines between the modular groups of $\bar{\psi}$ and $\bar{\varphi}$. This makes the proof of (i) complete.   □

It is noteworthy that due to formula (8.28) the proof of Theorem 4.5 is included in the previous proposition.

An *operational partition of unity* on a C*-algebra $\mathcal{A}$ is a k-tuple $\mathcal{V} = (V_1, V_2, \ldots, V_k)$ of elements of $\mathcal{A}$ such that $\sum_{i=1}^k V_i^* V_i = I$. Given an operational partition of unity, to each $1 \le i \le k$ one associates a completely positive mapping $\alpha_i : a \mapsto V_i^* a V_i$. $\mathcal{V}$ may be regarded as an instrument since $\gamma_{\mathcal{V}} \equiv \sum_i \alpha_i$ is unital. The operational partition of unity $\mathcal{V}$ is called φ-invariant if $\varphi \circ \gamma_{\mathcal{V}} = \varphi$. Let us introduce the concept of an *instrument* by the following probabilistic interpretation. If the state of system $\mathcal{A}$ at the time of the measurement is φ, then on the instrument one can find the outcome i with probability $p_i = \varphi \circ \alpha_i(I)$. The measurement changes the state of the system in such a way that just after the measurement its state is $p_i^{-1} \varphi \circ \alpha_i = p_i^{-1} \varphi(V_i^* \cdot V_i)$ with probability p_i. (Given $\mathcal{V}$, the correspondence $i \mapsto V_i^* V_i$ is a positive operator-valued measure and so it is an observation channel. The additional specificity of the concept of an instrument is in the statement on the state of the observed system after the measurement.) Thus the new state of the system is $\varphi \circ \gamma_{\mathcal{V}}$. Though it is not strongly related to the objective of our present discussion we note that the above probabilistic interpretation shows the importance of the invariant operational partitions of unity. If one intends to obtain information on a system by repeated measurements and operational partitions of unity are used for this purpose then only invariant operational partitions of unity should be applied.

Let $\mathcal{V} = (V_1, V_2, \ldots, V_k)$ be a φ-invariant operational partition of unity. Following *Lindblad* its entropy $S(\varphi, \mathcal{V})$ will be defined as the von Neumann entropy of the density matrix

$$\left(\varphi(V_i^* V_j)\right)_{i,j=1}^k.$$

Now let $\mathcal{A}$ be a matrix algebra $M_n(\mathbb{C})$. Any unital completely positive mapping $\alpha : M_n(\mathbb{C}) \to M_n(\mathbb{C})$ may be represented by an operational partition of unity $\mathcal{V} = (V_1, V_2, \ldots, V_k)$ in the form

$$\alpha(a) = \sum_{i=1}^k V_i^* a V_i. \tag{8.29}$$

If α leaves the faithful state φ invariant, set

$$H_\varphi^L(\alpha) = S(\varphi, \mathcal{V}).$$

Since the representation (8.29) is not at all unique we need to show that the definition is independent from it.

Performing the GNS construction with φ we obtain the action of our matrix algebra $\mathcal{A}$ on a Hilbert space $\mathcal{H}$ and a cyclic vector $\Phi \in \mathcal{H}$. Let J be the modular conjugation on $\mathcal{H}$ corresponding to the cyclic and separating vector Φ. The algebra $B(\mathcal{H})$ is linearly spanned by operators

$$aJbJ \qquad (a, b \in \mathcal{A}),$$

because due to Tomita's theorem JbJ is the typical element of the commutant of $\mathcal{A}$. Consider the Hilbert space $\mathcal{H} \otimes \mathcal{H}_k$ where $\mathcal{H}_k$ is a k-dimensional Hilbert space with orthonormal basis $\xi_1, \xi_2, \ldots, \xi_k$. The vector

$$\Omega = \sum_{l=1}^{k} V_l \Phi \otimes \xi_l \in \mathcal{H} \otimes \mathcal{H}_k$$

induces a pure state ω_{12} on $B(\mathcal{H}) \otimes B(\mathcal{H}_k) = B(\mathcal{H} \otimes \mathcal{H}_k)$ having marginals $\omega_1 \equiv \omega_{12}|B(\mathcal{H})$ and $\omega_2 \equiv \omega_{12}|B(\mathcal{H}_k)$. They have the same entropy according to Proposition 6.5.

Let (E_{ij}) be a system of matrix units in $B(\mathcal{H}_k)$ corresponding to the given basis. Then

$$\langle \Omega, I \otimes E_{ij} \Omega \rangle = \sum_{m} \langle V_m \Phi \otimes \xi_m, V_j \Phi \otimes \xi_i \rangle = \varphi(V_i^* V_j),$$

and we see that the entropy of ω_2 is just $S(\varphi, \mathcal{V})$. On the other hand,

$$\omega_1(aJbJ) = \sum_{i} \langle V_i^* a^* V_i \Phi, Jb\Phi \rangle = \langle\!\langle \alpha(a^*), b^* \rangle\!\rangle_\varphi, \qquad (8.30)$$

and the state ω_1 is expressed by α independently from the representing operational partition of unity $\mathcal{V}$. Formula (8.30) has some consequences. The self-polar form recovers the state φ in each of the two variables provided that the other one is fixed to be I. Hence from the subadditivity of the entropy we have

$$H_\varphi^L(\alpha) \le 2S(\varphi),$$

and equality holds if and only if $\alpha(a) = \varphi(a)I$ for every $a \in \mathcal{A}$. In other words, among the completely positive mappings leaving invariant a given state φ, the state itself possesses maximal Lindblad entropy. Simple analysis yields that $H_\varphi^L(\alpha) = 0$ if and only if α is an automorphism.

A further consequence of (8.30) is that the Lindblad entropy of α coincides with that of its transpose (with respect to the invariant state). Indeed, the states ω_1 and $\bar{\omega}_1 \equiv \omega_1(J \,.\, J)$ have the same entropy being antiunitarily equivalent. Moreover,

$$\bar{\omega}_1(aJbJ) = \langle\!\langle \alpha^{\star,\varphi}(a^*), b^* \rangle\!\rangle_\varphi$$

by simple verification.

We summarize the properties of the *Lindblad entropy* in a proposition.

Proposition 8.5 Let $\alpha : M_n(\mathbb{C}) \to M_n(\mathbb{C})$ be a completely positive mapping leaving a faithful state φ invariant. Then

(i) $H_\varphi^L(\alpha) \le 2S(\varphi)$,
(ii) $H_\varphi^L(\alpha) = H_\varphi^L(\alpha^{\star,\varphi})$,
(iii) $H_\varphi^L(\alpha)$ is a concave functional on the α-invariant states.

In quantum communication theory the signal transmission is represented by a scheme

$$\mathcal{C}_1 \xrightarrow{\ \gamma\ } \mathcal{A}_1 \xrightarrow{\ \alpha\ } \mathcal{A} \xrightarrow{\ \beta\ } \mathcal{C}$$

where $\mathcal{C}$ and $\mathcal{C}_1$ are abelian algebras corresponding to classical input and output systems, β is the "coding" map, γ is the "decoding" map and α performs the actual quantum sign transmission.

Before dealing with a concrete quantum model let us recall the intuitive formalism of signal transmission. We are given a set of symbols X to be sent and another set Y to be received. We may call X the input set and Y the output set. The two sets are connected by a stochastic matrix W, the entry $W(y|x)$ of which is interpreted as the probability that if $x \in X$ is sent then $y \in Y$ is received. If we want to send a message m we code it by an element $f(m)$ of X, then $f(m)$ is transmitted through the communication channel and an element $y \in Y$ is received. The receiver uses a dictionary from the input set Y to the set of messages and understands the coded information. Actually, the receiver recognizes the output with some uncertainty expressed by $W(y|x)$ and the communication channel coding problem consists in making the set of messages large while keeping the probability of error small.

Suppose that the signal transmission device operates by successively transmitting "letters" of a given "alphabet", one transmission at each time unit. In the course of n time units the input consists of "words" $(x_1, x_2, \ldots, x_n)$ of length n and the output is a word $(y_1, y_2, \ldots, y_n)$ also of length n. The usual solution of the coding problem deals with the asymptotic sense $(n \to \infty)$.

We restrict our discussion to the simplest pure quantum signal transmission model in which the the transmission is performed by a channel $\alpha : M_m(\mathbb{C}) \to \mathbb{C}^k$. Such an α is equivalently given by states $\varphi_1, \varphi_2, \ldots, \varphi_k$ of the matrix algebra $M_m(\mathbb{C})$. After sending the word $(x_1, x_2, \ldots, x_n)$ of the alphabet $\{1, 2, \ldots, k\}$, the state

$$\varphi_{(x_1, x_2, \ldots, x_n)} \equiv \varphi_{x_1} \otimes \varphi_{x_2} \otimes \cdots \otimes \varphi_{x_n}$$

of the nth tensor power $\mathcal{A}_n$ of $M_m(\mathbb{C})$ is received. The meaning of this output is understood by a measurement which plays the role of decoding. Formally, we define a code of length N as a system

$$\big((u_1, X_1), (u_2, X_2), \ldots, (u_N, X_N)\big) \tag{8.31}$$

where $u_1, u_2, \ldots, u_N$ are words of length n from the alphabet $\{1, 2, \ldots, k\}$ and $(X_1, X_2, \ldots, X_N)$ is a partition of unity. In other language $X_1, X_2, \ldots, X_N$ are positive operators whose sum is I. The number

$$\varphi_{(x_1, x_2, \ldots, x_n)}(X_j),$$

is regarded as the probability that the received message is the word u_j when the word $(x_1, x_2, \ldots, x_n)$ was sent. Since the aim is to decode the received message correctly we are interested in making $1 - \varphi_{u_j}(X_j)$ as small as possible. The probability of error of code (8.31) is defined as

$$\lambda(\mathbf{u}, \mathbf{X}) = \max\{1 - \varphi_{u_j}(X_j) : 1 \leq j \leq N\}.$$

We denote by $\lambda(n, N)$ the infimum of the error probabilities over all codes (8.31). The coding theorem announces a statement about the limit

$$\lim_{n \to \infty} \lambda(n, [2^{nR}]) \tag{8.32}$$

as a function of the parameter R. The computation of the critical value of R (in a theoretical sense) needs the introduction of a new information quantity. Consider a channel $\alpha \equiv (\varphi_1, \varphi_2, \ldots, \varphi_k) : \mathcal{A} \to \mathbb{C}^k$. Set

$$J(\alpha) = \sup\{I(\mu; \alpha \circ \delta) : \delta : \mathcal{C} \to \mathcal{A} \text{ is a}$$
$$\text{measurement and } \mu \text{ is an input state}\}, \tag{8.33}$$

where the supremum is over all δ's with a finite dimensional abelian output algebra and $\mu = (\mu_1, \mu_2, \ldots, \mu_k)$ is a state of $\mathbb{C}^k$. (Note that under these conditions the mutual entropy $I(\mu; \alpha \circ \delta)$ in (8.33) is expressed explicitly in the fashion of (8.11).)

It is quite trivial that the sequence

$$J_n(\alpha) = J(\alpha^{(1)} \otimes \alpha^{(2)} \otimes \ldots \otimes \alpha^{(n)})$$

satisfies the subadditivity condition $J_{n+m}(\alpha) \leq J_n(\alpha) + J_m(\alpha)$, and so the limit

$$C \equiv \lim_{n \to \infty} \frac{1}{n} J_n(\alpha) \tag{8.34}$$

exists (cf. Lemma 10.4). It is remarkable that the simplifying condition $C = J_1 = n^{-1} J_n$ holds in the classical case, that is, for a commutative $\mathcal{A}$. The following result of Holevo shows that C is the quantity what may be intuitively assigned to be the capacity of the transmission channel.

Proposition 8.6 Let $\alpha : M_m(\mathbb{C}) \to \mathbb{C}^k$ be a quantum channel and let C be given by (8.34). Then, for $R < C$ the limit (8.32) of minimal error probabilities is 0 but for $R > C$ this is not true.

We close this chapter with a short discussion on a generalized mutual entropy quantity. Mutual entropy quantities might be useful to examine the complexity of a system.

Let $\alpha : \mathcal{A}_1 \to \mathcal{A}$ be a channel with input state φ and output state $\psi \equiv \varphi \circ \alpha$. There are several states ω on the tensor product $\mathcal{A}_1 \otimes \mathcal{A}$ satisfying the following conditions. They will be called quasi-compound states.

(i) $\omega(a_1 \otimes I) = \psi(a_1)$ for any $a_1 \in \mathcal{A}_1$.
(ii) $\omega(I \otimes a) = \varphi(a)$ for any $a \in \mathcal{A}$.

For instance, the product $\omega_0 \equiv \psi \otimes \varphi$ is a trivial quasi-compound state. If a quasi-compound state indicates the correspondence between each elementary component (pure state) of φ and that of ψ, it is a "true" compound state. The compound state might play a similar role as the joint probability measure. Let μ be an extremal decomposition measure of φ. (The set of all those measures is denoted by M_φ.) Then a compound state ω_μ of φ and ψ is

$$\omega_\mu = \int (\nu \circ \alpha) \otimes \nu \, d\mu(\nu). \tag{8.35}$$

This state obviously satisfies (i) and (ii). Moreover, ω_μ contains the classical expression of the compound measure as a special case. Using the mutual entropy with respect to a decomposing measure $\mu \in M_\varphi$

$$I_\mu(\varphi; \alpha) = S(\omega_\mu, \omega_0)$$

we introduce the following variant of mutual entropy.

$$J(\varphi; \alpha) = \sup\{I_\mu(\varphi; \alpha) : \mu \in M_\varphi\}. \tag{8.36}$$

Monotonicity of the relative entropy implies the relation $J(\varphi; \alpha) \leq S(\varphi)$ for the generalized mutual entropy. The above definition have concerned decompositions in the full state space. When a reference system, that is, a compact convex subset $\mathcal{S}$ of the state space is chosen, the mutual entropies may be regarded in an even more general form, similarly to the mixture entropy.

Notes and Remarks. Complete positivity is the proper notion for the positivity of a mapping in operator algebras both from the mathematical and physical viewpoint. Some useful references are Chapter IV of [Takesaki 1979] and the monograph [Paulsen 1986]. The idea of identifying observables with positive operator-valued measures arose in [Davies and Lewis 1970], and the concept of instrument (without the requirement of complete positivity) originates from the same paper. The books [Davies 1976] and [Holevo 1982] treat the subject in detail.

The algebra of the canonical commutation relation and the quasi-free mappings are detailed in [Evans and Lewis 1977] and [Petz 1990]. The definition of *entropy defect* and the entropy of a channel together with the related results are from [Connes, Narnhofer and Thirring 1987]. In the quantum context the mutual entropy was discussed in [Ohya 1983].

The *self-polar form* associated to a state was used by several authors, for example, [Connes 1974] and [Accardi and Cecchini 1982]. [Petz 1984] studied systematically the *transpose mapping* (called *dual* there) with respect to a weight in the place of a state. In the language of this chapter, the transpose of a quasi-free channel with respect to a quasi-free state (on the algebra of the canonical commutation relation) is again a quasi-free mapping. This was observed in [Frigerio 1983]. Proposition 8.4 is from [Petz 1988b]. Mean ergodic theory in von Neumann algebras was created in [Kovács and Szűcs 1966], further references are [Kümmerer and Nagel 1979] and [Bratteli and Robinson 1979].

The *Lindblad entropy* of a channel is from [Lindblad 1979a], however, formula (8.30) seems to be new.

Our sketch on the signal transmission is based on [Csiszár and Körner 1981], the quantum aspects are emphasised in [Holevo 1973c]. Proposition 8.6 was achieved in [Holevo 1979] and its proof is a reduction to measure theoretic coding theory. [Holevo 1991] is a complete review of several related fields with a good bibliography.

The *compound state* (8.35) is due to [Ohya 1983a] and is linked to a quantum Markov chain in [Accardi and Ohya 1999]. In the quantum context the mutual entropy was first introduced in [Ohya 1983a] and its generalization to a C*-system was discussed in [Ohya 1985, 1989]. The mutual entropy can be used to define quantum ε-entropy and fractal dimension of states, see [Ohya 1991b]. An optical communication model is treated in Chapter 18.

Finally, we briefly mention the study of *entropy production* which was recently advanced by [Ojima, Hasegawa, Ichiyanagi 1988] and [Ojima 1989].

Comments. In this chapter (as well as in the whole monograph) a communication channel α is defined as a completely positive mapping $\mathcal{B} \to \mathcal{A}$ (between unital C*-algebras). If ψ is a linear functional on $\mathcal{A}$, then the channel output is $\psi \circ \alpha$. In many cases, especially in the finite dimensional situation when states are given by density matrices, the dual of α is considered. The dual is called the *state transformation* and it is a completely positive and trace preserving mapping. Let $\mathcal{E} : M_k(\mathbb{C}) \to M_m(\mathbb{C})$ be a state transformation. Then $\mathcal{E}$ can be written in the form

$$\mathcal{E}(D) = \sum_p A_p D A_p^*, \tag{8.37}$$

where the *operator coefficients* satisfy $\sum_p A_p A_p^* = I$. (This is the *Kraus representation* of a completely positive mapping between matrix algebras.)

If a state transformation is implemented by a quantum mechanical device, then it can be used to transmit classical information. To each classical input message x there corresponds a signal state D_x of the quantum communication system. The densities D_x are functioning as code words of the messages. The message x has a probability p_x. After the action of the state transformation $\mathcal{E}$, the distribution of the output is determined by a *measurement*, which is nothing else but another word for decoding in this context. To each message x there corresponds an obsevable A_x on the output Hilbert space. It is customary to assume that $0 \le A_x$ and $\sum_x A_x = \text{id}$ (id stands for the identity operator).

$$p_{yx} := p_x \text{Tr} \mathcal{E}(D_x) A_y$$

is the probability that the message x is sent and y is received. The joint distribution p_{yx} yields marginal probability distributions p_x and q_y on the set of messages. According to *Shannon*, the *mutual information*

$$I = \sum_{x,y} p_{yx} \log \frac{p_{yx}}{p_x q_y}$$

measures the amount of information going through the classical channel, from Alice to Bob. The so-called *Holevo bound* provides an upper bound on the amount of information accessible to Bob:

$$I \le \Delta := S(\mathcal{E}(D)) - \sum_x p_x S(\mathcal{E}(D_x)) = \sum_x p_x S(\mathcal{E}(D_x), \mathcal{E}(D)),$$

where $D = \sum_x p_x D_x$. (Note that the inequality $I \le \Delta$ follows from the monotonicity of the relative entropy or it may be regarded as an example of the monotonicity of the quantum mutual information.)

The basic problem of communication theory is to maximize the amount of information received by Bob from Alice asymptotically when long messages are sent. In the course of n time units the input consists of messages $(x_1, x_2, \ldots, x_n)$ of length n and the output has also length n. The message $(x_1, x_2, \ldots, x_n)$ appears with probability $p_{x_1} \times p_{x_2} \times \ldots \times p_{x_m}$ and coded by $D_{x_1} \otimes D_{x_2} \ldots \otimes D_{x_n}$. For the n-shot channel the mutual information can be computed as above and the supremum over all measurements A and over all codings (p_x, D_x) gives the n-shot capacity $C_n(\mathcal{E})$. The *classical capacity* of the channel is defined as

$$C(\mathcal{E}) := \lim_{n \to \infty} \frac{C_n(\mathcal{E})}{n} \tag{8.38}$$

This limit exists due to the superadditivity property

$$C_n(\mathcal{E}) + C_m(\mathcal{E}) \le C_{n+m}(\mathcal{E}).$$

The classical capacity comes from the optimization of classical mutual information. Similarly, one can define the *purely quantum capacity* by quantum mutual information. Let $\overline{C}_n(\mathcal{E})$ be the supremum of

$$\sum_x p_x S(\mathcal{E}^{n\otimes}(D_x), \mathcal{E}(D)^{n\otimes})$$

over all quantum codes $\{p_x, D_x\}$. Similarly to the classical capacity, the *quantum capacity* could be defined as

$$\overline{C}(\mathcal{E}) := \lim_{n\to\infty} \frac{\overline{C}_n(\mathcal{E})}{n}. \tag{8.39}$$

It follows from the Holevo bound that $C(\mathcal{E}) \leq \overline{C}(\mathcal{E})$.

A basic result about classical communication via quantum channels is that the classical mutual information can achieve the bound Δ when Alice prepares long code words and Bob uses decoding observables not necessarily of product type while the probability of error could be kept arbitrarily low, see [Holevo 1998b] and [Schumacher and Westmoreland 1997]. (Proposition 8.6 due to Holevo was an earlier result towards this aim.)

The quantum capacity $\overline{C}_1(\mathcal{E})$ can be viewd in terms of information geometry, it is nothing else but the relative entropy radius of the range of $\mathcal{E}$, that is,

$$\overline{C}_1(\mathcal{E}) = \inf_D \left\{ \sup_{D'} S\big(\mathcal{E}(D'), \mathcal{E}(D)\big) \right\},$$

see [Ohya, Petz and Watanabe 1997].

The capacity quantities $C_1, \overline{C}_1, C$ and $\overline{C}$ are not the only ones appearing in the literature. Heuristically, capacity is a bound for performance in an information transmission protocal and there is wide space for optimizing performance, another example is the concept of *entanglement-assisted classical capacity* ([Bennett et al 2002], or [Holevo 2002]) whose definition is based on a different protocol.

9 Sufficient Channels and Measurements

In an operator algebraic approach to quantum mechanics one considers observables of a system associated to elements of an operator algebra $\mathcal{A}$. If a subalgebra $\mathcal{A}_0$ is selected for experimental observation then the determination through $\mathcal{A}_0$ provides only partial information about the states of the system. The subalgebra $\mathcal{A}_0$ is sufficient for decision within a family $(\varphi_\theta)_\theta$ of states if the subalgebra contains all the information about the mutual relationship between these states. In that case one can select with certainty the appropriate member of the family of states $(\varphi_\theta)_\theta$ even if the behaviour of the family is known only on the subalgebra $\mathcal{A}_0$. For example, taking the simplest von Neumann algebra $\mathcal{A} = \mathcal{B}(\mathcal{H})$ let $(\varphi_\theta)_\theta$ be a family of vector states, $\varphi_\theta = \langle \xi_\theta, . \xi_\theta \rangle$ $(\xi_\theta \in \mathcal{H})$. If $\mathcal{H}_0 \subset \mathcal{H}$ is a closed subspace containing all the vectors ξ_θ and P is the corresponding orthogonal projection in $B(\mathcal{H})$, then the subalgebra $\mathcal{A}_0 = PB(\mathcal{H})P + \mathbb{C}P^\perp$ must be sufficient.

Now we discuss a parametric problem of statistics in connection with a measurement in order to motivate the concept of sufficiency. The observer of the physical system obtains information by measurements. In the mathematical model *measurements* are measures taking positive operator values in the operator algebra. If the system is in a state φ and $E(H)$ is the measure of the set H then $\mu_\varphi(H) := \varphi(E(H))$ is interpreted as the probability that the measured quantity has its value in the set H. Decision between two statistical hypotheses is the following problem. Suppose that we have to choose between two states φ and ω on the algebra. Performing a measurement we receive two probability distributions and we decide between them by applying standard methods of mathematical statistics. This may yield a satisfactory result only in the case in which the performed measurement extracted all the information on the mutual relation of the two states. The information contained in the measures μ_φ and μ_ω is typically less than the information carried by the couple φ and ω. In the fortunate case the two information quantities coincide and we call such a measurement sufficient. As it is intuitively expected, sufficient measurement does not exist for a generic couple of states. Below we formulate this statement mathematically in order to prove it. This seemingly awkward feature of measurements has a consolation. Sufficient measurement always exist in a weaker sense of mean. Hence asymptotically and theoretically the problem of discrimination between two statistical hypotheses may be solved always in the following way. First we perform sufficient measurements and then we apply classical statistics.

Before we give the formal definition of sufficiency, we review quickly the analogous concept in classical statistics. Let $\xi_1, \xi_2, \ldots, \xi_n$ be a random sample of size n from a distribution that has probability density function $f(\,.\,; \theta)$, where θ is a certain parameter. In other words, $\xi_1, \xi_2, \ldots, \xi_n$ are independent random variables with probability density $f(\,.\,; \theta)$. Any function $u : \mathbb{R}^n \to \mathbb{R}$ induces a new random variable $u(\xi_1, \xi_2, \ldots, \xi_n)$ possessing a probability density $g(\,.\,; \theta)$. If the function u is such that for every fixed value $y = u(x_1, x_2, \ldots, x_n)$ the quotient

$$\frac{f(x_1; \theta)\, f(x_2; \theta)\, \ldots\, f(x_n; \theta)}{g(u(x_1, x_2, \ldots, x_n); \theta)}$$

does not depend upon θ, then the function u (termed statistic) is called *sufficient*. This means that $u(\xi_1, \xi_2, \ldots, \xi_n)$ exhausts all the information about θ that is contained in the sample.

For example, let $f(x; \theta) = (2\pi)^{-1/2} \exp(-(x - \theta)^2/2)$ and

$$u(x_1, x_2, \ldots, x_n) = \frac{x_1 + x_2 + \ldots + x_n}{n}$$

then u is a *sufficient statistic*. Approaching sufficient statistics in a more abstract way, we may consider the joint probability μ_θ of $(\xi_1^\theta, \xi_2^\theta, \ldots, \xi_n^\theta)$ on $\mathbb{R}^n$ and the sub-σ-algebra $\mathcal{S}_0$ generated by the statistic u. Let us call $(\mathbb{R}^n, \mu_\theta)$ a statistical experiment. The σ-algebra $\mathcal{S}_0$ is sufficient if for every measurable set B there exists a $\mathcal{S}_0$-measurable function f_B such that

$$\int_{H_0} f_B \, d\mu_\theta = \mu_\theta(B \cap H_0)$$

for every value of θ and for all $\mathcal{S}_0$-measurable H_0. Roughly speaking, this is the case, if the conditional expectation $E(\mu_\theta | \mathcal{S}_0)$ is independent of θ.

Returning to the setting of operator algebras, we make the following formal definition. A *statistical experiment* $(\mathcal{A}, \varphi_\theta)$ consists of a C*-algebra $\mathcal{A}$ and a family φ_θ of states $(\theta \in \Theta)$. The C*-subalgebra $\mathcal{A}_0$ is called *sufficient* for $(\mathcal{A}, \varphi_\theta)$ if there exists a completely positive unital mapping α of $\mathcal{A}$ into $\mathcal{A}_0$ such that $\varphi_\theta \circ \alpha = \varphi_\theta$ for all θ. If $\mathcal{A}$ happens to be a von Neumann algebra, then both φ_θ and α are assumed to be normal.

The aim of this chapter is to sketch a theory on sufficient subalgebras and to analyze examples. Moreover, the concept of sufficiency will be extended to channels and discussed in an asymptotic sense as well.

Let $\mathcal{M} \subset B(\mathcal{H})$ be a von Neumann algebra and $\varphi_\theta(\,.\,) = \langle \xi_\theta, \,.\, \xi_\theta \rangle$ be a set of vector states $(\theta \in \Theta)$. If P is a projection in $\mathcal{M}$ such that $P\xi_\theta = \xi_\theta$ for all θ, then the subalgebra $M_0 = P\mathcal{M}P + \mathbb{C}\,P^\perp$ is sufficient for $\{\varphi_\theta\}$ because $\alpha(a) = PaP + \omega(a)\,P^\perp$ with any normal state ω leaves φ_θ invariant.

In the definition of a sufficient subalgebra we intentionally avoided conditional expectations. One reason for that is the empirical fact that conditioning is always a difficult problem in quantum probability. (Concerning conditional expectations in operator algebras see Chapter 4.)

The next proposition gives an example of sufficient subalgebra.

Proposition 9.1 Let G be a group of automorphisms of the von Neumann algebra $\mathcal{M}$, $\{\varphi_\theta : \theta \in \Theta\}$ the set of all G-invariant normal states, and let $\mathcal{M}_0$ be the fixed point algebra of G. Then $\mathcal{M}_0$ is sufficient for $\{\varphi_\theta\}$.

Proof. This result is a simple consequence of the *Kovács-Szűcs theorem* An easy reduction shows how one can assume that the family $\{\varphi_\theta\}$ is separating, that is, for every $0 \neq a \in \mathcal{M}^+$ there is a θ such that $\varphi_\theta(a) > 0$. Under this condition the classical *Kovács-Szűcs ergodic theorem* provides the existence of a G-invariant faithful normal condition expectation $E : \mathcal{M} \to \mathcal{M}_0$ which gives the sufficiency of $\mathcal{M}_0$. □

In the characterization of sufficient subalgebras *Raggio's transition probability* will play an essential role. Let $\mathcal{M}$ be a von Neumann algebra acting in the standard way and let φ, ω be normal states. Their transition probability is the inner product $\langle \Phi, \Omega \rangle$ where Φ and Ω are the vector representatives in the natural positive cone (see (7.6)). The advantage of the transition probability from the viewpoint of our proof and in comparison with other quasi-entropies – in particular, with the relative entropy – is its boundedness. Since a crucial use of modular theory is required, we try to localize the most delicate technicalities to the following lemma. Recall that $\Delta(./.)$ and $D(.,.)$ stand for the spatial derivative and for the lineal, respectively.

Lemma 9.2 Let $\mathcal{M}_0 \subset \mathcal{M} \subset B(\mathcal{H})$ be von Neumann algebras with commutants $\mathcal{M}' \subset \mathcal{M}_0'$. Let φ and ω_0' be normal states of $\mathcal{M}$ and $\mathcal{M}_0'$, respectively. Write φ_0 and ω' for their restrictions. Then, for $\eta' \in D(\mathcal{H}, \omega_0')$ the condition

$$\langle \eta', \Delta(\varphi/\omega')^{1/2}\eta' \rangle = \langle \eta', \Delta(\varphi_0/\omega_0')^{1/2}\eta' \rangle$$

implies

$$\Delta(\varphi/\omega')^{it}\eta' = \Delta(\varphi_0/\omega_0')^{it}\eta'$$

for every $t \in \mathbb{R}$.

Proof. For the sake of simplicity we write Δ and Δ_0 for the spatial derivatives $\Delta(\varphi/\omega')$ and $\Delta(\varphi_0/\omega_0')$, respectively. Let

$$\int_0^\infty \lambda \, dE_\lambda$$

be the spectral decomposition of Δ. We approximate Δ by the truncated operators

$$H_n = \int_0^n \lambda \, dE_\lambda \qquad (n \in \mathbb{N}).$$

So $(t + H_n)^{-1} \to (t + D)^{-1}$ strongly for every positive $t \in \mathbb{R}$. Due to the inequality $\Theta^{\omega'_0}(\eta) \geq \Theta^{\omega'}(\eta)$ we have

$$||\Delta_0^{1/2}\eta'||^2 = \varphi_0(\Theta^{\omega'_0}(\eta)) \geq \varphi_0(\Theta^{\omega'}(\eta)) = ||\Delta^{1/2}\eta'||^2 \geq ||H_n^{1/2}\eta'||^2 \quad (9.1)$$

for every $\eta' \in D(\mathcal{H}, \omega'_0)$. Let p'_0 be the support of ω'_0. Recall that it projects to the closure of $D(\mathcal{H}, \omega'_0)$. Since $D(\mathcal{H}, \omega'_0)$ is a core fore $\Delta_0^{1/2}$, we deduce from (9.1) that

$$\Delta_0 \geq p'_0 H_n p'_0 . \quad (9.2)$$

This implies

$$(t + \Delta_0)^{-1} \leq (t + p'_0 H_n p'_0)^{-1} \leq p'_0(t + H_n)^{-1}p'_0 .$$

Letting $n \to \infty$ we infer

$$(t + \Delta_0)^{-1} \leq p'_0(t + \Delta)^{-1}p'_0 . \quad (9.3)$$

Now we are going to benefit from the following integral representations:

$$\langle \eta', \Delta^{1/2}\eta' \rangle = \pi^{-1} \int_0^\infty \langle \eta', (\lambda^{-1/2} - \lambda^{-1/2}(\lambda + \Delta)^{-1})\eta' \rangle \, d\lambda$$

and similarly

$$\langle \eta', \Delta_0^{1/2}\eta' \rangle = \pi^{-1} \int_0^\infty \langle \eta', (\lambda^{-1/2} - \lambda^{-1/2}(\lambda + \Delta_0)^{-1})\eta' \rangle \, d\lambda .$$

The equality of the values of the integrals and the inequality between the integrands (thanks to (9.3)) give the equality

$$\langle \eta', (t + \Delta)^{-1}\eta' \rangle = \langle \eta', (t + \Delta_0)^{-1}\eta' \rangle \quad (9.4)$$

for almost all positive t which is easily extended by continuity for all $t > 0$. Since

$$p'_0(t + \Delta_0)^{-1}p'_0 \leq p'_0(t + \Delta)^{-1}p'_0 ,$$

(9.4) implies

$$p'_0(t + \Delta)^{-1}\eta' = (t + \Delta_0)^{-1}\eta' \quad (9.5)$$

for every $t > 0$. Differentiating this equality with respect to t we get

$$p'_0(t + \Delta)^{-2}\eta' = (t + \Delta_0)^{-2}\eta' . \quad (9.6)$$

Now we show by estimating the norms that the projection p'_0 in (9.5) may be omitted. Indeed,

$$||p'_0(t + \Delta)^{-1}\eta'||^2 = ||(t + \Delta_0)^{-1}\eta'||^2$$
$$= \langle \eta', (t + \Delta_0)^{-2}\eta' \rangle = \langle \eta', (t + \Delta_0)^{-2}\eta' \rangle = ||(t + \Delta)^{-1}\eta'||^2$$

and

$$(t + \Delta)^{-1}\eta' = (t + \Delta_0)^{-1}\eta'$$

for ever real $t > 0$. To complete the proof we refer to the Stone-Weierstrass approximation theorem leading to

$$f(\Delta)\eta' = f(\Delta_0)\eta'$$

for any bounded continuous function f.   □

The next theorem characterizes *sufficient subalgebras* for a pair of states. Recall that P_A stand for the transition probability.

Theorem 9.3 Let $\mathcal{M}_0 \subset \mathcal{M}$ be von Neumann algebras and φ, ω be faithful normal states on $\mathcal{M}$ and write E_φ, E_ω for the corresponding generalized conditional expectations from $\mathcal{M}$ into $\mathcal{M}_0$. The following conditions are equivalent.

  (i) $\varphi \circ E_\omega = \varphi$.
  (ii) $\mathcal{M}_0$ is a sufficient subalgebra for $\{\varphi, \omega\}$.
  (iii) $P_A(\varphi, \omega) = P_A(\varphi|\mathcal{M}_0, \omega|\mathcal{M}_0)$.
  (iv) $[D\varphi, D\omega]_t = [D(\varphi|\mathcal{M}_0), D(\omega|\mathcal{M}_0)]_t$ for every real t.
  (v) $[D\varphi, D\omega]_t \in \mathcal{M}_0$ for every real t.
  (vi) $E_\varphi = E_\omega$.

Proof. The equivalence will be shown by the following implications.

$$\text{(i)} \Rightarrow \text{(ii)} \Rightarrow \text{(iii)} \Rightarrow \text{(iv)} \Rightarrow \text{(v)} \Rightarrow \text{(iv)} \Rightarrow \text{(vi)} \Rightarrow \text{(i)}.$$

(i)⇒(ii) is immediate from the definition of the sufficiency. (The mapping E_ω can do the job of α.)

(ii)⇒(iii) is based on the monotonicity of the transition probability under completely positive mappings. $P_A(\varphi, \omega) \leq P_A(\varphi|\mathcal{M}_0, \omega|\mathcal{M}_0)$ is always true and $P_A(\varphi, \omega) \geq P_A(\varphi \circ \alpha, \omega \circ \alpha)$ for every $\alpha : \mathcal{M} \to \mathcal{M}_0$. When α leaves φ and ω invariant, (iii) follows.

(iii)⇒(iv) is the substantial part of the proof. Let us work in the standard representation and assume that Ω is a cyclic and separating vector inducing the state ω. Then (iv) is equivalent to

$$[D\varphi, D\omega]_t\Omega = [D(\varphi|\mathcal{M}_0), D(\omega|\mathcal{M}_0)]_t\Omega\,,$$

or in terms of spatial derivatives,

$$\Delta(\varphi/\omega')^{it}\Delta(\omega/\omega')^{it}\Omega = \Delta(\varphi_0/\omega_0')^{it}\Delta(\omega_0/\omega_0')^{it}\Omega\,, \tag{9.7}$$

where $\varphi_0, \omega_0, \omega_0'$ and ω' are the same as in the previous lemma. Choosing ω_0' to be the vector state associated to Ω (on the commutant of $\mathcal{M}_0$) the equation (9.7) simplifies essentially. Namely, we have

$$\Delta(\varphi/\omega')^{it}\Omega = \Delta(\varphi_0/\omega_0')^{it}\Omega\,.$$

This is exactly the statement of Lemma 9.2 while its hypothesis is an alternative form of condition (iii).

(iv)$\Rightarrow$(v) is trivial and the converse is a piece of modular theory. Let $\mathcal{N}$ be the von Neumann subalgebra generated by the cocycle $[D\varphi, D\omega]_t$ ($t \in \mathbb{R}$). Then $\mathcal{N}$ is stable under the modular group of ω, and this fact has two consequences. On the one hand, $[D\varphi, D\omega]_t$ is a cocyle with respect to the modular group of ω_0, and on the other hand, there exists a conditional expectation $F : \mathcal{M} \to \mathcal{N}$ preserving ω. The converse of Connes' theorem tells us that there exists a faithful normal weight f on $\mathcal{N}$ such that

$$[Df, D\omega_0]_t = [D\varphi, D\omega]_t \text{ for every real } t.$$

However,

$$[D(\varphi \circ F), D\omega]_t = [Df, D\omega_0]_t = [D\varphi, D\omega]_t \qquad (t \in \mathbb{R})$$

and from the identity of the cocycles with respect to ω we obtain $\varphi = f$. So we have arrived at (iv).

Finally, we show (iv)$\Rightarrow$(v), that is, from the identity of two cocycles we have to deduce the identity of the generalized conditional expectations E_φ and E_ω. Let us assume that $(\mathcal{M}, \mathcal{H}, J, \mathcal{P})$ and $(\mathcal{M}_0, \mathcal{H}_0, J_0, \mathcal{P}_0)$ are the standard representations of $\mathcal{M}$ and $\mathcal{M}_0$, respectively. (In spite of the assumption $\mathcal{M}_0 \subset \mathcal{M}$, we may represent the two algebras on different Hilbert spaces but we omit the denotation of the representations to alleviate our formulas. Let us recall the following formulas.

$$E_\varphi(a) = J_0 V_\varphi^* JaJV_\varphi J_0 \quad \text{and} \quad E_\omega(a) = J_0 V_\omega^* JaJV_\omega J_0 \qquad (a \in \mathcal{M}_0),$$

where

$$V_\varphi a\xi_{\varphi_0} = a\xi_\varphi \quad \text{and} \quad V_\varphi a\xi_{\omega_0} = a\xi_\omega .$$

In order to show $E_\varphi = E_\omega$ we prove $V_\varphi = V_\omega$ by means of analytic continuation. Fixing an $a \in \mathcal{M}_0$ consider the function

$$F(t) = V_\omega a[D\varphi_0, D\omega_0]_t\xi_{\omega_0} \equiv V_\omega a\Delta(\varphi_0, D\omega_0)^{it}\xi_{\omega_0} ,$$

which admits an analytic extension to the strip $\{z \in \mathbb{C} : -1/2 \le \operatorname{Im} z \le 0\}$. Since $\Delta(\varphi_0, D\omega_0)^{1/2}\xi_{\omega_0} = \xi_{\varphi_0}$, we have

$$F(-i/2) = V_\omega a\xi_{\varphi_0} .$$

Similarly,

$$G(t) \equiv a[D\varphi, D\omega]_t\xi_\omega \equiv a\Delta(\varphi, D\omega)^{it}\xi_\omega$$

has an analytic extension to the same strip and

$$G(-i/2) = a\xi_\varphi .$$

According to our hypothesis $F(t) = G(t)$ for every real t, and this implies the identity of the continuations. In particular,

$$V_\omega a\xi_{\varphi_0} = a\xi_\varphi ,$$

which is nothing else but $V_\varphi = V_\omega$. $\qquad\qquad\square$

It is a consequence of the theorem that the von Neumann subalgebra $\mathcal{N}$ generated by the cocycle $[D\varphi, D\omega]_t$ ($t \in \mathbb{R}$) is the smallest sufficient subalgebra for the couple $\{\varphi, \omega\}$. In analogy with the finite dimensional case, two states φ and ω are said to be *commuting* in modular theory if their *Connes cocyle* is a one-parameter group, which is the same as saying that $[D\varphi, D\omega]_t$ commutes with $[D\varphi, D\omega]_u$ for every t and u. The theorem shows that for non-commuting states there exist no *sufficient abelian subalgebras*.

In the characterization of a sufficient subalgebra the quasi-entropy P_A played a central role but the relative entropy may be used similarly. Since condition (i) in Theorem 9.3 implies

$$S(\varphi, \omega) = S(\varphi | \mathcal{M}_0, \omega | \mathcal{M}_0) \,, \tag{9.8}$$

via double application of the monotonicity of the relative entropy, any of the conditions of this theorem implies (9.8). Vice versa, assuming (9.8) one can prove (iv) by a small alteration of the above proof, provided the common value of the two entropy quantities of (9.8) is finite. If we replace the square root function by the logarithm in Lemma 9.2, then the following integral formula could be used:

$$\log x = \int_0^\infty (1+t)^{-1} - (x+t)^{-1} \, dt \,.$$

Hence (9.8) is equivalent to conditions (i)–(vi), provided $S(\varphi, \omega)$ or $S(\omega, \varphi)$ is finite. Under this condition we have the following interpretation of the entropic extension of Theorem 9.3. Performing a simple measurement (which corresponds to a projection-valued measure) we necessarily lose information concerning the mutual relation of non-commuting states. Indeed, a simple measurement is in fact a multiplicative observation channel α, which has, of course, a commutative range $\mathcal{A}_0$. The entropy equality

$$S(\omega, \varphi) = S(\omega \circ \alpha, \varphi \circ \alpha)$$

would imply the sufficiency of the subalgebra $\mathcal{A}_0$, but this would be impossible because the cocycle of non-commuting states cannot lie in a commutative algebra. The fact that any measurement loses information with respect to non-commuting state is somewhat depressing. (Although the above proof works only for simple measurements, it will be shown below that the same conclusion holds for an arbitrary measurement.) The way out will be an averaging approach. One can obtain information in the mean, arbitrarily close to the relative information of two states even if they do not commute. This will be treated in detail after the characterization of sufficient channels in the present chapter.

Before going on we show an example of a sufficient subalgebra in connection with the algebra of the *canonical commutation relation*. Let $\mathcal{A}_0 \subset \mathcal{A}$ be finite quantum systems and let φ be a faithful state on $\mathcal{A}$ which admits the existence of a conditional expectation $E : \mathcal{A} \to \mathcal{A}_0$. Endow $\mathcal{A}$ with the Hilbert-Schmidt inner product

$$\langle a, b \rangle = \varphi(a^* b),$$

and consider the C*-algebras $\text{CCR}(\mathcal{A}_0) \subset \text{CCR}(\mathcal{A})$ (with the understanding that the symplectic form is the imaginary part of the inner product). There exists a state ϱ on $\text{CCR}(\mathcal{A})$ such that

$$\varrho(W(a)) = \exp(-\tfrac{1}{4}\varphi(a^* a)) \qquad (a \in \mathcal{A}).$$

We denote by F the φ-preserving conditional expectation of $\mathcal{A}$ onto the centralizer of φ. The centralizer is the fixed point algebra of the modular group and hence admits the existence of a conditional expectation. Set a real linear subspace

$$K = \{a - F(a) : a \in \mathcal{A} \text{ and } a = a^*\}$$

and let the subalgebra $\mathcal{B}$ generated by the unitaries $\{W(a) : a \in K\}$. (The fluctuations of an observable $a \in K$ converge to the field operator $B(a)$ affiliated with $\mathcal{B}$ in the central limit theorem, see Chapter 14 for more detail.) The symplectic form

$$\sigma(a, b) = \text{Im} \langle a, b \rangle$$

restricted to K is non-degenerate and we have $\mathcal{B} \equiv \text{CCR}(K, \sigma)$. Performing GNS-construction on $\mathcal{B}$ with the state ϱ we shall get the von Neumann algebra $\mathcal{M} = \pi_\varrho(\text{CCR}(\mathcal{A})'')$ and the cyclic vector Ω. For the sake of simplicity we think the algebra $\mathcal{B}$ acts on the GNS Hilbert space. One can see that the group

$$\alpha_t : W(a) \mapsto W(\sigma_t^\varphi(a)) \qquad (a \in K,\, t \in \mathbb{R})$$

of *Bogoliubov transformations* satisfies the KMS condition and so determines the modular group of the vector state associated with Ω on $\mathcal{M}$. In particular, Ω is separating and we denote by ϱ the corresponding faithful normal state of $\mathcal{M}$. Now we describe a family of faithful normal states of the von Neumann algebra $\mathcal{M}$. Setting

$$K_0 = \{a - F(a) : a \in \mathcal{A}_0 \text{ and } a = a^*\}$$

let

$$\varrho_\theta(W(a)) = \varrho(W(\theta)W(a)W(-\theta)) \qquad (\theta \in K_0,\, a \in K). \tag{9.9}$$

Since ϱ_θ is the composition of ϱ with an inner automorphism, it is easy to find their Connes cocycle.

$$[D\varrho_\theta, D\varrho]_t = \pi_\rho(W(\theta - \sigma_t^\varphi(\theta)) \exp(\tfrac{1}{2} \varphi([\theta, \sigma_t^\varphi(\theta)])). \tag{9.10}$$

Let $\mathcal{M}_0$ be the von Neumann subalgebra of $\mathcal{M}$ generated by the Weyl unitaries $\{W(b) : b \in K_0\}$. We claim that $\mathcal{M}_0$ is a sufficient subalgebra with respect to the family $(\varrho_\theta)_\theta$. In the light of the previous theorem this follows from formula (9.10). Since K_0 is stable under the modular group of φ, we see that the Connes cocycle $[D\varrho_\theta, D\varrho]_t$ lies in $\mathcal{M}_0$ for every $\theta \in K_0$.

In this example it is not necessary to pass to the generated von Neumann algebra. We may consider instead the C*-algebra $\mathcal{B}$ with the family (9.9) of states. The subalgebra $\mathcal{B}_0$ generated by $\{W(b) : b \in K_0\}$ is sufficient. One can show this by constructing a completely positive mapping $\beta : \mathcal{B} \to \mathcal{B}_0$ preserving all the states ϱ_θ. Let β be the quasi-free mapping which is the transpose of the embedding $\mathcal{B}_0 \subset \mathcal{B}$ with respect to the quasi-free state ϱ. One verifies by computation that this β leaves invariant all the ϱ_θ's.

In the next proposition we characterize the *sufficient subalgebras* for a state and for its inner perturbation. Perturbation theory of states is treated in detail in Chapter 12. Here we recall only the most necessary formulas. Let $\mathcal{M}$ be a von Neumann algebra with a faithful normal state φ. In order to use relative modular operators it is convenient to assume that $\mathcal{M}$ acts standardly. The *inner perturbation* $[\varphi^h]$ of φ by a self-adjoint $h \in \mathcal{M}$ may be determined by the Connes cocycle $[D[\varphi^h], D\varphi]_t$ as follows.

$$
\begin{aligned}
[D[\varphi^h], D\varphi]_t &= \Delta([\varphi^h], \varphi)^{it} \Delta(\varphi, \varphi)^{-it} \\
&= \exp\big(it(\log \Delta(\varphi, \varphi) - h + c(\varphi, h))\big)\Delta(\varphi, \varphi)^{-it} \qquad (9.11) \\
&= e^{itc(\varphi, h)} \exp\big(it(\log \Delta(\varphi, \varphi) - h)\big)\Delta(\varphi, \varphi)^{-it},
\end{aligned}
$$

where $c(\varphi, h)$ is some normalization constant (see Theorem 12.6). Using the perturbation expansion of $\exp\big(it(\log \Delta(\varphi, \varphi) - h)\big)$ we have also

$$
\begin{aligned}
[D[\varphi^h], D\varphi]_t &= e^{itc(\varphi, h)} \\
&\times \sum_{n=0}^\infty (-\mathrm{i})^n \int_0^t dt_1 \int_0^{t_1} dt_2 \dots \int_0^{t_{n-1}} dt_n\, \sigma_{t_n}^\varphi(h) \dots \sigma_{t_1}^\varphi(h). \qquad (9.12)
\end{aligned}
$$

After this recollection we are in a position to prove the following.

Proposition 9.4 Let $\mathcal{M}$ be a von Neumann algebra with a faithful normal state φ and let $h \in \mathcal{M}$ be self-adjoint. The a subalgebra $\mathcal{M}_0$ is sufficient for the pair $\{\varphi, [\varphi^h]\}$ if and only if the orbit of h under the modular group of φ remains in $\mathcal{M}_0$.

Proof. Assume that $\mathcal{M}_0$ is sufficient. Then

$$
\sigma_s^\varphi\big([D[\varphi^h], D\varphi]_t\big) \in \mathcal{M}_0 \qquad (s, t \in \mathbb{R})
$$

according to Theorem 9.3. Differentiation at $t = 0$ of the formula (9.11) proves that

$$
\sigma_s^\varphi(h) \in \mathcal{M}_0 \quad \text{for every } s \in \mathbb{R}. \qquad (9.13)
$$

Conversely, if (9.13) holds, then the expansion (9.12) makes clear that the cocycle $[D[\varphi^h], D\varphi]_t$ $(t \in \mathbb{R})$ is in the subalgebra, and we refer to Theorem 9.3 again. $\qquad \square$

Theorem 9.3 gives several equivalent conditions for the sufficiency of a subalgebra but its start hypothesis is too restrictive. From the viewpoint of applications it is desirable to extend it to non-faithful states. This will be done next.

Theorem 9.5 Let $(\mathcal{M}, \varphi_\theta)$ be a statistical experiment and let $\mathcal{M}_0 \subset \mathcal{M}$ be von Neumann algebras. Assume that there exist a sequence $(\theta_i)_i$ and a state

$$\omega = \sum_{i=1}^\infty \lambda_i \varphi_{\theta_i}$$

such that $\omega | \mathcal{M}_0$ is faithful. Then the following conditions are equivalent.

(i) $\mathcal{M}_0$ is sufficient for (φ_θ).
(ii) $[D\varphi_\theta, D\omega]_t p = [D(\varphi_\theta | \mathcal{M}_0), D(\omega | \mathcal{M}_0)]_t p$ for every real t and for every θ, where p is the support of ω.
(iii)   The generalized conditional expectation $E_\omega : \mathcal{M} \to \mathcal{M}_0$ leaves all the states φ_θ invariant.

Proof. We can follow the proof of Theorem 9.3 with some additional care with regard to support projections, which makes the arguments more technical. Although the statement concerns a family φ_θ of states, we have to compare each element φ of $\{\varphi_\theta : \theta \in \Theta\}$ with ω separately. The algebra $\mathcal{M}$ is supposed to act standardly.

Condition (i) implies that $P_A(\varphi, \omega) = P_A(\varphi_0, \omega_0)$ in the usual notation (that is, ω_0 and φ_0 are the restrictions). Let Ω and Φ be the vector representatives of φ and ω in the positive cone. The vector state induced by Ω on $\mathcal{M}_0'$ is not necessarily faithful hence we choose a faithful positive functional ω_0' on $\mathcal{M}_0'$ such that

$$\omega_0'([\mathcal{M}_0\Omega]a_0'[\mathcal{M}_0\Omega]) = \langle a_0'\Omega, \omega \rangle \quad \text{for every} \quad a_0' \in \mathcal{M}_0'. \tag{9.14}$$

Although Lemma 9.2 was stated for states it holds for unnormalized positive functionals as well. So let us apply it to ω_0'. The hypothesis is satisfied because

$$\langle \Omega, \Delta(\varphi/\omega')^{1/2}\Omega \rangle = \langle \Omega, \Delta(\varphi, \omega)^{1/2}\Omega \rangle$$
$$= P_A(\varphi, \omega) = P_A(\varphi_0, \omega_0) = \langle \Omega, \Delta(\varphi_0/\omega_0')^{1/2}\Omega \rangle.$$

It follows from the lemma that

$$[D\varphi_\theta, D\omega]_t \Omega = [D(\varphi_\theta | \mathcal{M}_0), D(\omega | \mathcal{M}_0)]_t \Omega. \tag{9.15}$$

Since Ω is not separating, we cannot conclude the equality of the cocycles, but we obtain

$$[D\varphi_\theta, D\omega]_t p = [D(\varphi_\theta | \mathcal{M}_0), D(\omega | \mathcal{M}_0)]_t p,$$

which was to be proven.

Now assume (ii). Then we have (9.15), and by means of analytic continuation it follows that

$$\Delta(\varphi/\omega')^{1/2}\Omega = \Delta(\varphi_0/\omega_0')^{1/2}\Omega\,.$$

Denote by Q the projection onto the subspace $[\mathcal{M}_0\Omega]$ and perform the following computation:

$$\begin{aligned}
\varphi(E_\omega(a)) &= \langle \Delta(\varphi_0/\omega_0')^{1/2}\Omega, J_0QJaJQJ_0\Delta(\varphi_0/\omega_0')^{1/2}\Omega\rangle \\
&= \langle JaJ\Delta(\varphi_0/\omega_0')^{1/2}\Omega, \Delta(\varphi_0/\omega_0')^{1/2}\Omega\rangle \\
&= \langle JaJ\Delta(\varphi_0/\omega_0')^{1/2}\Omega, \Delta(\varphi/\omega')^{1/2}\Omega\rangle \\
&= \langle \Delta(\varphi_0/\omega_0')^{1/2}\Omega, a\Delta(\varphi/\omega')^{1/2}\Omega\rangle = \varphi(a)\,,
\end{aligned}$$

where $\Delta(\varphi/\omega')^{1/2}\Omega$ is the vector representative Φ of φ in the positive cone, and therefore the modular conjugation J leaves it untouched.

This completes the proof of (ii)⇒(iii), while (iii)⇒(i) is trivial due to the definition of a sufficient subalgebra.   □

We have concentrated on sufficient subalgebras. Now we turn to more general coarse graining and extend some of the previous results to channeling transformations.

In quantum theory the observation channels have great importance because the information on a system is obtained through measurements. Let $\mathcal{A}$ be a C*-algebra supposed to describe a system and let $\alpha_1 : \mathcal{C}_1 \to \mathcal{A}$ and $\alpha_2 : \mathcal{C}_2 \to \mathcal{A}$ be observations. We say that α_1 is a refinement of α_2 if there exists a channel $\beta : \mathcal{C}_2 \to \mathcal{C}_1$ such that $\alpha_1 \circ \beta = \alpha_2$. In this case we regard that performing α_1 in any state one obtains more valuable information than performing α_2. To make the idea more clear, suppose that α_1 corresponds to a partition of unity $(a_1, a_2, \ldots, a_n)$ and α_2 is given by $(b_1, b_2, \ldots, b_m)$. The observation α_1 leads to the probability vector

$$p = \big(\varphi(a_1), \varphi(a_2), \ldots, \varphi(a_n)\big)$$

in a state φ and similarly α_2 yields

$$q = \big(\varphi(b_1), \varphi(b_2), \ldots, \varphi(b_m)\big)\,.$$

α_1 is a refinement of α_2 if in any state φ the vector q is related to p through a stochastic matrix.

Let $\alpha : \mathcal{B} \to \mathcal{A}$ be a channel and $\{\varphi_\theta : \theta \in \Theta\}$ be a family of states on $\mathcal{A}$. The channel α is said to be *sufficient* for $\{\varphi_\theta\}$ if there exists a channel $\beta : \mathcal{A} \to \mathcal{B}$ such that

$$\varphi_\theta \circ \alpha \circ \beta = \varphi_\theta \quad \text{for every} \quad \theta \in \Theta\,. \tag{9.16}$$

Let us shed more light on this definition. The most perfect channel is the identity $\iota : \mathcal{A} \to \mathcal{A}$ both in the intuitive sense that it does not destroy any information and in the sense that any other channel $\alpha : \mathcal{B} \to \mathcal{A}$ is a coarse graining of ι. The existence of a channel $\gamma : \mathcal{A} \to \mathcal{B}$ such that $\alpha \circ \gamma = \iota$ would mean that α is a refinement of ι, and then they would give stochastically equivalent data in every state of $\mathcal{A}$. In the concept of sufficiency we do not require that α and ι should give stochastically equivalent data in any state but rather only in the states $\{\varphi_\theta\}$. This heuristic argument leads to the concept of sufficient channel. We shall have a more general version of the definition which is tailored to an asymptotic theory of measurements.

Let $\alpha_i : \mathcal{B}_i \to \mathcal{A}$ be a net of channels $(i \in I)$. We say that (α_i) is sufficient for a family of states $\{\varphi_\theta : \theta \in \Theta\}$ on $\mathcal{A}$ if for every $i \in I$ there exists a channel $\beta_i : \mathcal{A} \to \mathcal{B}_i$ such that

$$\lim_i \varphi_\theta(\alpha_i \circ \beta_i(a)) = \varphi_\theta(a)$$

for every $\theta \in \Theta$ and for every $a \in \mathcal{A}$. In the category of von Neumann algebras we require that all the involved channels and states be normal. It is easy to make connection with the previous definition (9.16). When the limit α of the channels (α_i) exists in certain sense (and (β_i) converges as well), then α is sufficient in the sense of the first definition. A reason for the second definition is the following. In infinite dimensional case one cannot expect the existence of a sufficient channel defined on a finite dimensional space but a sufficient net of such channels may exist.

Our next aim is to prove that for noncommuting states it is typical that a sufficient net of observation channels does not exist. Before proving the main results we state some auxiliary lemmas. The first one generalizes the well-known fact that a contraction and its adjoint have the same fixed vectors.

Lemma 9.6 Let I be a directed set and for every $i \in I$ let $V_i : \mathcal{H}_i \to \mathcal{K}_i$ be a contraction between Hilbert spaces. Assume that $\lim_i \|f_i\| = \lim_i \|u_i\|$ for some $u_i \in \mathcal{H}_i$ and $f_i \in \mathcal{K}_i$. If $\|V_i^* f_i - u_i\| \to 0$, then $\|V_i u_i - f_i\| \to 0$.

The second lemma is obtained by the refinement of the method used already in the proof of Lemma 9.2.

Lemma 9.7 Let ω and φ be faithful normal states of the von Neumann algebra $\mathcal{M}$ such that $\lambda\varphi \leq \omega \leq \mu\varphi$ with some positive numbers λ and μ. Let $(\mathcal{M}_i)$ be a net of von Neumann algebras with completely positive unital mappings $\alpha_i : \mathcal{M}_i \to \mathcal{M}$ so that $\varphi \circ \alpha_i$ and $\omega \circ \alpha_i$ are faithful and normal. If

$$\lim_i S(\omega \circ \alpha_i, \varphi \circ \alpha_i) = S(\omega, \varphi),$$

then

$$\alpha_i\big([D(\omega \circ \alpha_i), D(\varphi \circ \alpha_i)]_t\big) \to [D\omega, D\varphi]_t$$

for every $t \in \mathbb{R}$ in the strong operator topology.

Proof. Let us start with some remarks. The mutual majorization of φ and ω implies that ω is an inner perturbation of φ by a certain operator $h = h^* \in \mathcal{M}$ (see Chapter 12, in particular, Theorem 12.10). Combining this fact (namely, the existence of a relative Hamiltonians) with the monotonicity of the relative entropy, we obtain

$$0 \leq S(\omega \circ \alpha_i, \varphi \circ \alpha_i) \leq S(\omega, \varphi) \leq 2\|h\|.$$

It is convenient to assume that all von Neumann algebras act standardly. Let Φ_i, Ω_i, Φ, Ω be the vector representatives, from the natural positive cones, of $\varphi_i \equiv \varphi \circ \alpha_i$, $\omega_i \equiv \omega \circ \alpha_i$, φ, ω. Using the abbreviations $\Delta \equiv \Delta(\varphi, \omega)$ and $\Delta_i \equiv \Delta(\varphi_i, \omega_i)$ for the relative modular operators we recall that

$$S(\omega_i, \varphi_i) = -\langle \log \Delta_i \Omega_i, \Omega_i \rangle = \int_0^\infty \langle (\Delta_i + t)^{-1} \Omega_i, \Omega_i \rangle - (1+t)^{-1} \, dt,$$

$$S(\omega, \varphi) = \int_0^\infty \langle (\Delta + t)^{-1} \Omega, \Omega \rangle - (1+t)^{-1} \, dt.$$

Let us introduce the contraction

$$V_i : a_i \Omega_i \mapsto \alpha_i(a_i)\Omega \qquad (a_i \in \mathcal{M}_i).$$

It is not difficult to see by an extension of (9.3) that

$$V_i(t + \Delta_i)^{-1} V_i^* \leq (t + \Delta)^{-1}, \tag{9.17}$$

which implies

$$\langle (t + \Delta_i)^{-1} \Omega_i, \Omega_i \rangle \leq \langle (t + \Delta)^{-1} \Omega, \Omega \rangle. \tag{9.18}$$

Now a comparison of the assumption in the form of integral representation with inequality (9.18) yields that

$$\lim_i \langle (t + \Delta_i)^{-1} \Omega_i, \Omega_i \rangle = \langle (t + \Delta)^{-1} \Omega, \Omega \rangle \tag{9.19}$$

for every positive t.

Now we estimate benefiting from (9.17) as follows.

$$\begin{aligned}
\|(t + \Delta)^{-1}\Omega &- V_i(t + \Delta_i)^{-1}\Omega_i\|^2 \\
&= \|((t + \Delta)^{-1} - V_i(t + \Delta_i)^{-1} V_i^*)\Omega\|^2 \\
&\leq \|((t + \Delta)^{-1} - V_i(t + \Delta_i)^{-1} V_i^*)^{1/2}\| \\
&\quad \times \langle ((t + \Delta)^{-1} - V_i(t + \Delta_i)^{-1} V_i^*)\Omega, \Omega \rangle.
\end{aligned}$$

Since the operator is majorized by $\sqrt{2/t}$ and the inner product goes to 0, we arrive at

$$\lim_i V_i(t + \Delta_i)^{-1}\Omega_i = (t + \Delta)^{-1}\Omega. \tag{9.20}$$

The function $F(z) = \Delta^z \Omega$ is analytic on the strip $\{z \in \mathbb{C} : 0 < \operatorname{Re} z < 1/2\}$ because $\Omega \in \mathcal{D}(\Delta^{1/2})$. The majorization $\omega \leq \mu\varphi$ implies that $\Omega = a\Phi = JaJ\Phi$ for some $a \in \mathcal{M}$ and so $\Omega \in \mathcal{D}(\Delta^{-1/2})$. Therefore $F(z)$ is analytic on the strip $\{z \in \mathbb{C} : -1/2 < \operatorname{Re} z < 0\}$. In this way we conclude that $F(z)$ is analytic on $S = \{z \in \mathbb{C} : -1/2 < \operatorname{Re} z < 1/2\}$. The same argument yields the analyticity of $F_i(z) = \Delta_i^z \Omega_i$ on S. Note that both $F(z)$ and $F_i(z)$ are bounded and admit continuous extension to the closed strip. Our aim is to show that

$$V_i F_i(z) \to F(z) \text{ for every } z \in S. \tag{9.21}$$

In order to prove this claim it suffices to prove the convergence for $z = s$ with $0 < s < 1/2$. Since Ω_i is in the domain of Δ_i we have

$$V_i F_i(s) = \frac{\sin s\pi}{\pi} \int_0^\infty t^{s-1} V_i \Delta_i (t + \Delta_i)^{-1} \Omega_i \, dt. \tag{9.22}$$

It follows from (9.20) that

$$\begin{aligned}
V_i \Delta_i (t + \Delta_i)^{-1} \Omega_i &= V_i \Omega_i - t V_i (t + \Delta_i)^{-1} \Omega_i \\
&\to \Omega - t(t + \Delta)^{-1} \Omega = \Delta(t + \Delta)^{-1} \Omega
\end{aligned}$$

and hence

$$V_i \Delta_i^s \Omega_i \to \Delta^s \Omega \qquad (0 < s < \tfrac{1}{2}). \tag{9.23}$$

The relation (9.21) has been proven and in particular we have

$$V_i \Delta_i^{it} \Omega_i = V_i [D\varphi_i, D\omega_i]_t \Omega_i = \alpha_i([D\varphi_i, D\omega_i]_t)\Omega \to \Delta^{it}\Omega = [D\varphi, D\omega]_t \Omega.$$

Since on bounded sets of a von Neumann algebra the strong operator topology coincides with the corresponding Hilbert space norm topology the lemma is proven. □

The detailed proofs of the following two lemmas will be omitted here. Both are based on the following property of completely positive mappings. If for a completely positive mapping γ the relation $\gamma(a^*a) = \gamma(a)^*\gamma(a)$ holds, then $\gamma(xa) = \gamma(x)\gamma(a)$ for every x in the domain of γ.

Lemma 9.8 Let I be a directed set and for every $i \in I$ let $\alpha_i : \mathcal{A}_i \to B(\mathcal{H})$ be a completely positive unital operator-valued mapping defined on a C*-algebra. Assume that $\lim_i \alpha_i(a_i) = a$ and $\lim_i \alpha_i(u_i) = u$ in the strong* topology for some $a_i \in \mathcal{A}_i$, $a \in B(\mathcal{H})$ and for some unitaries $u_i \in \mathcal{A}_i$, $u \in B(\mathcal{H})$. Then $\lim_i \alpha_i(a_i u_i) = au$ in the strong* topology.

Lemma 9.9 Let $\mathcal{A} \subset B(\mathcal{H})$ be a C*-algebra and $(\beta_i : \mathcal{A} \to B(\mathcal{H}))_i$ be a net of completely positive unital mappings (over a directed set I). Let $u \in \mathcal{A}$ be a unitary and $\Phi \in \mathcal{H}$ such that $\langle \Phi, \beta_i(a)\Phi \rangle = \langle \Phi, a\Phi \rangle$ for every $a \in \mathcal{A}$. Then

$$\lim_i \langle u\Phi, \beta_i(a)u\Phi \rangle = \langle u\Phi, au\Phi \rangle$$

for every $a \in \mathcal{A}$ if $\beta_i(u) \to u$ in the weak operator topology.

Recall that two normal states on a von Neumann algebra are said to commute with each other if the corresponding modular groups do so. This is the case if and only if their Connes cocycle is a one-parameter group or equivalently, consists of commuting operators. (In case of a finite quantum system the commutativity of two states is equivalent to the commutativity of their densities.)

Theorem 9.10 Let ω and φ be faithful normal states of the von Neumann algebra $\mathcal{M}$ such that $\lambda\varphi \leq \omega \leq \mu\varphi$ with some positive numbers λ and μ. Suppose that φ and ω do not commute. Then there exists no sufficient net of observation channels.

Proof. We argue by contradiction and assume the existence of a sufficient net (only in the C*-algebraic sense). Then we can have a net $(\alpha_i : \mathcal{C}_i \to \mathcal{M})_i$ of observation channels such that $\mathcal{C}_i$ is finite dimensional and

$$\lim_i S(\omega \circ \alpha_i, \varphi \circ \alpha_i) = S(\omega, \varphi).$$

Now, Lemma 9.7 tells us that

$$\alpha_i\big([D(\omega \circ \alpha_i), D(\varphi \circ \alpha_i)]_t\big) \to [D\omega, D\varphi]_t$$

for every $t \in \mathbb{R}$ in the strong operator topology. Let us write u_t for $[D\omega, D\varphi]_t$ and u_t^i for $[D(\omega \circ \alpha_i), D(\varphi \circ \alpha_i)]_t$. Then

$$\alpha_i(u_t^i) \to u_t \quad \text{and} \quad \alpha_i((u_t^i)^* u_t^i) = I = u_t^* u_t \quad (t \in \mathbb{R}).$$

Lemma 9.9 yields that in this case the net $(\alpha_i)_i$ is asymptotically multiplicative on u_t^i's:

$$\lim_i \alpha_i(u_t^i u_s^i) = \lim_i u_t^i \lim_i u_s^i = u_t u_s$$

for every $t, s \in \mathbb{R}$. This is a contradiction, because $\mathcal{A}_i$ being commutative, the unitaries u_t^i and u_s^i commute for a given i and for every $t, s \in \mathbb{R}$, while we know that u_t and u_s do not commute for any $t, s \in \mathbb{R}$. $\qquad\square$

Theorem 9.11 Let ω and φ be faithful normal states of the von Neumann algebra $\mathcal{M}$ such that $\lambda\varphi \leq \omega \leq \mu\varphi$ with some positive numbers λ and μ. Let $(\mathcal{M}_i)$ be a net of von Neumann algebras with completely positive unital mappings $\alpha_i : \mathcal{M}_i \to \mathcal{M}$ so that $\varphi \circ \alpha_i$ and $\omega \circ \alpha_i$ are faithful and normal. If $(\alpha_i)_i$ is sufficient for the couple $\{\varphi, \omega\}$ then

$$\lim_i \alpha_i \circ \alpha_i^{*,\varphi}\big([D\omega, D\varphi]_t\big) = [D\omega, D\varphi]_t$$

in the strong* topology for every $t \in \mathbb{R}$.

Proof. Let $(\beta_i)_i$ be the net coming from the definition of sufficiency, that is, let

$$\lim_i \psi(\alpha_i \circ \beta_i(a)) = \psi(a)$$

for $\psi = \varphi, \omega$ and for every $a \in \mathcal{A}$. From the monotonicity property of the relative entropy we have

$$S(\omega, \varphi) \geq \limsup_i S(\omega \circ \alpha_i, \varphi \circ \alpha_i)$$
$$\geq \limsup_i S(\omega \circ \alpha_i \circ \beta_i, \varphi \circ \alpha_i \circ \beta_i).$$

According to the weak* lower semi-continuity we have

$$S(\omega, \varphi) \leq \liminf_i S(\omega \circ \alpha_i \circ \beta_i, \varphi \circ \alpha_i \circ \beta_i).$$

Therefore

$$S(\omega, \varphi) = \lim_i S(\omega \circ \alpha_i, \varphi \circ \alpha_i)$$

and now Lemma 9.7 may be applied. It gives that

$$\alpha_i([D(\omega \circ \alpha_i), D(\varphi \circ \alpha_i)]_t) \to [D\omega, D\varphi]_t$$

for every $t \in \mathbb{R}$ in the strong operator topology. We write u_t again for $[D\omega, D\varphi]_t$ and u_t^i for $[D(\omega \circ \alpha_i), D(\varphi \circ \alpha_i)]_t$. So

$$\alpha_i(u_t^i) \to u_t \qquad (t \in \mathbb{R}). \tag{9.24}$$

We may repeat what we have done, exchanging the role of φ and ω. So

$$\alpha_i([D(\varphi \circ \alpha_i), D(\omega \circ \alpha_i)]_t) \to [D\varphi, D\omega]_t$$

may be obtained, too. Since

$$[D(\varphi \circ \alpha_i), D(\omega \circ \alpha_i)]_t \equiv (u_t^i)^* \quad \text{and} \quad [D\varphi, D\omega]_t \equiv u_t^*,$$

the limit relation (9.24) holds in the strong* topology.

The linear transformation $\alpha_i : \mathcal{M}_i \to \mathcal{M}$ is a contraction with respect to the self-dual form inner product. This fact is based on an interpolation argument and it means that

$$\|\alpha_i(a_i)\|_{(1/4)} \leq \|a_i\|_{(1/4)}$$

for every $a_i \in \mathcal{M}_i$. In order to use Lemma 9.6 we verify that (9.24) holds in the norm $(1/4)$ (with respect to the states φ and $\varphi \circ \alpha_i$) and

$$\|u_t^i\|_{(1/4)} \to \|u_t\|_{(1/4)}. \tag{9.25}$$

The convergence (9.24) in the the norm $(1/4)$ follows readily from the strong* convergence and (9.25) will be seen by the analytic continuation. Below we write σ and σ^i for the modular group of φ and $\varphi \circ \alpha_i$, respectively. The limit

$$\begin{aligned}
\lim_i \varphi \circ \alpha_i \big((u_t^i)^* \sigma_s^i(u_t^i)\big) &= \lim_i \varphi \circ \alpha_i \big((u_t^i)^* (u_s^i)^* u_{t+s}^i\big) \\
&= \lim_i \langle \Phi, \alpha_i\big((u_t^i)^* (u_s^i)^* u_{t+s}^i\big) \Phi \rangle \\
&= \lim_i \langle \Phi, \alpha_i(u_t^i)^* \alpha_i(u_s^i)^* \alpha_i(u_{t+s}^i) \Phi \rangle \\
&= \langle \Phi, u_t^* u_s^* u_{t+s} \Phi \rangle = \varphi\big(u_t^* \sigma_s(u_t)\big)
\end{aligned}$$

follows from Lemma 9.8 for every real s. Analytic continuation to $-i/2$ yields (9.25). Now we are in a position to conclude from Lemma 9.6 that

$$\|\alpha_i^{*,\varphi}(u_t) - u_t^i\|_{(1/4)} \to 0, \tag{9.26}$$

The relations (9.26) and (9.24) give

$$\alpha_i \circ \alpha_i^{\star,\varphi}(u_t) \to u_t$$

for every $t \in \mathbb{R}$ in the norm (1/4). Hence

$$\alpha_i \circ \alpha_i^{\star,\varphi}(u_t)\Phi \to u_t\Phi \tag{9.27}$$

weakly in the standard Hilbert space. Since $||\alpha_i \circ \alpha_i^{\star,\varphi}(u_t)\Phi|| \leq 1$ and $||u_t\Phi|| = 1$, we have (9.27) in the norm of the standard Hilbert space as well, which in turn is equivalent to

$$\alpha_i \circ \alpha_i^{\star,\varphi}(u_t) \to u_t \quad \text{strongly.}$$

The reasoning for the adjoint net is completely similar.   □

The definition of the sufficiency of a net (α_i) for a pair of states was given by means of another net (β_i) of mappings which behaves asymptotically as the right inverse of (α_i) as far as the given states are concerned. The information content associated with the two states is given by their relative entropy and sufficiency means that all this information is gained during the application of the net of channels. More formally, we have the following result.

Theorem 9.12 Let ω and φ be faithful normal states of the von Neumann algebra $\mathcal{M}$ such that $\lambda\varphi \leq \omega \leq \mu\varphi$ with some positive numbers λ and μ. Let $(\mathcal{M}_i)$ be a net of von Neumann algebras with normal completely positive unital mappings $\alpha_i : \mathcal{M}_i \to \mathcal{M}$. Then $(\alpha_i)_i$ is sufficient for the couple $\{\varphi, \omega\}$, if and only if

$$\lim_i S(\omega \circ \alpha_i, \varphi \circ \alpha_i) = S(\omega, \varphi). \tag{9.28}$$

Proof. Assume first that $(\alpha_i)_i$ is sufficient. In fact, it was shown in the first part of Theorem 9.11 that (9.28) holds. This proof used only the lower semi-continuity and the monotonicity of the relative entropy.

Conversely, if we assume (9.28), then according to Theorem 9.11

$$\lim_i \alpha_i \circ \alpha_i^{\star,\varphi}(u_t) = u_t$$

in the strong operator topology with $u_t = [D\omega, D\varphi]_t$. By the previous lemma

$$\lim_i \langle u_t\Phi, \alpha_i \circ \alpha_i^{\star,\varphi}(a)u_t\Phi \rangle = \langle u_t\Phi, au_t\Phi \rangle \tag{9.29}$$

for every $a \in \mathcal{M}$ and $t \in \mathbb{R}$. In (9.29) both sides have an analytic extension to the strip $S = \{z \in \mathbb{C} : -1/2 < \text{Re}\, z < 1/2\}$. (Note that the real t in (9.29) corresponds to $z = it$ in the strip.) By analytic continuation we have

$$\lim_i \langle \Omega, \alpha_i \circ \alpha_i^{\star,\varphi}(a)\Omega \rangle = \langle \Omega, a\Omega \rangle,$$

where $\Omega = \Delta(\omega, \varphi)^{1/2}\Phi$ is the vector representative of ω in the positive cone. In this way we arrive at

$$\lim_i \omega\big(\alpha_i \circ \alpha_i^{*,\varphi}(a)\big) = \omega(a) \tag{9.30}$$

which shows that $\alpha_i^{*,\varphi}$ may play the role of β_i in the definition of sufficiency for $\{\varphi, \omega\}$.   □

Yet another type of information quantity similar to relative entropy can be defined as follows.

$$S_{\mathrm{cp}}(\omega, \varphi) = \sup\{\textstyle\sum_i \omega(a_i)(\log \omega(a_i) - \log \varphi(a_i))$$
$$: a_i \in \mathcal{A}^+, \quad \textstyle\sum_i a_i = I\}.$$

The subscript "cp" might need some explanation. Clearly,

$$S_{\mathrm{cp}}(\omega, \varphi) = \sup\{S(\omega \circ \alpha, \varphi \circ \alpha) : \alpha : \mathcal{C} \to \mathcal{A} \text{ is a unital positive}$$
$$\text{mapping from a commutative algebra of finite dimension}\}.$$

The inequality $S_{\mathrm{cp}}(\omega, \varphi) \leq S(\omega, \varphi)$ is a consequence of the Uhlmann monotonicity theorem. (Note that S_{cp} appeared in (7.31).) Under the conditions of the above theorems $S_{\mathrm{cp}}(\omega, \varphi) < S(\omega, \varphi)$ is a strict inequality.

Assume that we may observe a physical system in n independent copies, that is, we may perform measurements on a composite system consisting of noninteracting subsystems identical with the initial system. (Of course, we do not claim that this is always possible.) Formally, instead of the C*-algebra $\mathcal{A}$ we consider its nth tensor power $\mathcal{A}^{(n)}$ with product states $\varphi^{(n)}$ and $\omega^{(n)}$. The sequence of observations $\alpha_n : \mathcal{C}_n \to \mathcal{A}^{(n)}$ is said to be sufficient in the mean if

$$\lim_n \frac{1}{n} S(\omega^{(n)} \circ \alpha_n, \varphi^{(n)} \circ \alpha_n) = S(\omega, \varphi).$$

Our definition means that in the mean all the relative information of the states φ and ω is extracted by the sequence $(\alpha_n)_n$ of observations.

Theorem 9.13 Let $\mathcal{M}$ be an injective von Neumann algebra with normal states φ and ω. Then there exists a sequence of finite observation channels which is sufficient in the mean.

Proof. Using Theorem 5.30 we choose a sequence $(\gamma_n : \mathcal{A}_n \to \mathcal{M})_n$ such that

$$\lim_n S(\omega \circ \gamma_n, \varphi \circ \gamma_n) = S(\omega, \varphi)$$

and the algebras $\mathcal{A}_n$ are finite dimensional. Apply Lemma 7.12 we can find for each $n \in \mathbb{N}$ a $k(n) \in \mathbb{N}$ and an observation channel $\delta_n : \mathcal{C}_n \to \mathcal{A}^{k(n)\otimes}$ such that

$$\frac{1}{k(n)}\big(S(\omega_n, \varphi_n) - S(\omega_n \circ \delta_n, \varphi_n \circ \delta_n)\big) \leq \varepsilon_n$$

where we wrote ω_n for $(\omega \circ \gamma_n)^{k(n)\otimes}$ and φ_n for $(\varphi \circ \gamma_n)^{k(n)\otimes}$. Setting $\beta_n : \mathcal{C}_n \to \mathcal{M}^{k(n)\otimes}$ as $\beta_n = \gamma_n^{k(n)\otimes} \circ \delta_n$ we have

$$\lim_n \frac{1}{k(n)} S(\omega^{k(n)\otimes} \circ \beta_n, \varphi^{k(n)\otimes} \circ \beta_n) = S(\omega, \varphi)$$

provided $\varepsilon_n \to 0$.   □

It is worth noting that by a careful analysis the dimension of the commutative finite dimensional algebras $\mathcal{C}_n$ appearing in the previous proof may be chosen to be two. An observation channel on a two dimensional algebra is called a test in statistics. Therefore there exists a sequence of tests which is sufficient in the mean. If the given von Neumann algebra is not only injective but hyper-finite then each test may be chosen to be simple.

It is easy to modify Theorem 9.13 as well as its proof for arbitrary states of nuclear C*-algebras.

At the end of this chapter we attempt to make a bridge between sufficiency and certain approximation properties of operator algebras. Let us call a channel to be finite if its domain is of finite dimension. A von Neumann algebra $\mathcal{M}$ is injective if there exist nets $\alpha_i : \mathcal{B}_i \to \mathcal{M}$ and $\beta_i : \mathcal{A} \to \mathcal{B}_i$ of channels $(i \in I)$ such that $\mathcal{B}_i$ is finite dimensional, β_i is normal and

$$\lim_i \psi(\alpha_i \circ \beta_i(a)) = \psi(a)$$

for every $a \in \mathcal{M}$ and for every $\psi \in \mathcal{M}_*$. Using the terminology of sufficient channels, we have that $\mathcal{M}$ is injective if and only if there exists a sufficient net of finite channels with respect to the whole (normal) state space. The methods of this chapter allow to prove that injective von Neumann algebras possess an approximation property in the strong operator topology. (Recall that on bounded subsets the strong operator topology is independent of the representation of the von Neumann algebra.) Let $\mathcal{M}$ be an injective von Neumann algebra. Then there exist nets $\alpha_i : \mathcal{B}_i \to \mathcal{M}$ and $\beta_i : \mathcal{A} \to \mathcal{B}_i$ of channels $(i \in I)$ such that $\mathcal{B}_i$ is finite dimensional, β_i is normal and

$$\lim_i \alpha_i \circ \beta_i(a)) = a \quad \text{for every } a \in \mathcal{M} \tag{9.31}$$

strongly.

Now we introduce two definitions. Let $(\alpha_i : \mathcal{A}_i \to \mathcal{M})$ be a net of unital completely positive mappings with finite dimensional algebras $(\mathcal{A}_i)_i$. We shall call $(\alpha_i)_i$ a weak* approximating net if for each i there exists a unital completely positive mapping $\beta_i : \mathcal{M} \to \mathcal{A}_i$ such that (9.31) is true in the weak* (i.e., ultra-weak) topology. If the normality of β_i is required and the convergence (9.31) holds in the strong operator topology then we speak of a strong approximating net. Observe that the existence of a weak* approximating net is slightly less than the above requirement for injectivity (because of the normality of β_i's in the definition of injectivity) while the existence of a strong approximating net is somewhat more (because of the difference of the topologies).

Theorem 9.14 Let $\mathcal{M}$ be a von Neumann algebra admitting a faithful normal state. Then any weak* approximating net for $\mathcal{M}$ is a strong approximating net for $\mathcal{M}$.

Proof. Let φ be a faithful normal state on $\mathcal{M}$ and assume that $(\alpha_i : \mathcal{A}_i \to \mathcal{M})$ is a weak* approximating net. If ω is a normal state on $\mathcal{M}$ then $(\alpha_i)_i$ is sufficient for $\{\varphi, \omega\}$. If $\lambda\varphi \leq \omega \leq \mu\varphi$ for some positive numbers λ and μ then Theorem 9.11 tells us that

$$\lim_i \alpha_i \circ \alpha_i^{*,\varphi}([D\omega, D\varphi]_t) = [D\omega, D\varphi]_t$$

in the strong* topology for every $t \in \mathbb{R}$. Let $\mathcal{M}_0$ be the von Neumann subalgebra generated by the set

$$\{[D\omega, D\varphi]_t : t \in \mathbb{R}, \ \lambda\varphi \leq \omega \leq \mu\varphi \text{ for some positive numbers } \lambda \text{ and } \mu\}.$$

Then

$$\lim_i \alpha_i \circ \alpha_i^{*,\varphi}(a) = a$$

holds in the strong* topology for every $a \in \mathcal{M}_0$. To show that $\mathcal{M}_0$ coincides with $\mathcal{M}$ we can argue using the φ-preserving conditional expectation. Therefore (α_i) is a strong approximating net. $\qquad\square$

The last theorem of the chapter will become important when the theory of dynamical entropy is developed in injective von Neumann algebras. The assumption on the existence of a faithful normal state is totally superfluous, it made the proof simpler.

Notes and Remarks. The notion of a *sufficient subalgebra* in the context of operator algebras appeared first in [Umegaki 1962] and was discussed later in [Hiai, Ohya and Tsukada 1981]. Here conditional expectation was used. *Sufficiency of a channel*, as it appears in this chapter, was defined in [Petz 1989] and [Petz 1991c]. For a general theory of statistical decisions, see [Holevo 1978]. Proposition 9.1 is essentially from [Emch and Wolfe 1974], the Kovács-Szűcs theorem was published in [Kovács and Szűcs 1966].

One can find something about sufficient σ-algebras in most books of mathematical statistics, we suggest [Heyer 1982] which contains a chapter "Information and sufficiency". The definition we used in the operator algebraic setting is close to the concept of sufficiency in the sense of Blackwell (in classical statistics). Due to the Kullback-Leibler-Csiszár theorem every quasi-entropy with a strictly convex f may be used for the characterization of sufficient subalgebras. In our treatment the relative entropy and Raggio's transition probability were applied and it is not clear whether any quasi-entropy with an operator convex f could be used.

The intensive use of modular theory cannot be avoided when one deals with sufficiency in infinite dimensional operator algebras. Concerning modular theory [Strătilă 1981] is the comprehensive reference but many things are summarized in [Bratteli and Robinson 1979] as well.

The example of a sufficient subalgebra of a CCR-algebra is from [Petz 1989] and it is in the context of the algebra of fluctuations introduced in [Goderis, Verbeure and Vets 1989a], see also Chapter 14.

Most results on sufficiency in the mean and asymptotic sufficiency of nets are from [Petz 1994a]. The remarks after Theorem 9.13 are based on the paper [Hiai and Petz 1991a].

The equivalence of injectivity with the existence of a weak* approximating net (for the identity) was proved in [Choi and Effros 1979]. We refer to the reviews [Effros 1978] and [Lance 1982]. It folows from the above result that a von Neumann algebra admitting a strong approximating net must be injective. The fact that a weak* approximating net is necessarily a strongly approximating one appeared in [Petz 1994a].

Comments.

Theorem 9.3 characterizes the sufficient subalgebras with respect to a pair of faithful states. This theorem extends to more general channels under the mutual majorization condition (in Theorem 9.11). Let ω and φ be faithful normal states of the von Neumann algebra $\mathcal{M}$. Let $\mathcal{M}_0$ be another von Neumann algebra and let a completely positive unital mapping $\alpha : \mathcal{M}_0 \to \mathcal{M}$ be given. Assume also that $\varphi \circ \alpha$ and $\omega \circ \alpha$ are faithful and normal. Then the following conditions are equivalent.

(i) $\varphi \circ \alpha^{\star,\omega} = \varphi$.
(ii) There exists a completely positive mapping $\beta : \mathcal{M} \to \mathcal{M}_0$ such that $\varphi \circ \alpha \circ \beta = \varphi$ and $\omega \circ \alpha \circ \beta = \omega$.
(iii) $P_A(\varphi, \omega) = P_A(\varphi \circ \alpha, \omega \circ \alpha)$.
(iv) $[D\varphi, D\omega]_t = \alpha\big([D(\varphi \circ \alpha), D(\omega \circ \alpha)]_t\big)$ for every real t.
(v) $[D\varphi, D\omega]_t = \alpha \circ \alpha^{\star,\varphi}\big([D\varphi, D\omega]_t\big)$ for every real t.
(vi) $\alpha^{\star,\varphi} = \alpha^{\star,\omega}$.

Moreover, if $S(\varphi, \omega)$ is finite, then $S(\varphi, \omega) = S(\varphi \circ \alpha, \omega \circ \alpha)$ is another equivalent condition. Note that Theorem 9.11 extends condition (v) for a net. (The original publication about the equivalent conditions is [Petz 1988b].)

The above results can be applied to analyse the condition of equality in the *strong subadditivity* of entropy. Following the lines of the proof of Proposition 1.9, the embedding

$$\alpha(b \otimes c) = I \otimes b \otimes c$$

plays a role and

$$[D\varphi_{123}, D(\varphi_{12} \otimes \tau_3)]_t = [D\varphi_{23}, D(\varphi_2 \otimes \tau_3)]_t$$

for all real t is the necessary and sufficient condition. In terms of density matrices this reads as

$$D_{123}^{it} D_{12}^{-it} = D_{23}^{it} D_2^{-it} .$$

Taking the derivative we obtain

$$\log D_{123} - \log D_{12} = \log D_{23} - \log D_2 ,$$

which is still a necessary and sufficient condition. The exact structure of a density D_{123} with this property was established in [Hayden et al 2003].

10 Dynamical Entropy

The main goal of this chapter is to introduce the entropy $h_\varphi(\gamma)$ of an automorphism γ of a C*-algebra $\mathcal{A}$ with respect to an invariant state φ. The entropy $h_\varphi(\gamma)$ is the non-commutative extension of the Kolmogorov-Sinai entropy of a measure preserving transformation. If T is a homeomorphism of a compact metric space X then $\gamma : f \mapsto f \circ T$ is an automorphism of the C*-algebra $C(X)$ of continuous functions. When T preserves the Borel probability measure μ, the automorphism γ will leave invariant the state $\varphi : f \mapsto \int f \, d\mu$. In this special case our dynamical entropy $h_\varphi(\gamma)$ will reduce to the Kolmogorov-Sinai entropy of the transformation T with respect to μ. The traditional definition of the Kolmogorov-Sinai entropy involves a kind of partition calculus. If the topological space X is connected then there exists no partition which consists of open sets. Therefore, if one insisted on the computation of the entropy in a strictly topological way then the tool of measurable partitions would not be suitable. This argument shows that already in the abelian case there might be a need for another approach to entropy.

Below we shall not assume any knowledge of the entropy of measure preserving transformations. Nevertheless, readers familiar with the measure theoretic notion will observe that the quantum dynamical entropy is not an obvious algebraization of its forefather in all details, although the definitions are globally somewhat similar. Let us give a brief summary of the operator algebraic definition of the dynamical entropy: The notion of partition will be replaced by a completely positive map of a finite dimensional algebra. Shortly speaking, one could say that the *dynamical system* $(\mathcal{A}, \gamma, \varphi)$ is tested by finite channels $\alpha : \mathcal{A}_1 \to \mathcal{A}$ (with a finite dimensional $\mathcal{A}_1$) and a *mean entropy* $h_\varphi(\gamma; \alpha)$ is observed in course of the iteration of γ. The entropy is the optimum of the tests, that is,

$$h_\varphi(\gamma) = \sup\{h_\varphi(\gamma; \alpha) : \alpha \text{ is a channel into } \mathcal{A}\}. \tag{10.1}$$

In quantum thermodynamics one has a notion of mean entropy (entropy density) which refers in a crucial manner to a sequence of finite subsystems. Having, for example, a lattice system in mind, one can not be sure that the mean entropy reflects only the relation of a translation invariant state to the space translation automorphism. The thermodynamic limit, even if it is understood as by *van Hove*, seems to be an essential ingredient of the definition. The dynamical entropy does not refer to a specific sequence of finite subsystems, so it is a more conceptual improvement of the thermodynamic limit and it gives a *conjugation invariant* of an automorphism.

The entropy $S_\alpha(\varphi)$ of a channel $\alpha : \mathcal{A}_1 \to \mathcal{A}$ has been defined with respect to an input state φ. Now this definition will be extended to n-tuples of channels as a first step towards the dynamical entropy. Below we suppose that the C*-algebra $\mathcal{A}$ and its state φ are fixed. Although our discussion is not related to communication theory we shall use the terms input and output.

An n-tuple of channels with common input algebra will be termed *multiple-channel* (of order n). Let $(\alpha_i : \mathcal{A}_i \to \mathcal{A})_{i=1}^n$ be a multiple-channel and let φ be a fixed input state (of $\mathcal{A}$). We shall speak of an *abelian model* of the system $\mathcal{S} = ((\alpha_i)_{i=1}^n, \varphi)$. An abelian model consists of an abelian finite dimensional algebra $\mathcal{C}$, a state μ of $\mathcal{C}$, a positive unital mapping P of $\mathcal{A}$ into $\mathcal{C}$ such that $\mu \circ P = \varphi$, and finally n subalgebras $(\mathcal{C}_i)_{i=1}^n$ of $\mathcal{C}$ such that they generate the whole $\mathcal{C}$. Note that this concept of abelian model is independent of the mappings α_i and depends only on $\varphi, \mathcal{A}$ and n. Given the abelian model $\mathcal{M} = ((\mathcal{C}_i)_{i=1}^n, \mathcal{C}, P, \mu)$ we set ϱ_i for the composition of $P \circ \alpha_i$ and the μ-preserving conditional expectation $E_i : \mathcal{C} \to \mathcal{C}_i$. Hence the following diagram is commutative by definition.

$$
\begin{array}{ccc}
(\mathcal{A}_i, \varphi \circ \alpha_i) & \xrightarrow{\alpha_i} & (\mathcal{A}, \varphi) \\
\varrho_i \downarrow & & \downarrow P \\
(\mathcal{C}_i, \mu_i) & \xleftarrow{E_i} & (\mathcal{C}, \mu)
\end{array}
$$

Writing ι_i for the embedding of $\mathcal{C}_i$ into $\mathcal{C}$ we have another commutative diagram.

$$
\begin{array}{ccc}
(\mathcal{A}_i, \varphi \circ \alpha_i) & \xrightarrow{\alpha_i} & (\mathcal{A}, \varphi) \\
\varrho_i \downarrow & & \downarrow P \\
(\mathcal{C}_i, \mu_i) & \xrightarrow{\iota_i} & (\mathcal{C}, \mu)
\end{array}
$$

This picture shows that the multiple-channel $(\iota_i : \mathcal{C}_i \to \mathcal{C})_{i=1}^n$ is a coarse graining of the multiple-channel $(\alpha_i)_i$. This is the idea behind the abelian model.

The *entropy of the abelian model* is defined as

$$
\mathrm{Ent}\,(\mathcal{M}, \mathcal{S}) \equiv S(\mu) - \sum_{i=1}^n (S(\mu_i) - I(\mu_i; \varrho_i)), \tag{10.2}
$$

where $\mu_i \equiv \mu|\mathcal{C}_i$. (For convenience abelian algebras will be consistently denoted by $\mathcal{C}$ and $\mathcal{C}_i$ in this chapter.) Sometimes we shall refer to the first term $S(\mu)$ of (10.2) as the classical part and the second term

$$\sum_{i=1}^{n}(S(\mu_i) - I(\mu_i\,;\varrho_i)) \geq 0 \qquad (10.3)$$

will be called the correction term. Since

$$S(\mu) - \sum_{i=1}^{n} S(\mu_i) \leq 0 \quad \text{and} \quad I(\mu_i\,;\varrho_i) \leq S(\varphi \circ \alpha_i)$$

we arrive at the upper estimate

$$\text{Ent}\,(\mathcal{M},\mathcal{S}) \leq \sum_{i=1}^{n} S(\varphi \circ \alpha_i)\,. \qquad (10.4)$$

This bound for the entropy of the abelian model does not depend on the model chosen. Now define $H_\varphi(\alpha_1,\ldots,\alpha_n)$, *the entropy of the multiple-channel,* as the least upper bound of the quantities $\text{Ent}\,(\mathcal{M},\mathcal{S})$ when $\mathcal{M}$ ranges over all abelian models of $((\alpha_i)_{i=1}^{n},\varphi)$. The next proposition tells about the properties of the entropy of multiple-channels.

Proposition 10.1 Let $(\alpha_i : \mathcal{A}_i \to \mathcal{A})_{i=1}^{n}$ be a multiple-channel and let φ be a state of $\mathcal{A}$. The entropy $H_\varphi(\alpha_1,\alpha_2,\ldots,\alpha_n)$ possesses the following properties.

(i) $H_\varphi(\alpha_1 \circ \beta_1, \alpha_2 \circ \beta_2, \ldots, \alpha_n \circ \beta_n) \leq H_\varphi(\alpha_1,\alpha_2,\ldots,\alpha_n)$ when β_i are channels with input algebras $\mathcal{A}_i$ $(i = 1,2,\ldots,n)$.

(ii) $H_{\varphi'}(\gamma\circ\alpha_1,\gamma\circ\alpha_2,\ldots,\gamma\circ\alpha_n) \leq H_\varphi(\alpha_1,\alpha_2,\ldots,\alpha_n)$ for every completely positive unital mapping $\gamma : \mathcal{A} \to \mathcal{A}'$ such that $\varphi = \varphi' \circ \gamma$ for a state φ' of $\mathcal{A}'$.

(iii) $H_\varphi(\alpha_1,\alpha_2,\ldots,\alpha_n) = H_\varphi(\alpha_{\pi(1)},\alpha_{\pi(2)},\ldots,\alpha_{\pi(n)})$ for every permutation π of the set $\{1,2,\ldots,n\}$.

(iv) $H_\varphi(\alpha_1,\alpha_2,\ldots,\alpha_k) \leq H_\varphi(\alpha_1,\alpha_2,\ldots,\alpha_n)$ if $k \leq n$.

(v) $H_\varphi(\alpha_1,\alpha_2,\ldots,\alpha_n) \leq H_\varphi(\alpha_1,\alpha_2,\ldots,\alpha_k)+H_\varphi(\alpha_{k+1},\ldots,\alpha_n)$ if $1 \leq k < n$.

Proof. The common scheme of the proof of the above properties is the following. For a given abelian model of the left-hand side one constructs an abelian model for the right-hand side so that the entropy changes increasingly.

To prove (i) we consider an abelian model $\mathcal{M} \equiv ((\mathcal{C}_i)_{i=1}^{n},\mathcal{C},P,\mu)$ for $((\alpha_i \circ \beta_i)_{i=1}^{n},\varphi)$. Of course $\mathcal{M}$ is an abelian model for $((\alpha_i)_{i=1}^{n},\varphi)$ as well. To compare the two entropies with respect to this abelian model it suffices to have

$$I(\mu_i\,;E_i \circ P \circ \alpha_i \circ \beta_i) \leq I(\mu_i\,;E_i \circ P \circ \alpha_i) \qquad (1 \leq i \leq n)$$

which follows from the monotonicity of the mutual entropy (see (8.8)). Hence (i) is proven.

In order to show (ii) let $\mathcal{M}_1 \equiv ((\mathcal{C}_i)_{i=1}^{n},\mathcal{C},P,\mu)$ be an abelian model for $\mathcal{S}_1 \equiv ((\gamma \circ \alpha_i)_{i=1}^{n},\varphi')$. Then $\mathcal{M}_2 \equiv ((\mathcal{C}_i)_{i=1}^{n},\mathcal{C},\gamma \circ P,\mu)$ is an abelian model for $\mathcal{S}_2 \equiv ((\alpha_i)_{i=1}^{n},\varphi)$. A term by term comparison

$$\mathrm{Ent}\,(\mathcal{M}_1 , \mathcal{S}_1) = \mathrm{Ent}\,(\mathcal{M}_2 , \mathcal{S}_2)$$

yields inequality (ii).

(iii) is obvious and it is based on the invariance of (10.2) under the permutation of the subscripts.

Let $((\mathcal{C}_i)_{i=1}^k , \mathcal{C} , P , \mu) \equiv \mathcal{M}_k$ be a model for $((\alpha_i)_{i=1}^k , \varphi) \equiv \mathcal{S}_k$. By trivial extension we make a model $\mathcal{M}_n$ from $\mathcal{M}_k$ $(k < n)$. Define $\mathcal{C}_i = \mathbb{C} \cdot I$ for $k < i \le n$. Then

$$\mathrm{Ent}\,(\mathcal{M}_k , \mathcal{S}_k) = \mathrm{Ent}\,(\mathcal{M}_n , \mathcal{S}_n)$$

because the terms $S(\mu_i)$ and $I(\mu_i; E_i \circ P \circ \alpha_i)$ vanish for $k < i \le n$. Hence (iv) is obtained.

In the proofs of properties (i)–(iv) we have manipulated with the correction part (10.3) of the entropy of the abelian model. In contrast to this the emphasis in property (v) is on the classical part. Let $\mathcal{M} \equiv ((\mathcal{C}_i)_{i=1}^n , \mathcal{C} , P , \mu)$ be an abelian model for $((\alpha_i)_{i=1}^n , \varphi)$. Set $\mathcal{C}_B = \vee\{\mathcal{C}_i : 1 \le i \le k\}$, $\mathcal{C}_E = \vee\{\mathcal{C}_i : k < i \le n\}$, $\mu_B = \mu|\mathcal{C}_B$, $\mu_E = \mu|\mathcal{C}_E$. Set F_B and F_E for the μ-preserving conditional expectations of $\mathcal{C}$ onto $\mathcal{C}_B$ and $\mathcal{C}_E$. Then $((\mathcal{C}_i)_{i=1}^k , \mathcal{C}_B , F_B \circ P , \mu_B) \equiv \mathcal{M}_B$ and $((\mathcal{C}_i)_{i=k+1}^n , \mathcal{C}_E , F_E \circ P , \mu_E) \equiv \mathcal{M}_E$ are abelian models for $\mathcal{S}_B \equiv ((\alpha_i)_{i=1}^k , \varphi)$ and $\mathcal{S}_E \equiv ((\alpha_i)_{i=k+1}^n , \varphi)$, respectively. The difference

$$\mathrm{Ent}\,(\mathcal{M}_B , \mathcal{S}_B) + \mathrm{Ent}\,(\mathcal{M}_E , \mathcal{S}_E) - \mathrm{Ent}\,(\mathcal{M} , \mathcal{S}) = S(\mu_B) + S(\mu_E) - S(\mu)$$

is nonnegative due to the subadditivity of the classical entropy. This proves property (v). $\qquad\square$

The concept of an abelian model shows very well the intention of reducing the non-abelian situation to the abelian one. However, in the definition of the entropy of a multiple-channel one can proceed in a slightly different (but mathematically equivalent) way, putting the emphasis on certain decompositions of the given state.

Let $((\mathcal{C}_i)_{i=1}^n , \mathcal{C} , P , \mu)$ be an abelian model for the multiple-channel $(\alpha_i : \mathcal{A}_i \to \mathcal{A})_{i=1}^n$ with an input state φ. We choose an integer $N \in \mathbb{N}$ so that each of the abelian algebras $\mathcal{C}_1, \mathcal{C}_2, \ldots, \mathcal{C}_n$ have at most N minimal projections. We write

$$p_{i,1}, p_{i,2}, \ldots, p_{i,N}$$

for the sequence of minimal projections of $\mathcal{C}_i$. (If it has only $m < N$ minimal projections, we set $p_{i,l} = 0$ for $l > m$.) For any sequence $\bar{l} = (j_1, j_2, \ldots, j_n) \in \{1, 2, \ldots, N\}^n$ the projection

$$p(\bar{l}) = p_{1,j_1} p_{2,j_2} \cdots p_{n,j_n}$$

is either minimal in $\mathcal{C}$ or it is 0. Let $\nu(\bar{l}) = \mu(p(\bar{l}))$ so $\nu(\bar{l})$ is a probability distribution on the product set $\Omega = \{1, 2, \ldots, N\}^n$. To each $\bar{l} \in \Omega$ there is a state $\varphi(\bar{l})$ of $\mathcal{A}$ which is the corresponding component of P, that is

$$P(a) = \sum_{\bar{l}} p(\bar{l})\varphi(\bar{l})(a) \quad \text{and} \quad \varphi = \sum_{\bar{l}} \nu(\bar{l})\varphi(\bar{l}).$$

Hence an abelian model is simply a multi-indexed decomposition

$$\varphi = \sum_{\bar{l}} \psi(\bar{l}) \tag{10.5}$$

where $\psi(\bar{l}) = \nu(\bar{l})\varphi(\bar{l})$ are a positive linear functionals. With

$$\psi_m^k = \sum \{\psi(\bar{l}) : j_k = m\} \quad (m = 1, 2, \ldots, N)$$

the entropy of the abelian model (or equivalently that of the decomposition (10.5)) is

$$S(\nu) + \sum_{k=1}^{n} \sum_{m=1}^{N} S(\psi_m^k \circ \alpha_k, \varphi \circ \alpha_k). \tag{10.6}$$

Therefore, $H_\varphi(\alpha_1, \alpha_2, \ldots, \alpha_n)$ is the supremum of the quantities (10.6) over all decompositions (10.5) of φ.

It is easy to explain (10.6) in the case $n = 2$. Then the decomposition (10.5) can be visualized in a matrix arrangement.

$$\psi(1,1) \quad \psi(1,2) \quad \ldots \quad \psi(1,N)$$
$$\psi(2,1) \quad \psi(2,2) \quad \ldots \quad \psi(2,N)$$
$$\cdot \qquad\qquad \cdot \qquad \ldots \qquad \cdot$$
$$\psi(N,1) \quad \psi(N,2) \quad \ldots \quad \psi(N,N)$$

Evaluating each functional of this arrangement on the identity we obtain a probability distribution ν and its entropy is the first term in (10.6). The functional ψ_m^1 is the sum of the mth raw and similarly ψ_m^2 is that of the mth column $(m = 1, 2, \ldots, N)$.

Turning to another special case, assume that φ is a faithful normal state of a von Neumann algebra. Then any decomposition (10.5) of ψ corresponds to a partition of unity in the von Neumann algebra. The formula

$$\psi(\bar{l})(\,.\,) = \varphi(x(\bar{l})\,.\,)$$

determines the density $x(\bar{l})$ of $\psi(\bar{l})$ and we have

$$I = \sum_{\bar{l}} x(\bar{l}) \quad \text{and} \quad x_m^k = \sum \{x(\bar{l}) : j_k = m\} \quad (m = 1, 2, \ldots, N).$$

Suppose moreover that the channels $\alpha_1, \alpha_2, \ldots, \alpha_n$ are embeddings of finite dimensional subalgebras $\mathcal{N}_1, \mathcal{N}_2, \ldots, \mathcal{N}_n$ and denote by E_k the φ-preserving conditional expectation onto $\mathcal{N}_k$. Then we may rewrite the ingredients of (10.6) as follows.

$$S(\nu) = \sum_{\bar{l}} \eta(\varphi(x(\bar{l}))) \quad \text{and} \quad S(\psi_m^k \circ \alpha_k, \varphi \circ \alpha_k) = -\varphi(\eta(E_k(x_m^k))).$$

Now it is time to turn to the *multiple-channel entropy* in full generality.

Proposition 10.2 The quantity $H_\varphi(\alpha_1, \alpha_2, \ldots, \alpha_n)$ depends only on the set $\{\alpha_1, \alpha_2, \ldots, \alpha_n\}$.

Proof. After properties (iii) and (iv) of the previous proposition it remains to show that

$$H_\varphi(\alpha_1, \alpha_2, \ldots, \alpha_{n-1}, \alpha_n, \alpha_n) \leq H_\varphi(\alpha_1, \alpha_2, \ldots, \alpha_{n-1}, \alpha_n).$$

Let $((\mathcal{C}_i)_{i=1}^{n+1}, \mathcal{C}, P, \mu) \equiv \mathcal{M}_L$ be an abelian model for the left-hand side. If

$$\overline{\mathcal{C}}_i = \begin{cases} \mathcal{C}_i & \text{for} \quad 1 \leq i < n \\ \mathcal{C}_n \vee \mathcal{C}_{n+1} & \text{for} \quad i = n \end{cases}$$

then $((\overline{\mathcal{C}}_i)_{i=1}^{n}, \mathcal{C}, P, \mu) \equiv \mathcal{M}_R$ is a model for the right-hand side. We have

$$\begin{aligned} \text{Ent}\,&(\mathcal{M}_R, (\alpha_1, \alpha_2, \ldots, \alpha_{n-1}, \alpha_n; \varphi)) \\ &-\text{Ent}\,(\mathcal{M}_L, (\alpha_1, \alpha_2, \ldots, \alpha_{n-1}, \alpha_n; \alpha_n \varphi)) \\ &= s_{\mu_n}(E_n \circ P \circ \alpha_n) + s_{\mu_{n+1}}(E_{n+1} \circ P \circ \alpha_n) - s_{\overline{\mu}_n}(F \circ P \circ \alpha_n) \end{aligned} \qquad (10.7)$$

where $\overline{\mu}_n \equiv \mu | \overline{\mathcal{C}}_n$ and F is the μ-preserving conditional expectation onto $\overline{\mathcal{C}}_n = \mathcal{C}_n \vee \mathcal{C}_{n+1}$. Since $E_n = E_n \circ F$ and $E_{n+1} = E_{n+1} \circ F$ Lemma 8.1 is applicable and yields that (10.7) is nonnegative. $\qquad \square$

The entropy $H_\varphi(\alpha_1, \alpha_2, \ldots, \alpha_n)$ measures the uncertainty contained in the multiple-channel $(\alpha_1, \alpha_2, \ldots, \alpha_n)$ if the whole system is in a state φ. Since $H_\varphi(\alpha_1, \alpha_2, \ldots, \alpha_n) \leq H_\varphi(\text{id}, \text{id}, \ldots, \text{id})$ for the identity mapping id of the input algebra, we naturally have

$$H_\varphi(\alpha_1, \alpha_2, \ldots, \alpha_n) \leq S(\varphi). \qquad (10.8)$$

Note that if $\mathcal{A} \subset \tilde{\mathcal{A}}$ and $\varphi = \tilde{\varphi} | \mathcal{A}$ then the entropies $H_\varphi(\alpha_1, \alpha_2, \ldots, \alpha_n)$ and $H_{\tilde{\varphi}}(\alpha_1, \alpha_2, \ldots, \alpha_n)$ may be different and that they are equal when there exist a conditional expectation of $\tilde{\mathcal{A}}$ onto $\mathcal{A}$ which preserves the state $\tilde{\varphi}$. This can be deduced from Proposition 10.1.

On the set of multiple-channels one has a simple *join operation*

$$(\alpha_1, \alpha_2, \ldots, \alpha_n) \vee (\beta_1, \beta_2, \ldots, \beta_m) = (\alpha_1, \alpha_2, \ldots, \alpha_n, \beta_1, \beta_2, \ldots, \beta_m)$$

and (v) in Proposition 10.1 tells us that the multiple-channel entropy is *sub-additive* with respect to this join. If two multiple-channels are independent then entropy is additive:

Proposition 10.3 Let $(\alpha_i : \mathcal{A}_i \to \mathcal{A})_{i=1}^{n}$ and $(\beta_j : \mathcal{A}'_i \to \mathcal{A}')_{j=1}^{m}$ be multiple-channels and let $\psi = \varphi \otimes \varphi'$ be a product state on $\mathcal{A} \otimes \mathcal{A}'$. Then

$$\begin{aligned} H_\psi&(\bar{\alpha}_1, \bar{\alpha}_2, \ldots, \bar{\alpha}_n, \bar{\beta}_1, \bar{\beta}_2, \ldots, \bar{\beta}_m) \\ &= H_\varphi(\alpha_1, \alpha_2, \ldots, \alpha_n) + H_{\varphi'}(\beta_1, \beta_2, \ldots, \beta_m) \end{aligned}$$

where $\bar{\alpha}_i(a_i) = \alpha_i(a_i) \otimes I$ and $\bar{\beta}_j(a_j) = I \otimes \beta_j(a_j)$.

Proof. Due to subadditivity (that is, property (v) in Proposition 10.1) we shall prove only $\geq$. Let $\varphi = \sum_{\bar{n}} \varphi(\bar{n})$ and $\varphi' = \sum_{\bar{m}} \varphi'(\bar{m})$ be multi-indexed decompositions for the right-hand side. Then

$$\psi = \sum_{\bar{n}, \bar{m}} \varphi(\bar{n}) \otimes \varphi'(\bar{m})$$

is a decomposition of ψ such that for the entropies of the corresponding abelian models we have additivity according to the additivity of entropy and relative entropy under tensor product. $\square$

The proof of the following observation is elementary.

Lemma 10.4 If $(x_i) \subset \mathbb{R}^+$ is a sequence satisfying $x_{n+m} \leq x_n + x_m$ for every $n, m \in \mathbb{N}$ then

$$\lim_{n \to \infty} \frac{1}{n} x_n = \inf\{n^{-1} x_n : \quad n \in \mathbb{N}\}.$$

Let $\mathcal{A}$ be a C*-algebra, φ a state on $\mathcal{A}$ and γ a completely positive unital mapping of $\mathcal{A}$ which preserves φ. We define $h_\varphi(\gamma)$, the entropy of γ with respect to the invariant state φ. First we set

$$h_\varphi(\gamma; \alpha) = \lim_{n \to \infty} \frac{1}{n} H_\varphi(\alpha, \gamma \circ \alpha, \dots, \gamma^{n-1} \circ \alpha) \tag{10.9}$$

for a channel $\alpha : \mathcal{A}_1 \to \mathcal{A}$ with input algebra $\mathcal{A}$. The sequence $x_n = H_\varphi(\alpha, \gamma \circ \alpha, \dots, \gamma^{n-1} \circ \alpha)$ is subadditive. Indeed, according to (v) of Proposition 10.1

$$x_{n+k} \leq x_n + H_\varphi(\gamma^n \circ \alpha, \dots, \gamma^{n+k-1} \circ \alpha)$$

and property (ii) gives that the second term of the right hand side is majorized by x_k. The existence of the limit is provided by Lemma 10.4.

The entropy $h_\varphi(\gamma)$ is the supremum of $h_\varphi(\gamma; \alpha)$ for all possible testing channels $\alpha : \mathcal{A}_1 \to \mathcal{A}$ with a finite dimensional output algebra $\mathcal{A}_1$.

The dynamical entropy is a notion for infinite systems. If $S(\varphi) < \infty$, then the bound (10.8) gives $h_\varphi(\gamma; \alpha) = 0$ for every channel $\alpha : \mathcal{A}_1 \to \mathcal{A}$ and we have $h_\varphi(\gamma) = 0$. Although the definition of the dynamical entropy was formulated for a completely positive mapping γ, the primary interest is in endomorphisms. For example, the dynamical entropy of a conditional expectation is zero.

The following properties of the dynamical entropy are direct consequences of the definition.

Proposition 10.5 Let γ and β be automorphisms of a C*-algebra $\mathcal{A}$ and let φ be a state such that $\varphi \circ \gamma = \varphi$. Then

(i) $h_\varphi(\gamma) = h_\varphi(\gamma^{-1})$,
(ii) $h_\varphi(\gamma) = h_{\varphi \circ \beta}(\beta^{-1} \circ \gamma \circ \beta)$.

Proof. Property (i) follows from

$$H_\varphi(\alpha, \gamma \circ \alpha, \gamma^2 \circ \alpha, \ldots, \gamma^n \circ \alpha) = H_\varphi(\gamma^{-n} \circ \alpha, \gamma^{-n+1} \circ \alpha, \ldots, \alpha)$$

and (ii) is based on

$$H_{\varphi \circ \beta}(\alpha, (\beta^{-1} \circ \gamma \circ \beta) \circ \alpha, \ldots, (\beta^{-1} \circ \gamma \circ \beta)^n \circ \alpha)$$
$$= H_\varphi(\beta \circ \alpha, \gamma \circ (\beta \circ \alpha), \ldots, \gamma^n \circ (\beta \circ \alpha)).$$

This relation holds according to property (ii) in Proposition 10.1.   □

The automorphisms γ and $\beta^{-1} \circ \gamma \circ \beta$ are conjugate to each other. So property (ii) expresses the *conjugation invariance* of dynamical entropy.

Our definition is not suitable for the practical computation of the dynamical entropy $h_\varphi(\gamma)$; we need a theorem which restricts the class of the necessary testing channels. Such a theorem would allow to take the lowest upper bound of the quantities $h_\varphi(\gamma; \alpha)$ over a countable class of testing channels or the dynamical entropy would be obtained as a limit. In the measure theoretic case this problem is resolved by the Kolmogorov–Sinai theorem. theory. We should search for its operator algebraic generalization which is strongly related to approximation properties of operator algebras. Interrupting, for a while, the discussion of dynamical entropy we recall some facts on nuclear C*-algebras and injective von Neumann algebras.

Let $(\alpha_i : \mathcal{A}_i \to \mathcal{A})_i$ be a net of channels with finite dimensional output algebras $\mathcal{A}_i$. We shall call $(\alpha_i)_i$ a *norm approximating net* if for each i there exists a channel $\beta_i : \mathcal{A} \to \mathcal{A}_i$ such that $\|\alpha_i \circ \beta_i(a) - a\| \to 0$ $(a \in \mathcal{A})$. The class of *nuclear C*-algebras* is characterized by the existence of a norm approximating net. A very similar property is characteristic for *injective von Neumann algebras*. Let $\mathcal{M}$ be a von Neumann algebra. A net $(\alpha_i : \mathcal{A}_i \to \mathcal{M})_i$ of finite channels is called *weak* approximating net* if there exists a net $(\beta_i : \mathcal{M}_i \to \mathcal{A}_i)_i$ of normal channels such that $\psi(\alpha_i \circ \beta_i(a)) \to \psi(a)$ holds for every normal state ψ of $\mathcal{M}$ and for every $a \in \mathcal{M}$. A von Neumann algebra is injective if and only if it possesses a weak* approximating net of finite channels.

Proposition 10.6 Let $\mathcal{A}$ be a nuclear C*-algebra with an approximating net $(\alpha_i : \mathcal{A}_i \to \mathcal{A})_i$. Then for every state φ of $\mathcal{A}$

$$S(\varphi) = \lim_i H_\varphi(\alpha_i)$$

holds.

Proof. By the definition of entropy we can find a finite convex decomposition $\sum_{k=1}^n \lambda_k \varphi_k$ of φ for an $\varepsilon > 0$ so that

$$c \le \sum_{k=1}^n \lambda_k S(\varphi_k, \varphi)$$

for any given $c < S(\varphi)$. We have

$$S(\varphi_k, \varphi) \leq \liminf_i S(\varphi_k \circ \alpha_i, \varphi \circ \alpha_i) \qquad (k = 1, 2, \ldots, n)$$

due to Theorem 5.29. Hence

$$c \leq \sum_{k=1}^{n} \lambda_k S(\varphi_k \circ \alpha_i, \varphi \circ \alpha_i) \leq H_\varphi(\alpha_i)$$

for some sufficiently large i. Since $H_\varphi(\alpha_i) \leq S(\varphi)$, the proof is complete. $\square$

The next proposition contains a concrete estimate for the norm continuity of the multiple-channel entropy.

Proposition 10.7 Let $(\alpha_i : \mathcal{A}_i \to \mathcal{A})_{i=1}^{n}$ and $(\alpha_i' : \mathcal{A}_i \to \mathcal{A})_{i=1}^{n}$ be multiple-channels. Assume that $\mathcal{A}_i$ is finite dimensional and its tracial dimension has an upper bound d $(i = 1, 2, \ldots, n)$. If $\|\alpha_i - \alpha_i'\| \leq \varepsilon < 1/3$ $(i = 1, 2, \ldots, n)$ then for every input state φ the estimate

$$|H_\varphi(\alpha_1', \alpha_2', \ldots, \alpha_n')| - H_\varphi(\alpha_1, \alpha_2, \ldots, \alpha_n)| \leq n(\varepsilon \log d + \eta(\varepsilon))$$

holds.

Proof. It suffices to show that for any $\delta > 0$

$$H_\varphi(\alpha_1, \alpha_2, \ldots, \alpha_n) \geq H_\varphi(\alpha_1', \alpha_2', \ldots, \alpha_n') - n\varepsilon \log d - n\eta(\varepsilon) - \delta. \quad (10.10)$$

First we choose an abelian model $\mathcal{M} \equiv ((\mathcal{C}_i)_{i=1}^{n}, \mathcal{C}, P, \mu)$ for the system $\mathcal{S} \equiv (\alpha_1, \alpha_2, \ldots, \alpha_n; \varphi)$ such that

$$|\mathrm{Ent}\,(\mathcal{M}, \mathcal{S}) - H_\varphi(\alpha_1, \alpha_2, \ldots, \alpha_n)| \leq \delta.$$

In order to prove (10.10) we check that

$$|\mathrm{Ent}\,(\mathcal{M}, \mathcal{S}) - \mathrm{Ent}\,(\mathcal{M}, \mathcal{S}')| \leq n\varepsilon \log d + n\eta(\varepsilon)$$

where $\mathcal{S}' = (\alpha_1', \alpha_2', \ldots, \alpha_n'; \varphi)$. Keeping the notation used in the definition $\mathrm{Ent}\,(\mathcal{M}, \mathcal{S}')$ we have

$$\mathrm{Ent}\,(\mathcal{M}, \mathcal{S}) - \mathrm{Ent}\,(\mathcal{M}, \mathcal{S}') = \sum_{i=1}^{n} (I(\mu_i; \varrho_i) - I(\mu_i; \varrho_i')).$$

Since $\|\varrho_i - \varrho_i'\| = \|E_j \circ P \circ (\alpha_i - \alpha_i')\| \leq \|\alpha_i - \alpha_i'\| \leq \varepsilon$, a reference to Proposition 8.2 completes the proof. $\square$

Now we are in a position to prove the (norm) *approximating net theorem* for the dynamical entropy of a completely positive mapping of a nuclear C*-algebra.

Theorem 10.8 Let $\alpha_i : \mathcal{A}_i \to \mathcal{A}$ be a norm approximating net of finite channels for the nuclear C*-algebra $\mathcal{A}$. Then

$$h_\varphi(\gamma) = \lim_i h_\varphi(\gamma; \alpha_i)$$

for every completely positive mapping γ of $\mathcal{A}$ and for every state φ such that $\varphi \circ \gamma = \varphi$.

Proof. We have to show that

$$\liminf_i h_\varphi(\gamma; \alpha_i) \geq h_\varphi(\gamma; \alpha) \tag{10.11}$$

for any finite channel $\alpha : \mathcal{A}_0 \rightarrow \mathcal{A}$. Let $\theta_i = \alpha_i \circ \beta_i \circ \alpha$ where β_i is from the definition of the approximating net (α_i). Then $||\theta_i - \alpha|| \rightarrow 0$ because $\mathcal{A}_0$ is finite dimensional. By application of Proposition 10.7 we infer

$$\lim_i |h_\varphi(\gamma; \theta_i) - h_\varphi(\gamma; \alpha)| = 0 \,.$$

From (i) of Proposition 10.1 we see that $h_\varphi(\gamma; \theta_i) \leq h_\varphi(\gamma; \alpha_i)$ and so (10.11) must hold. □

The simplest example of a norm approximating net of channels is related to AF-algebras. Assume that the C*-algebra $\mathcal{A}$ contains an increasing sequence $\mathcal{A}_1 \subset \mathcal{A}_2 \subset \ldots$ of finite dimensional subalgebras such that $\cup_n \mathcal{A}_n$ is norm dense in $\mathcal{A}$. Then there exists a conditional expectation E_k^n of $\mathcal{A}_n$ onto $\mathcal{A}_k$ whenever $n \geq k$. One sees easily that any limit point of $(E_k^n)_{n \geq k}$ will be a conditional expectation of $\mathcal{A}$ onto $\mathcal{A}_k$. So compactness provides the existence of a conditional expectation E_k of $\mathcal{A}$ onto any $\mathcal{A}_k$. Let α_n be the embedding of $\mathcal{A}_n$ into $\mathcal{A}$. Obviously $\alpha_n \circ E_n(a) \rightarrow a$ if $a \in \cup\{\mathcal{A}_n : n \in \mathbb{N}\}$, therefore $||\alpha_n \circ E_n(a) - a|| \rightarrow 0$ for every a $\in \mathcal{A}$. In other words, (α_n) is a approximating sequence of finite channels. For an embedding α_n we write $h_\varphi(\gamma; \mathcal{A}_n)$ instead of $h_\varphi(\gamma; \alpha_n)$ and we prefer to write $H_\varphi(\mathcal{A}_n, \gamma(\mathcal{A}_n))$ for $H_\varphi(\alpha_n, \gamma \circ \alpha_n))$.

Corollary 10.9 With the above notation we have

$$h_\varphi(\gamma) = \lim_{n \to \infty} h_\varphi(\gamma; \mathcal{A}_n)$$

for an *AF*-algebra $\mathcal{A}$.

The approximating net theorem and the additivity of the multiple-channel entropy (see Proposition 10.3) allow the computation of the dynamical entropy of an *independent shift*. Let $\mathcal{B}_j$ be a copy of a nuclear C*-algebra $\mathcal{B}_0$ for every $j \in \mathbb{Z}$ and consider the infinite tensor product $\mathcal{B} \equiv \otimes_{j \in \mathbb{Z}} \mathcal{B}_j$. The algebra $\mathcal{B}$ is nuclear and for $J \subset \mathbb{Z}$ we write $\mathcal{B}_J$ for its corresponding subalgebra $\otimes_{j \in J} \mathcal{B}_j$. Now we are going to compute the entropy of the right shift automorphism γ of $\mathcal{B}$ with respect to a product state φ.

Since $\mathcal{B}_0$ is supposed to be nuclear, it has a norm approximating net (α_i) of finite channels. We set

$$\alpha_{(i,n)} \equiv \otimes_{s=-n}^n \alpha_i : \mathcal{B}_{[-n,n]} \rightarrow \mathcal{B}$$

and obtain a net (indexed by pairs) of finite channels for $\mathcal{B}$. Obviously, this is an approximating net. Due to the approximating net theorem we need to compute

$$\lim_{(i,n)} h_\varphi(\gamma; \alpha_{(i,n)}). \tag{10.12}$$

In order to arrive at an upper estimate we note first that there is a conditional expectation mapping of $\mathcal{B}$ onto the subalgebras $\mathcal{B}_J$ which leaves the state φ invariant. Therefore the multiple-channel entropy

$$H_\varphi(\alpha_{(i,n)}, \gamma \circ \alpha_{(i,n)}, \ldots, \gamma^k \circ \alpha_{(i,n)})$$

does not change when φ is replaced by $\varphi|\mathcal{B}_{[-n,n+k]}$ because the subalgebra $\mathcal{B}_{[-n,n+k]}$ contains the ranges of the mappings $\gamma^s \gamma \alpha_{(i,n)}$ for $0 \leq s \leq k$. Hence (10.8) tells us that

$$H_\varphi(\alpha_{(i,n)}, \gamma \circ \alpha_{(i,n)}, \ldots, \gamma^k \circ \alpha_{(i,n)}) \leq S(\varphi|\mathcal{B}_{[-n,n+k]})$$

and

$$S(\varphi|\mathcal{B}_{[-n,n+k]}) = (2n + 1 + k)S(\varphi_0)$$

where φ_0 is just $\varphi|\mathcal{B}_0$. So

$$h_\varphi(\gamma; \alpha_{(i,n)}) \leq S(\varphi_0) \tag{10.13}$$

for every pair (i, n).

Consider now $k = (2n+1)l$. Benefiting from point (iv) of Proposition 10.1 we have

$$H_\varphi(\alpha_{(i,n)}, \gamma \circ \alpha_{(i,n)}, \ldots, \gamma^{l(2n+1)} \circ \alpha_{(i,n)})$$
$$\geq H_\varphi(\alpha_{(i,n)}, \gamma^{2n+1} \circ \alpha_{(i,n)}, \ldots, \gamma^{l(2n+1)} \circ \alpha_{(i,n)})$$

which equals to

$$lH_\varphi(\alpha_{(i,n)}) \tag{10.14}$$

according to Proposition 10.3. In this way we obtain

$$h_\varphi(\gamma; \alpha_{(i,n)}) \geq \frac{1}{2n+1} H_{\varphi_0}(\alpha_i) \tag{10.15}$$

Taking the limit in i this lower bound goes to

$$\frac{1}{2n+1} S(\otimes_{-n}^n \varphi_0) = S(\varphi_0)$$

(cf. Proposition 10.6) and we conclude

$$h_\varphi(\gamma) = S(\varphi_0). \tag{10.16}$$

The norm approximating net theorem is not completely satisfactory because some important operator algebras are not nuclear. For example, $B(\mathcal{H})$ with an infinite dimensional Hilbert space $\mathcal{H}$ is not nuclear. On the other hand, in the category of von Neumann algebras and normal states the norm approximation is not natural. We need a more general form of Proposition 10.7 which is applicable von Neumann algebras. Let $\alpha : \mathcal{A}_1 \to \mathcal{A}$ and $\alpha' : \mathcal{A}_1 \to \mathcal{A}$ be channels. We shall use the following distance.

$$\|\alpha - \alpha'\|_\varphi = \sup\{ \|\alpha(x) - \alpha'(x)\|_\varphi : x \in \mathcal{A}_1, \|x\| \leq 1 \}$$

where $\|x\|_\varphi = \varphi(x^*x)^{1/2}$ with respect to an input state φ. The advantage of this norm is the inequality

$$\|\beta \circ \alpha - \beta \circ \alpha'\|_\varphi \leq \|\alpha - \alpha'\|_{\varphi \circ \beta} \tag{10.17}$$

for every completely positive unital mapping $\beta : \mathcal{A} \to \mathcal{A}'$.

Lemma 10.10 Given $\varepsilon > 0$, let $r(d,\varepsilon)$ be the minimal number of balls of radius $\varepsilon/2$ needed to cover the state space of $M_d(\mathbb{C})$. Let $(\alpha_i : \mathcal{A}_i \to \mathcal{A})_{i=1}^n$ be a multiple-channel so that the tracial dimension of $\mathcal{A}_i$ is at most $d \in \mathbb{N}$. Then for any $\delta > 0$ there is an abelian model $((\mathcal{C}_i)_{i=1}^n, \mathcal{C}, P, \mu)$ so that the tracial dimension of $\mathcal{C}_i$ is at most $r(d,\varepsilon)$ and the entropy of the model is larger than

$$H_\varphi(\alpha_1, \alpha_2, \ldots, \alpha_n) - n\varepsilon_1 - \delta$$

where $\varepsilon_1 = \varepsilon \log d + \eta(\varepsilon)$.

Proof. Let $((\mathcal{C}'_i)_{i=1}^n, \mathcal{C}', P', \mu')$ be an abelian model. (We regard that its entropy is close to the entropy of the given multiple-channel.) We may assume that $\mathcal{C}'_i$ is the algebra of functions on a finite set X_i and $\mathcal{C}'$ is the function algebra on the product $X_1 \times X_2 \times \ldots \times X_n$. We may assume also that $\mathcal{A}_i \equiv M_d(\mathbb{C})$. Let $\{U_j : 1 \leq j \leq r(d,\varepsilon)\}$ be a partition of the state space $\mathfrak{S}$ of $M_d(\mathbb{C})$ into sets with diameter smaller than ε. The linear mapping $\varrho'_i = E'_i \circ P' \circ \alpha_i : \mathcal{A}_i \to \mathcal{C}'_i$ is given by a mapping $\bar{\varrho}'_i : X_i \to \mathfrak{S}$. We set an equivalence relation by saying $x, y \in X_i$ are in relation if $\bar{\varrho}'_i(x)$ and $\bar{\varrho}'_i(y)$ belong to the same element of the partition (U_j). This equivalence determines a subalgebra $\mathcal{C}_i$ of $\mathcal{C}'_i$ and the tracial dimension of $\mathcal{C}_i$ (that is, the number of equivalence classes) will not exceed $r(d,\varepsilon)$. Let $\mathcal{C}$ be the subalgebra of $\mathcal{C}'$ generated by $\{\mathcal{C}_i : 1 \leq i \leq n\}$, $P = F' \circ P'$ where F' is the μ'-preserving conditional expectation of $\mathcal{C}'$ onto $\mathcal{C}$ and $\mu = \mu'|\mathcal{C}$. In this way we have constructed another model $((\mathcal{C}_i)_{i=1}^n, \mathcal{C}, P, \mu)$ which is strongly related to the given one distinguished by a prime.

Now we have to compare the entropies of the two models. A slight modification of the argument of Proposition 8.2 gives

$$|I(\mu_i, \varrho_i) - I(\mu'_i, \varrho'_i)| \leq \varepsilon \log d + \eta(\varepsilon). \tag{10.18}$$

On the other hand,

$$S(\mu) - \sum_{i=1}^n S(\mu_i) \geq S(\mu') - \sum_{i=1}^n S(\mu'_i) \tag{10.19}$$

is equivalent to

$$S(\mu'|\mathcal{C}, \bar{\mu}|\mathcal{C}) \leq S(\mu', \bar{\mu}) \tag{10.20}$$

where $\bar{\mu}$ is the product measure $\mu'_1 \otimes \mu'_2 \otimes \ldots \otimes \mu'_n$. Since (10.20) holds due to the monotonicity, we have (10.19). Together with (10.18) this inequality implies that the difference of the entropies of the given model and the constructed one is less than the required bound $n(\varepsilon \log d + \eta(\varepsilon))$.   □

Lemma 10.10 allows one to obtain an estimate for the continuity of the multiple-channel entropy in terms of the φ-norm.

Proposition 10.11 Given $d \in \mathbb{N}$ and $\varepsilon_0 > 0$ there is a $\delta_0 > 0$ such that if $(\alpha_i : \mathcal{A}_i \to \mathcal{A})_{i=1}^n$ and $(\alpha_i' : \mathcal{A}_i \to \mathcal{A})_{i=1}^n$ are multiple-channels so that $\|\alpha_i - \alpha_i'\|_\varphi < \delta_0$ and the tracial dimension of $\mathcal{A}_i$ is smaller than d $(1 \le i \le n)$, one has

$$|H_\varphi(\alpha_1, \alpha_2, \ldots, \alpha_n) - H_\varphi(\alpha_1', \alpha_2', \ldots, \alpha_n')| \le n\varepsilon_0.$$

Proof. First of all we may assume that the systems $\mathcal{S} = ((\alpha_i : \mathcal{A}_i \to \mathcal{A})_{i=1}^n, \varphi)$ and $\mathcal{S}' = ((\alpha_i' : \mathcal{A}_i \to \mathcal{A})_{i=1}^n, \varphi)$ are given with $\mathcal{A}_i \equiv M_d(\mathbb{C})$. Let $\mathcal{M} = ((\mathcal{C}_i)_{i=1}^n, \mathcal{C}, P, \mu)$ be an abelian model for both channels. According to the previous lemma for any $\delta > 0$ this model $\mathcal{M}$ might be chosen such a way that

$$H_\varphi(\alpha_1, \alpha_2, \ldots, \alpha_n) - n\varepsilon_1 - \delta \le \mathrm{Ent}(\mathcal{M}, \mathcal{S}) \equiv S(\mu) - \sum_{i=1}^n (S(\mu_i) - I(\mu_i, \varrho_i))$$

where $\varrho_i = E_i \circ P \circ \alpha_i$ is as in the definition (10.2) of the entropy for the abelian model. We may assume that the (tracial) dimension of $\mathcal{C}_i$ is at most $r(d, \varepsilon)$ where $\varepsilon < 1/3$ is so small that $\varepsilon_1 < \varepsilon_0/2$.

We have a similar formula for the entropy of $\mathcal{S}'$. Namely,

$$\mathrm{Ent}(\mathcal{M}, \mathcal{S}') = S(\mu) - \sum_{i=1}^n (s(\mu_i) - I(\mu_i, \varrho_i'))$$

where $\varrho_i' = E_i \circ P \circ \alpha_i'$. Property (10.17) tells us that

$$\|\varrho_i - \varrho_i'\|_{\mu_i} \le \|\alpha_i - \alpha_i'\|_\varphi.$$

The mappings ϱ_i and ϱ_i' map $M_d(\mathbb{C})$ into a finite dimensional space $\mathcal{C}_i$, its dimension is majorized by $r(d, \varepsilon)$. By a simple compactness argument we can have a constant C (depending on d and $r(d, \varepsilon)$) such that for any channels $\nu_1, \nu_2 : M_d(\mathbb{C}) \to \mathbb{C}^{r(d,\varepsilon)}$ and for any state ψ of $\mathbb{C}^{r(d,\varepsilon)}$ we have the estimate

$$\|\varrho_i - \varrho_i'\| \le C \|\varrho_i - \varrho_i'\|_{\mu_i}.$$

In particular,

$$\|\varrho_i - \varrho_i'\| \le C \|\varrho_i - \varrho_i'\|_{\mu_i}.$$

Let us choose δ so that $\|\alpha_i - \alpha_i'\|_\varphi < \delta$ implies $\|\varrho_i - \varrho_i'\|_{\mu_i} < \varepsilon$. Since

$$\mathrm{Ent}\,(\mathcal{M}, \mathcal{S}) - \mathrm{Ent}\,(\mathcal{M}, \mathcal{S}') = \sum_{i=1}^n (I(\mu_i, \varrho_i) - I(\mu_i, \varrho_i')),$$

we have

$$|\mathrm{Ent}\,(\mathcal{M}, \mathcal{S}) - \mathrm{Ent}\,(\mathcal{M}, \mathcal{S}')| \le n\varepsilon_1$$

according to basic estimate Proposition 8.2. This yields

$$H_\varphi(\alpha_1, \alpha_2, \ldots, \alpha_n) - n\varepsilon_1 - \delta$$
$$\le \mathrm{Ent}\,(\mathcal{M}, \mathcal{S}') + n\varepsilon_1 \le H_\varphi(\alpha_1', \alpha_2', \ldots, \alpha_n') + n\varepsilon_1.$$

Interchanging the role of $\mathcal{S}$ and $\mathcal{S}'$ we obtain

$$|H_\varphi(\alpha_1, \alpha_2, \ldots, \alpha_n) - H_\varphi(\alpha_1', \alpha_2', \ldots, \alpha_n')| \leq 2n\varepsilon_1$$

and the continuity of the multiple-channel entropy is proved.   □

Now we check that the von Neumann algebra definition of the dynamical entropy fits in with the C*-algebraic definition.

Proposition 10.12 Let $\mathcal{M}$ be a von Neumann algebra with a faithful normal state φ and a unital completely positive mapping $\gamma : \mathcal{M} \to \mathcal{M}$ so that $\varphi \circ \gamma = \varphi$. If $\mathcal{A} \subset \mathcal{M}$ is a sub-C*-algebra which is strongly dense in $\mathcal{M}$ and stable under γ, then

$$h_\varphi(\gamma) = h_{\varphi_0}(\gamma_0)$$

where $\varphi_0 = \varphi|\mathcal{A}$ and $\gamma_0 = \gamma|\mathcal{A}$.

Proof. Any testing channel $\alpha_i : \mathcal{A}_0 \to \mathcal{A}$ is a testing channel for $\mathcal{M}$ and the multiple-channels $(\alpha_i, \gamma \circ \alpha_i, \ldots, \gamma^n \circ \alpha_i)$ and $(\alpha_i, \gamma_0 \circ \alpha_i, \ldots, \gamma_0 \circ \alpha_i)$ have the same abelian models (because any $P : \mathcal{A} \to \mathcal{C}$ with $\mu \circ P = \varphi_0$ extends to $\overline{P} : \mathcal{M} \to \mathcal{C}$ with $\mu \circ \overline{P} = \varphi$). So

$$H_\varphi(\alpha_i, \gamma \circ \alpha_i, \ldots, \gamma^n \circ \alpha_i) = H_{\varphi_0}(\alpha_i, \gamma_0 \circ \alpha_i, \ldots, \gamma_0 \circ \alpha_i \circ \gamma_0^n) \qquad (10.21)$$

and

$$h_\varphi(\gamma) \geq h_{\varphi_0}(\gamma_0). \qquad (10.22)$$

To prove the converse inequality we shall use the Kaplansky density theorem on the algebra $M_n(\mathbb{C}) \otimes \mathcal{M}$. Consider an arbitrary finite testing channel $\alpha : M_n(\mathbb{C}) \to \mathcal{M}$. This can be viewed as a positive element of $M_n(\mathcal{M}) = M_n(\mathbb{C}) \otimes \mathcal{M}$ (the correspondence is given by $\alpha \mapsto (\alpha(e_{kl})_{kl}$ with the usual matrix units e_{kl} of $M_n(\mathbb{C})$) and it can be approximated by a net $(\alpha_i) \subset M_n(\mathcal{A})$ so that $\alpha_i(a) \to \alpha(a)$ strongly for every $a \in M_n(\mathbb{C})$. Since $M_n(\mathbb{C})$ is of finite dimension the convergence is uniform on the unit ball and one obtains $\|\alpha_i - \alpha\|_\varphi \to 0$. Proposition 10.11 yields that

$$\lim_i h_\varphi(\gamma; \alpha_i) = h_\varphi(\gamma; \alpha).$$

However, from (10.21) $h_\varphi(\gamma; \alpha_i) = h_{\varphi_0}(\gamma_0; \alpha_i)$ and we observe that the equality must hold in (10.22).   □

Now we introduce a third type of approximating net which will be suitable both for C*-algebras and for von Neumann algebras. Let $\mathcal{A}$ be a C*-algebra with a state φ. A net $(\alpha_i : \mathcal{A}_i \to \mathcal{A})$ of unital completely positive mappings with finite dimensional algebras $(\mathcal{A}_i)_i$ is called a φ-approximating net if for each i there exists a unital completely positive mapping $\beta_i : \mathcal{A} \to \mathcal{A}_i$ such that

$$\lim_i \|\alpha_i \circ \beta_i(a) - a\|_\varphi = 0$$

for every $a \in \mathcal{A}$.

First of all any norm approximating net is a φ-approximating one. Another example is supplied by a von Neumann algebra $\mathcal{M}$ which contains an ascending net $(\mathcal{N}_i)_i$ of finite dimensional subalgebras so that $\cup_i \mathcal{N}_i$ is strongly dense in $\mathcal{M}$. Let κ_i be the embedding of $\mathcal{N}_i$ into $\mathcal{M}$. Then $(\kappa_i)_i$ is a φ-approximating net for any faithful normal state φ of $\mathcal{M}$. Indeed, for the generalized conditional expectation $E_i : \mathcal{M} \to \mathcal{M}_i$ we have $\kappa_i \circ E_i \to$ id strongly due to the martingale convergence theorem (Theorem 4.6) and convergence in the strong operator topology implies φ-norm convergence. In the light of Theorem 9.14 we have the following more general result. Let $\mathcal{M}$ be an injective von Neumann algebra. It must possess a weak* approximating net $(\alpha_i : \mathcal{A}_i \to \mathcal{M})$ of unital completely positive mappings with finite dimensional algebras $(\mathcal{A}_i)_i$. Then $(\alpha_i)_i$ is a φ-approximating net for each faithful normal state φ.

The next result is the non-abelian extension of the Kolmogorov–Sinai theorem and it is crucial in the theory of dynamical entropy. It will be referred as approximating net theorem.

Theorem 10.13 Let $\mathcal{A}$ be a C*-algebra and let $\gamma : \mathcal{A} \to \mathcal{A}$ be a completely positive unital mapping leaving the state φ of $\mathcal{A}$ invariant. If $(\alpha_i : \mathcal{A}_i \to \mathcal{A})_i$ is a φ-approximating net of finite channels, then

$$h_\varphi(\gamma) = \lim_i h_\varphi(\gamma; \alpha_i).$$

Proof. We have to show that

$$\liminf_i h_\varphi(\gamma; \alpha_i) \geq h_\varphi(\gamma; \alpha) \qquad (10.23)$$

for any finite channel $\alpha : \mathcal{A}_0 \to \mathcal{A}$. Let $\theta_i = \alpha_i \circ \beta_i \circ \alpha$ where β_i is from the definition of the φ-approximating net $(\alpha_i)_i$. Then

$$\|\theta_i(a) - \alpha(a)\|_\varphi \to 0$$

for every $a \in \mathcal{A}_0$. Since $\mathcal{A}_0$ is finite dimensional, the convergence is uniform on the unit ball and we have

$$\|\theta_i - \alpha\|_\varphi \to 0.$$

Now Proposition 10.11 yields that

$$\lim_i |h_\varphi(\gamma; \theta_i) - h_\varphi(\gamma; \alpha)| = 0.$$

From (i) of Proposition 10.1 we see that $h_\varphi(\gamma; \theta_i) \leq h_\varphi(\gamma; \alpha_i)$ and so (10.23) must hold. $\qquad \square$

Corollary 10.14 Let a von Neumann algebra $\mathcal{M}$ contain an increasing net $(\mathcal{N}_i)_i$ of finite dimensional subalgebras so that $\cup_i \mathcal{N}_i$ is strongly dense in $\mathcal{M}$. Then

$$h_\varphi(\gamma) = \lim_i H_\varphi(\gamma; \mathcal{N}_i) \qquad (10.24)$$

for every normal state φ and for every completely positive unital mapping $\gamma : \mathcal{M} \to \mathcal{M}$ so that $\varphi \circ \gamma = \varphi$.

By means of this consequence of the approximation net theorem formula (10.16) may be extended to more general independent shifts. For example, the local algebras $\mathcal{B}_j$ can be $B(\mathcal{H})$ for some Hilbert space $\mathcal{H}$.

Let us consider now the dynamical entropy of an automorphism of a commutative C*-algebra. Let T be a homeomorphism of a compact metric space X. Then $\gamma : f \mapsto f \circ T$ is an automorphism of the commutative C*-algebra $C(X)$ of all continuous functions. When T preserves the Borel probability measure μ, the automorphism γ will leave invariant the state $\varphi : f \mapsto \int f \, d\mu$. The coincidence of the Kolmogorov–Sinai entropy of the transformation T with the dynamical entropy of the automorphism γ can be seen by reference to the von Neumann algebra L^∞. Choose in L^∞ an increasing sequence $(\mathcal{N}_i)_i$ of finite dimensional subalgebras generating this von Neumann algebra. $\mathcal{N}_i$ corresponds to a finite measurable partition ξ_i. The entropy

$$H_\varphi(\mathcal{N}_i, \gamma(\mathcal{N}_i), \dots, \gamma^k(\mathcal{N}_i))$$

is nothing else but the entropy of the partition $\xi_i \vee T^{-1}\xi_i \vee \dots \vee T^{-k}\xi_i$. Hence

$$H_\varphi(\gamma; \mathcal{N}_i) = H_\mu^{K-S}(\xi_i).$$

Now let $i \to \infty$. The previous corollary tells that the left hand side tends to the dynamical entropy $h_\varphi(\gamma)$ of the automorphism γ of the von Neumann algebra L^∞ and the Kolmogorov–Sinai theorem implies that the right hand side goes to the entropy of the measure preserving transformation T. Thanks to Proposition 10.12 the former dynamical entropy is the same as that of this automorphism restricted to $C(X)$. The dynamical entropy developed in this chapter is a considerable extension of the Kolmogorov–Sinai entropy of measure preserving transformations.

Theorem 10.15 Let $\mathcal{A}$ be a C*-algebra with a state φ. Assume that the pair $(\mathcal{A}, \varphi)$ admits a φ-approximating net and let $\gamma : \mathcal{A} \to \mathcal{A}$ be a completely positive unital mapping so that $\varphi \circ \gamma = \varphi$. Then $h_\varphi(\gamma^k) = k\, h_\varphi(\gamma)$ for every $k \in \mathbb{N}$.

Proof. The inequality $h_\varphi(\gamma^k) \leq k\, h_\varphi(\gamma)$ is almost obvious and does not need the hypothesis on the pair $(\mathcal{A}, \varphi)$. Given $\varepsilon > 0$ one can find a finite channel $\alpha : \mathcal{A}_0 \to \mathcal{A}$ so that for large n

$$h_\varphi(\gamma^k) \leq \frac{1}{n} H_\varphi(\alpha, \gamma^k \circ \alpha, \dots, \gamma^{(n-1)k} \circ \alpha) + \varepsilon$$

holds. The right hand side is majorized by

$$k\frac{1}{nk}H_\varphi(\alpha,\gamma\circ\alpha,\gamma^2\circ\alpha,\ldots,\gamma^{nk-1}\circ\alpha)+\varepsilon$$

and this can not exceed $k\,h_\varphi(\gamma)+\varepsilon$. So $h_\varphi(\gamma^k)\le kh_\varphi(\gamma)$ has been obtained.

The idea of the proof of the converse inequality is the following. Under the hypothesis any multiple finite channel $(\alpha_1,\alpha_2,\ldots,\alpha_k)$ may be approximately refined into a single finite channel δ. (In measure theory the analogous statement is totally trivial even without approximation; the join of finitely many finite partitions is again a finite partition.)

Let $(\alpha_i)_i$ be a φ-approximating net with a net $(\beta_i)_i$ of mappings in the reversed direction. We choose a finite channel α so that

$$h_\varphi(\gamma)\le\frac{1}{nk}H_\varphi(\alpha,\gamma\circ\alpha,\ldots,\gamma^{nk-1}\circ\alpha)+\varepsilon \tag{10.25}$$

for n big enough. Since

$$\|\alpha_i\circ\beta_i\circ\gamma^j\circ\alpha(a)-\gamma^j\circ\alpha(a)\|_\varphi\to0$$

holds for every $0\le j\le k-1$, we use that the domain of α is of finite dimension and observe that

$$\|\alpha_i\circ\beta_i\circ\gamma^j\circ\alpha-\gamma^j\circ\alpha\|_\varphi\le\varepsilon$$

for large i. So

$$\|\gamma^m\circ\alpha_i\circ\beta_i\circ\gamma^j\circ\alpha-\gamma^{m+j}\circ\alpha\|_\varphi\le\varepsilon \tag{10.26}$$

for every $m\in\mathbb{N}$, $0\le j\le k-1$ and for large i. Now we replace each channel in (10.25) by an approximant appearing in (10.26) and estimate the error of the entropy with the help of Proposition 10.11.

$$\begin{aligned}
&H_\varphi(\alpha,\gamma\circ\alpha,\ldots,\gamma^{nk-1}\circ\alpha)\\
&\le H_\varphi(\alpha_i\circ\beta_i\circ\alpha,\alpha_i\circ\beta_i\circ\gamma\circ\alpha,\ldots,\alpha_i\circ\beta_i\circ\gamma^{k-1}\circ\alpha,\\
&\qquad\gamma^k\circ\alpha_i\circ\beta_i\circ\alpha,\gamma^k\circ\alpha_i\circ\beta_i\circ\gamma\circ\alpha,\ldots,\gamma^k\circ\alpha_i\circ\beta_i\circ\gamma^{k-1}\circ\alpha,\\
&\qquad\ldots,\\
&\qquad\gamma^{(n-1)k}\circ\alpha_i\circ\beta_i\circ\alpha,\ldots,\gamma^{(n-1)k}\circ\alpha_i\circ\beta_i\circ\gamma^{k-1}\circ\alpha)+nk\varepsilon\\
&\le H_\varphi(\alpha_i,\alpha_i,\ldots,\alpha_i,\\
&\qquad\gamma^k\circ\alpha_i,\gamma^k\circ\alpha_i,\ldots,\gamma^k\circ\alpha_i,\\
&\qquad\ldots,\\
&\qquad\gamma^{(n-1)k}\circ\alpha_i,\gamma^{(n-1)k}\circ\alpha_i,\ldots,\gamma^{(n-1)k}\circ\alpha_i)+nk\varepsilon\\
&=H_\varphi(\alpha_i,\gamma^k\circ\alpha_i,\gamma^{2k}\circ\alpha_i,\ldots,\gamma^{(n-1)k}\circ\alpha_i)+nk\varepsilon\,.
\end{aligned}$$

Here the second inequality was obtained by the application of (i) in Proposition 10.1, that is, we have deleted parts of the long compositions. Then all completely positive mappings appearing with some multiplicity and Proposition 10.2 allowed to keep one copy of each of them. Summarizing the above chain of inequalities and taking the limit $n\to\infty$ we have

$$h_\varphi(\gamma) \le \frac{1}{k} h_\varphi(\gamma^k) + 2\varepsilon$$

which makes the proof complete since ε is arbitrary small.   □

It is not an essential point that the result $k\,h_\varphi(\gamma) = h_\varphi(\gamma^k)$ holds only under some hypothesis. On the one hand, in the important examples a φ-approximating net exist and the hypothesis is satisfied, on the other hand it would be easy to modify the definition of the dynamical entropy in order to get rid of the restrictive condition. If we take

$$H_\varphi(\gamma; \alpha_1, \alpha_2, \ldots, \alpha_k)$$
$$= \lim_{n \to \infty} \frac{1}{n} H_\varphi(\alpha_1, \alpha_2, \ldots, \alpha_k,$$
$$\gamma \circ \alpha_1, \gamma \circ \alpha_2, \ldots, \gamma \circ \alpha_k,$$
$$\ldots$$
$$\gamma^{n-1} \circ \alpha_1, \gamma^{n-1} \circ \alpha_2, \ldots, \gamma^{n-1} \circ \alpha_k)$$

then the supremum $\bar{h}_\varphi(\gamma)$ of $H_\varphi(\gamma; \alpha_1, \alpha_2, \ldots, \alpha_k)$ over all multiple finite channels $(\alpha_1, \alpha_2, \ldots, \alpha_k)$ is a quantity which makes the relation $k\,\bar{h}_\varphi(\gamma) = \bar{h}_\varphi(\gamma^k)$ true and under the existence of a φ-approximating net $h_\varphi(\gamma) = \bar{h}_\varphi(\gamma)$.

Now we prove a version of the previous theorem for a one-parameter-semigroup of completely positive mappings.

Proposition 10.16 Let $\mathcal{A}$ be a C*-algebra with a state φ. Assume that the pair $(\mathcal{A}, \varphi)$ admits a φ-approximating net and let $\gamma_t : \mathcal{A} \to \mathcal{A}$ be a semigroup of completely positive unital mappings such that $\varphi \circ \gamma_t = \varphi$ ($t \in \mathbb{R}^+$) and $t \mapsto \gamma_t(a)$ is continuous in the φ-norm for every $a \in \mathcal{A}$. Then $h_\varphi(\gamma_t) = t\,h_\varphi(\gamma_1)$ for every $t \in \mathbb{R}^+$.

Proof. Let $(\alpha_i : \mathcal{A}_i \to \mathcal{A})_i$ be a φ-approximating net. We fix positive real numbers t, s and for a given $\varepsilon > 0$ we choose i so large that

$$|h_\varphi(\gamma_t) - h_\varphi(\gamma_t; \alpha_i)| \le \varepsilon$$

which is possible due to the approximating net theorem. Choose ε_0 so that $\varepsilon_0 \log d + \eta(\varepsilon) \le \varepsilon$ for the tracial dimension d of $\mathcal{A}_i$. Using the compactness of the unit ball of $\mathcal{A}_i$ we find $\delta > 0$ so that

$$\|\gamma_{t+x} \circ \alpha_i - \gamma_t \circ \alpha_i\|_\varphi \le \varepsilon_0 \quad \text{for} \quad 0 \le x \le \delta.$$

This implies that

$$\|\gamma_u \circ \alpha_i - \gamma_v \circ \alpha_i\|_\varphi \le \varepsilon_0 \quad \text{for} \quad t \le u, v \quad \text{and} \quad |u - v| \le \delta. \tag{10.27}$$

Now choose $m \in \mathbb{N}$ big so that $s/m < \delta$ and consider the multiple-channel entropy

$$H_\varphi(\alpha_i, \gamma_t \circ \alpha_i, \gamma_{2t} \circ \alpha_i, \ldots, \gamma_{nt} \circ \alpha_i).$$

To each time subscript kt we can find some integer $l(k)$ such that $|kt - l(k)s/m| \le \delta$ and $t \le l(k)s/m \le nt$. So (10.27) gives $\|\gamma_{kt} \circ \alpha_i - \gamma_{l(k)s/m} \circ \alpha_i\|_\varphi \le \varepsilon_0$ and

$$H_\varphi(\alpha_i, \gamma_t \circ \alpha_i, \gamma_{2t} \circ \alpha_i, \ldots, \gamma_{nt} \circ \alpha_i)$$
$$\le n\varepsilon + H_\varphi(\alpha_i, \gamma_{l(1)s/m} \circ \alpha_i, \gamma_{l(2)s/m} \circ \alpha_i, \ldots, \gamma_{l(n)s/m} \circ \alpha_i)$$

follows from Proposition 10.7. The multiple-channel entropy on the right hand side is majorized if we include $\gamma_{js/m} \circ \alpha_i$ for all $0 \le j \le l(n)$ (due to the monotonicity property, (iv) in Proposition 10.1). Hence letting $n \to \infty$ we obtain

$$h_\varphi(\gamma_t; \alpha_i) \le \varepsilon + \frac{mt}{s} h_\varphi(\gamma_{s/m}; \alpha_i) \le \varepsilon + \frac{mt}{s} h_\varphi(\gamma_{s/m}).$$

Since ε can be arbitrarily small we have

$$h_\varphi(\gamma_t) \le \frac{tm}{s} h_\varphi(\gamma_{s/m}) = \frac{t}{s} h_\varphi(\gamma_s)$$

which implies easily our statement. □

Corollary 10.17 Let $\mathcal{A}$ be a nuclear C*-algebra with a state φ and let $\gamma_t : \mathcal{A} \to \mathcal{A}$ be a semigroup of completely positive unital mappings such that $\varphi \circ \gamma_t = \varphi$ $(t \in \mathbb{R}^+)$ and $t \mapsto \gamma_t(a)$ is norm continuous for every $a \in \mathcal{A}$. Then $h_\varphi(\gamma_t) = t\, h_\varphi(\gamma_1)$ for every $t \in \mathbb{R}^+$.

Corollary 10.18 Let $\mathcal{M}$ be an injective von Neumann algebra with a faithful normal state φ and let $\gamma_t : \mathcal{A} \to \mathcal{A}$ be a semigroup of completely positive unital mappings such that $\varphi \circ \gamma_t = \varphi$ $(t \in \mathbb{R}^+)$ and $t \mapsto \gamma_t(a)$ is continuous in the strong operator topology for every $a \in \mathcal{A}$. Then $h_\varphi(\gamma_t) = t\, h_\varphi(\gamma_1)$ for every $t \in \mathbb{R}^+$.

These results show that the dynamical entropy on a one-parameter semigroup is a continuous function of the parameter but it is not at all continuous on the completely positive mappings preserving a given state. A simple counterexample can be given in a finite factor with a faithful normal trace τ. Let $\alpha(\,.\,) = U^* . U$ be an inner automorphism. It follows from the spectral theorem that one can approximate U by a sequence $(U_n)_n$ of periodic unitaries in operator norm. For the automorphism $\alpha_n(\,.\,) = U_n^* . U_n$ we know that $h_\tau(\alpha_n)$ must be 0 or ∞. Consequently, h_τ can not be continuous at any inner automorphism α with non-vanishing finite entropy.

Notes and Remarks. The concept of entropy in measure theoretic ergodic theory was introduced by Kolmogorov in 1958 in order to prove that certain shifts are not conjugate. A year later Sinai improved the notion and introduced the conjugation invariant quantity used today. The *entropy of* *a measure preserving transformation* is treated in many books, for example [Walters 1982]. For the operator algebras the concept was first worked out for the case of an invariant tracial state in [Connes and Størmer 1975]. This paper contained already the idea that in contradiction to the measure theoretic definition the join of two subalgebras considered as analogues of partitions should be regarded as a pair of subalgebras rather than taking the generated subalgebra or something like that. This theory was built for finite von Neumann algebras and relied intensively on the existence of trace preserving conditional expectations. In the short announcement [Connes 1985] the emphasis was put on the relative entropy and the concept of the dynamical entropy for a non-tracial state of an AF C*-algebra and for an automorhism was first sketched. The general theory with full proofs was given in [Connes, Narnhofer and Thirring 1987a], this paper is the basis of the chapter. Here the use of finite dimensional subalgebras was replaced by completely positive mappings of finite dimensional algebras and a theory was made for nuclear C*-algebras and hyperfinite von Neumann algebras.

Theorem 10.13 is an operator algebraic analogue of the Kolmogorov–Sinai theorem, its first form appeared in [Connes and Størmer 1975]. In this form the theorem is a slight generalization of the result in [Connes, Narnhofer and Thirring 1987a]. In fact, Theorem 10.13 restricted to the measure preserving setting gives much less than the usual Kolmogorov–Sinai theorem which tells that the Kolmogorov–Sinai invariant equals the entropy of a particular generating partition. In most textbooks the proof of the Kolmogorov–Sinai theorem is given in a short combinatorial form. The proof of Theorem 10.13 is closer to the book [Shields 1973] where a continuity lemma is stated for $h_\varphi(\alpha; \varphi)$ with a different notation.

The equivalence of nuclearity with the existence of a norm approximating net (for the identity) was proved in [Choi and Effros 1978]. A similar characterization holds for injective von Neumann algebras in terms of an approximating net with respect to the weak* topology. We refer to the reviews [Effros 1978] and [Lance 1982]. It follows from the above mentioned result that a von Neumann algebra admitting a strong approximating net must be injective. The existence of a strong approximating net was pointed out in [Haagerup 1985], see also [Petz 1994a].

The concept of dynamical entropy has several different definitions in the literature. Let $\mathcal{M}$ be a von Neumann algebra with an automorphism α and with an invariant normal state φ. It was argued in [Emch 1974] that the mean entropy of the time development should be computed for only those finite dimensional subalgebras which "do not interact" with the whole algebra. Formally this requirement means that the subalgebra must be in the centralizer of the state. The disadvantage of this definition is that the set of finite dimensional subalgebras used in the supremum do depend on the invariant state itself. The corresponding dynamical entropy is known for a few examples only and the theory is rather limited. For a discussion of the various possible dynamical entropies, see [Emch 1992].

Another proposal for dynamical entropy is due to [Lindblad 1979a], see also [Lindblad 1988]. Here the finite partition of measure theory is replaced by an *operational partition of unity* leaving the given state invariant. (This invariance requirement is common with Emch's proposal.) The join of two operational partitions of unity is defined in a nice and natural way. The problem is again in finding enough testing objects, i.e. operational partitions of unity. Following Lindblad an entropy may be defined for a process. The Lindblad entropy of an independent shift over non-full matrix algebras is presumably different from the entropy introduced in the chapter. It is noteworthy that while information on the system is measured by comparison with abelian models in the concept of the Connes–Størmer–Narnhofer–Thirring entropy (so-called *CNT entropy*), comparison with the product of matrix algebras is the leading idea in Lindblad's proposal. An equivalent definition is given for the dynamical entropy in [Sauvageot and Thouvenot 1992]. The non-commutative dynamical system $(\mathcal{A}, \gamma, \varphi)$ is coupled to commutative ones $(\mathcal{C}, \beta, \mu)$. The coupling is a state ω on $\mathcal{A} \otimes \mathcal{C}$ which is invariant under the product action $\gamma \otimes \beta$. To each coupling an entropy is constructed in terms of classical conditional entropy and non-commutative mutual entropy. The dynamical entropy is reached as the supremum of the entropy of all possible couplings.

In [Størmer and Voiculescu 1990] the dynamical entropy of *Bogoliubov automorphisms* of the CAR algebra is given. In [Størmer 1992] the shift of the von Neumann group algebra associated to a *free group* generated by $\mathbb{Z}$ is found to have vanishing entropy with respect to the tracial state.

[Voiculescu 1991] contains a further proposal for entropy which is not an extension of the common Kolmogorov–Sinai invariant. [Hudetz 1993] is an attempt at the topological version of the dynamical entropy.

Comments. The paper [Narnhofer, Størmer and Thirring 1995] gives an example such that $h_\tau(\sigma) = 0$, but $h_{\tau \otimes \tau}(\sigma \otimes \sigma) = \log 2$. Hence the *additivity* of the dynamical entropy h does not hold, the subadditivity is obvious from the definition.

More results are available about the dynamical entropy $h_\tau(\alpha)$ if τ is a faithful normal trace on a von Neumann algebra $\mathcal{M}$ and α is a τ-preserving endomorphism. Assume that A_n is an inreasing sequence of finite dimensional subalgebras whose union is dense in $\mathcal{M}$. It was obtained in [Neshveyev and Størmer 2001] such that

$$h_\tau(\gamma) = \tfrac{1}{2} H_\tau(\mathcal{M}|\alpha(\mathcal{M})) + \tfrac{1}{2} \lim_{n \to \infty} H_\tau(Z(A_n))$$

under some regulaity condition on the sequence A_n. In the above formula the first term is the conditional entropy (7.49) and $Z(.)$ denotes the center.

In the paper [Alicki and Fannes 1994] another concept of quantum dynamical entropy was developed by means of operational partition of unities.

Let $(\mathcal{A}, \gamma, \varphi)$ be a non-commutative dynamical system and $\mathcal{A}_0$ be a dense sub-algebra of $\mathcal{A}$. Given an operational partition of unity $\mathcal{V} = V_1, V_2, \ldots, V_k \in \mathcal{A}_0$ and $n \in \mathbb{N}$, a state ψ_n can be defined on the n-fold tensor product $M_k(\mathbb{C}) \otimes \ldots \otimes M_k(\mathbb{C})$ as

$$\psi_n \big(E_{i_1 j_1} \otimes E_{i_2 j_2} \ldots \otimes E_{i_n j_n} \big) = \varphi \big(V_{i_1}^* V_{i_2}^* \ldots V_{i_n}^* V_{j_n} \ldots V_{j_2} V_{j_1} \big).$$

The states ψ_n are compatible and give a state ψ on the infinite tensor product $M_k(\mathbb{C}) \otimes M_k(\mathbb{C}) \otimes \ldots$ but this ψ is not translationally invariant in general. Nevertheless,

$$h_\varphi(\gamma, \mathcal{V}) := \limsup_{n \to \infty} \frac{1}{n} S(\psi_n)$$

might be considered. (When $\mathcal{V}$ is φ-invariant, ψ is translationally invariant, and the limit exists.) The entropy $h_\varphi^{ALF}(\gamma)$ is the supremum of $h_\varphi(\gamma, \mathcal{V})$'s over all $\mathcal{V}$ in $\mathcal{A}_0$. The spin half-chain and the right-shift constitute a symbolic model of the original quantum dynamical system. If $\mathcal{A}_0$ is a proper subalgebra, then the entropy $h_\varphi^{ALF}(\gamma)$ is not necessarily conjugate invariant, on the other hand, the use of a proper subalgebra makes easier the computation of $h_\varphi^{ALF}(\gamma)$.

If $\mathcal{A} = \ldots M_n(\mathbb{C}) \otimes M_n(\mathbb{C}) \ldots$ is the quantum spin algebra, γ is the shift, φ is a shift-invariant state, and $\mathcal{A}$ consists of the local elements, then $h_\varphi^{ALF}(\gamma)$ equals to the *mean entropy* of φ plus $\log n$. (Actually, $\mathcal{A}_0$ could be chosen slightly bigger, see [Fannes and Tuyls 2003].)

11 Stationary Processes

In this chapter a conceptual basis for processes will be formulated in the algebraic framework of the book. The obvious orientation is coming from probability theory where classical stochastic processes have been studied for many years. The quantum case we are interested in shows several new features and it seems impossible to incorporate all of them in a comprehensive general treatment. The basic notions should be somewhat flexible in order to adjust them to the actual model under study. After introducing a few general notions we shall restrict ourselves to a bulk of examples and no attempt is made to build a complete theory of quantum stationary processes.

A pair $(\mathcal{A}, \varphi)$ consisting of an algebra $\mathcal{A}$ and one of its state φ is called an *algebraic probability space*. An *algebraic random variable* is an embedding $j : \mathcal{B} \to \mathcal{A}$ of an algebra $\mathcal{B}$ into $\mathcal{A}$. The state $\varphi \circ j$ of $\mathcal{B}$ is called the *distribution* of the random variable j. Only random variables with identical domains can be compared. By an algebraic stochastic process we mean a family of algebraic random variables $j_t : \mathcal{B} \to \mathcal{A}$ indexed by a set T.

The simplest example of an algebraic stochastic process may be defined on a spin lattice system over the integers. To each $i \in \mathbb{Z}$ a copy $\mathcal{A}_i$ of the matrix algebra $M_m(\mathbb{C})$ is associated and $\mathcal{A}$ is the infinite tensor product $\otimes_i \mathcal{A}_i$. In this case the algebraic random variable $\mathcal{A}_i$ is the spin at the lattice site $i \in \mathbb{Z}$. (Note that $\mathcal{A}_i$ is not an observable of the spin system but it is determined by the different direction spin observables at the site i.) The quantum spin lattice system gives rise to a particularly simple algebraic stochastic process because $j_i(a)$ and $j_k(b)$ commute for any different $i, k \in \mathbb{Z}$ and for every $a, b \in M_m(\mathbb{C})$.

Let $j_t : \mathcal{B} \to \mathcal{A}$ be an algebraic stochastic process indexed by a set T. The correlations among the different random variables j_t are expressed by the following functionals which are called *correlation kernels*.

$$\mathbf{k}(a_1, a_2, \ldots, a_n \,;\, b_1, b_2, \ldots, b_n \,;\, t_1, t_2, \ldots, t_n)$$
$$\equiv \varphi\big(j_{t_1}(a_1)^* j_{t_2}(a_2)^* \ldots j_{t_n}(a_n)^* j_{t_n}(b_n) \ldots j_{t_2}(b_2) j_{t_1}(b_1)\big). \tag{11.1}$$

Often we choose a set of integers $\mathbb{Z}$ to index the processes. (Continuous time with $t \in \mathbb{R}$ (or $t \in \mathbb{R}^+$) would be another natural choice but for our purposes the discretized "time" will do.) Let $j_t : \mathcal{B} \to \mathcal{A}$ be an algebraic stochastic process indexed by $\mathbb{Z}$. This process is called *stationary* if its correlation kernels are translation invariant, that is

$$\mathbf{k}(a_1, a_2, \ldots, a_n \,; b_1, b_2, \ldots, b_n \,; t_1 + s, t_2 + s, \ldots, t_n + s)$$

is independent of s.

An algebraic stochastic process $j_t : \mathcal{B} \to \mathcal{A}$ determines a "localization" on the algebra $\mathcal{A}$. Similarly to spin lattice systems for any subset $J \subset \mathbb{Z}$ we set $\mathcal{A}_J$ for the subalgebra spanned by $\cup\{j_t(\mathcal{B}) : t \in J\}$. Many important qualitative properties of the process do not depend on the random variables j_t but only on the localization $\{\mathcal{A}_J : J \subset \mathbb{Z}\}$.

The heart of probability theory is the notion of *independence*. Let $j_1 : \mathcal{B} \to \mathcal{A}$ and $j_2 : \mathcal{B} \to \mathcal{A}$ be algebraic random variables over the probability space $(\mathcal{A}, \varphi)$. The independence of j_1 and j_2 must consist of two different kinds of conditions. On the one hand, the mutual position of the subalgebras $j_1(\mathcal{B})$ and $j_2(\mathcal{B})$ should be somewhat specific, and, on the other hand, the fixed state φ of $\mathcal{A}$ should fit in. To make a formal definition of independence we have to impose more on the randon variables. Namely, we require the existence of a conditional expectation onto their ranges.

A stationary algebraic stochastic process $j_t : \mathcal{B} \to \mathcal{A}$ $(t \in \mathbb{Z})$ on the probability space $(\mathcal{A}, \varphi)$ is called *Bernoulli process* if the following conditions are fulfilled.

(i) For every $J \subset \mathbb{Z}$ there exists a conditional expectation $E_J : \mathcal{A} \to \mathcal{A}_J$ which leaves the state φ invariant.

(ii) For any disjoint subsets J_1 and J_2 of $\mathbb{Z}$ and for every $a_1 \in \mathcal{A}_{J_1}$ and $a_2 \in \mathcal{A}_{J_2}$ the property $\varphi(a_1 a_2) = \varphi(a_1)\varphi(a_2)$ holds.

These conditions imply that the factorization of the state φ follows for arbitrary finite products as well.

$$\varphi(j_1(a_1)j_2(a_2)\ldots j_n(a_n)) = \varphi(j_1(a_1))\varphi(j_2(a_2))\ldots\varphi(j_n(a_n)). \tag{11.2}$$

A product state of the quantum spin lattice system provides a simple example for a Bernoulli process.

Sometimes an algebraic stochastic process $(\mathcal{A}, j_t, \varphi)$ is given by an algebra $\mathcal{A}$, its subalgebra $\mathcal{A}_0$, an automorphism γ of $\mathcal{A}$ and a state φ. In this case j_0 is the embedding of $\mathcal{A}_0$ and the other embeddings are defined as $j_t = \gamma^t \circ j_0$. The latter formula motivates us to call γ time shift. Now stationarity of the process is the invariance of φ under γ. For another state ω the *mean relative entropy* is defined as follows.

$$S_M(\omega, \varphi) = \limsup_n \frac{1}{n} S(\omega|\mathcal{A}_{[1,n]}, \varphi|\mathcal{A}_{[1,n]}). \tag{11.3}$$

We shall see that in (11.3) the limit exists for any invariant state ω when a stationary Bernoulli process is considered.

Let us recall that a state φ of a C*-algebra $\mathcal{A}$ is called *separating* if in the corresponding GNS-Hilbert space the cyclic vector Φ is separating for the generated von Neumann algebra. (It is known that a KMS-state with respect to a group of automorphisms is always a separating one.)

Now we state an abstract *strong superadditivity* for relative entropy.

Proposition 11.1 Let $\mathcal{A}$ be a C*-algebra with a separating state φ and sub-algebras $\mathcal{A}_{12}$ and $\mathcal{A}_{23}$. Assume that the φ-preserving conditional expectation E_{12} of $\mathcal{A}$ onto $\mathcal{A}_{12}$ exists. If $E_{12}(\mathcal{A}_{23}) = \mathcal{A}_2$ is a C*-subalgebra of $\mathcal{A}$ then for every state ω we have

$$S(\omega, \varphi) + S(\omega_2, \varphi_2) \geq S(\omega_{12}, \varphi_{12}) + S(\omega_{23}, \varphi_{23})$$

where the subscripts denote the restriction of states to the corresponding subalgebra.

Proof. This result is a consequence of Theorem 5.15. Observe that all the relative entropies may be computed in the von Neumann algebras obtained by the GNS-construction from φ. Performing this GNS-construction E_{12} and E_{23} determine faithful normal conditional expectations and application of Theorem 5.15 is allowed. Therefore we have

$$S(\omega, \varphi) = S(\omega_{12}, \varphi_{12}) + S(\omega, \omega_{12} \circ E_{12}) \tag{11.4}$$

and

$$S(\omega_2, \varphi_2) + S(\omega_{23}, \omega_2 \circ F) = S(\omega_{23}, \varphi_{23}) \tag{11.5}$$

where $F = E_{12}|\mathcal{A}_{23}$ is the conditional expectation of $\mathcal{A}_{23}$ onto $\mathcal{A}_2$ (which preserves φ_{23}). Since

$$S(\omega, \omega_{12} \circ E_{12}) \geq S(\omega_{23}, \omega_2 \circ F)$$

due to the monotonicity, we obtain the stated inequality by addition of (11.4) and (11.5).   □

Let $(\mathcal{A}, \mathcal{A}_0, \gamma, \varphi)$ be a stationary Bernoulli process and let ω be another state of $\mathcal{A}$. If φ is separating then we may apply Proposition 11.1 to the subalgebras $\mathcal{A}_{[1,n]}$ and $\mathcal{A}_{[n+1,n+m]}$ in place of $\mathcal{A}_{12}$ and $\mathcal{A}_{23}$, respectively. Since the intersection $\mathcal{A}_{[1,n]} \cap \mathcal{A}_{[n+1,n+m]}$ is trivial, the proposition tells that

$$S(\omega|\mathcal{A}_{[1,n+m]}, \varphi|\mathcal{A}_{[1,n+m]})$$
$$\geq S(\omega|\mathcal{A}_{[1,n]}, \varphi|\mathcal{A}_{[1,n]}) + S(\omega|\mathcal{A}_{[n+1,n+m]}, \varphi|\mathcal{A}_{[n+1,n+m]})$$

which is equivalent to

$$S(\omega|\mathcal{A}_{[1,n+m]}, \varphi|\mathcal{A}_{[1,n+m]}) \geq S(\omega|\mathcal{A}_{[1,n]}, \varphi|\mathcal{A}_{[1,n]}) + S(\omega|\mathcal{A}_{[1,m]}, \varphi|\mathcal{A}_{[1,m]})$$

when ω is assumed to be γ-invariant. Hence the sequence $S(\omega|\mathcal{A}_{[1,n]}, \varphi|\mathcal{A}_{[1,n]})$ is superadditive and the limsup in (11.3) is in fact a limit (cf. Lemma 10.4).

After the independent Bernoulli process we discuss Markov processes. A process $(\mathcal{A}, \mathcal{A}_0, \gamma, \varphi)$ is called a stationary *Markov process* (with conditional expectations) if

(i) $\varphi \circ \gamma = \varphi$,
(ii) the φ-preserving conditional expectation E_J of $\mathcal{A}$ onto $\mathcal{A}_J$ exists for every interval $J = [m, n]$ $(m \leq n)$,
(iii) $E_{[k,m]}(\mathcal{A}_{[m,n]}) = \mathcal{A}_m$ for every $k < m < n$.

Condition (iii) is the *Markov property* which possesses the usual interpretation: The future (events after the time m) does not depend on the past (events before the time m) but only on the present (events localized at time m). This property can be stated in a number of equivalent ways. Consider the following diagram

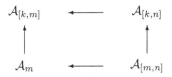

where the vertical arrows stand for embeddings and the horizontal ones for conditional expectations (preserving the state φ). Condition (iii) is equivalent to the commutativity of the above diagram. This means that the conditional expectation $E_{[k,m]}$ restricted to $\mathcal{A}_{[m,n]}$ yields the conditional expectation of $\mathcal{A}_{[m,n]}$ onto $\mathcal{A}_m$. Commutative diagrams which are built from two horizontal embeddings and two vertical conditional expectations like the one above are frequently called a *commuting squares* in the theory of operator algebras. A commuting square may be characterized in several equivalent ways. Here we mention only the *conditional independence* of future and past. Formally,

$$E_m(ab) = E_m(a)E_m(b) \quad \text{for any} \quad a \in \mathcal{A}_{[k,m]}, \ b \in \mathcal{A}_{[m,n]}. \tag{11.6}$$

Proposition 11.2 If $(\mathcal{A}, \mathcal{A}_0, \gamma, \varphi)$ is a stationary Markov process with a separating state φ and if ω is an arbitrary γ-invariant state then

$$S_M(\omega, \varphi) = \lim_n \frac{1}{n} S(\omega|\mathcal{A}_{[1,n]}, \varphi|\mathcal{A}_{[1,n]}).$$

Moreover, $\omega \mapsto S_M(\omega, \varphi)$ is an affine, weakly lower semicontinuous functional on the γ-invariant states.

Proof. By Proposition 11.1 we have

$$S(\omega|\mathcal{A}_{[1,n+k]}, \varphi|\mathcal{A}_{[1,n+k]}) + S(\omega|\mathcal{A}_{\{n\}}, \varphi|\mathcal{A}_{\{n\}})$$
$$\geq S(\omega|\mathcal{A}_{[1,n]}, \varphi|\mathcal{A}_{[1,n]}) + S(\omega|\mathcal{A}_{[n,n+k]}, \varphi|\mathcal{A}_{[n,n+k]}).$$

Due to the γ-invariance of φ and ω this may be written

$$S_{n+k} - S_0 \geq (S_n - S_0) + (S_k - S_0).$$

Therefore,

$$\lim_n \frac{1}{n} S_n = \sup \left\{ \frac{1}{n}(S_n - S_0) : n \in \mathbb{N} \right\}. \tag{11.7}$$

The affine property of $\omega \mapsto S_M(\omega, \varphi)$ is a consequence of Proposition 5.24. To show its lower semicontinuity note that according to the conditional expectation property

$$S_n - S_0 = S(\omega|\mathcal{A}_{[1,n]}, \omega \circ E|\mathcal{A}_{[0,n]})$$

where E is the φ-preserving conditional expectation from $\mathcal{A}$ onto $\mathcal{A}_0$, and we may refer to the weak lower semicontinuity of the relative entropy.   □

Stationary Markov processes with conditional expectations appear in connection with *dilations*. Let $\varphi_0 \otimes \psi$ be a state of the C*-algebra $\mathcal{A}_0 \otimes \mathcal{C}_0$ which is invariant under an automorphism β. Consider the infinite product

$$\mathcal{A}_0 \otimes_{j \in \mathbb{Z}} \mathcal{C}_j \equiv \mathcal{A} \tag{11.8}$$

where the indexed algebras $\mathcal{C}_j$ are all isomorphic to $\mathcal{C}_0$. On the algebra $\mathcal{A}$ we construct a *Markov process*. Its state φ is chosen to be a product state, the product of φ_0 on $\mathcal{A}_0$ and ψ on each factor $\mathcal{C}_j$. So $(\mathcal{A}, \mathcal{A}_0, \gamma, \varphi)$ becomes a process as soon as the automorphism γ is defined. Let $\tilde{\beta}$ be the trivial extension of β to $\mathcal{A}$, that is, it acts identically on the factors $\mathcal{C}_j$. Let the automorphism $\tilde{\alpha}$ be defined as the right shift on the factors $\mathcal{C}_j$ and the identity on $\mathcal{A}_0$. Finally, set

$$\gamma = \tilde{\beta} \circ \tilde{\alpha}. \tag{11.9}$$

Since both $\tilde{\beta}$ and $\tilde{\alpha}$ leave the product state φ invariant, we have $\varphi \circ \gamma = \varphi$. It is easy to see that

$$\mathcal{A}_{[0,n+1]} \equiv \mathcal{A}_0 \otimes \mathcal{C}_0 \otimes \mathcal{C}_1 \otimes \ldots \otimes \mathcal{C}_n \tag{11.10}$$

and there exists a φ-preserving conditional expectation onto every subalgebra $\mathcal{A}_{[0,n+1]}$. It follows from the γ-invariance that E_J exists for any interval $J \subset \mathbb{Z}$. A typical element of $\mathcal{A}_{[0,\infty]}$ is of the form $a_0 \otimes c_0 \otimes c_1 \otimes \ldots \otimes c_m$ and the conditional expectation $E_{[n,0]}$ sends it into $a_0 \prod_{i=0}^{m} \psi(c_i)$. Hence the Markov property $E_{[n,0]}(\mathcal{A}_{[0,\infty]}) = \mathcal{A}_0$ holds, indeed. The interesting feature of this example is the contradiction of the simple product structure of the algebra $\mathcal{A}$ with the tricky time shift γ. The automorphism γ determines a simple localization (11.10) which makes it possible to compute the dynamical entropy of γ at least in the case when both $\mathcal{A}_0$ and $\mathcal{C}_0$ are of finite dimension.

Due to the monotonicity of the multiple-channel entropy we have

$$H_\varphi(\mathcal{A}_{[0,n+1]}, \mathcal{A}_{[1,n+2]}, \ldots, \mathcal{A}_{[k,n+k+1]})$$
$$\leq H_\varphi(\mathcal{A}_{[0,n+k+1]}) = S(\varphi|\mathcal{A}_{[0,n+k+1]}) = S(\varphi_0) + nS(\psi).$$

This yields

$$h_\varphi(\gamma; \mathcal{A}_{[0,n+1]}) \leq S(\psi)$$

and from the approximating net theorem (Theorem 10.8) we infer the bound

$$h_\varphi(\gamma) \leq S(\psi). \tag{11.11}$$

For the subalgebra $\mathcal{C}_0$ we have $h_\varphi(\gamma; \mathcal{C}_0) = S(\psi)$ and in (11.11) the equality must hold. What we arrived at is rather similar to the Kolmogorov-Sinai entropy of a *classical Markov chain*. Both in the probabilistic and in the above operator algebraic example the entropy is easily computable and it is

$$h_\varphi(\gamma) = S(\varphi|\mathcal{A}_{[0,1]}) - S(\varphi|\mathcal{A}_0). \tag{11.12}$$

Let us make the example more concrete and consider two unitaries u, $v \in M_m(\mathbb{C})$. The tensor product $M_m(\mathbb{C}) \otimes \mathbb{C}^2$ is isomorphic to $M_m(\mathbb{C}) \oplus M_m(\mathbb{C})$ and

$$\beta : a \oplus b \mapsto uau^* \oplus vav^*$$

becomes an automorphism leaving invariant the product state $\tau \otimes \mu$ where τ is the tracial state of $M_m(\mathbb{C})$ and μ corresponds to the uniform distribution on $\mathbb{C}^2$. (So that it takes the value $(\tau(a)+\tau(b))/2$ at $a \oplus b$.) Then the construction of a Markov process may take place according to the above receipt. The automorphism β is the dilation of the mapping $a \mapsto (uau^* + vav^*)/2$ of $M_m(\mathbb{C})$ onto itself.

Now we are going to deal with translation invariant states of the quantum spin lattice system $\mathcal{A}$ over the integers. A translation invariant state φ is called a Markov state of the spin lattice if the following condition is satisfied.

(iii′) For every $n \in \mathbb{N}$ there exists a completely positive unital mapping F_n of $\mathcal{A}_{[0,n+2]}$ into $\mathcal{A}_{[0,n+1]}$ which preserves the state φ and leaves $\mathcal{A}_{[0,n]}$ pointwise invariant.

To make the relation of (iii′) to (iii) more transparent we note that (iii′) implies that

$$F_n(\mathcal{A}_{[n,n+1]}) \subset \mathcal{A}_n. \tag{11.13}$$

Indeed, this condition is equivalent to the commutation relation $[F_n(\mathcal{A}_{[n,n+1]}), \mathcal{A}_{[0,n-1]})] = 0$ and for $a \in \mathcal{A}_{[n,n+1]}$ and $b \in \mathcal{A}_{[0,n-1]})]$ we have

$$bF_n(a) = F_n(ba) = F_n(ab) = F_n(a)b$$

because b is a fixed point of F_n.

The spin lattice system with a *Markov state* is an algebraic stochastic process but it can not be a Markov process (with conditional expectations) except for the case of a product state. Namely, if φ is not a product then φ-preserving conditional expectations E_J of $\mathcal{A}$ onto $\mathcal{A}_J$ do not exists for intervals $J = [n, k]$.

The next lemma provides several equivalent formulations of property (iii′).

Lemma 11.3 Let $\mathcal{M}_1 \subset \mathcal{M}_2 \subset \mathcal{M}_3$ be von Neumann algebras and let ψ be a faithful normal state of $\mathcal{M}_3$. Then the following conditions are equivalent.

(a) The generalized conditional expectation $E_\psi : \mathcal{M}_3 \to \mathcal{M}_2$ with respect to ψ leaves $\mathcal{M}_1$ pointwise fixed.

(b) There exists a completely positive mapping $\alpha : \mathcal{M}_3 \to \mathcal{M}_2$ such that $\alpha(I) = I$, $\psi \circ \alpha = \psi$ and $\alpha(a_1) = a_1$ for every $a_1 \in \mathcal{M}_1$.

(c) There exists a von Neumann subalgebra $\mathcal{M}$ of $\mathcal{M}_3$ such that $\mathcal{M}_1 \subset \mathcal{M} \subset \mathcal{M}_2$ and there is a conditional expectation $E : \mathcal{M}_3 \to \mathcal{M}$ with the property $\psi \circ E = \psi$.

(d) The modular group (σ_t^3) of ψ maps $\mathcal{M}_1$ into $\mathcal{M}_2$.

(e) The action of the modular group σ_t^3 of ψ is identical with that of the modular group σ_t^2 of $\psi|\mathcal{M}_2$.

Moreover, if there is a conditional expectation E of $\mathcal{M}_3$ onto $\mathcal{M}_2$ then the above conditions are equivalent to the next one.

(f) For every $t \in \mathbb{R}$ there is a unitary $u_t \in \mathcal{M}_3 \cap \mathcal{M}_1'$ such that $u_t a u_t^* = \sigma_{-t}^3 \sigma_t^2(a)$ for all $a \in \mathcal{M}_2$.

The proof of this lemma is omitted. In case of the quantum lattice system all the algebras in the triplet $(\mathcal{A}_{[0,n]}, \mathcal{A}_{[0,n+1]}, \mathcal{A}_{[0,n+2]})$ are finite dimensional and the lemma applies to a Markov state φ provided its restriction to every subalgebra $\mathcal{A}_{[0,n]}$ is faithful. Such a state will be called locally faithful. We restrict our discussion to locally faithful Markov states.

Proposition 11.4 A locally faithful Markov state of the quantum spin algebra is separating.

Proof. Let $(\mathcal{H}, \pi, \Phi)$ be the GNS-triplet associated with the locally faithful Markov state φ. We must show that Φ is separating for the von Neumann algebra $\pi(\mathcal{A})''$, which is equivalent to Φ being cyclic for $\pi(\mathcal{A})'$.

We denote by $D_{[k,n]}$ the invertible density of the restriction of φ to $\mathcal{A}_{[k,n]}$ and set $\mathcal{A}_\infty = \cup\{\mathcal{A}_{[-n,n]} : n \in \mathbb{N}\}$. For $a \in \mathcal{A}_\infty$ we define an operator A as

$$A\pi(b)\Phi = \pi(ba)\Phi \quad (b \in \mathcal{A}_\infty).$$

We claim that A is bounded. If $a \in \mathcal{A}_{[0,n]}$ then

$$\langle A\pi(b)\Phi, A\pi(b)\Phi \rangle = \varphi(a^*b^*ba) = \varphi(a^*F_n(b^*b)a)$$

$$= \mathrm{Tr}\left(D_{[0,n+1]}^{1/2} F_n(b^*b) D_{[0,n+1]}^{1/2} |D_{[0,n+1]}^{1/2} a^* D_{[0,n+1]}^{-1/2}|^2\right)$$

$$\leq \|D_{[0,n+1]}^{1/2} a^* D_{[0,n+1]}^{-1/2}\|^2 \times \|\pi(b)\Phi\|^2$$

and we arrive at $\|A\| \leq \|D_{[0,n+1]}^{1/2} a^* D_{[0,n+1]}^{-1/2}\|$. (Recall that F_n is the completely positive mapping from (iii').) This shows that $A \in \pi(\mathcal{A})'$ and so $A \in \pi(\mathcal{A})'\Phi$ is dense in $\mathcal{H}$. □

Proposition 11.5 For a translation invariant state φ of the quantum spin algebra the following conditions are equivalent.

(a) φ is a Markov state.

(b) $S(\varphi|\mathcal{A}_{[0,n+2]}) + S(\varphi|\mathcal{A}_{n+1}) = S(\varphi|\mathcal{A}_{[0,n+1]}) + S(\varphi|\mathcal{A}_{[n+1,n+2]})$.

(c) $D_{[0,n+2]}^{it} D_{[0,n+1]}^{-it} \in \mathcal{A}_{[n+1,n+2]}$ for every $t \in \mathbb{R}$ if $D_{[k,n]}$ stands for the density of $\varphi|\mathcal{A}_{[k,n]}$.

(d) With the previous notation $D_{[0,n+2]}^{1/2} D_{[0,n+1]}^{-1/2} \in \mathcal{A}_{[n+1,n+2]}$ holds.

Proof. In showing (a)$\Rightarrow$(c) we use property (f) of Lemma 11.3. Let $u_t \in \mathcal{A}_{[n+1,n+2]}$ be the unitary used there. Then

$$I = D^{it}_{[0,n+2]} D^{-it}_{[0,n+1]} a D^{it}_{[0,n+1]} D^{-it}_{[0,n+2]} \quad \text{for all} \quad a \in \mathcal{A}_{[0,n]}$$

yields (c). Due to simple analytic continuation (c)$\Rightarrow$(d).

Consider the generalized conditional expectation E_φ from $\mathcal{A}_{[0,n+2]}$ into $\mathcal{A}_{[0,n+1]}$ with respect to the state φ. This has the form

$$E_\varphi(a) = E_{[0,n+1]}(D^{-1/2}_{[0,n+1]} D^{1/2}_{[0,n+2]} a D^{1/2}_{[0,n+2]} D^{-1/2}_{[0,n+1]}), \tag{11.14}$$

where $E_{[0,n+1]}$ stands for the trace preserving conditional expectation. It is clear from (11.14) that under the assumption of (d) the subalgebra $\mathcal{A}_{[0,n]}$ is left elementwise fixed by E_φ. This proves (d)$\Rightarrow$(a).

To complete the proof we have to deal with (b)$\Longleftrightarrow$(c). Proposition 11.1 tells us that

$$S(\varphi|\mathcal{A}_{[0,n+2]}, \tau|\mathcal{A}_{[0,n+2]}) + S(\varphi|\mathcal{A}_{n+1}, \tau|\mathcal{A}_{n+1})$$
$$\geq S(\varphi|\mathcal{A}_{[0,n+1]}, \tau|\mathcal{A}_{[0,n+1]}) + S(\varphi|\mathcal{A}_{[n+1,n+2]}, \tau|\mathcal{A}_{[n+1,n+2]}) \tag{11.15}$$

(with τ denoting the tracial state on $\mathcal{A}_{[0,n+2]}$). Our condition (b) is exactly the same as the equality in this formula. From the proof of Proposition 11.1 is easy to analyze the necessary and sufficient condition for the equality. We see that in (11.15) the equality holds if and only if

$$S(\varphi \circ E_{[n+1,n+2]}, \varphi|\mathcal{A}_{[0,n+2]}) = S(\varphi \circ E_{[n+1,n+2]}|\mathcal{A}_{[n+1,n+2]}, \varphi|\mathcal{A}_{[n+1,n+2]})$$

holds with the trace preserving conditional expectation $E_{[n+1,n+2]} : \mathcal{A}_{[0,n+2]} \to \mathcal{A}_{[n+1,n+2]}$. In the language of Chapter 9, the latest condition is the sufficiency of the subalgebra $\mathcal{A}_{[n+1,n+2]}$ of $\mathcal{A}_{[0,n+2]}$ with respect to the two states $\varphi \circ E_{[n+1,n+2]}$ and $\varphi|\mathcal{A}_{[0,n+2]}$. According to Theorem 9.3 the sufficiency of $\mathcal{A}_{[n+1,n+2]}$ is equivalent to

$$[D(\varphi \circ E_{[n+1,n+2]}), D(\varphi|\mathcal{A}_{[0,n+2]})]_t \in \mathcal{A}_{[n+1,n+2]}$$

for every real t. In our simple finite dimensional situation this condition reads as (c) in the proposition. $\qquad\square$

Property (b) provides an entropic characterization of the Markov states. Thanks to the translation invariance (b) is written as

$$S(\varphi|\mathcal{A}_{[0,n+1]}) - S(\varphi|\mathcal{A}_{[0,n]}) = S(\varphi|\mathcal{A}_{[0,1]}) - S(\varphi|\mathcal{A}_0). \tag{11.16}$$

So a Markov state has a constant entropy increment at each step. The mean entropy is

$$s(\varphi) = S(\varphi|\mathcal{A}_{[0,1]}) - S(\varphi|\mathcal{A}_0) \tag{11.17}$$

which yields also the dynamical entropy of the translation γ with respect to φ. This is the content of the next proposition.

Proposition 11.6 For a Markov state φ we have $h_\varphi(\gamma) = S(\varphi|\mathcal{A}_{[0,1]}) - S(\varphi|\mathcal{A}_0)$.

Proof. Since $h_\varphi(\gamma) \leq s(\varphi)$ holds for any γ-invariant state, we have to show a finite dimensional subalgebra $\mathcal{N}$ for any $\varepsilon > 0$ such that

$$h_\varphi(\gamma; \mathcal{N}) \geq s(\varphi) - \varepsilon.$$

The practical estimation of $H_\varphi(\mathcal{N}, \gamma(\mathcal{N}), \ldots, \gamma^t(\mathcal{N}))$ is easier for expected subalgebras. Hence we choose $\mathcal{A}_{[0,n]} \subset \mathcal{N}_n \subset \mathcal{A}_{[0,n+1]}$ in such a way that it is the range of a φ-preserving conditional expectation of $\mathcal{A}$ (cf. condition (c) in Lemma 11.3). Then

$$H_\varphi(\mathcal{N}_n, \gamma(\mathcal{N}_n), \ldots, \gamma^{t(n+2)}(\mathcal{N}_n)) \geq H_\varphi(\mathcal{N}_n, \gamma^{n+2}(\mathcal{N}_n), \ldots, \gamma^{t(n+2)}(\mathcal{N}_n))$$

by the monotonicity of the multiple-channel entropy. Observe that the subalgebras $\mathcal{N}_n$, $\gamma^{n+2}(\mathcal{N}_n)$, $\ldots$ $\gamma^{t(n+2)}(\mathcal{N}_n)$ are pairwise commuting, all of them are expected. Let $\mathcal{M}_t$ be the subalgebra generated by them. Then

$$H_\varphi(\mathcal{N}_n, \gamma^{n+2}(\mathcal{N}_n), \ldots, \gamma^{t(n+2)}(\mathcal{N}_n)) = H_\varphi(\mathcal{M}_t) = S(\varphi|\mathcal{M}_t)$$

because $\mathcal{M}_t$ is an expected subalgebra, too. Denoting by τ the tracial state of $\mathcal{A}$ we express the latest von Neumann entropy by relative entropy:

$$S(\varphi|\mathcal{M}_t) = -S(\varphi|\mathcal{M}_t, \tau|\mathcal{M}_t) + \log d^t$$

where d is the tracial dimension of $\mathcal{N}_n$ and so d^t is the tracial dimension of $\mathcal{M}_t$. Let m stand for the tracial dimension of $\mathcal{A}_0$. Then from the monotonicity of the relative entropy

$$S(\varphi|\mathcal{M}_t) \geq -S(\varphi|\mathcal{A}_{[0,(n+2)(t+1)]}, \tau|\mathcal{A}_{[0,(n+2)(t+1)]}) + \log d^t$$
$$= S(\varphi|\mathcal{A}_{[0,(n+2)(t+1)]}) + \log d^t - \log m^{(n+2)(t+1)+1}.$$

Hence we have

$$h_\varphi(\gamma; \mathcal{N}_n) \geq \lim_{t \to \infty} \frac{1}{(n+2)t} \left(S(\varphi|\mathcal{A}_{[0,(n+2)(t+1)]}) + \log d^t - \log m^{(n+2)(t+1)+1} \right)$$

$$\geq s(\varphi) - \frac{1}{n+2} \log m.$$

Since n is arbitrary, the proof is complete.   □

The dynamical entropy of the right shift γ of the quantum spin algebra $\mathcal{A}$ is always majorized by the mean entropy (of a given translation invariant state). In the above example of a Markov state we have the equality of the two quantities and one expects that it is so if the translation invariant state φ is close to a product (in a sense made precise below). However, this seems to be a difficult problem. Since

$$S(\varphi|\mathcal{A}_{[1,n]}) \geq H_\varphi(\mathcal{A}_{[1,n]}),$$

we have in general

$$s(\varphi) = \lim_{n\to\infty} \frac{1}{n} S(\varphi|\mathcal{A}_{[1,n]}) \geq \limsup_{n\to\infty} \frac{1}{n} H_\varphi(\mathcal{A}_{[1,n]}) \geq h_\varphi(\gamma).$$

Our next aim is to show that under certain conditions the first inequality here is an equality. Before proceeding with the formulation of the exact result we prove the continuity of the entropy quantity $H_\omega(\mathcal{B})$.

Lemma 11.7 Let $\mathcal{B}$ be a subalgebra of the finite dimensional C*-algebra $\mathcal{C}$ and let $\omega, \overline{\omega}$ be states on $\mathcal{C}$. Then there exists a constant C such that

$$|H_\omega(\mathcal{B}) - H_{\overline{\omega}}(\mathcal{B})| \leq C\big(\varepsilon^{1/3}\log d + \eta(\varepsilon^{1/3})\big)$$

where d is the tracial dimension of $\mathcal{B}$ and $\varepsilon = ||\omega - \overline{\omega}||$.

Proof. We consider $\mathcal{C}$ in a standard form. Let Ω and $\overline{\Omega}$ be vector representatives of ω and $\overline{\omega}$, respectively. Then $||\Omega - \overline{\Omega}||^2 \leq \varepsilon$. We choose a decomposition $\omega = \sum_i \lambda_i \omega_i$ such that

$$H_\omega(\mathcal{B}) = S(\omega|\mathcal{B}) - \sum_i \lambda_i S(\omega_i|\mathcal{B}).$$

We can find operators $x_i \in \mathcal{C}$ such that $\sum_i x_i^* x_i = I$ and ω_i is induced by the vector $x_i\Omega/||x_i\Omega||$. Note that $\lambda_i = ||x_i\Omega||^2$. Now we are going to construct a decomposition of $\overline{\omega}$. Let us write

$$||x_i(\Omega - \overline{\Omega})||^2 = \varepsilon_i ||x_i\Omega||^2 \quad \text{and} \quad J = \{i : \varepsilon_i \leq \varepsilon^{2/3}\}.$$

Let $\overline{\omega}_i$ be the vector state induced by the unit vector $x_i\overline{\Omega}/||x_i\overline{\Omega}||$ for $i \in J$ and let $\overline{\nu}$ be defined as

$$\overline{\nu}(a) = \sum_{i\notin J} \kappa^{-1}\langle x_i\overline{\Omega}, a x_i\overline{\Omega}\rangle, \quad \kappa = \sum_{i\notin J}\langle x_i\overline{\Omega}, x_i\overline{\Omega}\rangle.$$

Then

$$\overline{\omega} = \sum_{i\in J} ||x_i\overline{\Omega}||^2 \overline{\omega}_i + \kappa\overline{\nu}$$

is a decomposition of $\overline{\omega}$ and we have

$$H_\omega(\mathcal{B}) - H_{\overline{\omega}}(\mathcal{B}) \leq S(\omega|\mathcal{B}) - \sum_{i\in J}\lambda_i S(\omega_i|\mathcal{B}) - \sum_{i\notin J}\lambda_i S(\omega_i|\mathcal{B})$$

$$-S(\overline{\omega}|\mathcal{B}) + \sum_{i\in J} ||x_i\overline{\Omega}||^2 S(\overline{\omega}_i|\mathcal{B}) + \kappa S(\overline{\nu}|\mathcal{B})$$

$$\leq |S(\omega|\mathcal{B}) - S(\overline{\omega}|\mathcal{B})| + \sum_{i\in J}\lambda_i|S(\omega_i|\mathcal{B}) - S(\overline{\omega}_i|\mathcal{B})|$$

$$+ \sum_{i\in J}\big|\lambda_i - ||x_i\overline{\Omega}||^2\big| S(\overline{\omega}_i|\mathcal{B}) + \kappa S(\overline{\nu}|\mathcal{B}).$$

Now we estimate each term on the end of the above chain of inequalities. From the Fannes estimate (Proposition 1.8) we have

$$|S(\omega|\mathcal{B}) - S(\overline{\omega}|\mathcal{B})| \le \varepsilon \log d + \eta(\varepsilon)$$

for the first term. If $i \in J$ we estimate as

$$\left\| \frac{x_i \Omega}{||x_i \Omega||} - \frac{x_i \overline{\Omega}}{||x_i \overline{\Omega}||} \right\| \le \frac{||x_i \Omega - x_i \overline{\Omega}||}{||x_i \Omega||} + ||x_i \overline{\Omega}|| \left| \frac{||x_i \Omega|| - ||x_i \overline{\Omega}||}{||x_i \Omega|| \, ||x_i \overline{\Omega}||} \right|$$

$$\le \varepsilon_i^{1/2} + \frac{||x_i(\Omega - \overline{\Omega})||}{||x_i \Omega||} = 2\varepsilon_i^{1/2} \le 2\varepsilon^{1/3}$$

and obtain for the second term

$$\sum_{i \in J} \lambda_i |S(\omega_i|\mathcal{B}) - S(\overline{\omega}_i|\mathcal{B})| \le 4\varepsilon^{1/3} \log d + \eta(4\varepsilon^{1/3}) \,.$$

In case of the third term the entropy part is estimated trivially by the logarithm of the dimension and we have for the coefficients

$$\left| \lambda_i - ||x_i \overline{\Omega}||^2 \right| = \left| \langle \Omega, x_i^* x_i (\Omega - \overline{\Omega}) \rangle + \langle \Omega - \overline{\Omega}, x_i^* x_i \Omega \rangle \right.$$

$$\left. + \langle \Omega - \overline{\Omega}, x_i^* x_i (\overline{\Omega} - \Omega) \rangle \right| \le 2\varepsilon_i^{1/2} ||x_i \Omega||^2 + \varepsilon_i ||x_i \Omega||^2 \le 3\varepsilon^{1/3} ||x_i \Omega||^2 \,.$$

Hence

$$\sum_{i \in J} \left| \lambda_i - ||x_i \overline{\Omega}||^2 \right| S(\overline{\omega}_i|\mathcal{B}) \le 3\varepsilon^{1/3} \log d \,.$$

We proceed by finding bounds as

$$\sum_{i \notin J} ||x_i \Omega||^2 \le \varepsilon^{-2/3} \sum_{i \notin J} \varepsilon_i ||x_i \Omega||^2$$

$$\le \varepsilon^{-2/3} \sum_{i \notin J} ||x_i(\overline{\Omega} - \Omega)||^2 \le \varepsilon^{-2/3} \varepsilon = \varepsilon^{1/3}$$

and

$$\kappa = \sum_{i \notin J} \langle x_i \overline{\Omega}, x_i \overline{\Omega} \rangle = \langle \overline{\Omega}, \textstyle\sum_{i \notin J} x_i^* x_i \overline{\Omega} \rangle$$

$$\le \langle \Omega, \textstyle\sum_{i \notin J} x_i^* x_i \Omega \rangle + 2|\langle (\overline{\Omega} - \Omega), \textstyle\sum_{i \notin J} x_i^* x_i \overline{\Omega} \rangle| \le \varepsilon^{1/3} + 2\varepsilon^{1/2} \le 3\varepsilon^{1/3} \,,$$

which yields

$$\kappa S(\overline{\nu}|\mathcal{B}) \le 3\varepsilon^{1/3} \log d$$

for the last term. In this way we arrive at

$$H_\omega(\mathcal{B}) - H_{\overline{\omega}}(\mathcal{B}) \le 11\varepsilon^{1/3} \log d + \eta(4\varepsilon^{1/3})$$

if ε is small enough. Due to the symmetry in ω and $\overline{\omega}$ the proof is complete.

$\square$

Lemma 11.8 Let $\mathcal{C}_1 \equiv M_n(\mathbb{C})$ and $\mathcal{C}_2 \equiv M_d(\mathbb{C})$ be full matrix algebras and let ω be a state of $\mathcal{C}_1 \otimes \mathcal{C}_2$. Then

$$H_\omega(\mathcal{C}_1) \ge S(\omega|\mathcal{C}_1) - \log d \,.$$

Proof. Let $\omega = \sum_i \lambda_i \omega_i$ be an extremal orthogonal decomposition. Due to the definition we have

$$H_\omega(\mathcal{C}_1) \geq \sum_i \lambda_i S(\omega_i|\mathcal{C}_1, \omega|\mathcal{C}_1) = S(\omega|\mathcal{C}_1) - \sum_i \lambda_i S(\omega_i|\mathcal{C}_1).$$

The application of purification (Lemma 6.4) gives

$$S(\omega_i|\mathcal{C}_1) = S(\omega_i|\mathcal{C}_2) \leq \log d$$

and our statement follows.   □

The state φ of $\mathcal{A}$ is said to cluster if for every $\varepsilon > 0$ there are sequences $k(n)$ and $m(n)$ in $\mathbb{N}$ so that $k(n)/m(n) \to 0$ as $n \to \infty$ and there is a unitary

$$U_n \in \mathcal{A}_{[-m(n)-k(n),-m(n)+k(n)]\cup[m(n)-k(n),m(n)+k(n)]} \tag{11.18}$$

such that

$$\|\varphi(U_n^* \cdot U_n) - \varphi|(\mathcal{A}_{[-m(n),m(n)]} \otimes \mathcal{A}_{[-m(n),m(n)]^c})\| < \varepsilon \tag{11.19}$$

if n is big enough. (For an interval $J \subset \mathbb{Z}$ we denoted its complement $\mathbb{Z} \setminus J$ by J^c. This notation will be kept below.) The content of the cluster condition is the following. One can find an arbitrary large interval $J(n) = [-m(n), m(n)]$ and a unitary U_n localized close to the boundary of $J(n)$ such that the state $\varphi(U_n^* \cdot U_n)$ is factorizing approximately.

Proposition 11.9 If the states φ satisfies the cluster condition then

$$s(\varphi) = \limsup_{n\to\infty} \frac{1}{n} H_\varphi(\mathcal{A}_{[1,n]}).$$

Proof. For typographical reasons the subalgebra $\mathcal{A}_J$ of $\mathcal{A}$ will be written here as $\mathcal{A}(J)$. First we fix some notation. Write $J(n)$ for the interval $[-m(n), m(n)]$ and $M(n)$ for $(-m(n) + k(n), m(n) - k(n))$

Now we start with the trivial estimate

$$\begin{aligned} 0 \leq \; &S(\varphi|\mathcal{A}(M(n))) - H_\varphi(\mathcal{A}(M(n))) \\ \leq \; &|S(\varphi|\mathcal{A}(M(n))) - H_{\varphi|\mathcal{A}(J(n))\otimes\varphi|\mathcal{A}(J(n)^c)}(\mathcal{A}(M(n)))| \\ &+ |H_{\varphi|\mathcal{A}(J(n))\otimes\varphi|\mathcal{A}(J(n)^c)}(\mathcal{A}(M(n))) - H_\varphi(\mathcal{A}(M(n)))|. \end{aligned} \tag{11.20}$$

Since the subalgebra $\mathcal{A}(M(n))$ is invariant under the action of U_n we have

$$H_\varphi(\mathcal{A}(M(n))) = H_\psi(\mathcal{A}(M(n))), \quad \psi = \varphi(U_n^* \cdot U_n)$$

and we can estimate the second term on the right hand side of (11.20) by (11.19) and Lemma 11.7.

$$\begin{aligned} &|H_{\varphi|\mathcal{A}(J(n))\otimes\varphi|\mathcal{A}(J(n)^c)}(\mathcal{A}(M(n))) - H_\varphi(\mathcal{A}(M(n)))| \\ &\leq C(\varepsilon^{1/3} m(n) \log d + \eta(\varepsilon^{1/3})). \end{aligned} \tag{11.21}$$

To get a bound for the first term on the right hand side of (11.20) we need a bit longer argument. Since there is a conditional expectation of $\mathcal{A}$ onto $\mathcal{A}(J(n))$ which preserves the product state $\varphi|\mathcal{A}(J(n)) \otimes \varphi|\mathcal{A}(J(n)^c)$ we have

$$H_{\varphi|\mathcal{A}(J(n)) \otimes \varphi|\mathcal{A}(J(n)^c)}\big(\mathcal{A}(M(n))\big) = H_{\varphi|\mathcal{A}(J(n))}\big(\mathcal{A}(M(n))\big)\,. \tag{11.22}$$

Let $L(n) = [-m(n), -m(n)+k(n))$ and $R(n) = (m(n)-k(n), m(n)]$. So $J(n) = L(n) \cup M(n) \cup R(n)$ and $\mathcal{A}(J(n)) = \mathcal{A}(L(n)) \otimes \mathcal{A}(M(n)) \otimes \mathcal{A}(R(n))$ and by the application of Lemma 11.8 we obtain

$$H_{\varphi|\mathcal{A}(J(n))}\big(\mathcal{A}(M(n))\big) \geq S\big(\varphi|\mathcal{A}(M(n))\big) - 2k(n) \log d\,. \tag{11.23}$$

Finally, the first term on the right hand side of (11.20) is estimated as follows.

$$\begin{aligned}
S\big(\varphi|\mathcal{A}(M(n))\big) &- S\big(\varphi|\mathcal{A}(J(n))\big) \\
&\leq S\big(\varphi|\mathcal{A}(M(n))\big) - H_{\varphi|\mathcal{A}(J(n)) \otimes \varphi|\mathcal{A}(J(n)^c)}\big(\mathcal{A}(M(n))\big) \\
&\leq S\big(\varphi|\mathcal{A}(M(n))\big) - S\big(\varphi|\mathcal{A}(J(n))\big) + 1k(n) \log d\,.
\end{aligned}$$

In this way from (11.20) we arrive at

$$\frac{1}{m(n)}S\big(\varphi|\mathcal{A}(M(n))\big) - \frac{1}{m(n)}H_\varphi\big(\mathcal{A}(M(n))\big) \leq C\varepsilon^{1/3}\log d + \varepsilon$$

for n large enough. This is sufficient to conclude the proposition.   □

Now we discuss correlations of states of the quantum spin algebra $\mathcal{A}$. Set

$$\mathcal{A}_- = \cup\{\mathcal{A}_{[-n,-1]} : n \in \mathbb{N}\} \quad \text{and} \quad \mathcal{A}_+ = \cup\{\mathcal{A}_{[0,n]} : n \in \mathbb{N}\}\,.$$

For a state ψ of $\mathcal{A}$ and for $x \in \mathcal{A}_+$ we call the application

$$C_x : a \mapsto \psi(ax) \qquad (a \in \mathcal{A}_-) \tag{11.24}$$

a correlation functional. Clearly, a translation invariant state is known if and only if all of its correlation functionals are given. The state ψ is called finitely correlated if the linear space $\cup\{C_x : x \in \mathcal{A}_+\}$ of correlation functionals is of finite dimension. Markov states are examples of finitely correlated states. Indeed, for a Markov state φ we have

$$C_x(a) = \varphi(aF(x)) \qquad (a \in \mathcal{A}_-, x \in \mathcal{A}_+)\,, \tag{11.25}$$

where F is a positive mapping of $\mathcal{A}_+$ into $\mathcal{A}_0$. F is a "quasiconditional" expectation, $F(ab) = aF(b)$ holds for every $a \in \mathcal{A}_-$ and $b \in \mathcal{A}$ but it is not a projection. Formula (11.25) shows that the correlation functionals may be parametrized by $\mathcal{A}_0$ and so the dimension of the space of correlation functionals is at most the dimension of $\mathcal{A}_0$.

Let the quantum lattice algebra be built from matrix algebras $M_d(\mathbb{C})$, that is $\mathcal{A}_0 = M_d(\mathbb{C})$. Suppose that a completely positive unital mapping $\mathcal{E} : M_d(\mathbb{C}) \otimes M_m(\mathbb{C}) \to M_m(\mathbb{C})$ and a state ω of $M_m(\mathbb{C})$ are given so that

$$\omega(\mathcal{E}(I \otimes A) = \omega(A) \qquad (A \in M_m(\mathbb{C}))\,. \tag{11.26}$$

There is then a unique translation invariant state ψ of $\mathcal{A}$ such that for every $a_i \in \mathcal{A}_i$ $(0 \leq i \leq n)$ we have

$$\psi(a_0 a_1 \ldots a_n) = \omega(\mathcal{E}(a_0 \otimes \mathcal{E}(\ldots \mathcal{E}(a_{n-1} \otimes \mathcal{E}(a_n \otimes I))\ldots))). \qquad (11.27)$$

(Here I denotes the unit element in the auxiliary matrix algebra $M_m(\mathbb{C})$.) The translation invariance of ψ is a plain consequence of the hypothesis (11.26). To see that (11.27) determines a finitely correlated state ψ we set

$$\mathcal{F}[a] : M_m(\mathbb{C}) \to M_m(\mathbb{C}), \quad \mathcal{F}[a](X) = \mathcal{E}(a, X)$$

and we obtain

$$\psi(a_{-n} a_{-n+1} \ldots a_0 a_1 \ldots a_n)$$
$$= \omega(\mathcal{F}[a_{-n}] \circ \mathcal{F}[a_{-n+1}] \circ \ldots \circ \mathcal{F}[a_0] \circ \mathcal{F}[a_1] \circ \ldots \circ \mathcal{F}[a_n](I)).$$

Now the correlation functionals are parametrized by the linear transformations of the auxiliary algebra $M_m(\mathbb{C})$ and ψ is a finitely correlated state. We call ψ algebraic state generated by $(\mathcal{E}, \omega)$.

Notes and Remarks. The fundamental paper on the general theory of *algebraic stochastic processes* is [Accardi, Frigerio and Lewis 1982]; [Kümmerer 1988] is a suggested survey, it explains the connection of Markov processes with dilation theory, our formulation of Bernoulli processes follows it. Commuting squares are treated comprehensively in the monograph [Goodman, de la Harpe and Jones 1989]. Proposition 11.1 and 11.2 are from [Petz 1991].

Independence is a delicate point in quantum probability, in the setting of field theory [Summers 1990] is a suggested reference. Recently the notion of free independence has attracted a lot of attention, see [Voiculesu 1986] and [Speicher 1990], for example.

Markov states of the quantum spin algebra were introduced in [Accardi 1975], [Accardi and Frigerio 1983] is a suitable reference, too. Lemma 11.3 contains a part of equivalent conditions which appeared in [Cecchini and Petz 1989a]. [Cecchini 1992] is looking for analogues of the Markov property in an abstract operator algebraic setting. [Nachtergaele 1990] discusses the Markov property by comparing the probabilistic and quantum cases. An example for a Markov process shows up also in Chapter 17.

Lemma 11.7 and Proposition 11.9 are from [Narnhofer and Thirring 1988] where the third law of thermodynamics is discussed as well. It is not clear whether the dynamical entropy of the shift of a one-dimensional quantum lattice system is identical with the mean entropy when the equilibrium state of an interaction of finite range is considered.

States determined as in (11.27) were introduced in [Accardi 1981], at least under the restriction $d = m$, that is, by a restrictive choice of the auxiliary algebra. In [Accardi and Frigerio 1983] such state is called generalized quantum Markov chain and $\mathcal{E}$ is called a transition expectation. *Finitely correlated* and algebraic states were studied in [Fannes, Nachtergaele and Werner

1992a]. Their importance comes from the fact that the ground state of certain models is of this form.

Comments. A translation invariant state of the quantum spin algebra may be constructed by means of a conditional density amplitude which is an operator $K \in M_m(\mathbb{C}) \otimes M_m(\mathbb{C})$ such that $\mathrm{Tr}_2 KK^* = I$, Tr_2 denotes the partial trace over the second factor. Given a density matrix $W \in M_m(\mathbb{C})$ such that $\mathrm{Tr}_1 K^*(W \otimes I)K = W$, a translation invariant state φ can be constructed in the following way.

$$E : M_m(\mathbb{C}) \otimes M_m(\mathbb{C}) \to M_m(\mathbb{C}), \qquad E(x) := \mathrm{Tr}_2 KxK^*$$

and

$$\varphi(x_i \otimes x_{i-1} \ldots \otimes x_k) := \mathrm{Tr}\big(WE(x_i \otimes E(x_{i-1} \otimes E(x_{i-2} \otimes \ldots E(x_k \otimes I)))))\big).$$

This state φ is called quantum Markov chain and it was shown in [Park 1994] that the Connes–Narnhofer–Thirring entropy of the shift is the mean entropy.

Part IV

Perturbation Theory

12 Perturbation of States

For a finite quantum system $\mathcal{A}$ with Hamiltonian $H \in \mathcal{A}^{sa}$ the *free energy functional* at inverse temperature β is defined as

$$F(\omega) = \omega(H) - \frac{1}{\beta}S(\omega)$$

for a state ω. The *canonical state* φ minimizing the free energy functional possesses the density

$$\frac{e^{-\beta H}}{\operatorname{Tr} e^{-\beta H}}.$$

The *perturbed state* $[\varphi^h]_\beta$ is the canonical state for the Hamiltonian $H + h$. Hence $[\varphi^h]_\beta$ represents the equilibrium state of the perturbed physical system in which the energy of each state σ has been increased by $\sigma(h)$. The density of $[\varphi^h]_\beta$ is

$$\frac{e^{-\beta(H+h)}}{\operatorname{Tr} e^{-\beta(H+h)}}.$$

The state $[\varphi^h]_\beta$ is the normalization of the (nonnormalized) perturbed functional which appeared already in Chapter 1:

$$[\varphi^h]_\beta = \varphi^{-\beta h}/\varphi^{-\beta h}(I).$$

The aim of this chapter is to develop a theory of perturbed states. Perturbation of states has great importance in quantum statistical mechanics in connection with equilibrium states and their stability. For example, the *Gibbs condition* for equilibrium will be formulated in terms of perturbed state in the Chapter 15. The perturbation may be considered from several aspects. Ours is thermodynamic, the definition will be given by a variational formula. Since the Hamiltonian in physics may be unbounded and it can even take infinite values, we will allow very general perturbations.

In the finite quantum case $[\varphi^h]_\beta$ minimizes the functional

$$\omega \mapsto \beta^{-1}S(\omega, \varphi) + \omega(h)$$

and following Donald we shall use this property as a definition in the general case. This thermodynamic definition will be shown to be equivalent to the kinematical one, and continuity properties of the perturbation will be treated as well. By the help of the simple transformation

$$[\varphi^h]_\beta = [\varphi^{\beta h}]_1$$

the value $\beta = 1$ can often be fixed and we shall write $[\varphi^h]$ instead of $[\varphi^h]_1$.

Let $\mathcal{M}$ be a von Neumann algebra acting on a Hilbert space $\mathcal{H}$. The normal state space of $\mathcal{M}$ will be denoted by $\mathfrak{S}_*(\mathcal{M})$. We are going to deal with self-adjoint operators affiliated with $\mathcal{M}$ and bounded from below. Let us recall that a self-adjoint operator is affiliated with $\mathcal{M}$ if all its spectral projections belong to $\mathcal{M}$. It will be allowed that $+\infty$ be an eigenvalue of the operator. More formally, an *extended-valued lower-bounded self-adjoint operator* affiliated with $\mathcal{M}$ is an affine mapping $h : \mathfrak{S}_*(\mathcal{M}) \to \mathbb{R} \cup \{+\infty\}$ such that

   (i)   the set $h(\mathfrak{S}_*(\mathcal{M}))$ is bounded from below,
   (ii) h is weakly lower semi-continuous.

We use the notation $\mathcal{M}^{\text{ext}}$ for the set of all such generalized operators. One can see that to each $h \in \mathcal{M}^{\text{ext}}$ a unique spectral decomposition

$$\int_c^\infty \lambda dE_\lambda + \infty \times p \tag{12.1}$$

is associated where $c \in \mathbb{R}$, p is a certain projection in $\mathcal{M}$ and E_λ is a spectral measure with values in the von Neumann algebra $p^\perp \mathcal{M} p^\perp$ and given on the interval $[c, \infty)$. The generalized operator $h \in \mathcal{M}^{\text{ext}}$ with spectral decomposition (12.1) takes the value

$$\int_c^\infty \lambda d\sigma(E_\lambda) + \infty \times \sigma(p) \tag{12.2}$$

at $\sigma \in \mathfrak{S}_*(\mathcal{M})$. In other language, for $\sigma(p) \neq 0$ we have $h(\sigma) = +\infty$, otherwise $h(\sigma)$ is given by the first term of (12.2). ($0 \times \infty = 0$ convention.) Obviously, $\mathcal{M}^{\text{ext}}$ is closed under pointwise addition. A pleasant feature of $\mathcal{M}^{\text{ext}}$ is that it is closed under increasing limit as well. If $(h_i)_i$ is an increasing net of operators from $\mathcal{M}^{\text{ext}}$, then

$$h(\sigma) = \lim_i h_i(\sigma) = \sup\{h_i(\sigma) : i\} \qquad (\sigma \in \mathfrak{S}_*(\mathcal{M}))$$

determines an element of $\mathcal{M}^{\text{ext}}$. It is clear from the spectral decomposition (12.1) that every element of $\mathcal{M}^{\text{ext}}$ is the increasing limit of a sequence of commuting self-adjoint operators in $\mathcal{M}$.

If $Q \in \mathcal{M}$ is a projection, we set

$$QhQ(\sigma) = \begin{cases} 0 & \text{if} \quad \sigma(Q) = 0, \\ h(\sigma_Q) & \text{otherwise,} \end{cases} \qquad (\sigma \in \mathfrak{S}_*(\mathcal{M})),$$

where

$$\sigma_Q(a) = \frac{\sigma(QaQ)}{\sigma(Q)} \qquad (a \in \mathcal{M}).$$

This notation is justified by the fact that for $h \in \mathcal{M}^{sa}$ the above defined QhQ is the product of the three operators in the common sense.

Now we are ready to define $[\varphi^h]_\beta$ for a normal state φ of $\mathcal{M}$ and for a generalized operator $h \in \mathcal{M}^{\text{ext}}$. If

$$d \equiv \inf\{F(\omega) \equiv \beta^{-1}S(\omega, \varphi) + h(\omega) : \omega \in \mathfrak{S}_*(\mathcal{M})\}$$

is finite, then, due to Theorem 5.26, there is a unique state $[\varphi^h]_\beta$ of $\mathcal{M}$ satisfying

$$S([\varphi^h]_\beta, \varphi) + h([\varphi^h]_\beta) = d.$$

Hence the perturbed state $[\varphi^h]_\beta$ is determined as the unique minimizer of the weakly lower semi-continuous functional F. If $d = \infty$ then $[\varphi^h]_\beta$ is not defined. Note that if $h(\varphi)$ is finite, then $[\varphi^h]_\beta$ exists, and thanks to $S([\varphi^h]_\beta, \varphi) < +\infty$ the state $[\varphi^h]_\beta$ is always normal. Moreover,

$$\text{supp}\,[\varphi^h] \leq \text{supp}\,\varphi. \tag{12.3}$$

It follows also from the definition that

$$[\varphi^h] = [\varphi^{QhQ}], \tag{12.4}$$

whenever Q is a projection in $\mathcal{M}$ with $Q \geq \text{supp}\,\varphi$. Property (12.4) has the consequence that one can easily pass to an algebra on which φ is faithful. One should only replace h by $(\text{supp}\,\varphi)h(\text{supp}\,\varphi)$.

Before proving general theorems we discuss a simple example. Let $\mathcal{A}$ be a finite quantum system and let $p \in \mathcal{A}$ be a projection. If $k \in \mathcal{A}^{sa}$ is such that $\text{supp}\,k \perp p$ then $k + \infty \times p$ is a lower-bounded extended-valued operator. Set e^{-H} for the density of a state φ of $\mathcal{A}$ and define

$$\psi(a) = \frac{\text{Tr}\,aqe^{-qHq-k}}{\text{Tr}\,qe^{-qHq-k}} \qquad (a \in \mathcal{A})$$

where $q = I - p$. We are going to show that $[\varphi^{k+\infty p}] = \psi$. One computes

$$S(\psi, \varphi) = -\log \text{Tr}\,qe^{-qHq-k} - \psi(k).$$

Hence it suffices to show the inequality

$$S(\omega, \varphi) + \omega(k) + \infty \times \omega(p) \geq -\log \text{Tr}\,qe^{-qHq-k} \tag{12.5}$$

for every state ω of $\mathcal{A}$. This holds trivially if $\omega(p) \neq 0$, so we may assume $\text{supp}\,\omega \leq q$. According to Proposition 1.11 we have

$$S(\omega, \varphi) + \omega(h) \geq -\log \text{Tr}\,e^{-H-h} \tag{12.6}$$

for every $h \in \mathcal{A}^{sa}$. Now replace h here by $-H + qHq + k + np$ $(n \in \mathbb{N})$ and (12.6) becomes

$$S(\omega, \varphi) + \omega(k) \geq -\log \operatorname{Tr} e^{-qHq-k} e^{-np} = -\log \operatorname{Tr}\left(qe^{-qHq-k} + e^{-n}p\right).$$

Taking the limit $n \to \infty$ we conclude (12.5). Notice in this example that $\operatorname{supp}[\varphi^{k+\infty p}] = q$ and does not need to be equal to $\operatorname{supp}\varphi$.

Although the concept of *perturbed state* was formulated in the the setting of von Neumann algebras, the same definition works if ψ is an arbitrary state of a C*-algebra $\mathcal{A}$ and $h = h^* \in \mathcal{A}$. Observe that $[\psi^h] = \psi$ for pure states ψ.

We define

$$c_\beta(\varphi, h) = \inf\{\tfrac{1}{\beta}S(\omega, \varphi) + h(\omega) : \omega \in \mathfrak{S}_*(\mathcal{M})\} \tag{12.7}$$

given a state $\varphi \in \mathfrak{S}_*(\mathcal{M})$ and an extended-valued lower-bounded operator h affiliated with the von Neumann algebra $\mathcal{M}$. Then $c_\beta(\varphi, h) < +\infty$ is equivalent to the claim that $[\varphi^h]_\beta$ exists, and in this case

$$c_\beta(\varphi, h) = \frac{1}{\beta}S([\varphi^h], \varphi) + [\varphi^h](h). \tag{12.8}$$

We write $c(\varphi, h)$ with the understanding that $\beta = 1$.

Lemma 12.1 If $c(\varphi, h) < +\infty$, then

$$S(\omega, \varphi) + \omega(h) \geq c(\varphi, h) + S(\omega, [\varphi^h])$$

for every $\omega \in \mathfrak{S}_*(\mathcal{M})$.

Proof. We may assume that $S(\varphi, \omega) + \omega(h)$ is finite. By the definition of $[\varphi^h]$ we have

$$S(\lambda\omega + (1-\lambda)[\varphi^h], \varphi) + (\lambda\omega + (1-\lambda)[\varphi^h])(h) \geq S([\varphi^h], \varphi) + [\varphi^h](h)$$

for every $0 < \lambda < 1$. Proposition 5.22 gives

$$S(\lambda\omega + (1-\lambda)[\varphi^h], \varphi) = \lambda S(\omega, \varphi) + (1-\lambda)S([\varphi^h], \varphi)$$
$$-\lambda S(\omega, \lambda\omega + (1-\lambda)[\varphi^h]) - (1-\lambda)S([\varphi^h], \lambda\omega + (1-\lambda)[\varphi^h]).$$

Combining these two equations and (12.8) we arrive at the inequality

$$S(\omega, \varphi) + \omega(h) \geq c(\varphi, h)$$
$$+S(\omega, \lambda\omega + (1-\lambda)[\varphi^h]) + \frac{1-\lambda}{\lambda}S([\varphi^h], \lambda\omega + (1-\lambda)[\varphi^h]).$$

Now $\lambda \to 0$ completes the proof. $\square$

Lemma 12.2 If $c(\varphi, h) < +\infty$ and there is a constant $\mu > 0$ such that $\omega \leq \mu[\varphi^h]$, then

$$S(\omega, \varphi) + \omega(h) = c(\varphi, h) + S(\omega, [\varphi^h]).$$

Proof. By assumption we can write $[\varphi^h]$ in the form $\lambda\omega + (1-\lambda)\sigma$ with some $\sigma \in \mathfrak{S}_*(\mathcal{M})$ and $0 < \lambda < 1$. By means of Proposition 5.22 we have

$$\begin{aligned}
S([\varphi^h], \varphi) + [\varphi^h](h) &= \lambda(S(\omega, \varphi) + \omega(h)) + (1-\lambda)(S(\sigma, \omega) + \sigma(h)) \\
&\quad -\lambda S(\omega, [\varphi^h]) - (1-\lambda)S(\sigma, [\varphi^h]).
\end{aligned}$$

On the other hand, (12.8) yields

$$\begin{aligned}
S([\varphi^h], \varphi) + [\varphi^h](h) &= \lambda(c(\varphi, h) + S(\omega, [\varphi^h])) \\
&\quad +(1-\lambda)(c(\varphi, h) + S(\sigma, [\varphi^h])) - \lambda S(\omega, [\varphi^h]) - (1-\lambda)S(\sigma, [\varphi^h]).
\end{aligned}$$

It follows that

$$\begin{aligned}
\lambda(S(\omega, \varphi) + \omega(h)) &+ (1-\lambda)(S(\sigma, \varphi) + \sigma(h)) \\
&= \lambda(c(\varphi, h) + S(\omega, [\varphi^h])) + (1-\lambda)(c(\varphi, h) + S(\sigma, [\varphi^h])),
\end{aligned}$$

since $S(\omega, [\varphi^h])$ and $S(\sigma, [\varphi^h])$ are finite. According to Lemma 12.1 the second term on the right hand side is majorized by $(1-\lambda)(S(\sigma, \varphi) + \sigma(h))$ and we obtain

$$S(\omega, \varphi) + \omega(h) \leq c(\varphi, h) + S(\omega, [\varphi^h]).$$

The opposite inequality is contained in Lemma 12.1.   □

We note that it follows from Lemma 12.1 that $\varphi(h) < \infty$ implies

$$\text{supp}\,[\varphi^h] = \text{supp}\,\varphi. \tag{12.9}$$

Indeed, in this case $S([\varphi^h], \varphi)$ is finite and $\text{supp}\,\varphi \leq \text{supp}\,[\varphi]^h$ must hold. This with (12.3) gives (12.9). Below we shall see that for a bounded $h \in \mathcal{M}^{sa}$ the statement of Lemma 12.2 holds for every $\sigma \in \mathfrak{S}_*(\mathcal{M})$. Now our goal is to obtain an analytic expression for the perturbed state, at least for bounded perturbations.

The next result concerns the *strong continuity of perturbations*.

Theorem 12.3 Let $(k_n)_n$ be a sequence of self-adjoint elements of the von Neumann algebra $\mathcal{M}$ and $\varphi \in \mathfrak{S}_*(\mathcal{M})$. If $k_n \to k$ strongly then $[\varphi^{k_n}](k_n) \to [\varphi^k](k)$, $c(\varphi, k_n) \to c(\varphi, k)$ and $[\varphi^{k_n}] \to [\varphi^k]$ in norm.

Proof. The simple estimate

$$S([\varphi^h], \varphi) \leq \varphi(h) - [\varphi^h](h) \leq 2\|h\|$$

shows that $S([\varphi^{k_n}], \varphi)$ is bounded. Hence the set $\mathcal{W} = \{[\varphi^k]\} \cup \{[\varphi^{k_n}] : n \in \mathbb{N}\}$ is weakly relatively compact in $\mathfrak{S}_*(\mathcal{M})$ (cf. Proposition 5.27). This implies that there exists a $\sigma \in \mathfrak{S}_*(\mathcal{M})$ such that for any given $\varepsilon > 0$ there exists a $\delta > 0$ with the property $|\rho(a)| < \varepsilon$ ($\rho \in \mathcal{W}$) whenever $\|a\| \leq 1$ and $\sigma(a^*a + aa^*) < \delta$. (Concerning relative compactness in the predual space of a von Neumann algebra we refer to Theorem 5.4 of Chapter III in [Takesaki 1979].)

We choose a weakly convergent subnet $([\varphi^{k_i}]) \to \psi$ of $([\varphi^{k_n}])$. We show that $[\varphi^{k_i}](k_i) \to \psi(k)$. Indeed, $|[\varphi^{k_i}](k) - \psi(k)| < \varepsilon$ if i is big enough and the relative compactness yields $|[\varphi^{k_i}](k_i) - [\varphi^{k_i}](k)| < \varepsilon$ if i is sufficiently large.

If we prove

$$S([\varphi^k], \varphi) + [\varphi^k](k) \geq S(\psi, \varphi) + \psi(k), \tag{12.10}$$

then the definition of the perturbed state provides $\psi = [\varphi^k]$. Lemma 12.1 yields

$$S([\varphi^k], \varphi) + [\varphi^k](k_i) \geq S([\varphi^{k_i}], \varphi) + [\varphi^{k_i}](k_i) + S([\varphi^k], [\varphi^{k_i}]). \tag{12.11}$$

Here the limit of the left hand side becomes the left hand side of (12.10) and the liminf of the right hand side is minorized by $S(\varphi, \psi) + \psi(k)$. This yields (12.10). So $\psi = [\varphi^k]$ is established, and

$$S([\varphi^k], \varphi) \geq \limsup_i S([\varphi^{k_i}], \varphi) \tag{12.12}$$

also follows from (12.10). Combining this with the lower semi-continuity we have

$$S([\varphi^k], \varphi) = \lim_i S([\varphi^{k_i}], \varphi).$$

Once more taking the limit of (12.11) as $i \to \infty$ we infer

$$\lim_i S([\varphi^k], [\varphi^{k_i}]) = 0,$$

and the norm estimate of Theorem 5.5 gives

$$[\varphi^{k_i}] \to [\varphi^k] \quad \text{in norm.}$$

What we have proved holds for an arbitrary convergent subnet of $([\varphi^{k_n}])$, hence it holds for the sequence $([\varphi^{k_n}])_n$ itself. $\qquad \square$

Let φ be a faithful normal state of the von Neumann algebra $\mathcal{M}$ and k be a self-adjoint element. We assume that $\mathcal{M}$ is in standard form and write $\Delta(\varphi)$ for the modular operator of φ. For $n \in \mathbb{N}$ we define

$$k_n = \sqrt{\frac{n}{\pi}} \int_{-\infty}^\infty e^{-nt^2} \tau_t(k) \, dt, \tag{12.13}$$

where $\tau_t(a) = \Delta(\varphi)^{-it} a \Delta(\varphi)^{it}$ ($t \in \mathbb{R}$, $a \in \mathcal{M}$) is the modular group of φ. It is well-known that $k_n \to k$ strongly as $n \to \infty$. The operator k_n is analytic for the group (τ_t) in the following sense. The function

$$z \mapsto k_n(z) = \sqrt{\frac{n}{\pi}} \int_{-\infty}^\infty e^{-n(t-z)^2} \tau_t(k) \, dt \tag{12.14}$$

is an entire $\mathcal{M}$-valued analytic function on the complex plane and for real s

$$k_n(s) = \tau_s(k_n).$$

Set

$$T_m(z,k,n) \equiv \int_0^1 \int_0^{t_1} \cdots \int_0^{t_{m-1}} k_n(t_m z)...k_n(t_1 z)\, dt_m...dt_1 \tag{12.15}$$

$(z \in \mathbb{C},\, k, n \in \mathbb{N})$. Since

$$\|k_n(z)\| \le 2e^{n \operatorname{Im} z}\|k\|, \tag{12.16}$$

one obtains

$$\|T_m(z,k,n)\| \le 2^m \|k\|^m e^{n \operatorname{Im} z}/m! . \tag{12.17}$$

This shows that the power series

$$\Gamma_n(z) \equiv 1 + \sum_{m=1}^{\infty} (iz)^m T_m(z,k,n) \tag{12.18}$$

converges for every $z \in \mathbb{C}$ and $\Gamma_n(z)$ is an entire analytic function. The perturbation expansion of $\exp it(-\log \Delta(\varphi) + k_n)$ is exactly $\Gamma_n(t)\exp(-it\log \Delta(\varphi))$. From this we recognize that

$$e^{iz(-\log \Delta(\varphi) + k_n)} = \Gamma_n(z)e^{-iz\log \Delta(\varphi)} \tag{12.19}$$

including the equality of the domains. It follows also (by analytic continuation from real z) that

$$\Gamma_n(\bar{z})^* = \Gamma_n(z)^{-1} \qquad (z \in \mathbb{C}).$$

Let ξ_φ be the vector representative of φ in the natural positive cone. (So ξ_φ is a cyclic and separating vector for $\mathcal{M}$.) We define

$$\xi_n = \Gamma_n(i/2)\xi_\varphi \equiv e^{-\frac{1}{2}(-\log \Delta(\varphi) + k_n)}\xi_\varphi.$$

$\Gamma_n(i/2)$ being invertible, the vector ξ_n is cyclic for $\mathcal{M}$.

Lemma 12.4 ξ_n is in the natural positive cone.

Proof. Due to the self-duality of the natural positive cone it suffices to show that

$$\langle aJaJ\xi_\varphi,\, \xi_n \rangle \ge 0$$

for every $a \in \mathcal{M}$. This is obtained as follows.

$$\begin{aligned}
\langle aJaJ\xi_\varphi,\, \Gamma_n(i/2)\xi_\varphi \rangle &= \langle Ja J\xi_\varphi,\, J\Delta(\varphi)^{1/2}\Gamma_n(i/2)a\xi_\varphi \rangle \\
&= \langle e^{-\frac{1}{2}(-\log \Delta(\varphi) + k_n)}a\xi_\varphi,\, a\xi_\varphi \rangle
\end{aligned}$$

which is obviously nonnegative. □

Let ψ_n be the vector state generated by the vector $\eta_n^0 \equiv \eta_n/\|\eta_n\|$ and write c_n for $-\log \|\eta_n\|^2$. For $a, b \in \mathcal{M}$ we compute

$$\langle \Delta(\psi_n, \varphi)^{1/2} a\xi_\varphi, b\xi_\varphi \rangle = \langle Ja^* \xi_n^0, b\xi_\varphi \rangle$$
$$= \langle Ja^* \Gamma_n(i/2) e^{\frac{1}{2}c_n} \xi_\varphi, b\xi_\varphi \rangle$$
$$= \langle \Delta(\varphi)^{1/2} \Gamma_n(i/2)^* a\xi_\varphi, b\xi_\varphi \rangle e^{\frac{1}{2}c_n}$$
$$= \langle a\xi_\varphi, \Gamma_n(i/2) \Delta(\varphi)^{1/2} b\xi_\varphi \rangle e^{\frac{1}{2}c_n}$$
$$= \langle a\xi_\varphi, e^{-\frac{1}{2}(-\log \Delta(\varphi) + k_n - c_n)} b\xi_\varphi \rangle$$

and arrive at

$$\Delta(\psi_n, \varphi)^{1/2} | \mathcal{M}\xi_\varphi = e^{-\frac{1}{2}(-\log \Delta(\varphi) + k_n - c_n)} | \mathcal{M}\xi_\varphi. \tag{12.20}$$

$\mathcal{M}\xi_\varphi$ being a core for $\Delta(\psi_n, \varphi)^{1/2}$, we obtain that the two operators in (12.20) are actually equal, and taking logarithms we have

$$\log \Delta(\psi_n, \varphi) = \log \Delta(\varphi) - k_n + c_n. \tag{12.21}$$

Lemma 12.5 For every faithful normal state σ of $\mathcal{M}$

$$\log \Delta(\psi_n, \sigma) = \log \Delta(\varphi, \sigma) - k_n + c_n \tag{12.22}$$

holds.

Proof. Observe that for $\sigma = \varphi$ our statement is just (12.21) which tells us that

$$[D\varphi, D\psi_n]_t = \Delta(\varphi, \varphi)^{it} \Delta(\psi_n, \varphi)^{-it} = \Gamma_n(-t)^* e^{itc_n}.$$

On the other hand we have

$$\Delta(\psi_n, \sigma)^{it} \Delta(\varphi, \sigma)^{-it} = [D\psi_n, D\varphi]_t = [D\varphi, D\psi_n]_t^*,$$

and arrive at

$$\Delta(\psi_n, \sigma)^{it} = \Gamma_n(-t) e^{itc_n} \Delta(\varphi, \sigma)^{it} \tag{12.23}$$

for every $t \in \mathbb{R}$. Our aim is to show that (12.23) may be differentiated at $t = 0$ and applied to at every $\zeta \in \mathcal{D}(\log \Delta(\varphi, \sigma))$. An obvious estimation gives

$$\left\| \left(\frac{1}{it}(\Delta(\psi_n, \sigma)^{it} - I) - \log \Delta(\varphi, \sigma) + k_n - c_n \right)\zeta \right\|$$
$$\leq \left\| \Gamma_n(-t) e^{itc_n} \left(\frac{1}{it}(\Delta(\varphi, \sigma)^{it} - I) - \log \Delta(\varphi, \sigma) \right)\zeta \right\|$$
$$+ \left\| (\Gamma_n(-it) e^{itc_n} - I) \log \Delta(\varphi, \sigma)\zeta \right\|$$
$$+ \left\| \left(\frac{1}{it}(\Gamma_n(-t) e^{itc_n} - I) + k_n - c_n \right)\zeta \right\|.$$

The first and the second terms of the right hand side converge to 0 trivially as $t \to 0$. A brief look at (12.18) and (12.15) shows that the third term converges to 0 as well. In this way we have established

$$\log \Delta(\psi_n, \varphi) \supset \log \Delta(\varphi, \sigma) - k_n + c_n \,,$$

which must be an equality, because the right hand side is a self-adjoint operator with domain $\mathcal{D}(\log \Delta(\varphi, \sigma))$. □

In the lemma (12.22) is stated as an equality of operators. However, one can see that it may well be interpreted as an equality of forms, too. This comes from the boundedness of the operator $k_n + c_n$. Hence

$$\langle \xi_\sigma, \log \Delta(\psi_n, \sigma)\xi_\sigma \rangle = \langle \xi_\sigma, \log \Delta(\varphi, \sigma)\xi_\sigma \rangle - \langle \xi_\sigma, k_n \xi_\sigma \rangle + c_n$$

or

$$c_n + S(\sigma, \psi_n) = S(\sigma, \varphi) + \sigma(k_n).$$

We see that ψ_n minimizes the functional $\sigma \mapsto S(\sigma, \varphi) + \sigma(k_n)$. This implies

$$\psi_n = [\varphi^{k_n}] \quad \text{and} \quad c_n = c(\varphi, k_n).$$

Now we are ready to prove the following important result.

Theorem 12.6 Let φ be a faithful state and σ an arbitrary normal state on the von Neumann algebra $\mathcal{M}$ and let $k \in \mathcal{M}^{sa}$. Then

$$P \log \Delta([\varphi^k], \sigma) = P \log \Delta(\varphi, \sigma) - Pk + c(\varphi, k)P$$

where $P = [\mathcal{M}\xi_\sigma]$, and $\mathcal{M}$ is supposed to be in standard form.

Proof. Recall that $P \in \mathcal{M}'$ and $\operatorname{supp} \Delta(\varphi, \sigma) = \operatorname{supp} \Delta([\varphi^k], \sigma) = P$. First the case of a faithful σ will be treated. From the discussion preceding the theorem we have

$$\log \Delta([\varphi^{k_n}], \sigma) = \log \Delta(\varphi, \sigma) - k_n + c(\varphi, k_n) \,, \tag{12.24}$$

where k_n is the regularization of k defined by (12.13). The idea of the proof is to take the limit of (12.24) in the strong resolvent sense as $n \to \infty$. Proposition 12.3 provides $c(\varphi, k_n) \to c(\varphi, k)$ and $[\varphi^{k_n}] \to [\varphi^k]$ in norm. The latter implies

$$\lim_{n \to \infty} \log \Delta([\varphi^{k_n}], \sigma) = \log \Delta([\varphi^{k_n}], \sigma) \,,$$

which is the key for the limit $n \to \infty$ in (12.24).

When σ is not faithful we approximate it by setting

$$\sigma_n = n^{-1}\varphi + (1 - n^{-1})\sigma \quad (n \in \mathbb{N}) \,.$$

So $\sigma_n \to \sigma$ in norm and we have

$$\log \Delta([\varphi^k], \sigma_n) = \log \Delta(\varphi, \sigma_n) - k + c(\varphi, k) \,, \tag{12.25}$$

since the theorem is already proven for faithful σ_n. Moreover, Proposition 4.8 yields the strong limits

$$(1 + \Delta([\varphi^k], \sigma_n)^{1/2})^{-1} P \to (1 + \Delta([\varphi^k], \sigma)^{1/2})^{-1} P$$
$$(1 + \Delta(\varphi, \sigma_n)^{1/2})^{-1} P \to (1 + \Delta(\varphi, \sigma)^{1/2})^{-1} P. \qquad (12.26)$$

We infer

$$
\begin{aligned}
(i + \log \Delta([\varphi^k], \sigma))^{-1} P &= \lim_n (i + \log \Delta([\varphi^k], \sigma_n))^{-1} P \\
&= \lim_n (i + \log \Delta(\varphi, \sigma_n) - k + c(\varphi, k))^{-1} P \\
&= (i + \log \Delta(\varphi, \sigma) - k + c(\varphi, k))^{-1}.
\end{aligned}
$$

Here we have first applied (12.26) as well as Lemma 4.10 with the choice $f(x) = (i + 2 \log x)^{-1}$, followed by (12.25) and finally by a combination of (12.26), Lemma 4.10 and Lemma 4.11. Since P commutes with k and with the relative modular operators, our statement follows. $\qquad \square$

Corollary 12.7 Suppose that $\mathcal{M}$ is in standard form and $k \in \mathcal{M}^{sa}$. Then the vector representative of the perturbed state $[\varphi^k]$ in the natural positive cone is of the form

$$e^{\frac{1}{2}(\log \Delta(\varphi, \varphi) - pkp + c(\varphi, k))} \xi_\varphi$$

with $p = \operatorname{supp} \varphi$.

Proof. This corollary is based on the fact $\xi_\psi = \Delta(\psi, \varphi)^{1/2} \xi_\varphi$. $\qquad \square$

The corollary provides an analytic determination of the perturbed state. The vector functional induced by the *perturbed vector*

$$\xi_\varphi^k = e^{\frac{1}{2}(\log \Delta(\varphi, \varphi) + pkp)} \xi_\varphi$$

is denoted by φ^k. The change in the sign of the exponent in the relation

$$[\varphi^k] = e^{c(\varphi, k)} \varphi^{-k} \qquad (12.27)$$

comes from the different conventions of mathematicians and physicists. In statistical mechanics β corresponds to the inverse temperature and is therefore nonnegative, but mathematicians take β to be arbitrary and set $\beta = -1$.

The perturbed vector possesses the following *expansion*

$$\xi_\varphi^k = \sum_{n=0}^\infty \int_0^{1/2} dt_1 \int_0^{t_1} dt_2 \ldots \int_0^{t_{n-1}} dt_n \, k(t_n) k(t_{n-1}) \ldots k(t_1) \, \xi_\varphi \qquad (12.28)$$

where $k(t) = \Delta(\varphi, \varphi)^t k \Delta(\varphi, \varphi)^{-t}$. The series (12.28) is absolutely convergent due to the estimate

$$\| \Delta^{s_1} k^{(1)} \Delta^{s_2} k^{(2)} \ldots \Delta^{s_n} k^{(n)} \xi_\varphi \| \le \| \xi_\varphi \| \prod_{i=1}^n \| k^{(i)} \| \qquad (12.29)$$

($k^{(i)} \in \mathcal{A}^{sa}$, $0 \le s_i$ and $\sum_i s_i \le 1/2$).

Corollary 12.8 If φ and σ are normal states on $\mathcal{M}$ and $k \in \mathcal{M}^{sa}$, then

$$S(\sigma, [\varphi^k]) = S(\sigma, \varphi) + \sigma(k) - c(\varphi, k).$$

Proposition 12.9 Let φ be a normal state of the von Neumann algebra $\mathcal{M}$, and let (h_n) be an increasing sequence of extended valued self-adjoint operators affiliated with $\mathcal{M}$. Let h stand for $\sup\{h_n : n \in \mathbb{N}\}$ and assume that $c(\varphi, h)$ is finite. Then $c(\varphi, h_n) \to c(\varphi, h)$, $S([\varphi^{h_n}], \varphi) \to S([\varphi^h], \varphi)$ and $[\varphi^{h_n}] \to [\varphi^h]$ in norm.

Proof. First of all, the relation

$$c(\varphi, h) = S([\varphi^h], \varphi) + [\varphi^h](h)$$
$$\geq S([\varphi^h], \varphi) + [\varphi^h](h_n)$$

shows that $c(\varphi, h_n)$ is finite (and hence $[\varphi^{h_n}]$ is defined). It follows also that $S([\varphi^{h_n}], \varphi)$ has a finite upper bound.

From this point we can follow the proof of Proposition 12.3. The set $\{[\varphi^{h_n}] : n \in \mathbb{N}\}$ is weakly relatively compact and any convergent subnet from it minimizes the functional

$$\psi \mapsto S(\psi, \varphi) + \psi(h).$$

Therefore, $[\varphi^{h_n}]$ converges to $[\varphi^h]$ pointwise and $c(\varphi, h_n) \to c(\varphi, h)$. The norm convergence of $([\varphi^{h_n}])_n$ is established through

$$S([\varphi^{h_n}], [\varphi^h]) \to 0$$

as in Theorem 12.3.   □

The relation $[\varphi^{h+k}] = [[\varphi^h]^k]$ is very plausible. It is referred to as the chain rule. Our goal is to prove it for an unbounded h affiliated with the given von Neumann algebra.

Theorem 12.10 Let $k = k^* \in \mathcal{M}$ and let h be an extended-valued lower-bounded self-adjoint operator affiliated with the von Neumann algebra $\mathcal{M}$. Let φ be a normal state of $\mathcal{M}$ such that $c(\varphi, h) < \infty$. Then $[\varphi^{h+k}] = [[\varphi^h]^k]$, $c(\varphi, h + k) = c([\varphi^h], k) + c(\varphi, h)$, and for every $\sigma \in \mathfrak{S}_*(\mathcal{M})$ the equality

$$S(\sigma, [\varphi^h]) + \sigma(k) + c(\varphi, h) = c(\varphi, h + k) + S(\sigma, [\varphi^{h+k}]) \qquad (12.30)$$

holds.

Proof. Without restricting generality we may assume that φ is faithful and $\mathcal{M}$ is in standard form. In the first part of the proof we shall suppose that h is bounded, and the case of a general h will be treated by the approximation provided by Proposition 12.9.

Corollary 12.8 tells us that

$$c(\varphi, h + k) + S(\sigma, [\varphi^{h+k}]) = S(\sigma, \varphi) + \sigma(h + k)$$
$$= S(\sigma, [\varphi^h]) + c(\varphi, h) + \sigma(k)$$
$$= S(\sigma, [[\varphi^h]^k]) + c([\varphi^h], k) + c(\varphi, h)$$

for any faithful normal state σ. Putting first $\sigma = [\varphi^h]^k$ and then $\sigma = [\varphi^{h+k}]$ we obtain

$$c(\varphi, h + k) + S([[\varphi^h]^k], [\varphi^{h+k}]) = c([\varphi^h], k) + c(\varphi, h)$$

and

$$c(\varphi, h + k) = S([\varphi^{h+k}], [[\varphi^h]^k]) + c([\varphi^h], k) + c(\varphi, h).$$

Therefore we arrive at

$$S([\varphi^{h+k}], [[\varphi^h]^k]) + S([[\varphi^h]^k], [\varphi^{h+k}]) = 0$$

which yields $[\varphi^{h+k}] = [[\varphi^h]^k]$ as well as the other stated relations for a bounded h.

When h is extended-valued we choose a sequence $(h_n) \subset \mathcal{M}^{sa}$ such that $h_n \to h$ increasingly. Let $q = \operatorname{supp}[\varphi^h]$.

According to the lower semi-continuity of the relative entropy we have

$$S([\varphi^{h+k}], [\varphi^h]) + S([\varphi^h], [\varphi^{h+k}])$$
$$\leq \liminf_{n \to \infty} S([\varphi^{h_n+k}], [\varphi_n^h]) + S([\varphi_n^h], [\varphi^{h_n+k}]) \leq 2\|k\|.$$

The finite upper bound yields

$$\operatorname{supp}[\varphi^{h+k}] = \operatorname{supp}[\varphi^h])$$

Let σ be a normal state of $\mathcal{M}$. If $\operatorname{supp}\sigma$ is not smaller than q, then $S(\sigma, [\varphi^h]) + \sigma(k)$ is infinite. Otherwise,

$$S(\sigma, [\varphi^h]) + \sigma(k) = S(\sigma, [\varphi^h]) + \sigma(qkq),$$

and we establish $[[\varphi^h]^k] = [[\varphi^h]^{qkq}]$. Similarly one gets that $[\varphi^{h+k}] = [\varphi^{h+qkq}]$. These relations show that we may assume $k = qkq$ in the rest of the proof.

By an application of Theorem 12.6 we have

$$P \log \Delta([\varphi^{h_n+k}], \sigma) = P \log \Delta([\varphi^{h_n}], \sigma) - Pk + c([\varphi^{h_n}], h_n + k)P$$
$$= P \log \Delta([\varphi^{h_n}], \sigma) - Pk + c(\varphi, h_n + k)P + c(\varphi, h_n)P,$$

where $P = [\mathcal{M}\xi_\sigma]$ is a projection in the commutant $\mathcal{M}'$. We know from Proposition 12.9 that $[\varphi^{h_n+k}] \to [\varphi^{h+k}]$ and $[\varphi^{h_n}] \to [\varphi^h]$ in norm as $n \to \infty$. These relations imply the strong resolvent convergence of the corresponding relative modular operators. Namely,

$$(1 + \Delta([\varphi^{h_n+k}], \sigma)^{1/2})^{-1} P \to (1 + \Delta([\varphi^{h+k}], \sigma)^{1/2})^{-1} P$$

and

$$(1 + \Delta([\varphi^{h_n}], \sigma)^{1/2})^{-1} P \to (1 + \Delta([\varphi^h], \sigma)^{1/2})^{-1} P$$

(see Proposition 4.8). Now the argument can be continued like in the proof
of Theorem 12.6. By means of Lemma 4.10 and 4.11 we deduce

$$(i + Pq \log \Delta([\varphi^{h+k}], \sigma))^{-1} Pq$$
$$(i + Pq \log \Delta([\varphi^h], \sigma) - Pk + c(\varphi, h+k)P + c(\varphi, h)P)^{-1} Pq$$

and hence

$$Pq \log \Delta([\varphi^{h+k}], \sigma) = Pq \log \Delta([\varphi^h], \sigma) - Pk + c(\varphi, h+k)P + c(\varphi, h)P.$$

Now take the functional $\langle \xi_\sigma, \cdot \xi_\sigma \rangle$ of both sides of this equality. Since $\xi_\sigma = P\xi_\sigma = q\xi_\sigma$ what we obtain is the required inequality (12.30). By the definition
of the perturbed state this implies the chain rule $[\varphi^{h+k}] = [[\varphi^h]^k]$. ☐

Suppose that the states φ and ω are given. If $\omega = [\varphi^h]$, then h is called
the relative Hamiltonian of ω with respect to φ. The discussion at the beginning of the chapter explains this terminology. Roughly speaking, if φ is the
equilibrium state for the Hamiltonian H, then ω is the equilibrium state for
the Hamiltonian $H + h$. The next result of Araki establishes the existence of
the relative Hamiltonian under a majorization condition.

Theorem 12.11 Let $\varphi, \omega \in \mathfrak{S}_*(\mathcal{M})$ such that $\lambda\varphi \leq \omega \leq \mu\varphi$ with some
positive constants λ and μ. Then there exists a $h \in \mathcal{M}^{sa}$ such that $[\varphi^h] = \omega$
and $-\log\mu \leq h \leq -\log\lambda$.

Proof. We may assume that φ and ω are faithful and the algebra $\mathcal{M}$ is in
standard form. The Connes cocycle

$$t \mapsto [D\omega, D\varphi]_t = \Delta(\omega, \varphi)^{it} \Delta(\varphi, \varphi)^{-it}$$

is a unitary operator-valued function on the real line and under the hypothesis
it admits an analytic extension to a strip. Differentiation at $t = 0$ shows that

$$\mathcal{D}(\log \Delta(\omega, \varphi)) \supset \mathcal{D}(\log \Delta(\varphi, \varphi))$$

must hold, and the densely defined operator

$$\log \Delta(\varphi, \varphi) - \log \Delta(\omega, \varphi)$$

has a bounded closure $h \in \mathcal{M}^{sa}$. The perturbed vector

$$\xi_\varphi^{-h} = \exp \tfrac{1}{2}(\log \Delta(\varphi, \varphi)) - h)\xi_\varphi = \Delta(\omega, \varphi)^{1/2}\xi_\varphi$$

is just the vector representative of ω in the natural positive cone. So the
relation $[\varphi^h] = \omega$ is readily concluded. The bounds for h follow from the
relations

$$\lambda\Delta(\varphi, \varphi) \leq \Delta(\omega, \varphi) \leq \mu\Delta(\varphi, \varphi)$$

(see the definition of the spatial derivative operator in Chapter 4). ☐

Let $\mathcal{A}$ be a C*-algebra with a state φ fixed for a while. The relationship between the functional $\omega \mapsto S(\omega, \varphi)$ defined on the state space and the

function $k \mapsto c(\varphi, k)$ defined on the self-adjoint part of $\mathcal{A}$ may be clarified by means of a little standard convex analysis. Let V and U be two linear spaces being in duality through the pairing $\langle . , . \rangle$. The spaces V and U will be endowed with the topologies $\sigma(V, U)$ and $\sigma(U, V)$, respectively. Let F be a function from V into $\mathbb{R} \cup \{+\infty\}$. Then the formula

$$F^*(u) = \sup\{\langle v, u \rangle - F(v) : v \in V\} \tag{12.31}$$

defines a function F^* from U into $\mathbb{R} \cup \{+\infty\}$ called the conjugate function or Legendre transform of F. It is obvious that in (12.31) we may confine ourselves to those v for which $F(v)$ is finite. This process can be repeated and leads to the second conjugate F^{**}, which is a function of V into $\mathbb{R} \cup \{+\infty\}$:

$$F^{**}(v) = \sup\{\langle v, u \rangle - F^*(u) : u \in U\} \tag{12.32}$$

A basic result in convex analysis states that $F = F^{**}$ if F is a lower semi-continuous convex function.

Denoting by $\mathcal{A}_h^*$ the real linear space of hermitian functionals on $\mathcal{A}$ we adopt $U = \mathcal{A}^{sa}$ and $V = \mathcal{A}_h^*$. With respect to the duality

$$\langle \nu, k \rangle = -\nu(k) \quad (k \in \mathcal{A}^{sa} \text{ and } \nu \in \mathcal{A}_h^*)$$

the conjugate function of

$$F(\omega) = \begin{cases} \beta^{-1} S(\omega, \varphi) & \text{if } \omega \text{ is a state} \\ +\infty & \text{otherwise} \end{cases}$$

is nothing else but $-c_\beta(\varphi, k)$. Since F is lower semi-continuous and convex, we have

$$\beta^{-1} S(\omega, \varphi) = \sup\{-\omega(k) + c_\beta(\varphi, k) : k \in \mathcal{A}^{sa}\}. \tag{12.33}$$

From the definition of $c(\varphi, k)$ it follows that

$$c(\varphi, k) \leq \varphi(k) \quad (k \in \mathcal{A}^{sa}), \tag{12.34}$$

which is a generalization of the *Peierls-Bogoliubov inequality*. For a finite quantum system (12.34) would read as

$$-\log \mathrm{Tr}\, \exp(-h - H) \leq \mathrm{Tr}\, h \exp(-H).$$

The estimate (12.34) is based on the concavity of the function $f(t) = c(\varphi, th)$. Since $k \mapsto c(\varphi, k)$ is concave, we have $f(1) \leq f'(0) \cdot 1 + f(0)$, and from (12.28) we obtain $f'(0) = \varphi(h)$. The lower estimate

$$-\log \varphi(\exp(-h)) \leq c(\varphi, h) \quad (h \in \mathcal{A}^{sa}) \tag{12.35}$$

is due to *Araki* and extends the *Golden-Thompson inequality*. (See (3.22) for finite quantum systems with a slightly different notation.) In principle, the proof of (12.35) is similar to that of Theorem 3.11. The only essential difference is that the required Trotter-type exponential formula is more sophisticated. The details are omitted here.

Proposition 7.13 provides another example of Legendre transform. Let φ be a fixed state of a finite quantum system $\mathcal{A}$ and consider the natural duality $\langle \omega, h \rangle \equiv \omega(h)$ between the hermitian parts of $\mathcal{A}$ and its dual. Then the conjugate convex function of $h \mapsto \log \varphi(e^h)$ is $\omega \mapsto S_{\text{co}}(\omega, \varphi)$.

Finally we devote some space to an analytic extension approach to state perturbation. In this approach it is more convenient to deal with the nonnormalized functional φ^h than with the normalized one $[\varphi^h]$. Let $\mathcal{M}$ be a von Neumann algebra in standard form and let φ be a faithful normal state of $\mathcal{M}$. According to Theorem 12.6 for $k \in \mathcal{M}^{sa}$ the operator $\log \Delta(\varphi, \varphi) + k$ has a self-adjoint closure which is nothing else but $\log \Delta(\varphi^k, \varphi)$. By means of the regularization (12.14) one can show that for a fixed $x \in \mathcal{M}$ the function

$$it \mapsto \langle \xi_\varphi, x \exp(it(\log \Delta(\varphi, \varphi) + k))\xi_\varphi \rangle$$

admits an analytic continuation to the strip $S = \{z \in \mathbb{C} : 0 \leq \operatorname{Re} z \leq 1\}$. We denote by $\varphi^{(z,k)}(x)$ the value of the continuation at $z \in \mathbb{C}$. By the simple computation

$$\langle \xi_\varphi, x \exp(\log \Delta(\varphi, \varphi) + k)\xi_\varphi \rangle = \langle \Delta(\varphi^k, \varphi)^{1/2} x^* \xi_\varphi, \xi_\varphi^k \rangle$$
$$= \langle \xi_\varphi^k, J\Delta(\varphi^k, \varphi)^{1/2} x^* \xi_\varphi \rangle = \langle \xi_\varphi^k, J\Delta(\varphi^k, \varphi)^{1/2} J\Delta(\varphi^k, \varphi)^{1/2} x \xi_\varphi \rangle$$
$$= \langle \xi_\varphi^k, \Delta(\varphi^k, \varphi)^{-1/2} \Delta(\varphi^k, \varphi)^{1/2} x \xi_\varphi \rangle = \varphi^k(x)$$

we obtain that

$$\varphi^{(1,k)}(x) = \varphi^k(x) \qquad (x \in \mathcal{M}). \tag{12.36}$$

In this way the functional φ^k may be regarded as a value of the analytic family $\varphi^{(z,k)} \in \mathcal{M}_*$ $(z \in S)$. It can be shown that

$$\|\varphi^{(1,k)}\| \leq \exp \|k\| \tag{12.37}$$

and

$$\|\varphi^{(1,k)} - \varphi^{(1,h)}\| \leq \|k - h\| \exp(\|k\| + \|h\|). \tag{12.38}$$

This technique facilitates the handling unbounded perturbations. Let H be a self-adjoint operator affiliated with the von Neumann algebra $\mathcal{M}$ such that

(i) $\xi_\varphi \in \mathcal{D}(H)$,
(ii) $\log \Delta[\varphi, \varphi) + H$ is densely defined and has a self-adjoint closure.

Then $\varphi^{(1,H)} \in \mathcal{M}_*$ could be defined by means of the analytic extension for $z \in S$ as it was done for a bounded k.

Notes and Remarks. Traditionally perturbation theory concerns semigroups and there is a huge literature on this topic. A standard reference is [Kato 1966], in particular IX.2. Most results from perturbation theory of semigroups do not apply directly to the modular group. Perturbation theory was adapted to the context of operator algebras in the fundamental paper

[Araki 1973a], see also [Bratteli and Robinson 1981]. The relative entropy approach to state perturbation was suggested in [Donald 1990]. The chapter is based on this paper. (The book [Bratteli and Robinson 1981] follows a much more dynamical approach, the perturbed time evolution is the starting point.)

The concept of the extended-valued self-adjoint operator affiliated with a von Neumann algebra appears in [Haagerup 1979], see also [Strătilă 1981]. Our finite dimensional example of perturbation is related to an entropy mimimization problem in [Raggio and Werner 1990]; the minimizer of the functional $\omega \mapsto S(\omega, \varphi)$ under the constraint $\operatorname{supp} \omega \leq p$ is the state $[\varphi^{\infty q}]$, where $q = I - p$.

The estimate (12.29) and Theorem 12.10 are from [Araki 1973a]. The presented proof of Theorem 12.10 is new and it uses the result of [Connes 1973] concerning the analytic extension of the Connes cocycle. (A different proof was sketched in [Donald 1990], where the result was more general. The majorization $\omega \leq \mu\varphi$ ensures the existence of an extended-valued relative Hamiltonian which becomes a semi-bounded self-adjoint operator for faithful states.)

The conjugate convex function is contained in many books on convex analysis, for example [Ekeland and Temam 1976]. The connection of relative entropy in operator algebras with Legendre transformation was first explicitly mentioned in [Petz 1988c]. Inequality (12.35) was proved in [Araki 1973b]. The analytic family $\varphi^{(z,k)}$ was defined in [Sakai 1986] and extended the analytic machinery of Araki. The estimates (12.37) and (12.38) are from [Sakai 1985]. The generalization to unbounded perturbation is in [Sakai 1987], where also the strong resolvent continuity of the mapping $H \mapsto \varphi^H$ is proved. The paper [Donald 1991] is devoted to the joint continuity of $(\varphi, k) \mapsto [\varphi^k]$. An example for the perturbation of a quasi-free state by unbounded (field) operators is worked out in [Honegger 1990].

13 Variational Expression
of Perturbational Limits

In mathematical physics one often meets the problem of the asymptotic behaviour, for large n, of a sequence

$$\text{trace } (\exp H_n) \qquad (n = 1, 2, \dots),$$

where (H_n) is a sequence of self-adjoint operators acting on some Hilbert
space. The problem arises, for example, in statistical mechanics of quantum
lattice systems. In case of a one-dimensional infinite lattice one has a copy
$\mathcal{B}_n$ of a C*-algebra associated to each lattice node $n \in \mathbb{Z}$. For any $\mathcal{J} \subset \mathbb{Z}$ we
introduce the notation $\mathcal{B}_{\mathcal{J}} = \otimes_{n \in \mathcal{J}} \mathcal{B}_n$ and the shortscript $\mathcal{B} = \mathcal{B}_{\mathbb{Z}}$. The n-
fold tensor product $\mathcal{B}_{[1,n]}$ represents a compound system made of n identical
neighbouring lattice nodes. (A more detailed discussion of quantum lattice
systems is found in Chapter 15.) We assume that a Hamiltonian H_n can
be associated with each finite subsystem $\mathcal{B}_{[1,n]}$. If these Hamiltonian can be
diagonalized, the limit

$$\lim_{n \to \infty} \frac{1}{n\beta} \log \text{Tr} \, e^{-\beta H_n} \tag{13.1}$$

may be treated by powerful probabilistic methods. Roughly speaking, the
sequence (H_n) is modelled by a sequence of random variables and the large
deviation technique may be invoked.

Let (μ_n) be a sequence of measures on a (sufficiently regular) topological
space X. The sequence is said to obey the *large deviation principle* with
constants (V_n) and rate function $\mathcal{L} : X \to \mathbb{R}^+ \cup \{\infty\}$ if the following conditions
are satisfied:

(i)   The level sets $\{x \in X : \mathcal{L}(x) \le t\}$ are compact for each $t \in \mathbb{R}^+$.
(ii) $\limsup_n V_n^{-1} \log \mu_n(F) \le -\inf\{\mathcal{L}(x) : x \in F\}$ whenever $F \subset X$ is a
      closed set.
(iii) $\liminf_n V_n^{-1} \log \mu_n(G) \ge -\inf\{\mathcal{L}(x) : x \in G\}$ whenever $G \subset X$ is an
      open set.

The strong attachment of the large deviation principle to the thermodynamic limit (13.1) will be transparent from the next result due to Varadhan.
The large deviation principle is equivalent to the limit relation

$$\lim_{n \to \infty} \frac{1}{V_n} \log \int_X e^{V_n f(x)} \, d\mu_n(x) = \sup\{f(x) - \mathcal{L}(x) : x \in X\} \tag{13.2}$$

for every continuous bounded function $f : X \to \mathbb{R}$. Let the local Hamiltonians be of the form

$$H_{[m,n]} = \sum_{i=m}^{n} h_i + \sum_{i=m}^{n} x_i \times \left(\frac{1}{n-m+1} \sum_{j=m}^{n} x_j \right) \tag{13.3}$$

where $h_i \in \mathcal{B}_i$ and $x_i \in \mathcal{B}_i$ are copies of some self-adjoint operators $h_1 \in \mathcal{B}_1$ and $x_1 \in \mathcal{B}_1$, respectively. The interaction (13.3) is of a mean-field type and it is one of the simplest example when the classical large deviation theory is not suitable for the determination of the limit (13.1). The expression "mean-field" reflects the second term of (13.3). (At the end of Chapter 15 one can find a short discussion of more general mean-field models.) If the single system is described by a finite dimensional C*-algebra $\mathcal{B}_1$ then the minimal free energy density at inverse temperature β is given by

$$F_n(\beta) = -\frac{1}{n\beta} \log \text{Tr} \exp \left(-\beta \sum_{i=1}^{n} h_i - \frac{\beta}{n} \sum_{i,j=1}^{n} x_i x_j \right). \tag{13.4}$$

Let φ be the state of $\mathcal{B}_1$ which possesses the density

$$D_\varphi = \frac{\exp(-\beta h)}{\text{Tr} \exp(-\beta h)}, \tag{13.5}$$

and let φ_∞ be the corresponding product state on $\mathcal{B}$. In the notation of the previous chapter we have

$$F_n(\beta) = \frac{1}{n} c_\beta \left(\varphi_\infty, \frac{1}{n} \sum_{i,j=1}^{n} x_i x_j \right). \tag{13.6}$$

It is not a real restriction if we put $\beta = 1$ in the sequel. In order to make contact with probability theory, let us assume for a while that $\mathcal{B}$ is abelian, or least $[h, x] = 0$. Then we have a classical mean-field picture and φ_∞ corresponds to a product measure which is the joint distribution of the sequence (x_i) of identically distributed independent random variables. In this probabilistic translation (13.4) becomes

$$\frac{1}{n} \log \int \exp \left(-n \left(\frac{1}{n} \sum_{i=1}^{n} x_i \right)^2 \right) d\mu. \tag{13.7}$$

If μ_n is the distribution of $\frac{1}{n} \sum_{i=1}^{n} x_i$ then the sequence (μ_n) obeys the large deviation principle with the rate function $\mathcal{L}$ given below:

$$L(u) = \int \exp u x_1 \, d\mu(u) \quad \text{and} \quad \mathcal{L}(x) = \sup\{ux - \log L(u) : u \in \mathbb{R}\} \tag{13.8}$$

(The large deviation result for the average of identically distributed independent random variables is a reformulation of a classical theorem proved by *Cramér*.) Typically $\mathcal{L}$ is a convex function vanishing at the expectation value m of x_1. One knows from the law of large numbers that if $F \subset \mathbb{R}$ is a closed set such that $m \notin F$ then

$$\mathrm{Prob}\left(\frac{1}{n}\sum_{i=1}^{n} x_i \in F\right) \to 0$$

as $n \to \infty$. Condition (ii) of the large deviation principle tells us that this convergence is exponentially fast and its speed is shown by the rate function.

We want to conclude from the above discussion that the thermodynamics of a mean-field picture is strongly related to the large deviation principle in the case $[h, x] = 0$. All this will serve as a motivation for us to develop an analogous theory in the pure quantum case $[h, x] \neq 0$. The functional c from (13.6) will take the place of "log $\int$ exp" here and our guideline will be version (13.2) of the large deviation principle. Although the primary aim of this chapter is to treat limit theorems of Varadhan type, this means the verification of the free energy minimum principle of thermodynamics for *mean-field models*.

Let γ denote the right shift automorphism of the C*-algebra $\mathcal{B}$. The limit of the averages

$$s_n(a) \equiv \frac{1}{n}\sum_{i=1}^{n-1}\gamma^i(a) \qquad (a \in \mathcal{B}^{sa},\ n \in \mathbb{N})$$

is the subject of ergodic theorems. The C*-algebraic ergodic theorem we are going to prove needs the following lemma.

Lemma 13.1 Let f be a polynomial and $K \in \mathbb{R}^+$. Then for each $\varepsilon > 0$ there exists a $\delta > 0$ such that

$$\|f(A) - f(B)\| < \varepsilon$$

whenever $A, B \in \mathcal{B}^{sa}$, $\|A - B\| < \delta$ and $\|A\|, \|B\| \leq K$.

Proposition 13.2 Let $g : [s, t] \to \mathbb{R}$ be a continuous function and $b \in \mathcal{B}^{sa}$ such that the spectrum of b is contained in $[s, t]$. Then

$$\lim_{n\to\infty} \omega\big(g(s_n(b))\big) \tag{13.9}$$

exists for every γ-invariant state ω of $\mathcal{B}$.

Proof. It is a plain consequence of the Weierstrass approximation theorem that it suffices to prove the existence of the limit (13.9) when g is a polynomial. A further possibility of reduction is based on Lemma 13.1. Since the local elements are norm dense in the algebra $\mathcal{B}$, we may assume that b is local, that is, there exists $k \in \mathbb{N}$ such that $b \in \mathcal{B}_{[1,k]}$.

First we consider the case $k = 1$. Then the sequence $s_n(b)$ is in the abelian C*-algebra $\ldots \otimes \mathcal{C} \otimes \mathcal{C} \otimes \ldots$, where $\mathcal{C}$ is the subalgebra generated by a in $\mathcal{B}_1$. According to the representation theorem of abelian C*-algebras, $\mathcal{C}$ may be viewed as an algebra of continuous functions on the spectrum of b, and ω restricted to $\ldots \otimes \mathcal{C} \otimes \mathcal{C} \otimes \ldots$ corresponds to an integration with respect to a measure μ. By the individual ergodic theorem $s_n(b)$ converges μ-almost everywhere, and so does $g(s_n(b))$ for any continuous bounded function g. The Lebesgue theorem tells us that

$$\int g(s_n(b)) \, d\mu \equiv \omega(g(s_n(b))$$

converges as well as $n \to \infty$. This completes the proof for $b \in \mathcal{A}_1^{sa}$.

The general case $b \in \mathcal{B}_{[1,k]}^{sa}$ will be reduced to the previously discussed abelian case by changing the localization. For an integer $l \geq k$ we set

$$b_l \equiv \sum_{i=0}^{l-k} \gamma^i(b) \in \mathcal{B}_{[1,l]}$$

and write the arbitrary n in the form $u \cdot l + r$ where u and r are integers and $0 \leq r < l$. Then

$$\left\| \sum_{j=0}^{n-1} \gamma^j(k) - \sum_{j=0}^{u-1} \gamma^{jl}(b_l) \right\| \leq (uk + r) \|b\| . \tag{13.10}$$

This gives that

$$\left\| s_n(b) - \frac{1}{u} \sum_{j=1}^{u-1} \gamma^{jl}(b_l/l) \right\|$$

can be arbitrary small if l is large enough (with respect to k) and n is large too. According to Lemma 13.1

$$\left\| f(s_n(b)) - f\left(\frac{1}{n} \sum_{j=1}^{n-1} \gamma^{jl}(b_l/l) \right) \right\|$$

is small as well. On the other hand, for fixed l

$$\omega\left(f\left(\frac{1}{n} \sum_{j=0}^{n-1} \gamma^{jl}(b_l/l) \right) \right)$$

is a Cauchy sequence. Indeed, the sequence $\gamma^{jl}(b_l/l)$ is built from pairwise commuting operators and the above abelization argument works. In this way we are able to conclude that $\omega(f(s_n(b)))$ is a Cauchy sequence, proving the convergence of (13.9). $\qquad \square$

At this point we can anticipate how the analogue of the variational formula (13.2) will look in the quantum case. Let $b \in \mathcal{B}_{[1,k]}^{sa}$ be a local operator in the quasi-local algebra $\mathcal{B}$. By the definition of the function c we have

$$c_\beta(\varphi_\infty, nf(s_n(b))) = c_\beta(\varphi_\infty|\mathcal{B}_{[1,n+k]}, nf(s_n(k)))$$

$$\leq \omega(nf(s_n(a))) + \frac{1}{\beta} S(\omega|\mathcal{B}_{[1,n+k]}, \varphi_\infty|\mathcal{B}_{[1,n+k]}) \tag{13.11}$$

for any state ω of $\mathcal{B}$. Assume now that ω is translation invariant and employ the superadditivity of the relative entropy (Corollary 5.21) as follows.

$$S(\omega|\mathcal{B}_{[1,n+m]}, \varphi_\infty|\mathcal{B}_{[1,n+m]})$$
$$\geq S(\omega|\mathcal{B}_{[1,n]}, \varphi_\infty|\mathcal{B}_{[1,n]}) + S(\omega|\mathcal{B}_{[n+1,n+m]}, \varphi_\infty|\mathcal{B}_{[n+1,n+m]})$$
$$= S(\omega|\mathcal{B}_{[1,n]}, \varphi_\infty|\mathcal{B}[1,n]) + S(\omega|\mathcal{B}_{[1,m]}, \varphi_\infty|\mathcal{B}_{[1,m]}).$$

This yields that $n^{-1}S(\omega|\mathcal{B}_{[1,n]}, \varphi_\infty|\mathcal{B}_{[1,n]})$ has a limit $S_M(\omega, \varphi_\infty)$ (see Lemma 10.4). The quantity $S_M(\omega, \varphi_\infty)$ will be called the *mean relative entropy* (or relative entropy density). Dividing the inequality (13.11) by n and letting $n \to \infty$ we arrive at the relation

$$\limsup_{n\to\infty} \frac{1}{n} c_\beta(\varphi_\infty, nf(s_n(b)), \beta)$$

$$\leq \lim_{n\to\infty} \omega(f(s_n(a))) + \frac{1}{\beta}S_M(\omega, \varphi_\infty), \tag{13.12}$$

which holds for every translation invariant state ω. It is our aim to show that

$$\lim_{n\to\infty} \frac{1}{n} c_\beta(\varphi_\infty, f(s_n(b)))$$

$$= \inf \left\{ \lim_{n\to\infty} \omega(f(s_n(b)) + \frac{1}{\beta}S_M(\omega, \varphi_\infty) : \omega \in \mathfrak{S}^\gamma(\mathcal{B}) \right\}, \tag{13.13}$$

where the notation $\mathfrak{S}^\gamma(\mathcal{B})$ stands for the translation invariant states of $\mathcal{B}$.

The permutation of the tensor factors according to a permutation of the set $\mathbb{Z}$ gives rise to an automorphism of $\mathcal{B}$. Since the local Hamiltonian (13.3) is invariant under permutation automorphisms the whole theory of mean-field (lattice) models is strongly related to permutations. We shall speak of a mean-field interaction on the quasi-local spin algebra in an extended sense in the case of highly permutation invariant (local) Hamiltonians. A state ω is called symmetric if it is invariant under all finite permutation automorphisms. The set $\mathfrak{S}^s(\mathcal{B})$ of all symmetric states is a compact convex subset of the state space $\mathfrak{S}(\mathcal{B})$ endowed with the weak* topology. Obviously, $\mathfrak{S}^s(\mathcal{B}) \subset \mathfrak{S}^\gamma(\mathcal{B})$. Let $\mathfrak{S}^s_p(\mathcal{B})$ stand for the set of all symmetric product states. The following result is due to *Størmer*.

Theorem 13.3 The closed extremal boundary of the compact convex set $\mathfrak{S}^s(\mathcal{B})$ is $\mathfrak{S}^s_p(\mathcal{B})$. Every $\omega \in \mathfrak{S}^s(\mathcal{B})$ admits an integral decomposition

$$\omega(b) = \int_{\mathfrak{S}^s(\mathcal{B})} \varrho(b) \, d\mu(\varrho) \qquad (b \in \mathcal{B})$$

with a unique Radon probability measure μ supported on $\mathfrak{S}^s_p(\mathcal{B})$.

Proof. $\mathfrak{S}^s_p(\mathcal{B})$ is obviously closed in $\mathfrak{S}(\mathcal{B})$. To prove the theorem it suffices to show the existence and uniqueness of the integral decomposition. We shall use the theory of completely positive definite functions on abelian semigroups.

Let S be the set of all positive contractions in $\cup\{\mathcal{B}_{[-n,n]} : n \in \mathbb{N}\}$ and let $\mathcal{S}$ be the free abelian semigroup generated by S. Its elements are formal expressions like

$$s \equiv b_1\hat{+}b_2\hat{+}\dots\hat{+}b_n \qquad (n \in \mathbb{N},\ b_i \in S). \tag{13.14}$$

Let u be the smallest natural number such that $b_j \in \mathcal{B}_{[-u,u]}$ for every $1 \le j \le n$. Set

$$G(b_1, b_2, \dots, b_n) = b_1\gamma^t(b_2)\dots\gamma^{(n-1)t}(b_n) \qquad (t = 2u + 1). \tag{13.15}$$

For a state $\omega \in \mathfrak{S}^s(\mathcal{B})$ define a function $F_\omega : \mathcal{S} \to [0,1]$ as follows. If $s \in \mathcal{S}$ is of the form (13.14), let

$$F_\omega(s) = \omega(G(b_1, b_2, \dots, b_n)). \tag{13.16}$$

Observe that thanks to the symmetry of ω the number $F_\omega(s)$ depends only on s (and it is independent of the order of the terms $b_1, b_2, \dots, b_n$).

Below we will show in a separate lemma that F_ω is completely positive definite, which means that

$$\sum_{i,j} \lambda_i\lambda_j F_\omega(s\hat{+}s_i\hat{+}s_j) \ge 0 \tag{13.17}$$

for every $\lambda_1, \lambda_2, \dots, \lambda_n \in \mathbb{R}$, $s, s_1, \dots, s_n \in \mathcal{S}$ and $n \in \mathbb{N}$. What we need to know about completely positive definite functions is the fact that they are unique mixtures of semicharacters. $\chi : \mathcal{S} \to [0,1]$ is called semicharacter if

$$\chi(0) = 1 \quad \text{and} \quad \chi(s_1\hat{+}s_2) = \chi(s_1)\chi(s_2) \quad (s_1, s_2 \in \mathcal{S}).$$

The semicharacters form a compact space K and the above result states that there exists a unique Radon measure on K such that

$$F_\omega(s) = \int_K \chi(s)\, d\mu(\chi) \qquad (s \in \mathcal{S}). \tag{13.18}$$

Each product state ϱ of $\mathcal{B}$ induces a semicharacter in a trivial way. Let K_0 be the set of all such semicharacters. We need to show that the measure μ appearing in (13.18) is actually supported on the set K_0. In another formulation, we need to show that for μ-almost every $\chi \in K$ there exists a product state ψ on $\mathcal{B}$ such that $\chi(b) = \psi(b)$ for every $b \in S$.

Choose $b_1, b_2 \in S$ so that $b_1 + b_2 = I$. Then

$$\begin{aligned}
\int (\chi(b_1) + \chi(b_2))^n\, d\mu(\chi) &= \sum_{k=0}^n \binom{n}{k} \int \chi(b_1)^k \chi(b_2)^{n-k}\, d\mu(\chi) \\
&= \sum_{k=0}^n \binom{n}{k} \int \chi(kb_1\hat{+}(n-k)b_2)\, d\mu(\chi) \\
&= \sum_{k=0}^n \binom{n}{k} F_\omega(kb_1\hat{+}(n-k)b_2) \\
&= \omega((a_1 + a_2)\gamma^l(a_1 + a_2)\dots\gamma^{(n-1)l}(a_1 + a_2)) \\
&= 1
\end{aligned}$$

with sufficiently large $l \in \mathbb{N}$. So we find that $\int (\chi(b_1) + \chi(b_2))^n \, d\mu(\chi) = 1$ for every $n \in \mathbb{N}$, which implies that

$$H^2(b_1, b_2) = \{\chi \in K : \chi(b_1) + \chi(b_2) = 1\}$$

is of measure 1. By a similar trick one obtains for $x_1, x_2, x_3 \in \mathcal{S}$ with $x_1 + x_2 + x_3 = I$ that the condition $\chi(x_1) + \chi(x_2) + \chi(x_3) = 1$ holds for μ-almost every $\chi \in K$. Since

$$H^3(x_1, x_2, x_3) = \{\chi \in K : \chi(x_1) + \chi(x_2) + \chi(x_3) = 1\}$$

is closed we conclude that

$$K_1 \equiv \cap\{H^2(b_1, b_2) : b_1 + b_2 = I\} \cap \{H^3(x_1, x_2, x_3) : x_1 + x_2 + x_3 = I\}$$

is a closed set of measure 1.

Now let a, b, c be elements of S so that $a + b + c = I$. Since $\chi(a) + \chi(b) + \chi(c) = 1 = \chi(a + b) + \chi(c)$ for every $\chi \in K_1$ we obtain $\chi(a + b) = \chi(a) + \chi(b)$, and in particular, χ is monotone, $\chi(0) = 1$ and $\chi(1) = 1$. By mathematical induction $\chi(\lambda a) = \lambda \chi(a)$ holds for every rational $0 < \lambda < 1$ and this property is extended by monotonicity for irrationals as well. Therefore $\chi|S$ is the restriction of a state ψ of $\mathcal{B}$ to S for every $\chi \in K_2$.

Choose positive contractions $b_i \in \mathcal{B}_i$ $(1 \le i \le n)$. Since

$$F_\omega(b_1 b_2 \ldots b_n) = F_\omega(b_1 \hat{+} b_2 \hat{+} \ldots \hat{+} b_n)$$

the condition

$$\chi(b_1 b_2 \ldots b_n) = \chi(b_1 \hat{+} b_2 \hat{+} \ldots \hat{+} b_n)$$

must hold for almost every $\chi \in K$. This implies that χ comes from a product state for almost every $\chi \in K_2$. $\square$

The following lemma was used in the proof of the previous theorem.

Lemma 13.4 F_ω is completely positive definite, that is, (13.17) holds.

Proof. By permuting the tensor factors of $\mathcal{B}$ one obtains that

$$F_\omega(s \hat{+} s_i \hat{+} s_j) = \omega\big(G(s)\gamma^{il}(G(s_i))\gamma^{(j+n)l}(G(s_j))\big)$$

if we choose $l \in \mathbb{N}$ big enough. (s and s_i are made from finitely many local operators which can be translated in such a way that they are supported disjointly on the infinite tensor product.) Therefore

$$\sum_{i,j=1}^{n} \lambda_i \lambda_j F_\omega(s \hat{+} s_i \hat{+} s_j) = \omega\Big(x_0 \Big(\sum_{i=1}^{n} \lambda_i x_i\Big) \gamma^{nl} \Big(\sum_{j=1}^{n} \lambda_j x_j\Big)\Big), \tag{13.19}$$

where $x_0 = G(s)$ and for $1 \le i \le n$ $x_i = \gamma^{il}(G(s_i))$. In order to prove that (13.19) is nonnegative we show that

$$\omega(xy\alpha^m(y^*)) \ge 0 \tag{13.20}$$

whenever $0 \le x \in \mathcal{B}_{[-t,0]}$ and $y \in \mathcal{B}_{[1,m]}$. Since

$$x\Big(\sum_{i=1}^{N} \gamma^{il}(y)\Big)\Big(\sum_{i=1}^{N} \gamma^{il}(y^*)\Big) \ge 0\,,$$

we have

$$0 \le \omega\Big(x\Big(\sum_{i=1}^{N} \gamma^{il}(y)\Big)\Big(\sum_{i=1}^{N} \gamma^{il}(y^*)\Big)\Big) = N\omega(xyy^*) + (N^2 - N)\omega(xy\gamma^l(y^*)).$$

Finally division by N^2 and the limit $N \to \infty$ gives (13.20).   □

The *symmetrization operator* $\mathbf{sym}_n : \mathcal{B} \to \mathcal{B}$ is the continuous linear extension of the mapping

$$\mathbf{sym}_n(b_{-k}b_{-k+1}\ldots b_0 b_1 \ldots b_n b_{n+1} \ldots b_k)$$
$$= \frac{1}{n!}\sum_{\pi} b_{-k}\ldots b_0 b_{\pi(1)}\ldots b_{\pi(n)}b_{n+1}\ldots b_k$$

where $k \ge n$, $b_i \in \mathcal{B}_i$ $(-k \le i \le k)$ and the summation is over all permutations π of the set $\{1,2,\ldots,n\}$. A translation invariant state of $\mathcal{B}$ is symmetric if and only if every symmetrization operator leaves it invariant. A symmetric sequence in $\mathcal{B}$ is a sequence (x_n) such that $x_n \in \mathcal{B}_{[1,n]}$ and

$$x_{n+k} = \mathbf{sym}_{n+k}x_n \qquad (n,k \in \mathbb{N}). \tag{13.21}$$

A sequence (x_n) is called approximately symmetric if $x_n = \mathbf{sym}_n x_n$ for n big enough and if for every $\varepsilon > 0$ there exists a symmetric sequence (y_n) such that $\|x_n - y_n\| < \varepsilon$ for n big enough. For a symmetric state ω of $\mathcal{B}$ and for an approximately symmetric sequence (x_n) the limit $\lim_n \omega(x_n)$ always exists.

Lemma 13.5 If $(x_n)_{n \ge m}$ and $(y_n)_{n \ge m}$ are approximately symmetric sequences then so is $(x_n y_n)_{n \ge 2m}$.

Proof. The lemma follows by approximation if it is proven for symmetric sequences. So we may assume that (x_n) and (y_n) are symmetric. One can see that $\mathbf{sym}_{l+k}(x_l \gamma^l(y_k))$ depends only on $l + k$. Therefore

$$z_n = \mathbf{sym}_{l+k}(x_l \gamma^l(y_k)) \qquad (l+k = n \ge 2m) \tag{13.22}$$

and since

$$\mathbf{sym}_{n+t}z_n = \mathbf{sym}_{n+t}\mathbf{sym}_{l+k}(x_l \gamma^l(y_k)) = \mathbf{sym}_{n+t}x_l\gamma^l(y_k)$$
$$= \mathbf{sym}_{n+t}x_l\gamma^l(\mathbf{sym}_{k+t}y_k) = \mathbf{sym}_{n+t}x_l\gamma^l(y_{k+t}) = z_{n+t}$$

$(z_n)_{n \ge 2m}$ is checked to be symmetric. To complete the proof of the lemma we show that

$$\|x_n y_n - z_n\| \le \frac{m^2}{n}\|x_m\|\,\|y_m\| \qquad (n \ge 2m). \tag{13.23}$$

Since x_n and y_n are symmetric we have

$$x_n y_n = (n!)^{-2} \sum_{\pi,\pi'} \alpha_\pi(x_m) \alpha_{\pi'}(y_m),$$ (13.24)

where $\pi \mapsto \alpha_\pi$ denotes the action of permutations of $\{1, 2, \ldots, n\}$ on $\mathcal{B}_{[1,n]}$. On the other hand,

$$z_n = (n!)^{-2} \sum_{\pi,\pi'} \alpha_\pi(x_m) \alpha_{\pi'}(y_m)$$ (13.25)

where the summation is over all permutations π and π' for which

$$\pi(\{1, 2, \ldots, m\}) \cap \pi'(\{1, 2, \ldots, m\}) = \emptyset.$$

The number of those pairs (π, π') is

$$\binom{n-m}{m} m!(n-m)!n! \equiv d(n,m)$$

and so $1 - d(n,m)/(n!)^2$ is to be estimated. This is the relative weight of the terms of (13.24) not appearing in (13.25). We claim that

$$1 - \frac{d(n,m)}{(n!)^2} = 1 - \prod_{i=0}^{m-1} \left(\frac{n-m-i}{n-i} \right) \leq \frac{m^2}{n}.$$

This is obvious for $m = 1$ and we proceed by induction over m.

$$\prod_{i=0}^{m} \left(\frac{n-m-i}{n-i} \right) = \frac{n-2m}{n-m} \prod_{i=0}^{m-1} \left(\frac{n-m-i}{n-i} \right)$$

$$\geq \frac{n-2m}{n-m} \left(1 - \frac{m^2}{n} \right) \geq 1 - \frac{(m+1)^2}{n}$$

as far as $m + 1 \leq n$. So we write arrive at the estimate (13.23). The sequence $(z_n)_{n \geq 2m}$ being symmetric we can conclude that $(x_n y_n)_{n \geq m}$ is approximately symmetric. $\square$

The previous lemma tells us that the approximately symmetric sequences form an algebra with the termwise multiplication. This property is an advantage of the approximately symmetric sequences against the strictly symmetric ones. It follows also that for an approximately symmetric self-adjoint sequence (x_n) and for a continuous function $f : \mathbb{R} \to \mathbb{R}$ the sequence $f(x_n)$ is approximately symmetric. This fact is deduced by a polynomial approximation of the function f and by the uniform norm continuity of the functional calculus (cf. Lemma 13.1). In particular, the sequence $(f(s_n(b)))_n$ arising in the statements about large deviation is approximately symmetric if $b \in \mathcal{B}_1^{sa}$.

Let ω be a translation invariant state of $\mathcal{B}$. Set $\omega_n = \omega|\mathcal{B}_{[1,n]}$ and $\varphi_n = \varphi_\infty|\mathcal{B}_{[1,n]}$. Recall that according to the superadditivity of the relative entropy the numbers $t_n = S(\omega_n, \varphi_n)$ form a superadditive sequence (that is, $t_{n+m} \geq t_n + t_m$) and so

$$S_M(\omega, \varphi_\infty) \equiv \lim_{n \to \infty} \frac{1}{n} S(\omega_n, \varphi_n) = \sup\left\{\frac{1}{n} S(\omega_n, \varphi_n) : n \in \mathbb{N}\right\}. \quad (13.26)$$

Since $S(\omega_n, \varphi_n)$ is a weak*-lower semicontinuous function of ω, it follows immediately that $S_M(\omega, \varphi_\infty)$ is lower semicontinuous.

Proposition 13.6 The mean relative entropy $\omega \mapsto S_M(\omega, \varphi_\infty)$ is a weak*-lower semicontinuous affine functional on $\mathfrak{S}^\gamma(\mathcal{B})$.

Proof. It remained to show the affine property. Proposition 5.24 tells us that

$$0 \geq S(\lambda\omega_n^1 + \mu\omega_n^2, \varphi_n) - \lambda S(\omega_n^1, \varphi_n) - \mu S(\omega_n^2, \varphi_n) \geq \lambda \log \lambda + \mu \log \mu$$

if $\lambda + \mu = 1$. Dividing by n and letting $n \to \infty$ we obtain that

$$S(\lambda\omega^1 + \mu\omega^2, \varphi_\infty) = \lambda S(\omega^1, \varphi_\infty) + \mu S(\omega^2, \varphi_\infty)$$

which makes the proof complete. □

Let ψ be a state of $\mathcal{B}_{[1,n]}$. By periodization ψ extends to a state

$$\tilde{\psi} \equiv \ldots \otimes \psi \otimes \psi \otimes \psi \otimes \ldots \quad (13.27)$$

of $\mathcal{B}$ which is invariant under γ^m. By averaging we transform it into a shift invariant state

$$\overline{\psi} \equiv \frac{1}{m} \sum_{i=0}^{m-1} \tilde{\psi} \circ \gamma^i. \quad (13.28)$$

The affine property of the mean relative entropy for the rescaled localization allows a simple computation of $S_M(\overline{\psi}, \varphi_\infty)$. Namely,

$$
\begin{aligned}
S_M(\overline{\psi}, \varphi_\infty) &= \frac{1}{m} \lim_{n \to \infty} \frac{1}{n} S(\overline{\psi}|\mathcal{B}_{[1,nm]}, \varphi_{nm}) \\
&= \frac{1}{m} \sum_{i=0}^{m-1} \frac{1}{m} \lim_{n \to \infty} \frac{1}{n} S(\overline{\psi} \circ \gamma^i|\mathcal{B}_{[1,nm]}, \varphi_{nm}) \quad (13.29) \\
&= \frac{1}{m^2} \sum_{i=0}^{m-1} \lim_{n \to \infty} \frac{1}{n} S(\overline{\psi}|\mathcal{B}_{[1,nm]}, \varphi_{nm}) = \frac{1}{m} S(\psi, \varphi_n).
\end{aligned}
$$

Now we are ready to prove a limit theorem which, in terms of quantum statistical mechanics, gives the validity of the *Gibbs variational principle* for a rather general class of *mean-field models*. The thermodynamic limit of the free energy density is obtained by minimizing the free energy density functional over the set of symmetric states of the infinite system. Although the formulation is done for a lattice system over $\mathbb{Z}$, as far as the limit theorems are concerned, there is no real difference between one-dimensional and higher dimensional lattices.

Theorem 13.7 For an approximately symmetric sequence $(x_n)_{n \geq m}$ the relation

$$\lim_{n \to \infty} \frac{1}{n} c(\varphi_n, n x_n) = \inf \left\{ \lim_{n \to \infty} \omega(x_n) + S_M(\omega, \varphi_\infty) : \omega \in \mathfrak{S}^s(\mathcal{B}) \right\}$$

holds. (Here $\varphi_n = \varphi_\infty | \mathcal{B}_{[1,n]}$.)

Proof. The upper estimate is fairly easy and it is a plain consequence of the inequality

$$c(\varphi_n, n x_n) \leq \omega(n x_n) + S(\omega_n, \varphi_n),$$

which is just the definition of the functional c. By the definition of the perturbed state we have

$$c(\varphi_n, n x_n) = [\varphi_n^{n x_n}](n x_n) + S([\varphi_n^{n x_n}], \varphi_n). \tag{13.30}$$

We shall say that $\nu \in \mathfrak{S}(\mathcal{B})$ is a limit point of the sequence $\psi_n \in \mathfrak{S}(\mathcal{B}_{[1,n]})$ $(n \in \mathbb{N})$, if for every local operator $a \in \cup \{\mathcal{B}_{[1,n]} : n \in \mathbb{N}\}$, every $\varepsilon > 0$ and $n_0 \in \mathbb{N}$ there is an $n \geq n_0$ such that $|\psi_n(a) - \nu(a)| < \varepsilon$. The idea is to show that any (stationary) limit point ν of the sequence $([\varphi_n^{n x_n}])_n$ satisfies

$$\liminf_{n \to \infty} \frac{1}{n} c(\varphi_n, n x_n) \geq \lim_{n \to \infty} \nu(x_n) + S_M(\nu, \varphi_\infty) \tag{13.31}$$

and therefore in (13.29) the equality operates. This is slightly more than the content of the theorem.

Let us write ψ_n for $[\varphi_n^{n x_n}]$ to have a more convenient notation. Since $\varphi_n \circ \mathbf{sym}_n = \varphi_n$ and $\mathbf{sym}_n(x_n) = x_n$, it follows that ψ_n is symmetric as well, that is, $\psi_n \circ \mathbf{sym}_n = \psi_n$. Now we consider the states $\overline{\psi}_n \in \mathfrak{S}(\mathcal{B})$ defined by (13.27) and (13.28). (In fact $\overline{\psi}_n \in \mathfrak{S}^\gamma(\mathcal{B})$.) The symmetry of ψ_n ensures that any translation invariant limit point ν of the sequence (ψ_n) is a weak* limit point of the sequence $(\overline{\psi}_n)$. Besides, ν must be in $\mathfrak{S}^s(\mathcal{B})$. From the lower semi-continuity of the mean relative entropy and from the relation (13.29) we obtain

$$S_M(\nu, \varphi_\infty) \leq \liminf_{n \to \infty} S_M(\overline{\psi}_n, \varphi_\infty) \leq \liminf_{n \to \infty} \frac{1}{n} S(\psi_n, \varphi_n).$$

In order to complete the proof of (13.31) we need to show that

$$\lim_{n \to \infty} \nu(x_n) \geq \liminf_{n \to \infty} \psi_n(x_n). \tag{13.32}$$

Since ν is symmetric, the limit on the left hand side certainly exists. Choose a symmetric sequence (y_n) such that $\|x_n - y_n\| < \varepsilon$ for $n \geq n_0$. There exists a state ψ_m so that $|\psi_m(y_{n_0}) - \nu(y_{n_0})| < \varepsilon$ and $m \geq n_0$. For $n \geq m$ we have

$$\nu(x_n) \geq \nu(y_n) - \varepsilon = \nu(y_{n_0}) - \varepsilon$$
$$\geq \psi_m(y_{n_0}) - 2\varepsilon = \psi_m(y_m) - 2\varepsilon \geq \psi_m(x_m) - 3\varepsilon.$$

Since $\varepsilon > 0$ is arbitrary we arrive at (13.32). $\qquad \square$

Here we make an interruption and explain that Theorem 13.7 contains the minimum free energy principle of thermodynamics (for a class of mean-field models). Assume that the local algebras are finite dimensional and let the density matrix of $\varphi_\infty | \mathcal{B}_n$ be $e^{-\beta h}$ ($n \in \mathbb{Z}$). The macroscopic thermodynamic functional

$$F_\beta(\omega) = \lim_{n \to \infty} \omega(x_n) + \frac{1}{\beta} S_M(\omega, \varphi_\infty)$$

$$= \lim_{n \to \infty} \frac{1}{n} \omega\left(n x_n + \sum_{i=1}^{n} h_i\right) + \frac{1}{\beta} \lim_{n \to \infty} \frac{1}{n} S(\omega | \mathcal{B}_{[1,n]})$$

is the difference of the specific internal energy functional and $1/\beta = kT$ times the specific (or mean) entropy functional. So $F_\beta(\omega)$ is the free energy functional. The variational principle expresses the minimum value of the free energy functional with the help of local minimal free energies through a thermodynamic limit.

Since for $b \in \mathcal{B}_1$ the sequence $(f(s_n(b)))_n$ is approximately symmetric we can deduce from Theorem 13.7 the variational formula

$$\lim_{n \to \infty} \frac{1}{n} c(\varphi_\infty, n f(s_n(b)))$$

$$= \inf\left\{ \lim_{n \to \infty} \omega(f(s_n(b))) + S_M(\omega, \varphi_\infty) : \omega \in \mathfrak{S}^s(\mathcal{B}) \right\}. \qquad (13.33)$$

It does not make any difference if the lower bound is over the stationary states. It was proven that there exists a symmetric state ν which minimizes

$$\omega \mapsto \lim_{n \to \infty} \omega\big(f(s_n(b))\big) + S_M(\omega, \varphi_\infty).$$

Størmer's theorem (i.e., Theorem 13.3) tells us that

$$\nu = \int_{\mathfrak{S}(\mathcal{B})} \varrho \, d\mu(\varrho)$$

with a Radon measure μ supported on the symmetric product states. Since

$$S_M(\nu, \varphi_\infty) = \int S(\varrho_1, \varphi_1) \, d\mu(\varrho),$$

and

$$\lim_{n \to \infty} \omega(f(s_n(b))) = \int f(\varrho_1(b)) \, d\mu(\varrho)$$

we have

$$\lim_{n \to \infty} \frac{1}{n} c(\varphi_\infty, n f(s_n(b)))$$

$$\geq \inf\left\{ \lim_{n \to \infty} \omega(f(s_n(b))) + S_M(\omega, \varphi_\infty) : \omega \in \mathfrak{S}^s(\mathcal{B}) \right\}.$$

The converse inequality is obvious and hence we arrive at the following

Proposition 13.8 For $b \in \mathcal{B}_1$ and a continuous function f the variational formula

$$\lim_{n \to \infty} \frac{1}{n} c(\varphi_\infty, nf(s_n(b)))$$

$$= \inf \left\{ \lim_{n \to \infty} \omega(f(s_n(b))) + S_M(\omega, \varphi_\infty) : \omega \in \mathfrak{S}^s(\mathcal{B}) \right\}.$$

holds.

This proposition is a typical example of what we shall call *perturbational limit theorems*. Let $\mathcal{A}$ and $\mathcal{C}$ be C*-algebras and ψ a state of $\mathcal{C}$. Assume that we are given positive unital mappings $\alpha_n : \mathcal{A} \to \mathcal{C}$ ($n \in \mathbb{N}$). Motivated by Proposition 13.8 (as well as large deviation theory, in particular *Varadhan's theorem*) the perturbational limit principle is said to hold if

$$\lim_{n \to \infty} \frac{1}{n} c(\psi, nf(\alpha_n(a)))$$

$$= \inf \left\{ \lim_{n \to \infty} \omega(f(\alpha_n(a))) + I(\omega) : \omega \in \mathfrak{S}(\mathcal{A}) \right\} \qquad (13.34)$$

for every $a = a^* \in \mathcal{A}$, for every continuous function $f : \mathbb{R} \to \mathbb{R}$ and for a certain weak* lower semi-continuous functional $I : \mathfrak{S}(\mathcal{A}) \to \mathbb{R}^+ \cup \{\infty\}$. The latter functional will be called the rate functional.

To consider Proposition 13.8 as a perturbational limit theorem one chooses $\mathcal{A} = \mathcal{B}_1$, $\mathcal{C} = \mathcal{B}$, $\psi = \varphi_\infty$ and

$$\alpha_n = n^{-1}(\mathrm{id} + \gamma + \ldots + \gamma^{n-1})|\mathcal{B}_1 \quad (n \in \mathbb{N}).$$

The rate function is the relative entropy in this case.

When the perturbational limit principle holds, the rate functional gives information on the sequence α_n. For example, if

$$\lim_{n \to \infty} \alpha_n(a) = \nu(a)$$

in norm for a state ν of $\mathcal{A}$ and for every $a \in \mathcal{A}$, then the perturbational limit principle holds trivially in the form

$$\lim_{n \to \infty} \frac{1}{n} c(\psi, nf(\alpha_n(a))) = \inf\{f(\nu(a)) + I(\omega) : \omega \in \mathfrak{S}(\mathcal{A})\},$$

where

$$I(\omega) = \begin{cases} 0 & \text{if } \omega = \nu \\ +\infty & \text{otherwise.} \end{cases}$$

In the proof of a perturbational limit theorem the inequality

$$|c(\psi, x) - c(\psi, y)| \leq \|x - y\|$$

allows some simplification. When

$$f \mapsto E_\omega^f(a) \equiv \lim_{n \to \infty} \omega(f(\alpha_n(a)))$$

is continuous with respect to the supremum norm, (13.34) is to be checked only for polynomials due to polynomial approximation. Similarly, if $a \mapsto E_\omega^f(a)$ is norm continuous then Lemma 13.1 yields that (13.34) holds for every a if and only if it holds on a norm dense subset.

The object $(\mathcal{B}, \mathcal{B}_1, \gamma, \varphi_\infty)$ studied above in Proposition 13.8 is called an independent process in the setting of C*-algebras, which might be viewed at different levels. When one is interested in the collective behaviour of the perturbational limit for every $b \in \mathcal{B}$, then the algebra $\mathcal{A}$ must be chosen larger and the rate functional must be defined on the state space of the enlarged algebra. Conversely, if a particular $b \in \mathcal{B}$ is viewed (as it is the case in statistical mechanics computing the free energy density for a given Hamiltonian), the variational right hand side should be contracted to a smaller domain. (Similarly to the classification of large deviation theorems Proposition 13.8 may be called a level-2 result while Theorem 13.11 below is of level-3.)

Lemma 13.9 Assume that

$$z_- \equiv \inf\{[\varphi^{tx}](x) - td : t \in \mathbb{R}\} < d < z_+ \equiv \sup\{[\varphi^{tx}](x) - td : t \in \mathbb{R}\}.$$

Then

$$\sup\{c(\varphi, tx) - td : t \in \mathbb{R}\} = \inf\{S(\psi, \varphi) : \psi(x) = d\}.$$

Proof. Let us recall that

$$\frac{d}{dt} c(\varphi, tx) = [\varphi^{tx}](x)$$

By the concavity of the functional $t \mapsto c(\varphi, tx)$ the function $t \mapsto [\varphi^{tx}](x)$ is monotone decreasing and due to continuity there exists a $t_0 \in \mathbb{R}$ such that

$$[\varphi^{tx}](x) = d.$$

It is easy to show that

$$\inf\{S(\psi, \varphi) : \psi(x) = d\} = S([\varphi^{t_0 x}], \varphi) = c(\varphi, t_0 x) - t_0 d,$$

which is the maximum of the function $t \mapsto c(\varphi, tx) - tu.$ □

The next perturbational limit theorem concerns a single $b \in \mathcal{B}_1$, so it is level-1.

Proposition 13.10 Under the conditions of Proposition 13.8 we have

$$\lim_{n \to \infty} \frac{1}{n} c(\varphi_\infty, nf(s_n(b))) = \inf\{f(u) + G(u) : u \in \mathbb{R}\},$$

where $g(u) = \sup\{c(\varphi, tb) - tu : u \in \mathbb{R}\}.$

Proof. Proposition 13.8 tells us that the stated limit equals to

$$\inf\{f(\psi(b)) + S(\psi,\varphi) : \psi \in \mathfrak{S}(\mathcal{B}_1)\}$$
$$= \inf\{f(u) + \inf\{S(\psi,\varphi) : \psi \in \mathfrak{S}(\mathcal{B}_1),\ \psi(b) = u\} : u \in \mathbb{R}\},$$

where we have used that

$$\inf\{S(\psi,\varphi) : \psi(b) = d\} = G(u)$$

for the essential values of u due to Lemma 13.9.   □

Whereas the previous proposition concerns the perturbational limit for a single $b = b^* \in \mathcal{B}_1$, in Proposition 13.8 the limit was formulated for the whole space $\mathcal{B}_1^{sa}$. In the next theorem the perturbational limit is established for the self-adjoint part of the full quasi-local algebra $\mathcal{B}$.

Theorem 13.11 For $b \in \mathcal{B}^{sa}$ and for a continuous real function f the variational formula

$$\lim_{n\to\infty} \frac{1}{n} c(\varphi_\infty, nf(s_n(b)))$$
$$= \inf\{E_\omega^f(b) + S_M(\omega,\varphi_\infty) : \omega \in \mathfrak{S}^\gamma(\mathcal{B})\}$$

holds with $E_\omega^f(a) = \lim_{n\to\infty} \omega(f(s_n(a)))$.

Proof. First we note that by the the norm continuity of both sides of the equality one may assume that for some $k \in \mathbb{N}$ we have $b \in \mathcal{B}_{[1,k]}$ and f is a polynomial. Inequality (13.12) already contains a part of the statement. It remains to show that the liminf of the left hand side is not smaller than

$$E_\omega^f(b) + S_M(\omega,\varphi_\infty) - \varepsilon$$

for an arbitrary small $\varepsilon > 0$ and with a certain stationary state ω.

Similarly to the proof of Proposition 13.2 we set

$$b_l \equiv \sum_{i=0}^{l-k} \gamma^i(b) \in \mathcal{B}_{[1,l]}$$

for $l > k$ and we write the arbitrary n in the form $u \cdot l + r$ where u and r are integers and $0 \le r < l$. Then

$$\left\| \frac{1}{n} \sum_{j=0}^{n-1} \gamma^j(b) - \frac{1}{u} \sum_{j=0}^{u-1} \gamma^{jl+t}(b_l/l) \right\| \le \delta$$

uniformly for every $0 \le t \le l$ if n is greater than $N_1 \in \mathbb{N}$. This yields

$$\left\| f(s_n(b)) - f\left(\frac{1}{u} \sum_{j=0}^{u-1} \gamma^{jl+t}(b_l/l) \right) \right\| \le \varepsilon_1 \tag{13.35}$$

for $n > N_1$. So

$$\frac{1}{n}c(nf(s_n(b)), \varphi_\infty) \geq \frac{1}{n}c\left(ulf\left(\frac{1}{u}\sum_{j=0}^{u-1}\gamma^{jl+t}(b_l/l)\right)\right) - \varepsilon_2$$

with a small ε_2. Now we use Proposition 13.8 with another localization. Taking the right shift γ^l the proposition tells that there exists a state ψ of $\mathcal{B}$ which is invariant under γ^l and

$$\frac{1}{u}c\left(ulf\left(\frac{1}{u}\sum_{j=0}^{u-1}\gamma^{jl+t}(b_l/l)\right)\right)$$

$$\geq \psi\left(lf\left(\frac{1}{u}\sum_{j=0}^{u-1}\gamma^{jl+t}(b_l/l)\right)\right) + \lim_{m\to\infty}\frac{1}{m}S(\psi|\mathcal{B}_{[1,ml]}, \varphi_\infty|\mathcal{B}_{[1,ml]}) - \varepsilon_3$$

whenever $u \geq u_0$. From the γ^l-invariant state ψ we obtain a γ-invariant one by setting

$$\omega \equiv \frac{1}{l}\sum_{j=0}^{l-1}\psi \circ \gamma^j. \tag{13.36}$$

It is seen that

$$S_M(\omega, \varphi_\infty) = \frac{1}{l}\lim_{m\to\infty}\frac{1}{m}S(\psi|\mathcal{B}_{[1,ml]}, \varphi_\infty|\mathcal{B}_{[1,ml]})$$

(to be compared with (13.29)). From the previous inequalities we have

$$\frac{1}{n}c(nf(s_n(b)), \varphi_\infty) \geq \psi\left(f\left(\frac{1}{u}\sum_{j=0}^{u-1}\gamma^{jl}(b_l/l)\right)\right) + S_M(\omega, \varphi_\infty) - \varepsilon_4$$

for n sufficiently large and for arbitrary small $\varepsilon_4 > 0$. We have to show that the first term on the right hand side is close to $\omega(f(s_n(b)))$. By application of (13.35) we obtain

$$\left|\omega(f(s_n(b))) - \omega\left(f\left(\frac{1}{u}\sum_{j=0}^{u-1}\gamma^{jl}(b_l/l)\right)\right)\right| \leq \varepsilon_1,$$

and

$$\left\|f\left(\frac{1}{u}\sum_{j=0}^{u-1}\gamma^{jl+s}(b_l/l)\right) - f\left(\frac{1}{u}\sum_{j=0}^{u-1}\gamma^{jl+t}(b_l/l)\right)\right\| \leq 2\varepsilon_1$$

for $0 \leq t, s \leq l$. Hence

$$\left|\psi\left(f\left(\frac{1}{u}\sum_{j=0}^{u-1}\gamma^{jl+t}(b_l/l)\right)\right) - \psi\left(f\left(\frac{1}{u}\sum_{j=0}^{u-1}\gamma^{jl}(b_l/l)\right)\right)\right| \leq 2\varepsilon_1.$$

From (13.36) we know that

$$\omega\Big(f\Big(\frac{1}{u}\sum_{j=0}^{u-1}\gamma^{jl}(b_l/l)\Big)\Big) = \frac{1}{l}\sum_{t=0}^{l-1}\psi\Big(f\Big(\frac{1}{u}\sum_{j=0}^{u-1}\gamma^{jl+t}(b_l/l)\Big)\Big).$$

Therefore we have

$$\Big|\omega\Big(f\Big(\frac{1}{u}\sum_{j=0}^{u-1}\gamma^{jl}(b_l/l)\Big)\Big) - \psi\Big(f\Big(\frac{1}{u}\sum_{j=0}^{u-1}\gamma^{jl}(b_l/l)\Big)\Big)\Big| \le 2\varepsilon_1$$

and we conclude

$$\Big|\omega(f(s_n(b))) - \psi\Big(f\Big(\frac{1}{u}\sum_{j=0}^{u-1}\gamma^{jl}(b_l/l)\Big)\Big)\Big| \le 3\varepsilon_1$$

that was the last thing to be proved.   □

Theorem 13.11 supplies us with a new example of Legendre transform (see (12.31) for the definition). Choose $U = \mathcal{B}^{sa}$ and let V be the real linear space of hermitian γ-invariant functionals of $\mathcal{B}$. With respect to the duality

$$\langle \nu, b\rangle = \lim_{n\to\infty} \nu(b)$$

the conjugate function of

$$F(\nu) = \begin{cases} S_M(\nu, \varphi_\infty) & \text{if } \nu \text{ is a } \gamma\text{-invariant state} \\ +\infty & \text{otherwise.} \end{cases}$$

is

$$F^*(b) = -\lim_{n\to\infty} \frac{1}{n} c(\varphi_\infty, n s_n(b)).$$

Notes and Remarks. Large deviation theory has an extensive literature. The small book [Varadhan 1984] contains most of the basic ideas and our short outline is close to [Ellis 1985], which puts the emphasis on classical statistical mechanical applications. Here the classical Curie–Weiss model is investigated in detail and the picture of phase transition in quantum mean-field models is not different, see [Fannes, Kossakowski and Verbeure 1991]. The papers [Lewis 1986] and [Lewis 1988] are expository accounts on large deviation methods in statistical mechanics. [Fannes, Spohn and Verbeure 1980] initiated the exploration of symmetrical states characterized by *Størmers's theorem*, [Størmer 1969]. This result had its origin in *de Finetti's* characterization of symmetrical sequences of random variables. Concerning related results and their application in (classical) particle sytems we refer to [Georgii 1979]. The proof of Theorem 13.3 is from [Petz 1990b] and its method is versatile enough to be used in different situations. Varadhan-type limit theorems in the quantum case motivated by the mean-field theory appeared in [Petz, Raggio and Verbeure 1989]. It should be noted that the local algebras could also be infinite dimensional so the limit theorems give more than required of statistical mechanics of mean-field models. Thermodynamics and lattice gases are treated in detail in [Sewell 1986] and [Israel 1979], see also Chapter 15. The useful concept of approximately symmetric sequences originates from [Raggio and Werner 1989]. The abstract formulation of the perturbation limit principle appears in [Petz 1990b]. In cases of inhomogeneous mean-field systems the Hamiltonian of the n-particle system depends on external parameters. This dependence destroys the permutation symmetry, like in BCS-theory ([Thirring and Wehrl 1967]). The basic idea of [Raggio and Werner 1991] is the introduction of an auxiliary algebra of functions on the space of external parameters taking values in the one-particle algebra of the system. The paper [Gerisch and Rieckers 1990] deals with mean-field models on multi-lattices where each local algebra $\mathcal{B}_n$ is a tensor product $\mathcal{B}_n^{(1)} \otimes \mathcal{B}_n^{(2)} \otimes \ldots \otimes \mathcal{B}_n^{(p)}$. Dynamics for mean-field models is considered in [Bóna 1989]. The paper [Werner 1992] contains a survey of large deviation and mean-field models. It is close to our presentation and contains further results and examples of several models in which the methods are applicable.

Part V

Miscellanea

14 Central Limit and Quasi-free Reduction

First we recall the simplest central limit theorem from probability theory. Let $\xi_1, \xi_2, \ldots$ be a sequence of independent identically distributed real random variables. For the sake of simplicity, assume that their expected value is 0 and variance is 1. The central limit theorem tells us that

$$\text{Prob}\left(\frac{\alpha}{\sqrt{n}} \leq \frac{\sum_{i=1}^{n} \xi_i}{n} \leq \beta\right) \rightarrow \frac{1}{\sqrt{2\pi}} \int_{\alpha}^{\infty} \exp\left(-\frac{u^2}{2}\right) du \tag{14.1}$$

as $n \rightarrow \infty$ for $0 < \alpha < \beta$. Another formulation of the result says that the distribition of

$$\frac{\xi_1 + \xi_2 + \ldots + \xi_n}{\sqrt{n}}$$

converges to the standard normal distribution. It is more clear from the form (14.1) that the central limit theorem concerns the fluctuations of the average of the independent variables. (The dependent case is more interesting than the independent and then it may happen that the fluctuations are not normal.)

In this chapter the simplest central limit theorems will be treated in a non-commutative setting. It will be shown that the fluctuations of the addition of independent spins lead naturally to the algebra of the canonical commutation relation with a quasi-free state.

A spin lattice system over the integers is a simple algebraic stochastic process indexed by the integers $\mathbb{Z}$. To each $i \in \mathbb{Z}$ one associates a copy $\mathcal{A}_i$ of the matrix algebra $M_n(\mathbb{C})$ and define $\mathcal{A}$ as the infinite tensor product $\otimes_i \mathcal{A}_i$. If also a product state φ of $\mathcal{A}$ is chosen then we have an independent (or Bernoulli) algebraic stochastic process at our disposal. The algebraic random variable $\mathcal{A}_i$ is the spin at the lattice site $i \in \mathbb{Z}$ and it will replace ξ_i of the starting probabilistic example. Note that $\mathcal{A}_i$ is not an observable of the spin system but it is determined by the different direction spin observables at the site i. Picking a copy $x_i \in \mathcal{A}_i$ of the self-adjoint matrix x from $M_n(\mathbb{C})$ we may consider the *fluctuations*

$$F_n(x) = \frac{\sum_{i=1}^{n}(x_i - \varphi(x_i))}{\sqrt{n}} \tag{14.2}$$

and their behaviour as $n \rightarrow \infty$ is completely determined by the above mentioned central limit theorem. Indeed, since x_i commutes with x_j for $i \neq j$, the sequence of fluctuation is diagonalizable and may be represented by random variables. The quantum central limit theorems ought to do more, they should handle all the self-adjoint x's together.

For the commutator of the fluctuations we have

$$[F_n(a), F_n(b)] = \frac{1}{n} \sum_{i=1}^{n} [a_i, b_i] \to cI$$

strongly as $n \to \infty$ where $c = \varphi([a_1, b_1])$ according to the ergodic theorem. If $F_n(a)$ converges to an operator A and $F_n(b)$ converges to an operator B in some sense then A and B obey the canonical commutation relation. This simple argument indicates that the limit of fluctuations is related to the canonical commutation relation. Next we will show that the Weyl form of the canonical commutation relation is asymptotically satisfied by the fluctuations.

For matrices a and b we introduce the notation

$$L(a, b) = \exp(ia) \exp(ib) - \exp(ia + ib) \exp(\tfrac{1}{2}[a, b]).$$

Inserting the power series expansion of the exponential function we find that only monomials (of a and b) with degree greater than 2 are present. Therefore the following lemma is straightforward. (It is nothing else but the approximate version of the so-called *Baker-Hausdorff formula*.)

Lemma 14.1 There exists a constant $C > 0$ such that for $\varepsilon > 0$ small enough

$$\|L(a, b)\| \leq C \varepsilon^3$$

provided that $\|a\| < \varepsilon$ and $\|b\| < \varepsilon$.

Lemma 14.2 If $[a_i, b_j] = [a_i, a_j] = [b_i, b_j] = 0$ for $i \neq j$ then

$$\|L(a_1 + a_2, b_1 + b_2)\| \leq \|L(a_1, b_1)\| + \|L(a_2, b_2)\| \exp(\tfrac{1}{2}\|[a_1, b_1]\|).$$

Proof. Under the hypothesis the equality

$$\begin{aligned} L(a_1 + a_2, b_1 + b_2) &= L(a_1, b_1) \exp(ia_2) \exp(ib_2) \\ &\quad + \exp(ia_1 + ib_1) \exp(\tfrac{1}{2}[a_1, b_1]) L(a_2, b_2) \end{aligned}$$

holds and provides the required estimate. □

Proposition 14.3 Let $a, b \in M_n(\mathbb{C})^{sa}$ then

$$\lim_{n \to \infty} \|L(F_n(a), F_n(b))\| = 0.$$

Proof. We represent n in the form $n = k \cdot m + l$ where $\log n \leq m < 1 + \log n$ (i.e., m is the integer part of $\log n$) and $0 \leq l < m$. Application of Lemma 14.2 yields

$$\|L(F_n(a), F_n(b))\| \leq \left\|L\left(\sqrt{k\,m\,n^{-1}}\,F_{km}(a), \sqrt{l\,m\,n^{-1}}\,F_{lm}(b)\right)\right\|$$
$$+ \left\|L\left(\sqrt{l\,n^{-1}}\,F_l(a), \sqrt{l\,n^{-1}}\,F_l(b)\right)\right\| \times \exp\left(k\,m\,\|a\|\,\|b\|\,n^{-1}\right).$$

First we concentrate on the second term. Since

$$\left\|\sqrt{l\,n^{-1}}\,F_l(a)\right\| \leq \frac{l\|a\|}{\sqrt{n}} \leq \frac{1 + \log n}{\sqrt{n}}\|a\|$$

(and similarly with b) the norm continuity of $L(.\,,.)$ tells us that the first factor converges to 0 while the second remains bounded. Hence the second term tends to 0 and we have to show that so does the first one. A repeated use of Lemma 14.2 gives that

$$\left\|L\left(\sqrt{k\,m\,n^{-1}}\,F_{km}(a), \sqrt{k\,m\,n^{-1}}\,F_{km}(b)\right)\right\|$$
$$\leq \left\|L\left(\sqrt{m\,n^{-1}}\,F_m(a), \sqrt{m\,n^{-1}}\,F_m(b)\right)\right\| \times \left(1 + C_n + \ldots + C_n^{k-1}\right)$$

where

$$C_n = \exp(m^2\,\|a\|\,\|b\|\,n^{-1}).$$

One can easily see that

$$1 + C_n + \ldots + C_n^{k-1} = O(k).$$

According to Lemma 14.1

$$\left\|L\left(\sqrt{m\,n^{-1}}F_m(a), \sqrt{m\,n^{-1}}\,F_m(a)\right)\right\| \leq C\frac{m^3}{n\sqrt{n}}$$

if n is big enough. Since

$$0 \leq \frac{k\,m^3}{n\sqrt{n}} \leq \frac{(1 + \log n)^2}{\sqrt{n}} \to 0$$

we arrive at the end of the proof.   □

The abstract way to study the canonical commutation relation is based on *symplectic structure*. Let H be a real linear space. A bilinear form σ of H is called a symplectic form if $\sigma(f,g) = -\sigma(g,f)$ for $f,g \in H$. By a symplectic space we mean a pair (H,σ) consisting of a linear space H and a symplectic form σ. The C*-algebra CCR(H,σ) of the *canonical commutation relation* associated to (H,σ) is related to the *Weyl form* of the canonical commutation relation:

$$W(f)W(g) = W(f+g)\exp\left(-\tfrac{i}{2}\sigma(f,g)\right). \tag{14.3}$$

$W(f)$ denotes a unitary element of the algebra CCR(H,σ) associated to $f \in H$. The essence of the relation of the central limit of a spin system to the Weyl relation (14.3) is the following.

$$\lim_{n\to\infty} [F_n(a), F_n(b)] = \sigma(a,b)$$

determines a symplectic form on the self-adjoint part of $M_m(\mathbb{C})$ and if one can give sense to

$$\lim_{n\to\infty} F_n(a) = W(a) \tag{14.4}$$

then Proposition 14.3 has the interpretation that the unitaries $W(f)$ obey (14.3). (This argument will turn out to be correct only if the limit in (14.4) is understood in expectation and not in some usual convergence of operators.)

The C*-algebra $\mathrm{CCR}(H,\sigma)$ is determined up to isomorphism by the following two conditions.

(i) $\mathrm{CCR}(H,\sigma)$ is generated by the unitary operators $W(f)$, $f \in H$, satisfying the canonical commutation relation (14.3).

(ii) If a C*-algebra $\mathcal{A}$ contains unitaries $W'(f)$ ($f \in H$) satisfying the relation (14.3) then there exists a homomorphism $\alpha : \mathrm{CCR}(H,\sigma) \to \mathcal{A}$ so that $\alpha(W(f)) = W'(f)$ ($f \in H$).

The symplectic form σ (and with it the space (H,σ)) is called non-degenerate if the condition $\sigma(f,g) = 0$ for every $f \in H$ implies $g = 0$. If (H,σ) is a non-degenerate symplectic space then $\mathrm{CCR}(H,\sigma)$ is determined uniquely by condition (i) of its definition. This result is known as the *Slawny theorem*.

Most often a symplectic space is induced by a (complex) Hilbert space. Let $\mathcal{H}$ be a complex Hilbert space. Then $\sigma(f,g) = \mathrm{Im}\,\langle f,g\rangle$ is a non-degenerate symplectic form and we write $\mathrm{CCR}(\mathcal{H})$ for the corresponding C*-algebra of the canonical commutation relation. Let us mention that in the extreme case when $\sigma \equiv 0$, the algebra $\mathrm{CCR}(H,\sigma)$ is isomorphic to the C*-algebra of all continuous functions on the compact space $\hat{H}$ where $\hat{H}$ is the compact dual group of the discrete abelian group H. Now we describe the Fock representation of $\mathrm{CCR}(\mathcal{H})$.

We consider the *full Fock space*

$$\mathcal{F}(\mathcal{H}) = \mathbb{C}\Omega \oplus \mathcal{H} \oplus (\mathcal{H} \otimes \mathcal{H}) \oplus (\mathcal{H} \otimes \mathcal{H} \otimes \mathcal{H}) \oplus \dots$$

and define for each $f \in \mathcal{H}$ two bounded operators $c(f)$ and $c^*(f)$.

$$\begin{aligned}
&c(f)\Omega = 0, \\
&c(f)(g_1 \otimes \dots \otimes g_n) = \langle f, g_1\rangle g_2 \otimes \dots \otimes g_n, \\
&c^*(f)\Omega = f, \\
&c^*(f)(g_1 \otimes \dots \otimes g_n) = f \otimes g_1 \otimes \dots \otimes g_n.
\end{aligned} \tag{14.5}$$

The correspondence $f \mapsto c^*(f)$ is (complex) linear and $c^*(\mathrm{f})$ is the adjoint of $c(f)$. Also, the relation

$$c(g)c^*(f) = \langle g, f\rangle I \qquad (g, f \in \mathcal{H}) \tag{14.6}$$

holds. The *number operator* N is defined to be a positive self-adjoint operator such that $\mathcal{H}^{(1)}\otimes\dots\otimes\mathcal{H}^{(n)}$ is an eigensubspace corresponding to the eigenvalue

n. Let P_+ be the projection of $\mathcal{F}(\mathcal{H})$ onto the *symmetric (Bose) Fock space* $\mathcal{F}_+(\mathcal{H})$ and set

$$a(f) = P_+ c(f) N^{1/2} P_+ \,,$$
$$a^+(f) = P_+ N^{1/2} c^*(f) P_+ \,. \qquad (14.7)$$

These are the (Bose) *annihilation and creation operators*, respectively. They satisfy the *canonical commutation relations*:

$$[a^+(f), a^+(g)] = [a(f), a(g)] = 0$$
$$[a(f), a^+(g)] = \langle f, g \rangle I \qquad (14.8)$$

for every $f, g \in \mathcal{H}$, the equalities being valid on dense subspace of vectors with a finite particle number. By means of the field operators

$$B(f) = \frac{1}{\sqrt{2}} \left(a(f) + a^+(f) \right) \qquad (f \in \mathcal{H}) \qquad (14.9)$$

one defines the Weyl unitaries

$$W(f) = \exp(iB(f)) \quad (f \in \mathcal{H}) \,. \qquad (14.10)$$

Since

$$W(f) W(g) = W(f + g) \exp \left(-\tfrac{i}{2} \operatorname{Im} \langle f, g \rangle \right) \quad (f, g \in \mathcal{H}) \qquad (14.11)$$

the *Weyl unitaries* acting on the Bose Fock space $\mathcal{F}_+(\mathcal{H})$ give us an irreducible representation of the abstract C*-algebra $\mathrm{CCR}(\mathcal{H}, \sigma)$ when the symplectic form is $\sigma(f, g) \equiv \operatorname{Im} \langle f, g \rangle$. Although $\mathrm{CCR}(\mathcal{H}, \sigma)$ admits different irreducible representations for an infinite dimensional Hilbert space $\mathcal{H}$, the Fock representation is the most important one.

Let ψ be a state of $\mathrm{CCR}(H, \sigma)$ and consider the corresponding GNS-representation π_ψ on the Hilbert space $\mathcal{H}_\psi$. If $t \mapsto \psi(W(tf))$ is continuous for every $f \in H$ then the state ψ is called regular and $t \mapsto \pi_\psi(W(tf))$ is a strongly continuous unitary group. Due to the Stone theorem it possesses an infinitesimal generator $B_\psi(f)$ which is a self-adjoint operator on $\mathcal{H}_\psi$. In most cases we restrict ourselves to states of $\mathrm{CCR}(\mathcal{H}, \sigma)$ such that $t \mapsto \varphi(W(tf))$ is at least twice differentiable. One can prove in this case that the cyclic vector $\Psi \in \mathcal{H}_\psi$ is in the common domain of all unbounded operators $B_\psi(f)$ with $f \in H$. If, in addition, the state ψ is even, (that is, $\psi(W(f)) = \psi(W(-f))$ for every $f \in H$) then by differentiating the relation

$$\langle \Psi, e^{itB_\psi(f)} \Psi \rangle = \langle \Psi, e^{-itB_\psi(t)} \Psi \rangle$$

we obtain

$$\langle \Psi, B_\psi(f) \Psi \rangle = 0 \qquad (f \in H) \,. \qquad (14.12)$$

Since the set of linear combinations of the Weyl unitaries $\{W(f) : f \in H\}$ are dense in $\mathrm{CCR}(H, \sigma)$, any state ψ is determined by the function $f \to G(f) \equiv \psi(W(f))$ $(f \in H)$, also called the *characteristic function* of the state

ψ. It can be proven that $f \mapsto G(f)$ is the characteristic function of a state of $\mathrm{CCR}(H, \sigma)$ if and only if $G(0) = 1$ and the kernel

$$(f, g) \to G(f - g) \exp\left(\tfrac{\mathrm{i}}{2}\sigma(f, g)\right) \tag{14.13}$$

is positive, that is, for any $f_1, f_2, \ldots, f_n \in H$ and $c_1, c_2, \ldots, c_n \in \mathbb{C}$ we have

$$\sum_{j,k=1}^{n} c_j, \bar{c}_k G(f_j - f_k) \exp\left(\tfrac{\mathrm{i}}{2}\sigma(f_j, f_k)\right) \geq 0.$$

In particular, let s be a symmetric positive bilinear form on H. There exists a state ω of $\mathrm{CCR}(H, \sigma)$ such that

$$\omega(W(f)) = \exp\left(-\tfrac{1}{2}s(f, f)\right) \tag{14.14}$$

if, and only if, $|\sigma(f, g)|^2 \leq s(f, f)\, s(g, g)$ for every $f, g \in H$. A state ω of the form (14.14) is called a quasi-free state.

After the short introduction to the algebra of the canonical commutation relation, we resume the study of the central limit for the quantum spin system. Recall that a product state φ of $\mathcal{A}$ was chosen and assume that φ is the product of factors $\varphi_0 \in \mathfrak{S}(M_m(\mathbb{C}))$. As it is customary the matrix algebra $M_m(\mathbb{C})$ is a Hilbert space if endowed with the scalar product

$$\langle a, b \rangle = \varphi_0(a^* b) \qquad (a, b \in M_m(\mathbb{C})).$$

On the self-adjoint part $M_m(\mathbb{C})^{sa}$ this induces the symplectic form

$$\sigma(a, b) = \operatorname{Im} \langle a, b \rangle = -\mathrm{i}\,\varphi_0([a, b]) \qquad (a, b \in M_m(\mathbb{C})^{sa}).$$

So we may consider the algebra of canonical commutation relation over the symplectic space $(M_m(\mathbb{C})^{sa}, \sigma)$. This algebra will be called the algebra of (normal) fluctuations of the spin system. The reason for the terminology is explained by the following *central limit theorems*.

Proposition 14.4 For each $x \in M_m(\mathbb{C})^{sa}$ the limit

$$\lim_{n \to \infty} \varphi\left(\exp \mathrm{i}\, F_n(x)\right)$$

exists and defines a quasi-free state ω on the algebra of fluctuations such that

$$\lim_{n \to \infty} \varphi\left(\exp \mathrm{i}\, F_n(x)\right) = \omega(W(x)) = \exp\left(-\tfrac{1}{2}s(x, x)\right)$$

where s is the real symmetric positive bilinear form on $M_m(\mathbb{C})^{sa}$ given by

$$s(x, y) = \operatorname{Re} \varphi_0\big((x - \varphi_0(x))(y - \varphi_0(y))\big).$$

Proof. In fact, the statement is an unusual reformulation of the standard probabilistic central limit theorem (for identically distributed independent random variables). Let us recall the argument. We may assume that $\varphi_0(x) = 0$. Since φ is a product state we have

$$\varphi(\exp(\mathrm{i}\, F_n(x)) = f(n)^n = \left(1 + \frac{1}{n}(n\, f(n) - n)\right)^n,$$

with the abbreviation

$$f(n) = \varphi_0\left(\exp\frac{\mathrm{i}\, x}{\sqrt{n}}\right).$$

It follows from the power series expansion of the exponential function that

$$n\, f(n) - n = -\tfrac{1}{2}\varphi_0(x^2) + O(n^{-1/2}).$$

Consequently,

$$\varphi(\exp(\mathrm{i}F_n(x)) \to \exp\left(-\tfrac{1}{2}\varphi_0(x^2)\right)$$

which is the value of $\omega(W(x))$. □

As a quasi-free state ω is regular and the Weyl operators $W(x)$ are represented by exponentials of the *Bose field operators* $B_\omega(x)$. In view of the identification of $B_\omega(x)$ with the "limit" of the fluctuations $F_n(x)$, the unbounded operators $B_\omega(x)$ are nothing else but the macroscopic fluctuations of the observables $x \in M_m(\mathbb{C})^{sa}$.

Now we prove other versions of Proposition 14.4. These results contain the genuine non-commutative feature, in contrast with the previous proposition where diagonalization of the operators was possible.

Theorem 14.5 Under the conditions of the previous proposition we have

$$\lim_{n\to\infty} \varphi(\exp(\mathrm{i}F_n(x^1))\exp(\mathrm{i}F_n(x^2))\ldots\exp(\mathrm{i}F_n(x^k))$$
$$= \omega\big(W(x^1 - \varphi_0(x^1))W(x^2 - \varphi_0(x^2))\ldots W(x^k - \varphi_0(x^k))\big).$$

for every $x^1, x^2, \ldots, x^k \in M_m(\mathbb{C})^{sa}$ and $k \in \mathbb{N}$.

Proof. We may assume that $\varphi_0(x^i) = 0$ ($1 \le i \le k$) and apply induction by k. Observe that the case $k = 1$ was stated in Proposition 14.4. We have to carry out the induction step.

Proposition 14.3 tells us that

$$\exp(\mathrm{i}F_n(x^k))\exp(\mathrm{i}F_n(x^{k+1}))$$
$$- \exp(\mathrm{i}F_n(x^k + x^{k+1}))\exp\left(\tfrac{1}{2}s_n([x^k, x^{k+1}])\right) \qquad (14.15)$$

converges to 0 in norm where

$$s_n(x) \equiv \frac{1}{n}\sum_{i=1}^{n} x_i \qquad (x \in M_m(\mathbb{C})).$$

Let us represent $\mathcal{A}$ on the GNS-Hilbert space $\mathcal{H}$ with cyclic vector Φ. We apply von Neumann's statistical ergodic theorem for the unitary $V : \mathcal{H} \to \mathcal{H}$ defined by the formula

$$V a\Phi = \gamma(a)\Phi \qquad (a \in \mathcal{A})$$

where γ is the right shift on $\mathcal{A}$. Set E for the projection onto the fixed point space of V. We claim that E is of rank one. It is easy to see that for local elements $a, b \in \mathcal{A}$ the relation

$$\varphi(b^* s_n(a)) \to \varphi(b^*)\varphi(a)$$

holds. Equivalently,

$$\langle b\Phi, n^{-1}(I + V + \ldots V^{n-1})a\Phi \rangle \to \langle \Phi, a\Phi \rangle \langle b\Phi, \Phi \rangle.$$

Therefore we have

$$\langle b\Phi, Ea\Phi \rangle = \langle \Phi, a\Phi \rangle \langle b\Phi, \Phi \rangle$$

for every $a, b \in \mathcal{A}$. The vector Φ being cyclic, the equality

$$Ea\Phi = \langle \Phi, a\Phi \rangle \Phi = \psi(a)\Phi$$

must hold and E is really of rank one.

Let $a \in M_m(\mathbb{C})$ and $(b_n) \subset \mathcal{A}$ be a bounded sequence. We show that

$$\lim_{n \to \infty} \varphi\big(b_n \exp(s_n(a))\big) = \exp \varphi_0(a) \lim_{n \to \infty} \varphi(b_n) \tag{14.16}$$

whenever the right hand side makes sense. Indeed,

$$\langle b_n \exp(s_n(a) - \varphi_0(a))\Phi, \Phi \rangle$$
$$= \langle b_n\Phi, \Phi \rangle + \langle s_n(a)\Phi - \varphi_0(a)\Phi, \sum_{k=0}^{\infty} \frac{(s_n(a^*) - \varphi_0(a^*))^k}{(k+1)!} b_n^*\Phi \rangle,$$

where the second term obviously converges to 0. Benefiting from the induction hypothesis and (14.16) we obtain

$$\lim_{n \to \infty} \varphi\big(\exp(\mathrm{i}\, F_n(x^1)) \ldots \exp(\mathrm{i}\, F_n(x^k)) \exp(\mathrm{i}\, F_n(x^{k+1})) \big)$$
$$= \lim_{n \to \infty} \varphi\big(\exp(\mathrm{i}\, F_n(x^1)) \ldots \exp(\mathrm{i}\, F_n(x^{k-1}))$$
$$\times \exp(\mathrm{i}\, F_n(x^k + x^{k+1})) \exp \big(\tfrac{1}{2} s_n([x^k, x^{k+1}]) \big) \big)$$
$$= \lim_{n \to \infty} \varphi\big(\exp(\mathrm{i}\, F_n(x^1)) \ldots \exp(\mathrm{i}\, F_n(x^{k-1}))$$
$$\times \exp(\mathrm{i}\, F_n(x^k + x^{k+1})) \big) \exp \big(-\tfrac{1}{2}\sigma(x^{k_1}, x^{k+1}) \big)$$
$$= \omega\big(W(x^1) \ldots W(x^{k-1}) W(x^k + x^{k+1}) \big) \exp \big(-\tfrac{1}{2}\sigma(x^k, x^{k+1}) \big)$$
$$= \omega\big(W(x^1) \ldots W(x^k) W(x^{k+1}) \big).$$

We have used the relationship

$$\varphi_0([x^k, x^{k+1}]) = -\mathrm{i}\, \sigma(x^k, x^{k+1})$$

as well as the basic commutation relation (14.3) of the Weyl operators.   □

In the next *central limit theorem* we need some further knowledge on quasi-free states. Let ω be a quasi-free state of the algebra $CCR(\mathcal{H})$. The cyclic vector Φ in the GNS-representation is in the domain of the field operators $B_\omega(f)$ ($f \in \mathcal{H}$) but much more is true. For any $f_1, f_2, \ldots, f_k \in \mathcal{H}$

$$\Phi \in \mathcal{D}\big(B_\omega(f_1)B_\omega(f_2)\ldots B_\omega(f_k)\big) .$$

So the scalar product $\langle \Phi, B_\omega(f_1)B_\omega(f_2)\ldots B_\omega(f_k)\Phi\rangle$ makes sense and we write it as $\omega(B_\omega(f_1)B_\omega(f_2)\ldots B_\omega(f_k))$. A characteristic property of the quasi-free states is the following. For k odd

$$\omega\big(B_\omega(f_1)B_\omega(f_2)\ldots B_\omega(f_k)\big) = 0 ,$$

and for k even we have

$$\omega\big(B_\omega(f_1)B_\omega(f_2)\ldots B_\omega(f_k)\big) = \sum \prod_{m=1}^{k/2} \omega\big(B_\omega(f_{n_m})B_\omega(f_{j_m})\big) , \tag{14.17}$$

where the summation is over all partitions $\{H_1, H_2, \ldots, H_{k/2}\}$ of $\{1, 2, \ldots, k\}$ such that $H_m = \{n_m, j_m\}$ with $n_m < j_m$ ($m = 1, 2, \ldots, k/2$).

Theorem 14.6 Under the conditions of Proposition 14.4 we have

$$\lim_{n\to\infty} \varphi\big(F_n(x^1)F_n(x^2)\ldots F_n(x^k)\big) = \omega(B(x^1 - \varphi_0(x^1))\ldots B(x^k - \varphi_0(x^k))$$

for $x^1, x^2, \ldots, x^k \in M_m(\mathbb{C})^{sa}$ and $k \in \mathbb{N}$.

Proof. The proof consists of an enjoyable combinatorial argument. First of all, we may assume that $\varphi_0(x^i) = 0$ for every $1 \leq i \leq k$. The idea is to group the terms in the product

$$n^{-k/2} F_n(x^1)F_n(x^2)\ldots F_n(x^k) \tag{14.18}$$

in a certain way. The general term in the expansion of (14.18) is of the form

$$\gamma^{n_1}(x^{m_1})\gamma^{n_2}(x^{m_2})\ldots\gamma^{n_k}(x^{m_k}) , \tag{14.19}$$

where γ denotes the right shift and x^i's are thought of as elements of $\mathcal{A}_1$. Due to the obvious commutations, we may reorder the monomial (14.19) as

$$\gamma^0(x^{i^0(1)}x^{i^0(2)}\ldots x^{i^0(m(0))})\ldots\gamma^{n-1}(x^{i^{n-1}(1)}\ldots x^{i^{n-1}(m(n-1))}) \tag{14.20}$$

where the sets

$$\{i^0(1),\ldots,i^0(m(0))\},\ldots,\{i^{n-1}(1),\ldots,i^{n-1}(m(n-1))\} \tag{14.21}$$

form a partition of the set $\{1, 2, \ldots, k\}$. Since n is going to be very large, many of the sets in (14.21) will be empty. We denote by l the number of the nonempty ones. The possible values of l are $1, 2, \ldots, k$. On the (tensor) product (14.19) φ takes the value

$$\varphi_0\big(x^{j^1(1)}\ldots x^{j^1(p(1))}\big)\ldots\varphi_0\big(x^{j^l(1)}\ldots x^{j^l(p(l))}\big) , \tag{14.22}$$

where the empty sets of the partition of $\{1, 2, \ldots, k\}$ are not counted any more and therefore the number of factors is exactly l. The multiplicity of (14.22) in the expression of

$$\varphi(F_n(x^1)F_n(x^2) \ldots F_n(x^k))$$

is $n(n-1)(n-2) \ldots (n-l+1)$. (You may arrive at this number by choosing l different values for an exponent of γ from the set $\{0, 1, \ldots, n-1\}$.) Therefore we have

$$\varphi(F_n(x^1) \ldots F_n(x^k)) = \sum_{l=0}^{k} \frac{n(n-1) \ldots (n-l+1)}{n^{k/2}}$$
$$\times \sum \varphi_0(x^{j^1(1)} \ldots x^{j^1(p(2))}) \ldots \varphi_0(x^{j^l(1)} \ldots x^{j^l(p(l))}), \tag{14.23}$$

where the second summation is over all partitions of $\{1, 2, \ldots, k\}$ into l sets

$$\{j^1(1), j^1(2), \ldots, j^1(p(1))\}, \ldots, \{j^l(1), j^l(2), \ldots, j^l(p(l))\}. \tag{14.24}$$

(Note that in each product

$$x^{j^t(1)} x^{j^t(2)} \ldots x^{j^t(p(t))}$$

the subscripts are ordered increasingly.) If a partition (14.24) contains a singleton then due to the assumption $\varphi_0(x^i) = 0$ the contribution of the corresponding term of (14.23) vanishes. If $l > k/2$ then any partition into l nonempty sets must contain at least one singleton. Hence we may neglect these values of l. On the other hand for $l < k/2$ the coefficient

$$\frac{n(n-1)(n-2) \ldots (n-l+1)}{n^{k/2}}$$

tends to 0 as $n \to \infty$ while

$$\sum \varphi_0(x^{j^1(1)} \ldots x^{j^1(p(1))}) \ldots \varphi_0(x^{j^l(1)} \ldots x^{j^l(p(l))})$$

remains bounded. The only value of l that may contribute to the limit is $k/2$, provided that k is even. So for k odd

$$\lim_{n \to \infty} \varphi(F_n(x^1) \ldots F_n(x^k)) = 0,$$

whereas for $k = 2l$

$$\lim_{n \to \infty} \varphi(F_n(x^1) \ldots F_n(x^k)) = \sum \varphi_0(x^{j^1(1)} x^{j^1(2)}) \ldots \varphi_0(x^{j^l(1)} x^{j^l(2)})$$

where the summation is over all partitions of $\{1, 2, \ldots, k\}$ into sets $\{j^1(1) < j^1(2)\}, \{j^2(1) < j^2(2)\}, \ldots, \{j^l(1) < j^l(2)\}$. In order to complete the proof of the theorem, we have to show that

$$\omega(B_\omega(x)B_\omega(y)) = \varphi_0(xy) \tag{14.25}$$

for $x, y \in M_m(\mathbb{C})$ with $\varphi_0(x) = \varphi_0(y) = 0$. This can be done by differentiation.

$$\omega(B_\omega(x)B_\omega(y)) = \frac{\partial^2}{\partial t \partial s} \langle \exp it B_\omega(x)\Omega, \exp is B_\omega(y)\Omega \rangle$$

$$= \frac{\partial^2}{\partial t \partial s} \omega\big(W\big((sy - tx)^2\big)\big) \exp i\sigma(tx, sy)$$

$$= \frac{\partial^2}{\partial t \partial s} \exp\big(-\tfrac{1}{2}s^2\varphi_0(y^2) - \tfrac{1}{2}t^2\varphi_0(x^2) + st\varphi_0(xy)\big)$$

$$= \varphi_0(xy).$$

The derivative was taken at $s = t = 0$ and Ω denoted the cyclic vector in the GNS-space of the quasi-free state ω.   □

What we proved in this theorem was a *convergence in law*. The sequence of fluctuations converge to Bose fields in the sense that all correlation functions converge. Let Λ be a set of parameter set. A process (indexed by Λ) means that for every $x \in \Lambda$ an operator $A(x)$ of an algebra $\mathcal{B}$ is given and a state ψ of $\mathcal{B}$ is fixed. We do not insist on the boundedness of $A(x)$ and allow that it be affiliated with the algebra $\mathcal{B}$. One can say that the sequence $(A_n, \mathcal{B}_n, \psi_n)$ of processes converges in law to a process $(A, \mathcal{B}, \psi)$ if for every $x^1, x^2, \ldots, x^k \in \Lambda$ and $k \in \mathbb{N}$ the limit relation

$$\lim_{n\to\infty} \psi_n(A_n(x^1)A_n(x^2)\ldots A_n(x^k)) = \psi(A(x^1)A(x^2)\ldots A(x^k))$$

holds. In this language Theorem 14.6 states that the fluctuation process converges to the process of Bose fields.

Now our goal is to formulate the central limit theorem as the convergence of states of the algebra of the canonical commutation relation. We define two operations for states. For $\varphi_1 \in \mathfrak{S}(\mathrm{CCR}(H, \sigma_1))$ and $\varphi_2 \in \mathfrak{S}(\mathrm{CCR}(H, \sigma_2))$ a state $\varphi_1 \hat{+} \varphi_2$ on $\mathrm{CCR}(H, \sigma_1 + \sigma_2)$ is determined by the formula

$$(\varphi_1 \hat{+} \varphi_2)(W_{12}(f)) = \varphi_1(W_1(f))\varphi_2(W_2(f)). \tag{14.26}$$

The characteristic function of $\varphi_1 \hat{+} \varphi_2$ is the product of the characteristic functions of φ_1 and φ_2. Since the pointwise product of positive kernels is positive again, according to (14.13) the formula (14.26) really defines a state.

The other operation associates a state $\varphi \cdot \lambda$ of the algebra $\mathrm{CCR}(H, |\lambda|^2\sigma)$ to $\varphi \in \mathfrak{S}(\mathrm{CCR}(H, \sigma))$ and $\lambda \in \mathbb{R}$. We set

$$\varphi \cdot \lambda\big(W_{|\lambda|^2\sigma}(f)\big) = \varphi\big(W_\sigma(\lambda f)\big). \tag{14.27}$$

It is readily checked that

$$(\varphi \hat{+} \psi) \cdot \lambda = \varphi \cdot \lambda \hat{+} \psi \cdot \lambda. \tag{14.28}$$

If $\varphi_i \in \mathfrak{S}(\mathrm{CCR}(H, \sigma))$ and $\sum_{i=1}^n |\lambda_i|^2 = 1$ then

$$\varphi_1 \cdot \lambda_1 \hat{+} \varphi_2 \cdot \lambda_2 \hat{+} \ldots \hat{+} \varphi_n \cdot \lambda_n \in \mathfrak{S}(\mathrm{CCR}(H, \sigma)) \tag{14.29}$$

and

$$(\varphi_1 \cdot \lambda_1 \hat{+} \varphi_2 \cdot \lambda_2 \hat{+} \ldots \hat{+} \varphi_n \cdot \lambda_n)(W(h)) = \prod_{i=1}^n \varphi_i(W(\lambda_i h)). \tag{14.30}$$

The state (14.29) might be called the convolution of $\varphi_1, \varphi_2, \ldots, \varphi_n$ with the weights $\lambda_1, \lambda_2, \ldots, \lambda_n$. We shall prove the convergence of the sequence

$$(\varphi\hat{+}\varphi\hat{+}\varphi\hat{+}\ldots\hat{+}\varphi)\frac{1}{\sqrt{n}} \qquad (n \text{ summands})$$

which is a kind of central limit. (The normalization $\sum_i \lambda_i^2 = 1$ in the definition was preferred to convex combination in order to emphasis the strong relationship with central limit.)

Proposition 14.7 Let φ be a twice differentiable even state of the algebra $CCR(H, \sigma)$ over the symplectic space (H, σ). Then the weighted convolution

$$(\varphi\hat{+}\varphi\hat{+}\ldots\hat{+}\varphi)\frac{1}{\sqrt{n}}$$

converges pointwise to a state φ_Q given by

$$\varphi_Q(W(f)) = \exp\left(-\tfrac{1}{2}\|B_\varphi(f)\Phi\|^2\right) \qquad (f \in H).$$

Proof. Since linear combinations of Weyl unitaries are dense in $CCR(H, \sigma)$, it is enough to show the pointwise convergence $\varphi_n \to \varphi_Q$ at $W(g)$ $(g \in H)$.

$$(\varphi\hat{+}\varphi\hat{+}\ldots\hat{+}\varphi)\frac{1}{\sqrt{n}}(W(g)) = \left(\varphi(W(n^{-1/2}g))\right)^n$$

and the argument of the proof of Proposition 14.4 works with

$$f(n) = \varphi\left(W(n^{-1/2}g)\right) = \langle \Phi, \exp\left(n^{-1/2}B_\varphi(g)\right)\Phi\rangle.$$

Now we have

$$nf(n) - n = -\tfrac{1}{2}\|B_\varphi(g)\Phi\|^2 + O(n^{-1/2})$$

and the proposition is concluded. $\square$

The state φ_Q appearing in the previous proposition is called the quasi-free reduction of the state φ. Below we shall study the convergence of the weighted convolution by means of entropy methods.

From now on we consider the algebra of the canonical commutation relation over a complex Hilbert space. Then the symplectic form $\sigma(f, g) = \operatorname{Im}\langle f, g\rangle$ is non-degenerate and $CCR(\mathcal{H})$ is simple. By means of the complex structure of $\mathcal{H}$ one can define gauge invariant states, creation and annihilation operators. The state φ of $CCR(\mathcal{H})$ is said to be *gauge invariant* if

$$\varphi(W(\lambda f)) = \varphi(W(f)) \quad (f \in \mathcal{H}, \; \lambda \in \mathbb{C}, \; |\lambda| = 1). \tag{14.31}$$

The creation and annihilation operators appear in the GNS-Hilbert space of a (regular) state φ:

$$a_\varphi(f) = \frac{1}{\sqrt{2}}(B_\varphi)(f) + iB_\varphi(it)$$

$$a_\varphi^+(f) = \frac{1}{\sqrt{2}}(B_\varphi(f) - iB_\varphi(it)). \tag{14.32}$$

The state

$$\varphi_F(W(f)) = \exp\left(-\tfrac{1}{4}||f||^2\right) \quad (f \in \mathcal{H}) \tag{14.33}$$

is a gauge invariant, infinitely differentiable state of $\mathrm{CCR}(\mathcal{H})$ and is called the Fock state because the corresponding GNS-Hilbert space is naturally isomorphic to the Bose-Fock space over the Hilbert space $\mathcal{H}$. The Fock state is a pure quasifree state. We shall mainly deal with twice differentiable gauge invariant states. Let φ be such a state and Φ be the cyclic vector of its GNS-representation. Then $\Phi \in \mathcal{D}(a_\varphi(f)) \cap \mathcal{D}(a_\varphi^+(f))$ for every $f \in \mathcal{H}$ and $\langle a_\varphi(g)\Phi, a_\varphi(f)\Phi \rangle$ is defined. Formally it can be written as $\varphi(a_\varphi^*(g)a_\varphi(f))$ which is called the two-point function of the state φ. The two-point function

$$t(f,g) = \varphi(a_\varphi^*(g)a_\varphi(f)) \tag{14.34}$$

is a positive sesquilinear form over the Hilbert space $\mathcal{H}$. To each positive sesquilinear form t there exists a state ω determined by the formula

$$\omega(W(f)) = \exp\left(-\tfrac{1}{4}||f||^2 - \tfrac{1}{2}t(f,f)\right) \quad (f \in \mathcal{H}) \tag{14.35}$$

which is a gauge invariant quasi-free state. From the relations

$$\omega(B_\omega(f)^2) = -\left.\frac{\partial^2 \omega(W(tf))}{\partial t^2}\right|_{t=0} = \frac{1}{2}||f||^2 + t(f,f) \tag{14.36}$$

and

$$B_\omega(f)^2 = \tfrac{1}{2}\left(a_\omega^*(f)^2 + a_\omega(f) + 2a_\omega^*(f)a_\omega(f) + ||f||^2 I\right)$$

follows that the two-point function of ω is $(f,g) \mapsto t(f,g)$. It is clear from Proposition 14.7 and formula (14.36) that ω is the quasi-free reduction of φ. So one can describe the quasi-free reduction as follows. A (twice differentiable gauge invariant) state φ of $\mathrm{CCR}(\mathcal{H})$ determines a two-point function (by (14.34)) and the quasi-free reduction of φ is the quasi-free state given by this two-point function, see (14.35).

Above the weighted convolution $\varphi_1 \cdot \lambda_1 \hat{+} \varphi_2 \cdot \lambda_2$ was defined for real coefficients λ_1 and λ_2. If the test function space $\mathcal{H}$ has a complex linear structure then $\lambda_1, \lambda_2 \in \mathbb{C}$ may be allowed. However, for a gauge invariant state φ_1, we have $\varphi_1 \cdot \lambda_1 = \varphi_2 \cdot |\lambda_1|$.

The next proposition contains the basic entropy inequality for weighted convolutions. The completeness of the Hilbert space $\mathcal{H}$ will not play any role, $\mathcal{H}$ could be a complex inner product space as well.

Proposition 14.8 Let $\varphi_1, \varphi_2 \in \mathfrak{S}(\mathrm{CCR}(\mathcal{H}))$ be gauge invariant states and set

$$\varphi = \frac{\varphi_1 \hat{+} \varphi_1 \hat{+} \ldots \hat{+} \varphi_1}{\sqrt{n+k}} \hat{+} \frac{\varphi_2 \hat{+} \ldots \hat{+} \varphi_2}{\sqrt{n+k}}$$

where $n, k \in N$ and φ_1 appears n times and φ_2 k times. Then

$$S(\varphi) \geq \frac{n}{n+k}S(\varphi_1) + \frac{k}{n+k}S(\varphi_2).$$

Proof. We consider $\mathcal{A}_1 = \mathcal{A}_2 = \mathrm{CCR}(\mathcal{H} \oplus \ldots \oplus \mathcal{H})$ with an $(n+k)$-fold direct sum. $\mathcal{A}_i$ can be identified with the $(n+k)$-fold tensor product

$$\mathrm{CCR}(\mathcal{H}) \otimes \mathrm{CCR}(\mathcal{H}) \otimes \ldots \otimes \mathrm{CCR}(\mathcal{H}))$$

under the identification

$$W_i(h_1 \oplus \ldots \oplus h_{n+k}) \mapsto W(h_1) \otimes W(h_2) \otimes \ldots \otimes W(h_{n+k}).$$

(The CCR-algebras are known to be nuclear, hence the C*-tensor product is unique.) On the algebra $\mathcal{A}_2$ we consider the product state

$$\psi = \varphi_1 \otimes \ldots \otimes \varphi_1 \otimes \varphi_2 \otimes \ldots \otimes \varphi_2$$

where φ_1 appears in n factors and φ_2 in k ones. According to the additivity of the entropy (see Theorem 6.15) we have

$$S(\psi) = nS(\varphi_1) + kS(\varphi_2). \tag{14.37}$$

If an $(n+k) \times (n+k)$ unitary matrix is given then a Bogoliubov isomorphism α arises between $\mathcal{A}_1$ and $\mathcal{A}_2$ as follows.

$$\alpha : W_1(h_1 \oplus \ldots \oplus h_{n+k}) \mapsto W_2\left(\sum_{i=1}^{n+k} a_{1,i}h_i \oplus, \ldots, \oplus \sum_{i=1}^{n+k} a_{n+k,i}h_i\right).$$

Let us compute the marginal of $\psi \circ \alpha$ on the j-th marginal $(1 \le j \le n+k)$.

$$\psi \circ \alpha(W_1(0 \oplus \ldots \oplus h^{(j)} \oplus \ldots \oplus 0)) = \psi(W_2(a_{1,j}h \oplus a_{2j} \oplus \ldots \oplus a_{n+k,j}h))$$

$$= \prod_{i=n+1}^{n} \varphi_1(W(a_{ij}h)) \prod_{i=n+1}^{n+k} \varphi_2(W(a_{ij}h)).$$

Choosing

$$(a_{ij}) = \frac{1}{\sqrt{n+k}} e^{2ij\pi/(n+k)} \quad (1 \le i,j \le n+k) \tag{14.38}$$

we have a unitary matrix (a_{ij}) and the j-th marginal is nothing else but φ. Refering to the subadditivity of the entropy (Theorem 6.16) we conclude

$$S(\psi) = S(\psi \circ \alpha) \le (n+k)S(\varphi)$$

which gives the proof with (14.37).   □

It φ be the same as in the previous proposition and define ψ analogously, that is

$$\psi = \frac{\psi_1 \hat{+} \psi_1 \hat{+} \ldots \hat{+} \psi_1}{\sqrt{n+k}} \hat{+} \frac{\psi_2 \hat{+} \psi_2 \hat{+} \ldots \hat{+} \psi_2}{\sqrt{n+k}}$$

(where the states ψ_1 and ψ_2 occur n- and k-times, respectively.) Following the proof of Proposition 14.8 one can obtain a relative entropy version. Namely,

$$S(\varphi, \psi) \le \frac{n}{n+k} S(\varphi_1, \psi_1) + \frac{k}{n+k} S(\varphi_2, \psi_2). \tag{14.39}$$

The direction of the inequality is the opposite because the relative entropy is superadditive under tensor product while the entropy is subadditive.

Another remark concerns the gauge invariance. If $n = k = 1$ then the matrix (14.38) is real and the proof works for even states.

Lemma 14.9 Let φ and ω be arbitrary states of $\mathrm{CCR}(\mathcal{H})$. Then

$$S(\psi, \varphi) = \sup\{S(\psi|\mathcal{A}, \varphi|\mathcal{A}) : \mathcal{A} = \mathrm{CCR}(\mathcal{K}),$$

$$\mathcal{K} \subset \mathcal{H} \text{ is finite dimensional subspace}\}.$$

Proof. Let $\mathcal{B} \subset \mathrm{CCR}(\mathcal{H})$ be the *-algebra of finite linear combinations of Weyl operators. Then $\mathcal{B}$ is a norm dense subalgebra of $\mathrm{CCR}(\mathcal{H})$. The relative entropy $S(\psi, \varphi)$ can be approximated arbitrarily well by an expression

$$\log m - \int_{1/m}^{\infty} t^{-1} \psi(y(t)^* y(t)) + t^{-2} \varphi(x(t)x(t)^*)\, dt \qquad (14.40)$$

due to Kosaki's formula (see Theorem 5.11 for the explanation of $x(t)$ and $y(t)$.) One can choose $x(t), y(t) \in \mathcal{B}$ and in fact, (14.40) is a finite sum. If we take the subspace $\mathcal{K}$ which is generated by the symbols of the Weyl unitaries appearing in (14.40) then $S(\psi|\mathrm{CCR}(\mathcal{K}), \varphi|\mathrm{CCR}(\mathcal{K}))$ is a good approximation of $S(\psi, \varphi)$. $\qquad\square$

Now everything is ready to prove a strengthened entropic central limit theorem.

Theorem 14.10 Let φ be a gauge invariant twice differentiable state of $\mathrm{CCR}(\mathcal{H})$ with quasi-free reduction ψ. If $S(\varphi, \psi)$ is finite then

$$S(\varphi_n, \psi) \to 0 \quad \text{and} \quad \|\varphi_n - \psi\| \to 0,$$

where

$$\varphi_n = \frac{\varphi \hat{+} \varphi \hat{+} \ldots \hat{+} \varphi}{\sqrt{n}} \qquad (\varphi \text{ is } n \text{ times}).$$

Proof. We show that $S(\varphi_n, \psi) \to 0$. Due to Lemma 14.9 it is sufficient to see that

$$S(\varphi_n|\mathrm{CCR}(\mathcal{K}), \psi|\mathrm{CCR}(\mathcal{K})) \to 0$$

for any finite dimensional subspace $\mathcal{K} \subset \mathcal{H}$. We fix $\mathcal{K} \subset \mathcal{H}$ and denote by $\bar{\varphi}_n$, $\bar{\varphi}$, $\bar{\psi}$ the restriction of φ_n, φ, ψ to $\mathcal{A} \equiv \mathrm{CCR}(\mathcal{K})$, respectively. Note that $\bar{\psi}$ is the quasi-free reduction of $\bar{\varphi}$ and $\bar{\varphi}_n$ is the weighted convolution sequence formed from $\bar{\varphi}$. Proposition 14.7 tells us that $\bar{\varphi}_n \to \bar{\psi}$ pointwise.

We view $\mathcal{A}$ in the Fock representation on the Hilbert space $\mathcal{F}_+(\mathcal{K})$. The quasi-free state $\bar{\psi}$ is normal with respect to the Fock representation, that is, it has a normal extension to $B(\mathcal{F}_+(\mathcal{K}))$ which has a density D. According to (14.39) the sequence $n \mapsto n S(\bar{\varphi}, \bar{\psi})$ is subadditive, in particular $S(\bar{\varphi}_n, \bar{\psi}) \leq S(\bar{\varphi}_n, \bar{\psi}) \leq S(\varphi, \psi) < +\infty$. (The subadditivity yields that $\lim_n S(\bar{\varphi}_n, \bar{\psi})$

exists and we are showing that it must be 0.) Since $S(\bar{\varphi}_n, \bar{\psi})$ is finite, $\bar{\varphi}_n$ also has a normal extension to $B(\mathcal{F}_+(\mathcal{K}))$ with a density D_n. We have

$$S(\bar{\varphi}_n, \bar{\psi}) = \operatorname{Tr} D_n(\log D_n - \log D_n) = -S(\bar{\varphi}_n) - \operatorname{Tr} D \log D.$$

We shall prove that $-\operatorname{Tr} D_n \log D = S(\bar{\psi})$ independently of n and $S(\bar{\varphi}_n) \to S(\bar{\psi})$. Having these relations proved, we arrive at $S(\bar{\varphi}_n, \bar{\psi}) \to 0$ and $S(\varphi_n, \psi) \to 0$.

The two-point function of $\bar{\psi}$ is given by a positive operator $T \in \mathcal{B}(\mathcal{K})$ in the form

$$\bar{\psi}(a^*(g)a(f)) = \langle f, Tg \rangle \quad (f, g \in \mathcal{K})$$

Assume that the positive eigenvalues of T are $\lambda_1 \geq \lambda_2 \geq \ldots \geq \lambda_n > 0$ with corresponding eigenvectors $f_1, f_2, \ldots, f_n$. Let

$$\lambda_i = \frac{e^{-S_i}}{1 - e^{-S_i}} \quad (1 \leq i \leq n)$$

and

$$H = \sum_{i=1}^{n} S_i \, a(f_i)^* a(f_i).$$

It is known that $D = C \exp(-H)$ where $C = 1/\operatorname{Tr} \exp(-H)$ is the normalization. From this it is clear that $\omega(\log D)$ depends only on the two-point function of ω, in particular $-\operatorname{Tr} D_n \log D = -\operatorname{Tr} D \log D = S(\bar{\psi})$. To show $S(\bar{\varphi}_n) \to S(\psi)$ we refer to the continuity of the entropy at bounded energy (Proposition 6.6). We have $\operatorname{Tr} \exp(-\beta H) < +\infty$ and

$$\bar{\varphi}_n(H) = \bar{\psi}(H) < +\infty$$

and the above cited result is applicable.

Now $S(\varphi_n, \psi) \to 0$ has been proved and the inequality

$$\|\varphi_n - \psi\|^2 \leq 2S(\varphi_n, \psi)$$

completes the theorem (cf. Proposition 5.23).   □

Since $S(\varphi, \varphi_Q) = -S(\varphi) + S(\varphi_Q) \geq 0$, the quasi-free state has the largest entropy among the states with a given two-point function. This is sometimes called as maximum entropy principle. The quasi-free reduction is a (non-affine) projection onto the set of quasi-free states.

Notes and Remarks. The pioneering works on quantum *central limit theorems* are [Cushen and Hudson 1971] and [Giri and Waldenfels 1978]. Recently several generalizations have been obtained. The algebra of fluctuation, that is a CCR-algebra over a C*-algebra, is originated from the paper [Goderis, Verbeure and Vets 1989a]. General references on the abstract CCR-algebra are [Bratteli and Robinson 1981], [Evans and Lewis 1977] (this booklet contains the proof of the nuclearity of the CCR-algebra), [Holevo 1982] and [Petz

1990a]. The *Slawny uniqueness theorem* was published in [Slawny 1971]. An instructive example is in [Waldenfels 1990]. The presentation of the chapter has benefited also from [Accardi and Bach 1989].

The setting of Proposition 14.4 is rather restrictive compared with the literature. There are central limit theorems on a higher dimensional lattice and for a state which satisfies a set of cluster conditions. (cf. [Goderis, Verbeure and Vets 1990]). Another direction of extension is the study of operator-valued mappings instead of the scalar-valued state. In this case the central limit gives rise to quasi-free completely positive mappings, for example, see [Quaegebeur 1984].

Theorem 14.10 is from [Petz 1992b] which was motivated by [Barron 1986] from the side of probability and [Streater 1987a] from the side of quantum mechanics. Concerning the principle of maximum entropy (in the setting of canonical ensembles) we refer to [Thirring 1983].

In the chapter central limit theorems were shown for the quantum spin system which was considered as an algebraic stochastic process ($j_n : M_n(\mathbb{C}) \to \mathcal{A}, \varphi$). The main feature of this process that $[j_n(a), j_m(b)] = 0$ if $n \neq m$ which is due to the tensor product structure of $\mathcal{A}$. One may consider more general commutation relations between $j_n(a)$ and $j_m(b)$ and the central limit will be different. (Nevertheless the limiting object always carries some reminiscent of Gaussianity.) In the anticommuting case the first result of this type is [Waldenfels 1978]. Recently central limit theorems have been obtained in the presence of "free independence" both in a probabilistic and algebraic setting, see [Voiculescu 1986], [Speicher 1990], [Maasen 1992] and [Speicher 1992].

15 Thermodynamics
of Quantum Spin Systems

Let an infinitely extended system of particles be considered in the simple cubic lattice $L = \mathbb{Z}^\nu$. It is thought that each lattice site is occupied by a finite number of particles; they interact with one another and jump from lattice site to lattice site. The observables confined to a lattice site $x \in \mathbb{Z}^\nu$ form the self-adjoint part of a finite dimensional C*-algebra $\mathcal{A}_x$ which is a copy of $M_d(\mathbb{C})$. It is assumed that the local observables in any bounded region $\Lambda \subset \mathbb{Z}^\nu$ are those of the finite quantum system

$$\mathcal{A}_\Lambda = \bigotimes_{x \in \Lambda} \mathcal{A}_x.$$

It follows from the definition that for $\Lambda \subset \Lambda'$ we have $\mathcal{A}_{\Lambda'} = \mathcal{A}_\Lambda \otimes \mathcal{A}_{\Lambda' \setminus \Lambda}$, where $\Lambda' \setminus \Lambda$ is the complement of Λ in Λ'. The algebra $\mathcal{A}_\Lambda$ and the subalgebra $\mathcal{A}_\Lambda \otimes \mathbb{C} I_{\Lambda' \setminus \Lambda}$ of $\mathcal{A}_{\Lambda'}$ have identical structure and we identify the element $A \in \mathcal{A}_\Lambda$ with $A \otimes I_{\Lambda' \setminus \Lambda}$ in $\mathcal{A}_{\Lambda'}$. If $\Lambda \subset \Lambda'$ then $\mathcal{A}_\Lambda \subset \mathcal{A}_{\Lambda'}$ and it is said that $\mathcal{A}_\Lambda$ is isotonic with respect to Λ. The definition also implies that if Λ_1 and Λ_2 are disjoint then elements of $\mathcal{A}_{\Lambda_1}$ commute with those of $\mathcal{A}_{\Lambda_2}$. The *quasi-local C*-algebra* $\mathcal{A}$ is the norm completion of the normed algebra $\mathcal{A}_\infty = \cup_\Lambda \mathcal{A}_\Lambda$, the union of all local algebras $\mathcal{A}_\Lambda$ associated with bounded (finite) regions $\Lambda \subset \mathbb{Z}^\nu$.

We denote by a_x the element of $\mathcal{A}_x$ corresponding to $a \in \mathcal{A}_0$ ($x \in \mathbb{Z}^\nu$). It follows from the definition that the algebra $\mathcal{A}_\infty$ consists of linear combinations of terms

$$a_{x_1}^{(1)} \ldots a_{x_k}^{(k)},$$

where $x_1, \ldots, x_k$ and $a^{(1)}, \ldots, a^{(k)}$ run through $\mathbb{Z}^\nu$ and $\mathcal{A}_0$, respectively. We define γ_x to be the linear transformation

$$a_{x_1}^{(1)} \ldots a_{x_k}^{(k)} \longmapsto a_{x_1+x}^{(1)} \ldots a_{x_k+x}^{(k)}.$$

γ_x corresponds to the space-translation by $x \in \mathbb{Z}^\nu$ and it extends to an automorphism of $\mathcal{A}$. Hence γ is a representation of the abelian group $\mathbb{Z}^\nu$ by automorphisms of the quasi-local algebra $\mathcal{A}$. Clearly, the covariance condition

$$\gamma_x(\mathcal{A}_\Lambda) = \mathcal{A}_{\Lambda+x}$$

holds, where $\Lambda + x$ is the space-translate of the region Λ by the displacement x. An important property of space-translations is that they are *asymptotically abelian*: If $|x|$ is large enough then

$$[\gamma_x(\mathcal{A}_{\Lambda_1}), \mathcal{A}_{\Lambda_2}] = 0 \tag{15.1}$$

for given finite regions $\Lambda_1, \Lambda_2 \subset \mathbb{Z}^\nu$.

Having described the kinematical structure of lattice systems we turn to the dynamics. The local Hamiltonian $H(\Lambda)$ is taken to be the total potential energy between the particles confined to Λ. This energy may come from *many-body interactions* of various orders. Most generally, we assume that there exists a global function Φ such that for any finite subsystem Λ the local Hamiltonian takes the form

$$H(\Lambda) = \sum_{X \subset \Lambda} \Phi(X). \tag{15.2}$$

Each $\Phi(X)$ represents the interaction energy of the particles in X. Mathematically, $\Phi(X)$ is a self-adjoint element of $\mathcal{A}_X$ and $H(\Lambda)$ will be a self-adjoint operator in $\mathcal{A}_\Lambda$. We restrict our discussion to translation-invariant interactions which satisfy the additional requirement

$$\gamma_x(\Phi(X)) = \Phi(X + x)$$

for every $x \in \mathbb{Z}^\nu$ and every region $X \subset \mathbb{Z}^\nu$. For a finite subset $\Lambda \subset \mathbb{Z}^\nu$ let $d(\Lambda)$ denote the largest distance between two points in Λ. $d(\Lambda)$ is called the diameter of Λ. An interaction Φ is said to be of *finite range* if there is a number $d_\Phi > 0$ such that $\Phi(\Lambda) = 0$ whenever $d(\Lambda) \geq d_\Phi$. The infimum of such numbers is termed the range of Φ.

If φ is a state of the quasi-local algebra $\mathcal{A}$ then it will induce a state φ_Λ on $\mathcal{A}(\Lambda)$, the finite system comprising the atoms in the bounded region Λ of $\mathbb{Z}^\nu$. The (local) *energy, entropy and free energy* of this finite system are given by the following formulas.

$$E_\Lambda(\varphi) = \mathrm{Tr}_\Lambda D_\Lambda H(\Lambda),$$
$$S_\Lambda(\varphi) = -\mathrm{Tr}_\Lambda D_\Lambda \log D_\Lambda, \tag{15.3}$$
$$F_\Lambda^\beta(\varphi) = E_\Lambda(\varphi) - \frac{1}{\beta} S_\Lambda(\varphi).$$

Here D_Λ denotes the density of φ_Λ with respect to the trace Tr_Λ of $\mathcal{A}_\Lambda$ and β denotes the inverse temperature. The functionals E_Λ, S_Λ and F_Λ^β are termed *local*. It is rather obvious that all three local functionals are continuous if the weak* topology is considered on the state space of the quasi-local algebra. The energy is affine, the entropy is concave (cf. Proposition 1.6) and consequently, the free energy is a convex functional.

The *time evolution* of the system occupying the finite region $\Lambda \subset \mathbb{Z}^\nu$ is represented by a one-parameter group of automorphisms of $\mathcal{A}_\Lambda$.

$$\sigma_t^\Lambda(A) = \exp(\mathrm{i}tH(\Lambda)) A \exp(-\mathrm{i}tH(\Lambda)). \tag{15.4}$$

The automorphism group associates to "the local observable" $A \in \mathcal{A}_\Lambda$ at time 0 the observable $\sigma_t^\Lambda(A)$ at time t.

The thermodynamic limit, "Λ tends to infinity" may be taken along lattice parallelepipeds. Let $a \in \mathbb{Z}^\nu$ with positive coordinates and define

$$\Lambda(a) = \{x \in \mathbb{Z}^\nu : 0 \le x_i < a_i, \quad i = 1, 2, \dots, \nu\}. \tag{15.5}$$

The volume of the parallelepiped $\Lambda(a)$ is

$$V(a) = \prod_{i=1}^{\nu} a_i$$

and we write $a \to \infty$ if $a_i \to \infty$ for all $1 \le i \le \nu$. When $a \to \infty$, $\Lambda(a)$ tends to infinity in a manner suitable for the study of thermodynamic limit: the boundary of the parallelepipeds is getting more and more negligible compared with the volume. The notion of *limit in the sense of van Hove* makes this idea more precise and physically more satisfactory. For the sake of simplicity we restrict ourselves to thermodynamic limit along parallelepipeds.

Denoting by $|\Lambda|$ the volume of Λ (or the number of points in Λ) we may define the *global energy, entropy and free energy* functionals of translationally invariant states to be

$$e(\varphi) = \lim_{\Lambda \to \infty} E_\Lambda(\varphi)/|\Lambda| \tag{15.6}$$

$$s(\varphi) = \lim_{\Lambda \to \infty} S_\Lambda(\varphi)/|\Lambda| \tag{15.7}$$

$$f^\beta(\varphi) = \lim_{\Lambda \to \infty} F_\Lambda^\beta(\varphi)/|\Lambda| \tag{15.8}$$

and assume that the *global time evolution* is given by the limit

$$\sigma_t = \lim_{\Lambda \to \infty} \sigma_t^\Lambda. \tag{15.9}$$

The existence of the limit in (15.7) is guaranteed by the strong subadditivity of entropy, while that of the limits in (15.6), (15.8) and (15.9) is assumed if the interaction is suitable tempered, as it certainly does if the interaction is of finite range. The *strong subadditivity* is an important feature of the local entropy functional. For $\Lambda_1, \Lambda_2 \subset \mathbb{Z}^\nu$ the inequality

$$S_{\Lambda_1 \cup \Lambda_2}(\varphi) + S_{\Lambda_1 \cap \Lambda_2}(\varphi) \le S_{\Lambda_1}(\varphi) + S_{\Lambda_2}(\varphi) \tag{15.10}$$

holds (see Proposition 1.9).

Proposition 15.1 If φ is a translationally invariant state of the quasi-local algebra $\mathcal{A}$ then the limit (15.7) exists and

$$s(\varphi) = \inf\{S_{\Lambda(a)}(\varphi)/|\Lambda(a)| : a \in \mathbb{Z}_+^\nu\}. \tag{15.11}$$

Moreover, the functional $\varphi \mapsto s(\varphi)$ is affine and upper semi-continuous when the state space is endowed with the weak* topology.

Proof. The existence of the limit is based on the ν-dimensional version of the following simple observation (probably due to G. Pólya). If (u_n) is a subadditive sequence, that is, $u_{n+m} \leq u_n + u_m$ $(n, m \in \mathbb{N})$, then

$$\lim_{n \to \infty} u_n/n = \inf\{u_n : n \in \mathbb{N}\} . \tag{15.12}$$

Denote by d the value of the infimum in (15.11) and choose a $b \in \mathbb{Z}_+^\nu$ such that

$$S_{\Lambda(b)}(\varphi)/|\Lambda(b)| \leq d + \varepsilon.$$

If $(a^{(n)})$ is a sequence of positive vectors in $\mathbb{Z}^\nu$ then we may write

$$a_i^{(n)} = p_i^{(n)} b_i + r_i^{(n)} \tag{15.13}$$

with $0 \leq r_i^{(n)} < b_i$ for every $1 \leq i \leq \nu$. Equation (15.13) may be visualized as $\Lambda(a^{(n)})$ being partitioned into $p_1^{(n)} p_2^{(n)} \ldots p_\nu^{(n)}$ translated copies of $\Lambda(b)$, and some parallelepipeds smaller than $\Lambda(b)$. The number of the latter ones is estimated by

$$\sum (p_{i_1}^{(n)} + 1) \ldots (p_{i_{\nu-1}}^{(n)} + 1) = g^{(n)} ,$$

where the summation is over all $(\nu - 1)$-element-subsets of $\{1, 2, \ldots, \nu\}$. It is easy to see that $g^{(n)}/|\Lambda(a^{(n)})| \to 0$ if $a_i^{(n)} \to \infty$ for every $1 \leq i \leq \nu$. There exists a constant C such that

$$S_{\Lambda(a)}(\varphi) \leq C$$

whenever $\Lambda(a) \subset \Lambda(b)$. Due to the subadditivity and the translational-invariance of the local entropy we have

$$S_{\Lambda(a^{(n)})}(\varphi) \leq \prod_{i-1}^\nu p_i^{(n)} S_{\Lambda(b)}(\varphi) + C g^{(n)}.$$

Dividing by $|\Lambda(a^{(n)})|$ we obtain

$$\limsup_{n \to \infty} S_{\Lambda(a^{(n)})}(\varphi)/|\Lambda(a^{(n)})| \leq S_{\Lambda(b)}(\varphi)/|\Lambda(b)| \leq d + \varepsilon.$$

Since obviously $\liminf \geq d$ and $\varepsilon > 0$ was arbitrary we arrive at (15.11).

S is upper semi-continuous because it is the infimum of continuous functionals. The affine property is a consequence of Proposition 1.6:

$$\lambda S_\Lambda(\varphi) + (1 - \lambda) S_\Lambda(\omega) + H(\lambda, 1 - \lambda)$$
$$\geq S_\Lambda(\lambda \varphi + (1 - \lambda)\omega) \geq \lambda S_\Lambda(\varphi) + (1 - \lambda) S_\Lambda(\omega).$$

The proof is complete. $\square$

In the treatment of the quantum spin system the set $\mathfrak{S}_\gamma$ of all translation-invariant states is essential. The global entropy functional s is a continuous affine function on $\mathfrak{S}_\gamma$ and physically it is a macroscopic quantity which does not have microscopic (that is, local) counterpart. Indeed, the local entropy functional is not an observable because it is not affine on the (local) state space. The local internal energy $E_\Lambda(\varphi)$ is microscopic observable and the energy density functional e of $\mathfrak{S}_\gamma$ is the corresponding global extensive quantity.

The translationally invariant interactions of finite range form a real vector space $\mathcal{F}$ which may be endowed with the norm

$$\|\Phi\| = \sum_{0 \in \Lambda} \|\Phi(\Lambda)\|/|\Lambda|. \tag{15.14}$$

The next result shows that $\|\,.\,\|$ is the natural norm for interactions.

Theorem 15.2 Let $\Phi \in \mathcal{F}$ and φ be a translationally invariant state of the quasi-local algebra $\mathcal{A}$. Then the thermodynamic limit (15.6) exists and the energy density is given by

$$e(\varphi) = \varphi(E_\Phi) \quad \text{and} \quad E_\Phi = \sum_{0 \in \Lambda} \frac{\Phi(\Lambda)}{|\Lambda|}.$$

Furthermore, $e(\varphi)$ is an affine weak* continuous functional of φ.

Proof. Upon elementary reasoning we may write

$$H_\Phi(\Lambda) = \sum_{x \in \Lambda} \sum_{\substack{x \in X \\ X \subset \Lambda}} \frac{\Phi(X)}{|X|},$$

and from invariance we have

$$\varphi(H_\Phi(\Lambda)) = \sum_{x \in \Lambda} \varphi(\gamma_x(E_\Phi)) = |\Lambda|\varphi(E_\Phi).$$

The proof of the other properties is even more obvious.   □

The normed space $\mathcal{F}$ of the interactions of finite range is not complete. Its completion $\mathcal{F}^-$ consist of the translationally invariant interactions Φ such that

$$\sum_{0 \in \Lambda} \frac{\|\Phi(\Lambda)\|}{|\Lambda|} < +\infty.$$

(Such Φ are sometimes called relatively short range interactions.) Since $\|E_\Phi\| \leq \|\Phi\|$, one can extend the previous theorem to $\Phi \in \mathcal{F}^-$ by a simple continuity argument.

Corollary 15.3 Under the assumption of the previous theorem the free energy density $f(\varphi)$ exists and it is an affine lower semi-continuous function of the translation-invariant state φ.

Theorem 15.4 Let $\Phi \in \mathcal{F}$ and let (σ_t^Λ) be the local time evolution (15.4). Then there exists a one-parameter group (σ_t) of automorphisms of the quasi-local algebra $\mathcal{A}$ such that

$$\lim_{t \to 0} \|\sigma_t((A) - A\| = 0 \qquad (A \in \mathcal{A})$$

and

$$\lim_{\Lambda \to \infty} \|\sigma_t^\Lambda(A) - \sigma_t(A)\| = 0 \qquad (A \in \mathcal{A}).$$

Proof. This group (σ_t) is termed the global time evolution (associated with the interaction Φ). In order to prove the thermodynamic limit of the local time evolution we expand $\sigma_t^\Lambda(A)$ in terms of commutators as

$$\sigma_t^\Lambda(A) = \sum_{n-0}^{\infty} \frac{(it)^n}{n!} [H(\Lambda), A]^{(n)}, \qquad (15.15)$$

where the commutators of higher order are defined recursively.

$$[B, A]^{(0)} = A \qquad \text{and} \qquad [B, A]^{(n+1)} = [B, [B, A]^{(n)}].$$

Formula (15.15) may be verified by taking the derivative of each sides with respect to t. Let A be a local element of $\mathcal{A}$, say $A \in \mathcal{A}_\Gamma$. It is known that

$$\|[H(\Lambda), A]^{(n)}\| \le \|A\| n! C_\Gamma C_\Phi^n, \qquad (15.16)$$

where C_Γ and C_Φ are certain constants depending on Γ and Φ, respectively. Observe that this estimate is independent of Λ. The numerical series

$$\sum_{n=0}^{\infty} |t|^n \|A\| C_\Gamma C_\Phi^n$$

converges if $|t| < (C_\Phi)^{-1}$. For such small t the thermodynamic limit of $\sigma_t^\Lambda(A)$ exists in norm and it will be denoted by $\sigma_t(A)$. Now we use norm continuity to allow arbitrary $A \in \mathcal{A}$ (at small t). To extend the limit relation (and with it σ_t) to general $t \in \mathbb{R}$ the group property $\sigma_t \circ \sigma_s = \sigma_{t+s}$ may be used. The strong continuity of σ_t^β follows from the absolute norm convergence of (15.15) as $\Lambda \to \infty$. $\qquad \square$

The above proof supplies us with some additional information not explicitly expressed in the theorem itself. For a given $A \in \mathcal{A}_\infty$ the operator-valued function $t \to \sigma_t^\beta(A)$ admits an analytical continuation to the strip $\{z \in \mathbb{C} : |\text{Im } z| < (C_\Phi)^{-1}\}$. Since $\mathcal{A}_\Lambda$ is of finite dimension, σ_t^Λ is the restriction of an entire analytical function $z \mapsto \sigma_z^\Lambda$ to the real line. For $|z| < (C_\Phi)^{-1}$ these operator-valued functions σ_z^Λ converge uniformly to an analytic function on the disc $\{z \in \mathbb{C} : |z| < (D_\Phi)^{-1}\}$. Application of the group property yields analyticity on the strip.

We consider now a finite local system $\mathcal{A}_\Lambda$. It was shown in Proposition 1.10 that the free energy functional

$$\omega \mapsto \omega(H(\Lambda)) - \frac{1}{\beta}S(\omega)$$

defined on the state space of $\mathcal{A}_\Lambda$ is minimized by the *canonical state*

$$\varphi_\Lambda^c(a) = \frac{\mathrm{Tr}_\Lambda\, a\, e^{-\beta H(\Lambda)}}{\mathrm{Tr}_\Lambda\, e^{-\beta H(\Lambda)}} \qquad (a \in \mathcal{A}_\Lambda). \tag{15.17}$$

The minimal value is given by

$$-\frac{1}{\beta} \log \mathrm{Tr}_\Lambda\, e^{-\beta H(\Lambda)}. \tag{15.18}$$

We shall call it canonical local free energy. This quantity depends on the inverse temperature β and on the interaction Φ. Our next goal is to show the existence of the thermodynamic limit of the local free energy for a finite range interaction.

Theorem 15.5 For a finite range interaction $\Phi \in \mathcal{F}$ and $0 < \beta < \infty$ the thermodynamic limit

$$\lim_{\Lambda \to \infty} \frac{1}{|\Lambda|} \log \mathrm{Tr}_\Lambda e^{-\beta H(\Lambda)} \equiv p(\beta, \Phi)$$

exists and satisfies the inequality

$$|p(\beta, \Phi) - p(\beta, \Psi)| \le \beta \|\Phi - \Psi\| \qquad (\Phi, \Psi \in \mathcal{F}).$$

Moreover, p is a convex functional of the interaction.

Proof. For the sake of simplicity, we treat the thermodynamic limit along cubes. Let $C(a)$ denote the cube (in $\mathbb{Z}^\nu$) with side a and centred at the origin. We prove that if a and a' are large enough then

$$\left| \frac{1}{|C(a)|} \log \mathrm{Tr}_{C(a)} \exp\left(-\beta H(C(a)) \right) \right.$$

$$\left. -\frac{1}{|C(a')|} \log \mathrm{Tr}_{C(a')} \exp\left(-\beta H(C(a')) \right) \right| \tag{15.19}$$

is small. We fill up the big cube $C(a)$ with translated images of a smaller cube $C(b)$ so that there is no interaction between two copies of $C(b)$. To reach this goal it suffices to translate $C(b)$ by vectors in $(b+d)\mathbb{Z}^\nu$. We write $\{\Lambda_i : 1 \le i \le n\}$ for the finite subset of

$$\{\gamma_x(C(b)) : x \in (b+d)\mathbb{Z}^\nu\}$$

which consists of translates contained in $C(a)$. Grouping the terms in $H(C(a))$ as

$$\sum_{X \subset C(a)} \Phi(X) = \sum_{i=1}^n \sum_{X \subset \Lambda_i} \Phi(X) + \sum_{\substack{X \not\subset \cup \Lambda_i \\ X \subset C(a)}} \Phi(X) \tag{15.20}$$

we observe that the contribution of $\{\Phi(X) : X \not\subset \cup \Lambda_i, \ X \subset C(a)\}$ to the local Hamiltonian is becoming negligible if a/b is large. Set

$$m(\Phi) = \sum_{0 \in X} \|\Phi(X)\| \tag{15.21}$$

and estimate

$$\left\| \sum_{\substack{X \not\subset \cup \Lambda_i \\ X \subset C(a)}} \Phi(X) \right\| \leq (|C(a)| - |\cup_i \Lambda_i|) m(\Phi) = (a^\nu - n(a,b)b^\nu) m(\Phi). \tag{15.22}$$

Application of inequality (3.21) yields

$$\left| \frac{1}{|C(a)|} \log \mathrm{Tr}_{C(a)} \exp(-\beta \sum_{X \subset C(a)} \Phi(X)) \right.$$

$$\left. - \frac{1}{|C(a)|} \log \mathrm{Tr}_{C(a)} \exp\left(-\beta \sum_{i=1}^{n(a,b)} \sum_{X \subset \Lambda_i} \Phi(X) \right) \right|$$

$$\leq \beta(1 - n(a,b)b^\nu a^{-\nu}) m(\Phi).$$

Since $n(a,b)$ is approximately $a^\nu(b+d)^{-\nu}$, the upper bound is arbitrary small if b and a/b are big enough. Due to the translation-invariance of the interaction we have

$$\frac{1}{|C(a)|} \log \mathrm{Tr}_{C(a)} \left(-\beta \sum_{i=1}^{n(a,b)} \sum_{X \subset \Lambda_i} \Phi(X) \right)$$

$$= \log t(1 - n(a,b)b^\nu a^{-\nu}) + n(a,b)a^{-\nu} \log \mathrm{Tr}_{C(b)} \exp\left(-\beta H(C(b)) \right),$$

where t is a constant (the trace of the identity in $\mathcal{A}_0$). It is easy to see that these estimates imply that (15.19) is small if a, a' (and a/b) are large enough. Consequently the thermodynamic limit along cubes exists.

Applying inequality (3.21) once more we infer

$$\left| \frac{1}{|\Lambda|} \log \mathrm{Tr}_\Lambda \exp(-\beta H^\Phi(\Lambda)) - \frac{1}{|\Lambda|} \log \mathrm{Tr}_\Lambda \exp(-\beta H^\Psi(\Lambda)) \right|$$

$$\leq \frac{\beta}{|\Lambda|} \|H^\Phi(\Lambda) - H^\Psi(\Lambda)\| = \frac{\beta}{|\Lambda|} \sum_{i \in \Lambda} \sum_{\substack{X \subset \Lambda \\ i \in X}} \frac{\|\Phi(X) - \Psi(X)\|}{|X|} = \beta\|\Psi - \Phi\|.$$

Letting $\Lambda \to \infty$ the stated inequality for $p(\beta, \Phi) - p(\beta, \Psi)$ follows.

The convexity of p is a direct consequence of Proposition 3.13. The limit of convex functions is convex again.   □

In accordance with the lattice gas interpretation of our model the global quantity p is termed pressure. By a simple continuity argument the previous theorem could be extended to the class $\mathcal{F}^-$ of relatively short range interactions.

As an analogue of the variational principle for finite quantum systems, the global free energy functional f_β attains an absolute minimum at a translationally invariant state, and the minimum value of f^β is equal to the thermodynamic limit of the canonical free energy densities of the local finite systems. In the next theorem this global variational principle will be formulated in a slightly different but equivalent way.

Theorem 15.6 For $\Phi \in \mathcal{F}$

$$p(\beta, \Phi) = \sup\{s(\omega) - \beta\, e(\omega) : \omega \text{ is a translationally invariant state on } \mathcal{A}\}.$$

Proof. The inequality $\geq$ is rather obvious. Let ω be a translationally invariant state of $\mathcal{A}$. By applying the variational principle to the finite quantum system $\mathcal{A}_\Lambda$, we see that

$$S(\omega|\mathcal{A}_\Lambda) - \beta\omega(H(\Lambda)) \leq \mathrm{Tr}_\Lambda \exp(-\beta H(\Lambda)). \tag{15.23}$$

Thus dividing by $|\Lambda|$ and passing to the thermodynamic limit we obtain

$$s(\omega) - \beta\, e(\omega) \leq p(\beta, \Phi). \tag{15.24}$$

In order to conclude the theorem we have to construct a translationally invariant state Ψ such that the equality sign holds here.

To this end we resolve the lattice $\mathbb{Z}^\nu$ into disjoint set of translated images Λ_j of a cube $C(a)$. We define ϱ_j^a to be the canonical state of the local algebra $\mathcal{A}_{\Lambda_j}$, which is by the density

$$\frac{\exp(-\beta H(\Lambda_j))}{\mathrm{Tr}_{\Lambda_j} \exp(-\beta H(\Lambda_j))}.$$

Now we set a periodic state by

$$\bar\psi_a = \underset{j}{\otimes}\, \varrho_j^a. \tag{15.25}$$

$\bar\psi_a$ is not translationally invariant yet. In order to convert it into a translationally invariant state, we average $\bar\psi_a$ over $C(a)$:

$$\psi_a = |C(a)|^{-1} \sum_{x \in C(a)} \bar\psi_a \circ \gamma_x. \tag{15.26}$$

This implies the identity

$$a^\nu \psi_a(\Phi(X)) - \varrho_1^a\left(\sum_{i+X \subset C(a)} \Phi(i+X) \right) = \sum_{\substack{i \in C(a) \\ i+X \not\subset C(a)}} \bar\psi_a(\Phi(i+X)) \tag{15.27}$$

for a fixed finite subset X of $\mathbb{Z}^\nu$ containing 0. The number of terms on the right hand side may be estimated by $a^\nu(a-d)^\nu$, where the d is the range of the interactions. Now divide equation (15.27) by $a^\nu|X|$ and sum up over X to get

$$\left| \psi_a\left(\sum_{0 \in X} \frac{\Phi(X)}{|X|} \right) - \varrho_1^a(H(C(a))) \right| \leq \frac{a^\nu - (a-d)^\nu}{a^\nu} \|\Phi\|.$$

Remember that

$$\psi_a\left(\sum_{0 \in X} \frac{\Phi(X)}{|X|} \right) = \psi_a(E_\Phi) = e(\psi_a)$$

and hence

$$|e(\psi_a) - \varrho_1^a(H(C(a)))| \leq \varepsilon \tag{15.28}$$

if a is large enough in which case we have

$$\left| p(\beta, \Phi) - \frac{1}{|C(a)|} \log \mathrm{Tr}_{C(a)} \exp\left(-\beta H(C(a)) \right) \right| \leq \varepsilon. \tag{15.29}$$

The affinity of the global entropy functional gives

$$s(\psi_a) = s(\bar\psi_a) - \frac{1}{C(a)} S(\varrho_1^a). \tag{15.30}$$

From (15.28)–(15.30) we conclude

$$p(\beta, \Phi) - s(\psi_a) + \beta e(\psi_a) \leq 2\varepsilon.$$

This is already enough to complete the proof of the global variational principle. However, thanks to the lower semi-continuity of $\omega \mapsto -s(\omega) + \beta e(\omega)$ we obtain that for any limit point ψ of the sequence $(\psi^a)_{a \in \mathbb{N}}$ the relation

$$p(\beta, \Phi) = S(\psi) - \beta e(\psi)$$

must hold.   □

Now we formulate abstract conditions which will characterize equilibrium states of the infinite system. Let ω be a translation-invariant state of the quasi-local algebra $\mathcal{A}$. It is said that ω is *globally thermodynamically stable* (at the inverse temperature β and with respect to the interaction Φ) if $p(\beta, \Phi) = s(\omega) - \beta e(\omega)$, that is, ω minimizes the global free energy functional. Another stability condition is formulated in terms of the dynamics of the infinite system.

A state ω of $\mathcal{A}$ satisfies the *KMS condition* (at inverse temperature β and with respect to the dynamics (σ_t)) if for every $a, b \in \mathcal{A}$ there exist a function $F(z)$ of a complex variable z on the closed strip

$$\{z \in \mathbb{C} : 0 \leq \mathrm{Im}\, z \leq \beta\}$$

such that $F(z)$ is bounded and continuous on the closed strip, holomorphic in the interior of the strip, and satisfies for all real t the following relation:

$$F(t) = \omega(a\sigma_t(b)) \quad \text{and} \quad F(t + \mathrm{i}\beta) = \omega(\sigma_t(b)a). \tag{15.31}$$

Let $\Lambda \subset \mathbb{Z}^\nu$ be a finite region. It is easy to see that the local canonical state φ_Λ^c (given by (15.17)) satisfies the KMS condition with respect to the local time evolution (defined by (15.4)). Indeed, the function

$$z \mapsto \varphi_\Lambda^c(a \exp(izH(\Lambda))b \exp(-izH(\Lambda))) \qquad (a, b \in \mathcal{A}_\Lambda)$$

is an entire analytic function and

$$\varphi_\Lambda^c(a \exp(i(t + i\beta)H(\Lambda))b \exp(-i(t + i\beta)H(\Lambda)))$$
$$= \frac{\text{Tr} \exp(-\beta H(\Lambda)) \exp(itH(\Lambda))b \exp(-itH(\Lambda))a}{\text{Tr} \exp(-\beta H(\Lambda))}$$
$$= \varphi_\Lambda^c(\exp(itH(\Lambda))b \exp(-itH(\Lambda))a).$$

Let $(\Lambda(n))_n$ be an increasing sequence of parallelepipeds suitable for thermodynamic limit. One can see by taking the limit of functions of type (15.31) that any state of the quasi-local algebra $\mathcal{A}$ which is an accumulation point of the sequence $(\varphi_{\Lambda(n)}^c)_n$ of local canonical states satisfies the KMS condition with respect to the limiting global dynamics established in Theorem 15.4.

The Gibbs condition provides a link between states of the quasi-local algebra $\mathcal{A}$ and the local canonical states. This condition is formulated by means of perturbation with a surface energy term. Given an interaction Φ and a finite volume $\Lambda \subset \mathbb{Z}^\nu$ the surface energy is defined as

$$W(\Lambda) = \sum \{\Phi(\Lambda_0) : \Lambda_0 \not\subset \Lambda \text{ and } \Lambda_0 \not\subset \Lambda^c\}. \tag{15.32}$$

It contains the contribution of all many-body interaction involving spins both from Λ and its complement Λ^c. We shall say that the state ω of $\mathcal{A}$ satisfies the *Gibbs condition* if the cyclic vector Ω of the corresponding GNS-representation is separating for the generated von Neumann algebra $\pi_\omega(\mathcal{A})''$ and if the perturbation of the faithful normal state $\overline{\omega}$ of $\pi_\omega(\mathcal{A})''$ induced by Ω has the following factorization property:

$$[\overline{\omega}^h](\pi_\omega(ab)) = \varphi_\Lambda^c(a) \times \psi(b) \tag{15.33}$$

Here $h = -\beta W(\Lambda)$, $a \in \mathcal{A}_\Lambda$, $b \in \mathcal{A}_{\Lambda^c}$, φ_Λ^c is the canonical state on $\mathcal{A}_\Lambda$, and ψ is certain state of $\mathcal{A}$.

A state satisfying one of the above three conditions might be considered as an equilibrium state of a model. The following theorem will be quoted without proof.

Theorem 15.7 Assume that the translation-invariant interaction Φ of the quantum spin system over the lattice $\mathbb{Z}^\nu$ is of finite range. Then, for a translation-invariant state φ of the quasi-local algebra $\mathcal{A}$, the following conditions are equivalent (at a fixed inverse temperature $0 < \beta$):

(i) φ satisfies the KMS condition.
(ii) φ satisfies the Gibbs condition.
(iii) φ is globally thermodynamically stable.

Let us use the notation $\mathfrak{S}_\Phi^\beta$ for the set of all translation-invariant states that satisfy the equivalent conditions of Theorem 15.7 and denote by $\mathfrak{S}_\gamma$ the set of all translation-invariant states. As it was noted above the convex set $\mathfrak{S}_\Phi^\beta$ is the set of equilibrium states of the infinite lattice system at inverse temperature β. The state $\omega \in \mathfrak{S}_\Phi^\beta$ is extremal if, and only if, the corresponding GNS-construction yields a von Neumann factor. The set $\mathcal{E}(\mathfrak{S}_\Phi^\beta)$ of all extremal states consists of extremal translation-invariant states, $\mathcal{E}(\mathfrak{S}_\Phi^\beta) \subset \mathcal{E}(\mathfrak{S}_\gamma)$. The "geometry" of $\mathfrak{S}_\Phi^\beta$ is one of the main issues in statistical mechanics. (Usually, the interaction Φ is fixed and β is varied.) $\mathfrak{S}_\Phi^\beta$ is a Choquet simplex, in the sense that the extremal decomposition of each $\varphi \in \mathfrak{S}_\Phi^\beta$ is unique. This decomposition corresponds to the physical separation of an equilibrium state into pure thermodynamic phases.

Lemma 15.8 Let $\varphi \in \mathcal{E}(\mathfrak{S}_\Phi^\beta)$ and $\psi \in \mathfrak{S}_\Phi^\beta$. If the relative entropy $S(\psi, \varphi)$ is finite, then $\varphi \equiv \psi$.

Proof. Let $(\mathcal{H}, \pi, \Omega)$ be the GNS-triplet associated with φ. By assumption (cf. Gibbs condition) Ω is a cyclic and separating vector for the von Neumann algebra $\pi(\mathcal{A})''$. By the C*-algebraic definition of the relative entropy the hypothesis $S(\psi, \varphi) < \infty$ implies that ψ is quasi-contained in φ and there exists a vector ξ in the natural positive cone of Ω such that $\psi(a) = \langle \xi, \pi(a)\xi \rangle$ for every $a \in \mathcal{A}$. Let $\bar\varphi$ and $\bar\psi$ be the vector states of $\pi(\mathcal{A})''$ induced by ξ and Ω, respectively. Since the KMS condition characterizes the modular group of automorphisms, it follows that the cocyle $[D\psi, D\varphi]_t$ belongs to the centre of $\pi(\mathcal{A})''$. The condition $\varphi \in \mathcal{E}(\mathfrak{S}_\Phi^\beta)$ yields that $\pi(\mathcal{A})''$ is a factor and hence ψ must coincide with φ.   □

Theorem 15.9 Let the translation-invariant interaction Φ of the quantum spin system over the lattice $\mathbb{Z}$ be of finite range. Then for every $\beta > 0$ the set $\mathfrak{S}_\Phi^\beta$ is a singleton.

Proof. Write φ_n^c for the canonical state of the local algebra $\mathcal{A}_{[-n,n]}$ and let φ_n be $\varphi_n^c \otimes \tau_n$, where τ_n is the tracial state of $\mathcal{A}_J$ with $J = (-\infty, -n) \cup (n, \infty)$. So φ_n is a state of $\mathcal{A}$. Since the state space is compact the sequence $(\varphi_n)_n$ has an accumulation point, say ψ. We know that ψ satisfies the KMS condition, hence $\psi \in \mathfrak{S}_\Phi^\beta$. Take an extremal point φ of $\mathfrak{S}_\Phi^\beta$. Our goal is to show that the relative entropy $S(\psi, \varphi)$ is finite. Then a reference to Lemma 15.8 will complete the proof.

Let $p \in \mathbb{N}$ be fixed for a while. Since the relative entropy is continuous in the first variable on finite dimensional algebras, we have

$$S(\psi | \mathcal{A}_{[-p,p]}, \varphi | \mathcal{A}_{[-p,p]}) = \lim_{n \to \infty} S(\varphi_n^c | \mathcal{A}_{[-p,p]}, \varphi | \mathcal{A}_{[-p,p]}) . \tag{15.34}$$

Now we use the fact that φ satisfies the Gibbs condition. Hence

$$\varphi_n^c | \mathcal{A}_{[-p,p]} = [\varphi^{-\beta h(n)}] | \mathcal{A}_{[-p,p]}$$

if $n \geq p$ and $h(n)$ is the surface energy term. Due to the condition $\Phi \in \mathcal{F}$ we have that $\sup\{\|h(n)\| : n \in \mathbb{N}\} \equiv C$ is finite. Then

$$S(\varphi_n^c|\mathcal{A}_{[-p,p]}, \varphi|\mathcal{A}_{[-p,p]}) = S([\varphi^{-\beta h(n)}]|\mathcal{A}_{[-p,p]}, \varphi|\mathcal{A}_{[-p,p]})$$
$$\leq S([\varphi^{-\beta h(n)}], \varphi) \leq 2\beta C. \tag{15.35}$$

In the latest estimate the definition of the perturbed state was used,

$$S([\varphi^k], \varphi) = \inf\{S(\omega, \varphi) + \omega(k) : \omega\} - [\varphi^k](k) \leq \varphi(k) - [\varphi^k](k) \leq 2\|k\|,$$

see Chapter 12 for more details. Comparing (15.34 with (15.35) we infer

$$S(\psi|\mathcal{A}_{[-p,p]}, \varphi|\mathcal{A}_{[-p,p]}) \leq 2\beta C.$$

Letting $p \to \infty$ we conclude that $2\beta C$ is an upper bound for $S(\psi, \varphi)$ (cf. (vi) in Proposition 5.23).   □

Now we show a few examples of spin-$\frac{1}{2}$ systems whose interactions are expressed by the following – so-called – *Pauli matrices*.

$$\sigma^1 = \begin{pmatrix} 0 & 1 \\ 1 & 0 \end{pmatrix}, \qquad \sigma^2 = \begin{pmatrix} 0 & -i \\ i & 0 \end{pmatrix}, \qquad \sigma^3 = \begin{pmatrix} 1 & 0 \\ 0 & -1 \end{pmatrix}. \tag{15.36}$$

The simplest interactions are translation-invariant and consist of one-body and two-body interactions.

$$\Phi(\{x\}) = h\gamma_x(\sigma^3), \qquad \Phi(\{x,y\}) = \sum_{i=1}^{3} j_i(x-y)\gamma_x(\sigma^i)\gamma_y(\sigma^i). \tag{15.37}$$

Here h is a real number and the j_i's are real functions on $\mathbb{Z}^\nu$. The one-particle terms correspond to the interaction of a particle with an external electro-magnetic field (of strength h), while the two-particle terms reflect the interaction of two different spins. The interaction Φ given by (15.37) is called the *anisotropic Heisenberg model* if j_1, j_2, j_3 are different from 0 and at least two of the three coupling constants differ; $X - Y$ *model* if $j_1, j_2 \neq 0$ but $j_3 = 0$; the *Ising model* if $j_1 = j_2 = 0$. The Ising model is well-understood in one and two dimensions.

Let us restrict ourselves to a lattice model over $\mathbb{Z}$. If the local Hamiltonians are of the form

$$H_{[-n,n]} = \sum_{i=-n}^{n} E_i + \frac{1}{2n} \sum_{\substack{i,j=-n \\ i \neq j}}^{n} V_i V_j \qquad (v \in \mathbb{N}),$$

then they can not be written as in (15.2) by means of an interaction Φ. However,

$$H_{[-n,n]} = \sum_{i=-n}^{n} (E_i + V_i \overline{V}_n) + o(n) \quad \text{where} \quad \overline{V}_n = \frac{1}{2n+1} \sum_{j=-n}^{n} V_j.$$

Since $\overline{V}_n$ is the mean "field", the similar models are called *mean-field type*.

The characteristic feature of the mean-field models is that every particle interacts with every other one in exactly the same way.

Notes and Remarks. The rigorous and comprehensive treatment of quantum lattice systems was one of the early successes of the algebraic approach to quantum statistical thermodynamics. The subject is well-summarized in [Bratteli and Robinson 1981], but we suggest [Israel 1979] as well. The book [Sewell 1986] contains more physics and has less in mathematical technicalities.

There are several other equivalent equilibrium conditions in the case of a translation-invariant finite range interaction, such as local thermodynamic stability, energy–entropy balance. They are contained with proofs and references in [Bratteli and Robinson 1981]; [Sewell 1980] and [Verbeure 1989] are reviews. (Concerning the proof of our Theorem 15.7 and the geometry of $\mathfrak{S}_{\varPhi}^{\beta}$ see Theorem 6.2.42 and 6.2.44 in [Bratteli and Robinson 1981]. There the results are more general.

Theorem 15.9 with its proof is from [Araki 1975]. A somewhat similar proof idea is followed in [Fröhlich and Pfister 1981] in a different context.

Concerning the thermodynamics of concrete interactions we refer to [Bratteli and Robinson 1981]. In the literature mean-field systems are considered mainly as approximations of realistic interactions. As in the treatment of quantum spin systems the translation-invariance plays an important role, in mean-field systems this role is taken over by invariance under the permutation of the lattice sites. Chapter 13 contains relevant material to general mean-field systems and the review [Werner 1992] gives many references.

16 Entropic Uncertainty Relations

In traditional or Hilbert space quantum mechanics a single, spinless, non-relativistic particle with one degree of freedom is treated in the Hilbert space $L^2(\mathbb{R})$. The (pure) states of the one-particle system are described by unit vectors of the complex Hilbert space $L^2(\mathbb{R})$ and the observables are described by self-adjoint operators. If A is an observable then the founders of quantum mechanics tell us that $\langle \xi, \chi^H(A)\xi \rangle$ is the probability that this observable has a value in the set $H \in \mathbb{R}$ (χ^H stands for the indicator function of H and if $\int \lambda \, dE_\lambda$ is the spectral decomposition of A then $\chi^H(A) = \int_H 1 \, dE_\lambda$). The *position operator* is the self-adjoint operator Q with domain

$$\mathcal{D}(Q) = \{f \in L^2(\mathbb{R}) : xf(x) \in L^2(\mathbb{R})\}$$

and it is defined by $(Qf)(x) = xf(x)$. The domain of the self-adjoint *momentum operator* P is the set of all absolutely continuous functions $f \in L^2(\mathbb{R})$ such that $f' \in L^2(\mathbb{R})$ and P is a differential operator: $(Pf)(x) = -i\hbar f'(x)$. The momentum and position operators obey the *canonical commutation relation*

$$(QP - PQ)f = i\hbar f \tag{16.1}$$

for all $f \in L^2(\mathbb{R})$ for which the left hand side is defined. The fundamental commutation rule (16.1) is the trade mark of quantum theory. The celebrated uncertainty principle says that "the more precisely the position of the electron is determined, the less precisely the momentum is known, and vice versa". This was expressed first by Heisenberg as

$$e(P)e(Q) \approx \hbar, \tag{16.2}$$

where $e(P)$ and $e(Q)$ are the imprecisions with which the values of P and Q are known. A measurement of position and momentum gives rise to probability distributions. The uncertainty principle intends to say, on the language of probability, that the probability distributions corresponding to these physical quantities can not be very concentrated at the same time.

In the above representation of the operators P and Q are related by the Fourier transform $\mathcal{F} : L^2(\mathbb{R}) \to L^2(\mathbb{R})$ defined by

$$(\mathcal{F}f)(x) = \frac{1}{\sqrt{2\pi\hbar}} \int f(t) \exp \frac{-itx}{\hbar} \, dt.$$

The inverse $\mathcal{F}^*$ of the unitary operator $\mathcal{F}$ is given by

$$(\mathcal{F}^* f)(x) = \frac{1}{\sqrt{2\pi\hbar}} \int f(t) \exp \frac{itx}{\hbar} \, dt.$$

It follows that $P = \mathcal{F}^* Q \mathcal{F}$ and one says that the Fourier transform of position is momentum. The general scheme for probabilistic interpretation yields that in the (vector) state $f \in L^2(\mathbb{R})$ the probability density of the distribution of the coordinate measurement is $|f|^2$. Due to the Fourier relationship the momentum wave function is $\mathcal{F}f$ and the corresponding density is $|\mathcal{F}f|^2$. In statistics there is no canonical way to attribute a characteristic width to a probability distribution. A possibility is to use the mean deviation $E(|\xi - E(\xi)|)$ of a random variable ξ for expressing the spread or dispersion of the values about the expectation. If we choose

$$f(x) = \frac{1}{\sqrt{2\pi\sigma}} \exp \frac{-x^2}{4\sigma^2}$$

for the position wave function then $|\mathcal{F}f|^2$ is the density of a Gaussian distribution with mean deviation $\sqrt{2\sigma^2\pi^{-1}}$. The distribution of momentum is Gaussian as well and possesses mean deviation $\hbar/\sqrt{2\sigma^2\pi^{-1}}$. We observe that the product of the mean deviations is just $\hbar$ and this can be the rigorous meaning of the relation (16.2) when we restrict ourselves to Gaussian wave functions. In fact, *Heisenberg* arrived at (16.2) by a similar argument.

Let $A = A^*$ be an arbitrary observable. Its standard deviation in a vector state corresponding to $f \in L^2(\mathbb{R})$ is understood as

$$\Delta(A, f) \equiv \sqrt{\langle f, (A - \langle f, Af \rangle)^2 f \rangle}.$$

Kennard was the first to choose standard deviation as a quantitative measure of uncertainty in connection with the position and momentum observables. He proved that

$$\Delta(P, f)\Delta(Q, f) \geq \tfrac{1}{2}\hbar \tag{16.3}$$

for every vector $f \in L^2(\mathbb{R})$. As a next step Robertson showed the relation

$$\Delta(A, f)\Delta(B, f) \geq \tfrac{1}{2}|\langle f, [A, B]f \rangle| \tag{16.4}$$

for an arbitrary pair (A, B) of observables. The inequality (16.4) has been very suitable for all textbooks of quantum mechanics. It demonstrates in a rather simple way that the mysterious uncertainty relation of quantum theory has its root in the presence of nontrivial commutation relations.

In the rest we fix $\hbar = 1$. Let u be a probability density function on the real line, that is, $u \geq 0$ and $\int u(x) \, dx = 1$. Beside the standard deviation the entropy measures the concentration of u on a set of small Lebesgue measure. The (differential) entropy is defined as

$$H(u) = -\int_{-\infty}^{\infty} u(x) \log u(x) \, dx \tag{16.5}$$

and a very small negative value of $H(u)$ corresponds to a high degree of concentration of the probability. Let $f \in L^2(\mathbb{R})$ be a wave function. The above discussions tell us that a lower estimate for $H(|f|^2) + H(|\mathcal{F}f|^2)$ may be regarded as an *entropic form of the momentum-position uncertainty*.

Theorem 16.1 If $f \in L^2(\mathbb{R})$ such that $\|f\|_2 = 1$ then

$$H(|f|^2) + H(|\mathcal{F}f|^2) \geq \log \pi e$$

provided that the left hand side is defined.

Proof. Note that even if the integrals of the form (16.5) are defined the inequality is meaningless if the left hand side is $\infty + (-\infty)$. The main ingredient of the proof is the *strengthened Hausdorff-Young inequality* derived by Beckner. This concerns the $L^p - L^q$ norm of the Fourier transform for conjugate exponents $(p^{-1} + q^{-1} = 1)$ and it says

$$\|\mathcal{F}f\|_q \leq \left(\frac{p}{2\pi}\right)^{1/2p} \left(\frac{q}{2\pi}\right)^{-1/2q} \|f\|_p \tag{16.6}$$

for $1 \leq p \leq 2$ and $f \in L^p(\mathbb{R})$.

First we assume that $f \in L^2(\mathbb{R}) \cap L^1(\mathbb{R})$. For each $x \in \mathbb{R}$

$$\frac{|f(x)|^2 - |f(x)|^p}{2 - p}$$

is increasing for $1 < p < 2$. Therefore, we have

$$\left.\frac{d\|f\|_p^p}{dp}\right|_{p=2} = -\frac{1}{2} H(|f|^2), \quad \left.\frac{d \log \|f\|_p}{dp}\right|_{p=2} = -\frac{1}{4} H(|f|^2)$$

and similar formulas hold for $\mathcal{F}f$ in place of f. Taking the derivative of the logarithm of (16.6) at $p = 2$ we obtain our statement.

The next step is to abandon the assumption $f \in L^1(\mathbb{R})$. If $\int |g(x)|^2 \, dx = c_g$ is not necessarily 1 then the proved relation reads

$$H(|g|^2) + H(|\mathcal{F}g|^2) \geq c_g \log \pi e - c_g \log c_g. \tag{16.7}$$

(Remember that $\int |\mathcal{F}g(x)|^2 \, dx = c_g$ as well.) Set

$$\omega_T(x) = \begin{cases} 1 - |x|/T & \text{if } -T \leq x \leq T, \\ 0 & \text{if } |x| > T, \end{cases}$$

where $T > 0$ is a parameter. The function $f \cdot \omega_T$ belongs to $L^1(\mathbb{R}) \cap L^2(\mathbb{R})$ and (16.7) applies for $g = f \cdot \omega_T$ for every $T > 0$. A careful inspection of the integrals (see [Hirschman 1957]) shows that

$$\lim_{T \to \infty} H(|f \cdot \omega_T|^2) = H(|f|^2), \quad \lim_{T \to \infty} H(|\mathcal{F}(f \cdot \omega_T)|^2) = H(|\mathcal{F}f|^2)$$

and the limit $T \to \infty$ yields the theorem for an arbitrary function $f \in L^2(\mathbb{R})$ with $\int |f(x)|^2 \, dx = 1$. $\qquad \square$

The entropic uncertainty relation of Theorem 16.1 is strong enough to provide the standard one (16.3). Let u be a probability distribution with mean $m = \int x u(x)\,dx$ and standard deviation $\sigma = (\int (x-m)^2 u(x)\,dx)^{1/2}$. The relative entropy of u with respect to the Gaussian

$$\frac{1}{\sigma\sqrt{2\pi}}\, e^{-(x-m)^2/2\sigma^2}$$

is nonnegative and this yields

$$H(u) \leq \log(\sigma\sqrt{2\pi e})\,. \tag{16.8}$$

Applying this estimate twice we infer from the theorem

$$2\pi e\,\Delta(f,P)\Delta(f,Q) \geq \exp\left(H(|f|^2) + H(|\mathcal{F}f|^2)\right) \geq \pi\,e$$

which is equivalent to (16.3) under the present convention $\hbar = 1$.

Let A and B be two self-adjoint operators on a finite dimensional Hilbert space. We assume that they are non-degenerate in the sense that their eigenvalues are free from multiplicities. If these observables share a common eigenvector and the system is prepared in the corresponding state, then the measurement of both A and B leads to a sharp delta distribution and one can not speak of uncertainty. In order to exclude this case, let (ξ_i) be an orthonormal basis consisting of eigenvectors of A and let (η_i) be a similar basis for B. We suppose that

$$c \equiv \sup\left\{|\langle \xi_i, \eta_j\rangle|: i,j\right\} \tag{16.9}$$

is strictly smaller than 1 and prove an entropic uncertainty relation. If $A = \sum_i \lambda_i^A P_i^A$ and $B = \sum_i \lambda_i^B P_i^B$ are the spectral decompositions then

$$H(A,\varphi) = \sum_i \eta(\varphi(P_i^A)) \quad \text{and} \quad H(B,\varphi) = \sum_i \eta(\varphi(P_i^B))$$

are the entropies of A and B in a state φ. When φ is pure and corresponds to a vector Φ then $\varphi(P_i^A) = |\langle \xi_i, \Phi\rangle|^2$ and $\varphi(P_i^B) = |\langle \eta_i, \Phi\rangle|^2$.

Theorem 16.2 With the notation above the uncertainty relation

$$H(A,\varphi) + H(B,\varphi) \geq -2\log c$$

holds.

Proof. Let n be the dimension of the underlying Hilbert space. First we assume that φ is a pure state corresponding to a vector Φ.

The $n \times n$ matrix $T \equiv (\langle \xi_i, \eta_j\rangle)_{i,j}$ is unitary and T sends the vector

$$f \equiv (\langle \xi_1, \Phi\rangle, \langle \xi_2, \Phi\rangle, \ldots, \langle \xi_n, \Phi\rangle)$$

into

$$Tf = (\langle \eta_1, \Phi\rangle, \langle \eta_2, \Phi\rangle, \ldots, \langle \eta_n, \Phi\rangle)\,.$$

The vectors f and Tf are elements of $\mathbb{C}^n$ and this space may be endowed with different L^p norms. Using interpolation theory we shall estimate the norm of the linear transformation T with respect to different L^p norms. Since T is a unitary

$$\|g\|_2 = \|Tg\|_2 \qquad (g \in \mathbb{C}^n).$$

With the notation (16.9) we have also

$$\|Tg\|_\infty \le c\|g\|_1 \qquad (g \in \mathbb{C}^n).$$

Let us set

$$N(p, p') = \sup\{\|Tg\|_p/\|g\|_{p'} : g \in \mathbb{C}^n, \quad g \ne 0\}$$

for $1 \le p \le \infty$ and $1 \le p' \le \infty$. The *Riesz–Thorin convexity theorem* says that the function

$$(t, s) \mapsto \log N(t^{-1}, s^{-1}) \tag{16.10}$$

is convex on $[0, 1] \times [0, 1]$ (where 0^{-1} is understood to be ∞). Application of convexity of (16.10) on the segment $[(0, 1), (1/2, 1/2)]$ yields

$$\|Tg\|_{2/\lambda} \le c^{1-\lambda}\|g\|_\mu \qquad (g \in \mathbb{C}^n),$$

where $0 < \lambda < 1$ and $\mu = (1 - \lambda/2)^{-1}$. This is rewritten by means of a more convenient parametrization in the form

$$\|Tg\|_p \le c^{1-2/p}\|g\|_q \qquad (g \in \mathbb{C}^n),$$

where $2 \le p < \infty$ and $p^{-1} + q^{-1} = 1$. Consequently

$$\log\|Tf\|_p \le \left(1 - \frac{2}{p}\right)\log c + \log\|f\|_q. \tag{16.11}$$

One checks easily that

$$\frac{d\log\|Tf\|_p}{dp}\bigg|_{p=2} = -\frac{1}{4}H(B, \varphi) \quad \text{and} \quad \frac{d\log\|f\|_q}{dp}\bigg|_{p=2} = \frac{1}{4}H(A, \varphi).$$

Hence dividing (16.11) by $p - 2$ and letting $p \searrow 2$ we obtain

$$-\tfrac{1}{4}H(B, \varphi) \le \tfrac{1}{2}\log c + \tfrac{1}{4}H(A, \varphi)$$

which proves the theorem for a pure state.

Concavity of the left hand side of the stated inequality in φ ensures the lower estimate for mixed states. $\qquad\qquad\square$

Let $\mathcal{H}$ be an n dimensional complex Hilbert space with orthonormal basis $(\xi_j)_j$. Setting

$$\eta_k = \frac{1}{\sqrt{n}}\sum_{j=1}^n e^{2\pi i/n}\xi_j \qquad (k = 1, 2, \ldots, n)$$

we have another basis (η_k) such that

$$|\langle \xi_j, \eta_k \rangle|^2 = n^{-1} \qquad (j, k = 1, 2, \ldots, n) . \tag{16.12}$$

Let A and B self-adjoint operators with eigenvectors (ξ_j) and (η_i), respectively and let φ be the pure state corresponding to ξ_1. Then $H(A, \varphi) = 0$ and $H(B, \varphi) = \log n$. Hence this example shows that the lower bound for the entropy sum in Theorem 16.2 is sharp. If (16.12) holds then the pair (A, B) of observables are called *complementary*. The position and momentum observables possess a property which is very much the continuous analogue of (16.12). Denote $E^P(H)$ and $E^Q(H)$ the spectral projections of the self-adjoint operators P and Q corresponding to the Borel set $H \subset \mathbb{R}$. The Fourier relationship allows to deduce

$$\text{Tr}\, E^P(H)\, E^Q(K)\, E^P(H) = \frac{1}{2\pi} \lambda(H)\lambda(K) \tag{16.13}$$

for arbitrary Borel sets $H, K \subset \mathbb{R}$. (λ stands for the Lebesgue measure of the real line.) One may regard $(2\pi)^{-1/2}$ as the replacement of the constant c of (16.9) and the continuous analogue of Theorem 16.2 for position and momentum would be

$$H(Q, \varphi) + H(P, \varphi) \geq \log 2\pi .$$

This lower bound is slightly worse than that of Theorem 16.1.

A particle which moves in a central, spherically symmetric field of force is described most conveniently in spherical polar coordinates. For the sake of simplicity we restrict our attention to the component of *angular momentum* in a given direction (say the z-axis) and to the angle of rotation around this axis. The operator of the angular momentum component is of the form

$$L \colon f(r, \phi, \theta) \mapsto -\mathrm{i} \frac{\partial}{\partial \phi} f(r, \phi, \theta) .$$

In the sequel we suppress the variables r and θ. The range of the angle variable ϕ is rather the unit circle than the real interval $[0, 2\pi)$. Any cut of the circle is unphysical and change the proximity relations rudely. For our operators we choose the underlying Hilbert space to be $L^2(\mathbb{T}, \mu)$ over the rotation invariant normalized measure μ on the unit circle $\mathbb{T}$. If (V_t) is the rotation group given by

$$(V_t f)(e^{\mathrm{i}\phi}) = f(e^{\mathrm{i}(\phi - t)}) \qquad (\phi, t \in \mathbb{R}) \tag{16.14}$$

then

$$V_t = e^{\mathrm{i}t\, L} \tag{16.15}$$

and L may be defined to be a self-adjoint operator. Due to the orthodox postulates of quantum theory the "angle observable" ought to be represented by a self-adjoint operator. We do not intend to insist on this postulate. Since the values of the angle lie on the circle $\mathbb{T}$ one is tempted to choose a unitary operator for the representation of the *angle observable*. In fact, if $U \colon L^2(\mathbb{T}, \mu) \to L^2(\mathbb{T}, \mu)$ is defined as

$$(Uf)(e^{i\phi}) = e^{i\phi}f(e^{i\phi}) \tag{16.16}$$

then

$$U^n V_t = e^{int}V_t U^n \qquad (n \in \mathbb{Z}, \, t \in \mathbb{R}) \tag{16.17}$$

which is completely analogous to the Weyl form of the canonical commutation relation. Loosening the above mentioned postulate we allow a unitary (or more generally a normal) operator to represent an observable. In this way the probabilistic interpretation will not break down. The unitary U admits a spectral decomposition

$$U = \int_{\mathbb{T}} \lambda \, dE_\lambda$$

and in the state φ the probability that the value of the angle is in $H \subset \mathbb{T}$ is given by

$$\int_H 1 \, d\varphi(e_\lambda) \, .$$

The concepts of mean and standard deviation do not apply for random variables with values in $\mathbb{T}$. In this case the use of entropy in an uncertainty relation is even more justified. For a normalized wave function $f \in L^2(\mathbb{T})$ we set

$$H(U, f) = - \int_{\mathbb{T}} |f|^2 \log |f|^2 \, d\mu \, .$$

Note that $-H(U, f)$ is the relative entropy of the density $|f|^2$ with respect to the background measure μ. So $H(U, f) \le 0$ and the equality holds if and only if $|f| \equiv 1$. The entropy of the angular momentum observable is expressed by means of the Fourier transform.

$$H(L, f) = - \sum_{n=-\infty}^{\infty} |\hat{f}|_n^2 \log |\hat{f}|_n^2 \, ,$$

where $f(e^{i\phi}) = \sum_{n=-\infty}^{\infty} (\hat{f})_n \, e^{in\phi}$ is a Fourier expansion. The quantity $-H(L, f)$ is again of relative entropy type, although the background measure is not normalized. It is the entropy of the distribution $|(\hat{f})_n|^2$ with respect to the counting measure on $\mathbb{Z}$.

Theorem 16.3 For every normalized wave function $f \in L^2(\mathbb{T}, \mu)$ the entropic uncertainty relation

$$H(U, f) + H(L, f) \ge 0$$

holds for the angular momentum and angle observables.

Proof. In the proof we can follow the pattern of Theorem 16.1. The Hausdorff-Young inequality tells that

$$\log \|\hat{f}\|_q \le \log \|f\|_p$$

for $1 < p < 2$ and $p^{-1} + q^{-1} = 1$. Dividing by $p - 2$ and letting $p \to 2$ we obtain the theorem.

The inequality is saturated by all eigenfunctions of the angular momentum observable. The canonically conjugate observables U and L are complementary. It is easy to compute that

$$\operatorname{Tr} E^U(H) E^L(K) = \mu(H) \#(K) \tag{16.18}$$

for $H \subset \mathbb{T}$ and $K \subset \mathbb{Z}$. Here $\#(K)$ is the number of integers in the set K.

Due to its importance we devote some more discussion to the position-momentum uncertainty. Recall that P and Q are self-adjoint operators on $\mathcal{H} = L^2(\mathbb{R})$ so that

$$(e^{ixP} f)(t) = f(t + x) \quad \text{and} \quad (e^{iyQ} f)(t) = e^{iyt} f(t) \tag{16.19}$$

for every $f \in L^2(\mathbb{R})$ and $x, y, t \in \mathbb{R}$. Let α be a vector of norm one and set

$$\alpha_{x,y} = e^{iyQ} e^{-ixP} \alpha \qquad (x, y \in \mathbb{R}).$$

The relations (16.19) yield

$$\alpha_{x,y}(t) = e^{iyt} \alpha(t - x) \qquad (t \in \mathbb{R}).$$

Let $E_{x,y}$ be the orthogonal projection of rank one associated with the vector $\alpha_{x,y} \in \mathcal{H}$. We show that for any normal state φ of $B(\mathcal{H})$ the function

$$\varrho(x, y) = \frac{1}{2\pi} \varphi(E_{x,y}) \tag{16.20}$$

is a probability density on the plane $\mathbb{R}^2$. Obviously we may restrict ourselves to pure states and assume that φ is a vector state associated with $f \in L^2(\mathbb{R})$. So

$$\int_{\mathbb{R}^2} \varrho(x, y) \, dx \, dy = \int_{\mathbb{R}^2} |(2\pi)^{-1/2} \langle f, \alpha_{x,y} \rangle|^2 \, dx \, dy.$$

In the equality

$$\frac{1}{\sqrt{2\pi}} \langle f, \alpha_{x,y} \rangle = \frac{1}{\sqrt{2\pi}} \int_{\mathbb{R}} e^{-iyt} \overline{f(t)} \alpha(t - x) \, dt$$

we recognize the Fourier transform. By the Plancherel theorem

$$\frac{1}{\sqrt{2\pi}} \int_{\mathbb{R}} |\langle f, \alpha_{x,y} \rangle|^2 \, dy = \int_{\mathbb{R}} |f(t)|^2 |\alpha(t - x)|^2 \, dt$$

and integrating in x we infer

$$\frac{1}{\sqrt{2\pi}} \int_{\mathbb{R}^2} |\langle f, \alpha_{x,y} \rangle|^2 \, dx \, dy = 1.$$

Our next goal is to estimate the L^p norm of the function $\varrho(x, y)$. We need the following result on convolution.

Lemma 16.4 Let $f \in L^s(\mathbb{R})$, $g \in L^t(\mathbb{R})$, $1 \le s,t \le \infty$ and $1 + r^{-1} = s^{-1} + t^{-1}$ with $r \ge 1$. Then

$$\|f * g\|_r \le C_s C_t C_r^{-1} \|f\|_s \|g\|_t,$$

where the constants are given by

$$C_p = p^{1/2p} q^{-1/2q} \qquad (p^{-1} + q^{-1} = 1). \tag{16.21}$$

Proposition 16.5 Let φ be a vector state of $B(L^2(\mathbb{R}))$ induced by $f \in L^2(\mathbb{R})$ and let $\varrho(x, y)$ be given by (16.20). Then, for $s \ge 1$

$$I_s \equiv \int_{\mathbb{R}^2} |\varrho(x, y)|^s \, dx \, dy \le (2\pi)^{1-s} s^{-1}.$$

Proof. The main ingredients of the proof are the inequality (16.6) for the Fourier transform and Lemma 16.4 concerning the norm of a convolution. Regarding $x \in \mathbb{R}$ as a parameter we set

$$g_x(t) = f(t) \alpha_{x,0}(t)$$

and observe

$$(F g_x)(-s) = \frac{1}{\sqrt{2\pi}} \langle f, \alpha_{x,s} \rangle \qquad (x, s \in \mathbb{R}).$$

It follows that

$$\|F g_x\|_q^q = (2\pi)^{-q/2} \int_{\mathbb{R}} |\langle f, \alpha_{x,s} \rangle|^q \, ds$$

and (16.6) yields

$$\int_{\mathbb{R}} |\langle f, \alpha_{x,s} \rangle|^q \, ds \le 2\pi C_p^q \|g_x\|_p^q \tag{16.22}$$

for $2 \le q \le \infty$ and $p^{-1} + q^{-1} = 1$. Since

$$\|g_x\|_p^q = (|f|^p * |\alpha|^p)^{q/p}$$

integrating (16.22) in x we may benefit from Lemma 16.4. Choosing $r = q/p$ and $s = t = 2/p$ we have

$$\left(\int_{\mathbb{R}} (|f|^p * |\alpha|^p)^{q/p} \, dx \right)^{p/q} \le C_{2/p}^2 C_{q/p}^{-1} \left(\int |f|^2 \, dx \right)^{p/2} \left(\int |\alpha|^2 \, dx \right)^{q/2}$$

$$= C_{2/p}^2 C_{q/p}^{-1}$$

and

$$\int_{\mathbb{R}^2} |\langle f, \alpha_{x,s} \rangle|^q \, ds \, dx \le 2\pi C_p^q (C_{2/p}^2 C_{q/p}^{-1})^{q/p} = 4\pi \, q^{-1}$$

which is equivalent to the statement $(q = 2s)$.   □

For a normal state φ of $B(\mathcal{H})$ and a parameter vector α of norm one *Wehrl* suggested to use the entropy quantity

$$H(\varphi, \alpha) = - \int_{\mathbb{R}^2} \varrho(x, y) \log \varrho(x, y) \, dx \, dy, \tag{16.23}$$

where ϱ is given (16.20). He called it classical entropy since it is the common entropy of the probability density $\varrho(x, y)$ on the plane $\mathbb{R}^2$ (which is regarded as the phase space here). Note that the integral in (16.23) is always well-defined thanks to $\varrho \leq 1/2\pi$. From Proposition 16.5 one can deduce easily the following *Wehrl-Lieb inequality*.

Corollary 16.6 For every normal state φ and $\alpha \in L^2(\mathbb{R})$ with $\|\alpha\| = 1$ the inequality

$$H(\varphi, \alpha) \geq \log 2\pi e$$

holds.

Proof. Due to the concavity of $H(\varphi, \alpha)$ in φ it suffices to show the lower bound for a pure state induced by $f \in L^2(\mathbb{R})$. From the proposition

$$\int \frac{\varrho^s - \varrho}{s - 1} \, dx \, dy \leq \frac{(2\pi)^{1-s} s^{-1} - 1}{s - 1}$$

for $s \geq 1$. Taking the limit $s \to 1+0$ we may benefit from the dominated convergence theorem of integrals provided that $\int \varrho \log \varrho \, dx \, dy$ is finite. Otherwise there is nothing to show. ∎

The example of the unitary angular momentum observable shows that it is worthwhile to think about observables as projection-valued measures. By means of the spectral theorem normal operators may be identified with such measures given on the Borel sets of the complex plain. Now we are going to take one more step towards the generality. Let Ω be a set with a σ-field $\mathcal{B}$. A positive operator-valued measure $A : \mathcal{B} \to B(\mathcal{H})$ will be called observable in the broad sense if $A(\Omega) = I$. Using the rank one projections $E_{x,y}$ associated with the vector $\alpha \in L^2(\mathbb{R})$ we obtain

$$A_\alpha(F) = \frac{1}{2\pi} \int_F E_{x,y} \, dx \, dy \qquad (F \subset \mathbb{R}^2) \tag{16.24}$$

to be an observable defined on the phase space $\mathbb{R}^2$. (In (16.24) the integral is meant in the weak operator topology and F runs on the Borel sets.) Obviously, the entropy $H(\varphi, \alpha)$ defined by (16.23) is connected with the observable A_α in the state φ. The marginal observables of A_α are defined as follows.

$$A_\alpha^1(E) = \frac{1}{2\pi} \int_{x \in E} \int_{\mathbb{R}} E_{x,y} \, dy \, dx \qquad (E \subset \mathbb{R}) \tag{16.25}$$

$$A_\alpha^2(F) = \frac{1}{2\pi} \int_{\mathbb{R}} \int_F E_{x,y} \, dy \, dx \qquad (F \subset \mathbb{R}). \tag{16.26}$$

The pure state associated with $f \in L^2(\mathbb{R})$ gives rise to the distribution $E \mapsto \langle A_\alpha^1(E)f, f \rangle$. Assume $\alpha(-t) = \alpha(t)$ $(t \in \mathbb{R})$ for the sake of simplicity. One obtains that

$$\langle A_\alpha^1(E)f, f \rangle = (2\pi)^{-1} \int_E \int_{\mathbb{R}} |\langle f, \alpha_{x,y} \rangle|^2 \, dy \, dx$$

$$= \int_{\mathbb{R}^2} |f(t)|^2 \chi_E(x) |\alpha(t-x)|^2 \, dt \, dx = \int_{\mathbb{R}} \chi_E(x)(|f|^2 * |\alpha|^2)(x) \, dx$$

and the density of the first marginal is $|f|^2 * |\alpha|^2$. Similar computation shows that the distribution $F \mapsto \langle A_\alpha^2(F)f, f \rangle$ has the density $|\mathcal{F}f|^2 * |\mathcal{F}\alpha|^2$. The first marginal A_α^1 is an approximate position observable because choosing

$$\alpha_\sigma(t) = (2\pi\sigma^2)^{-1/4} \exp(-t^2/4\sigma^2)$$

the distribution $|f|^2 * |\alpha|^2$ of this marginal converges to the "true distribution" $|f|^2$ of position as $\sigma \to 0$. (Symmetrically, if $\sigma \to +\infty$ then the second marginal distribution $|\mathcal{F}f|^2 * |\mathcal{F}\alpha|^2$ tends to $|\mathcal{F}f|^2$ which is the distribution of momentum at the wave function $f \in L^2(\mathbb{R})$.)

Using the subadditivity of the entropy one can infer the inequality

$$H(|f|^2 * |\alpha|^2) + H(|\mathcal{F}f|^2 * |\mathcal{F}\alpha|^2) \geq H(f, \alpha) \geq \log 2\pi e$$

from Corollary 16.6. A combination with (16.8) yields

$$\Delta(|f|^2 * |\alpha|^2) \Delta(|\mathcal{F}f|^2 * |\mathcal{F}\alpha|^2) \geq 1 \tag{16.27}$$

and an appropriate choice of α supplies the standard uncertainty relation (16.3) as a consequence.

Notes and Remarks. After a series of papers of Heisenberg, Born, Jordan and Dirac what we call today conventional quantum theory was mathematically summarized in [von Neumann 1932]. The uncertainty principle in the form (16.2) appeared in the paper [Heisenberg 1927]. In the same paper Heisenberg also introduced a similar relation for energy and time. [Kennard 1927] is the first appearance of the formulation (16.3) by means of the standard deviation. [Robertson 1929] proved the general relation (16.4) including already the commutator of the observables. Since that time this relation has been regarded as the expression of the uncertainty principle in nearly every textbook. Theorem 16.1 originated from [Hirschman 1957], where the lower bound $\log 2\pi$ was obtained and the actual sharp bound $1 + \log \pi$ was conjectured. The strengthed Hausdorff–Young inequality was derived in [Beckner 1975] and Theorem 16.1 was its consequence. (See also [Bialynicki–Birula and Mycielski 1975].) [Lieb 1990] contains generalizations of the Young inequality for the Fourier transform. The popular view that the Robertson inequality expresses the the validity of the uncertainty principle for any pair of non-commuting (selfadjoint) observables was criticized in [Deutsch 1983] and in [Hilgevoord and Uffink 1988]. Theorem 16.2 was conjectured in [Kraus 1987] and proven in [Maasen and Uffink 1988], see also [Maasen 1990]. The form of the entropic uncertainty relation of Theorem 16.2 is not the only possible one. One can deal with the uncertainty of the simultaneus measurment of two observables, see [Schroeck 1989].

The Riesz–Thorin convexity theorem may be found in [Triebel 1978] and in [Reed and Simon 1975], for example. The complementarity relation (16.13) is discussed in [Accardi 1984]. The use of a unitary operator for the representation of the angle variable was proposed in [Lévy–Leblond 1976]. The problem of uncertainty measures of a probability distribution over the unit circle is reviewed in [Breitenberger 1985]. The angular variance $\Delta(U, f) = \sqrt{1 - |\langle Uf, f \rangle|^2}$ or its functions may be used to replace the standard deviation of real random variables. For the details we refer to [Lévy–Leblond 1976], [Holevo 1982] and [Bersohn 1966]. The story of the entropic angle–angular momentum uncertainty relation is much simpler, Theorem 16.3 was shown by [Hirschman 1957]. A very readable account on measures of uncertainty and the uncertainty principle is the thesis [Uffink 1990]. It was also our source in the above historical sketch.

Phase space and coherent vectors have been widely used in the physics literature since the early days of quantum mechanics (see [Perelomov 1986]). In our simple treatement [Davies 1976] was followed. The entropy quantity (16.23) was proposed by [Wehrl 1979] under the name classical entropy of the statistical operator of φ. In this paper Corollary 16.6 was conjectured and the proof is due to [Lieb 1978]. (See [Carlen 1991] for another proof which tells about the case of equality.) The convolution inequality Lemma 16.4 was independently discoverd by [Beckner 1975] and [Brascamp and Lieb 1976]. In Wehrl's conjecture and in Lieb's proof the parameter function α was standard Gaussian. The possibility of allowing arbitrary α was observed in [Grabowski 1984]. The idea of identification of observables with positive operator-valued measures arose in [Davies and Lewis 1970] and [Holevo 1972] (see also [Holevo 1982]). The approximate position measurement is treated in [Davies 1976].

The theory of so-called hidden variables in quantum machanics is strongly related to the uncertainty. In a possible mathematical formulation the existence of hidden variables means the decomposition of quantum states into dispersion free (or less dispersed) states of a supersystem. One can think that the entropy may replace the standard deviation in this job as well. This idea is outlined in [Rédei 1988].

Comments. Relation (16.13) is the correct form of the complementarity, $E^P(H) E^Q(K)$ is not of trace class, hence $\operatorname{Tr} E^P(H) E^Q(K)$ is not defined.

When Lieb proved Wherl's conjecture concerning the Glauber coherent states in [Lieb 1978], he formulated a similar conjecture concerning Bloch coherent states. Inspite of the hope that the finite dimensional situation means significant simplification, this conjecture was partially proven in [Schupp 1999] much later.

17 Temperley-Lieb Algebras and Index

Shortly speaking the aim of this chapter is to show new examples of quantum Markov processes. Temperley and Lieb used projections to express the transfer matrices of the Potts model of statistical mechanics. These projections satisfied certain simple commutation rules and incidentally the same commutation relations appeared in the work of V. Jones when he developped an index theory for subfactors. It seems that the Temperley–Lieb relations determine a quantum Markov process and this is the common background of both cases. We stress once more that our goal is to present quantum Markov chains and to compute their entropy. The following brief discussion on the Potts model and on index theory is given only to show the full picture.

The q-state Potts model is a generalization of the Ising model. In the Potts model the spin at any lattice site can point in any of q directions of equal weight and significance. The value 2 of q corresponds to the Ising model. First we consider an array of $N + 1$ sites. Then a configuration of the system is a function $\sigma : \{1, 2, \ldots, N + 1\} \to \{1, 2, \ldots, q\}$, σ_i being the value of the spin at the lattice site i, $1 \leq i \leq N + 1$. Let us assume that only configurations with periodic boundary condition are allowed: $\sigma_1 = \sigma_{N+1}$. (So the space of all configurations is essentially $\{1, 2, \ldots, q\}^N$.) The energy comes from the interactions of different spins at neighbouring sites. The standard Potts model has a hamiltonian

$$H(\sigma) = -\sum_{i=1}^{N} J\delta(\sigma_i, \sigma_{i+1})$$

(where $\delta(s, t) = 1$ if $s = t$, otherwise $\delta(s, t) = 0$.) In order to calculate the expectation value of an observable, we need to compute the partition function.

$$Z = \sum_{\sigma} \exp(-\beta H(\sigma)) = \sum_{\sigma} \prod_{i=1}^{N-1} \exp\left(\beta J\delta(\sigma_i, \sigma_{i+1})\right).$$

Let T be a $q \times q$ matrix with entries

$$T_{k,l} = \exp(\beta J\delta(k, l)).$$

Then

$$Z = \mathrm{Tr}\, T^N,$$

(17.1)

so the partition function is expressed by the transfer matrix T. Similarly, the numerator in

$$\langle f \rangle_\beta = \frac{\sum_\sigma f(\sigma) \exp(-\beta H(\sigma))}{\sum_\sigma \exp(-\beta H(\sigma))} \tag{17.2}$$

can be expressed as $\mathrm{Tr}\, F_\beta T^N$, where F_β is a certain matrix in $M_q(\mathbb{C})$. Thus

$$\langle f \rangle_\beta = \frac{\mathrm{Tr}\, F_\beta T^N}{\mathrm{Tr}\, T^N}. \tag{17.3}$$

This is an interesting reduction of the one-dimensional classical Potts model to a "zero-dimensional model" set in the noncommutative algebra of $q \times q$ matrices.

Now we consider a square lattice on a rectangle with m atoms on the base and n atoms on the side. Each point of

$$\{1, 2, \ldots, m\} \times \{1, 2, \ldots, n\}$$

is again regarded as an atom with one of q possible spins. A configuration is now a double sequence $\sigma_j^{(k)}$ where $1 \leq j \leq m$, $1 \leq k \leq n$ and $\sigma_j^{(k)} \in \{1, 2, \ldots, q\}$. We assume that the energy comes from either horizontal or vertical nearest neighbour interactions and impose a vertical periodic boundary condition which is the most convenient in the following form. The atoms (j, n) and $(j, 1)$ are regarded as nearest neighbours $(1 \leq j \leq n)$. The convention $\sigma_{n+1}^{(k)} = \sigma_1^{(k)}$ will be used in accordance.

The horizontal contribution of some row with spins $\sigma_1, \sigma_2, \ldots, \sigma_m$ to the hamiltonian is

$$-K_h \sum_{j=1}^{m-1} \delta(\sigma_j, \sigma_{j+1}),$$

and the vertical contribution of two adjacent rows with spins $(\sigma_1, \sigma_2, \ldots, \sigma_m)$, $(\tau_1, \tau_2, \ldots, \tau_m)$ is

$$-K_v \sum_{j=1}^{m} \delta(\sigma_j, \tau_j).$$

We introduce two q^m-by-q^m matrices. V is diagonal and

$$V_{\sigma,\sigma} = \exp\left(\beta K_h \sum_{j=1}^{m-1} \delta(\sigma_j, \sigma_{j+1}) \right) \tag{17.4}$$

and it is related to the interaction of spins in one row. The other one contains the interaction between two neighbouring rows and it is given by

$$W_{\sigma,\varrho} = \exp\left(\beta K_v \sum_{j=1}^{m} \delta(\sigma_j, \varrho_j) \right). \tag{17.5}$$

Now the partition function is expressed as

$$\sum V_{\sigma^{(1)}\sigma^{(1)}} W_{\sigma^{(1)}\sigma^{(2)}} V_{\sigma^{(2)}\sigma^{(2)}} \ldots V_{\sigma^{(n)}\sigma^{(n)}} W_{\sigma^{(n)}\sigma^{(1)}} \, ,$$

where the summation is over all $\sigma^{(1)}, \sigma^{(2)} \ldots, \sigma^{(n)}$ corresponding to the rows of a configuration. In a more compact form,

$$Z = \mathrm{Tr}\,(V^{1/2} W V^{1/2})^n \, . \tag{17.6}$$

Now the matrices V and W or the single matrix $T \equiv V^{1/2} W V^{1/2}$ play a crucial role in the computations of this model.

It is convenient to introduce the following matrices for $1 \le i \le q$.

$$(E_{2i})_{\sigma,\tau} = \delta(\sigma_i, \sigma_{i+1}) \prod_{1 \le k \le m} \delta(\sigma_k, \tau_k) \, ,$$

$$(E_{2i-1})_{\sigma,\tau} = \frac{1}{q} \prod_{\substack{j \ne i \\ 1 \le j \le m}} \delta(\sigma_j, \tau_j) \, . \tag{17.7}$$

In particular, E_{2i} is diagonal. If we identify $M_{q^m}(\mathbb{C})$ with the mth tensor power of $M_q(\mathbb{C})$, then we have

$$E_{2i-1} = I^{(1)} \otimes I^{(2)} \otimes \ldots I^{(i-1)} \otimes g^{(i)} \otimes I^{(i+1)} \ldots \otimes I^{(m)} \, , \tag{17.8}$$

where g is a $q \times q$ matrix having the value q^{-1} at all entries. So g is a rank one projection onto the vector $(q^{-1/2}, q^{-1/2}, \ldots, q^{-1/2}) \in \mathbb{C}^q$ and E_{2i-1} is a projection, too. Similarly, E_{2i} is a projection of rank q. Namely,

$$E_{2i} = I^{(1)} \otimes I^{(2)} \otimes \ldots I^{(i-2)} \otimes f^{(i-1),(i)} \otimes I^{(i+1)} \ldots \otimes I^{(m)} \tag{17.9}$$

with $f = \sum_i h_i \otimes h_i \in M_q(\mathbb{C}) \otimes M_q(\mathbb{C})$, where h_i is the rank one projection onto $(\delta(1,i), \delta(2,i), \ldots, \delta(q,i)) \in \mathbb{C}^q$.

In this way we have obtained a sequence $E_1, E_2, \ldots, E_{2m-1}$ of projections.

$$E_j E_k = E_k E_j \quad \text{for} \quad |k - j| > 1 \tag{17.10}$$

is immediate from (17.8) and (17.9). The straightforward relations

$$g(f \otimes I)g = q^{-1}(f \otimes I) \quad \text{and} \quad (f \otimes I)g(f \otimes I) = q^{-1}(f \otimes f)$$

lead to the following formula:

$$E_j E_k E_j = q^{-1} E_j \quad \text{for} \quad |k - j| = 1 \, . \tag{17.11}$$

The next task is to express V and W in terms of the E_j's.

$$V = \exp\left(\beta K_h (E_2 + E_4 + \ldots + E_{2m-2})\right)$$

$$W = \prod_{i=1}^{m} (v_2 I + q E_{2i-1}) \quad \text{where} \quad v_2 = \exp(\beta K_v) - 1 \, , \tag{17.12}$$

or, in another form,

$$V = \prod_{i=1}^{m-1} (I + v_1 E_{2i})$$

$$W = v_2^m \exp \left(K(E_1 + E_3 + \ldots + E_{2m-1}) \right), \tag{17.13}$$

where

$$v_1 = \exp(\beta K_h) - 1 \quad \text{and} \quad \exp K = v_2 + q. \tag{17.14}$$

The relations (17.10) and (17.11) determine the maximal eigenvalue of the matrix VW (which is the same as that of $V^{1/2}WV^{1/2}$), and with it the large n behaviour of the partition function.

Above we have obtained a finite sequence $E_1, E_2, \ldots, E_{2m-1}$ of projections. The formulas (17.8) and (17.9) show that this could be continued to have an infinite sequence such that the relations (17.10) and (17.11) hold. (Of course, all these operators would then be defined in the infinite tensor power of $M_q(\mathbb{C})$.) The following result is somewhat surprising.

Proposition 17.1 Consider a sequence $e_1, e_2, \ldots$ of nonzero orthogonal projections on a complex Hilbert space such that the relations

$$e_j e_k = e_k e_j \quad \text{for} \quad |k - j| > 1$$
$$e_j e_k e_j = \kappa^{-1} e_j \quad \text{for} \quad |k - j| = 1 \tag{17.15}$$

hold for a real number $\kappa > 0$. Then either $\kappa \geq 4$ or $\kappa = 4\cos^2(\pi/p)$ for some integer $p \geq 3$.

The proposition will not be proven here. However, it is worthwile to introduce some notation for the set of possible κ's. Set

$$\mathbf{Ind} = \mathbf{Ind}_c \cup \mathbf{Ind}_d, \quad \mathbf{Ind}_c = \{t \in \mathbb{R} : t \geq 4\},$$
$$\mathbf{Ind}_d = \{4\cos^2(\pi/p + 2) : p \in \mathbb{N}\}. \tag{17.16}$$

The projections E_j satisfying the relations (17.10) and (17.11) were found by Temperley and Lieb in connection with the Potts model. Therefore we shall use the term Temperley–Lieb algebra for the C*-algebras generated by a finite or infinite sequence of projections possessing the property (17.15) with some $\kappa > 0$ Such algebras will also be encountered also in connection with the index theory of finite von Neumann algebras and provide new examples of noncommutative Markov processes. Returning to the Potts model, it is noteworthy that the values $q = 2$ and $q = 3$ lie in the discrete spectrum $\mathbf{Ind}_d$ while $q > 3$ is in the continuous one $\mathbf{Ind}_c$.

Let $\mathcal{N}$ be a von Neumann algebra acting on a Hilbert space $\mathcal{H}$ and having commutant $\mathcal{N}'$. Assume that there exist faithful normal tracial states τ on $\mathcal{N}$ and τ' on $\mathcal{N}'$. (So both $\mathcal{N}$ and $\mathcal{N}'$ are supposed to be finite.) For any vector $\xi \in \mathcal{H}$ the projection $[\mathcal{N}\xi]$ belongs to $\mathcal{N}'$ and similarly $[\mathcal{N}'\xi] \in \mathcal{N}$. The quotient

$$\dim_{\mathcal{N}}(\mathcal{H}) \equiv \frac{\tau'([\mathcal{N}\xi])}{\tau([\mathcal{N}'\xi])}$$

is known to be independent of the vector ξ and is called the coupling constant since the work of Murray and von Neumann. In a certain sense the coupling constant is the dimension of the Hilbert space $\mathcal{H}$ with respect to the von Neumann algebra $\mathcal{N}$. (When $\mathcal{N} \equiv \mathbb{C}I$ then the coupling constant is the usual dimension of $\mathcal{H}$, hence the notation $\dim_{\mathcal{N}}(\mathcal{H})$.) V. Jones used the coupling constant to define a size of a subfactor of a finite factor. He was inspired by the notion of the index of a subgroup of a group, he therefore called this relative size index.

Let $\mathcal{N}$ be a subfactor of a finite von Neumann factor $\mathcal{M}$ possessing a unique faithful normal tracial state τ. Denote by $L^2(\mathcal{M})$ the GNS Hilbert space obtained from τ and $\mathcal{M}$. The subalgebra $\mathcal{N}$ acts naturally on $L^2(\mathcal{M})$ and its index $[M : N]$ is defined as the coupling constant provided the commutant is finite. Else we put $[M : N] = \infty$ by definition. In fact, the index may be obtained in any terms of an arbitrary representation on a Hilbert space $\mathcal{H}$ as

$$[\mathcal{M} : \mathcal{N}] = \frac{\dim_{\mathcal{N}}(\mathcal{H})}{\dim_{\mathcal{M}}(\mathcal{H})}. \tag{17.17}$$

(When $\mathcal{M}$ acts standardly on $\mathcal{H}$ the denominator becomes 1 here.) The number $[M : N]$ is not always an integer, and the set of possible values of the index is exactly **Ind**.

By an evocative notation let $\mathcal{M}_0 \subset \mathcal{M}_1$ be finite von Neumann factors. The algebra $\mathcal{M}_1$ acts on $L^2(\mathcal{M}_1)$ by left multiplication and let Ω be the cyclic and separating trace vector in $L^2(\mathcal{M}_1)$. The projection $E_0 \equiv [\mathcal{M}_0\Omega]$ induces a trace-preserving conditional expectation $\mathcal{E}_0 : a \mapsto E_0 a E_0$ of $\mathcal{M}_1$ onto $\mathcal{M}_0$. By the definition of the index we have $[\mathcal{M}_1 : \mathcal{M}_0] = \tau(E_0)^{-1} \equiv \kappa$ where τ is the trace induced by Ω. Let $J_1 : a\Omega \mapsto a^*\Omega$ be the canonical conjugation on $L^2(\mathcal{M}_1)$. One can prove that the von Neumann algebra $\mathcal{M}_2 \equiv J_1 \mathcal{M}_0' J_1$ is genarated by $\mathcal{M}_1$ and E_0. So

$$[\mathcal{M}_2 : \mathcal{M}_1] = [J_1\mathcal{M}_2 J_1 : J_1\mathcal{M}_1 J_1] = [\mathcal{M}_0' : \mathcal{M}_1'] = \frac{\dim_{\mathcal{M}_1'}\left(L^2(\mathcal{M}_1)\right)}{\dim_{\mathcal{M}_0'}\left(L^2(\mathcal{M}_1)\right)}$$

$$= \frac{\langle \Omega, [\mathcal{M}_1'\Omega]\Omega \rangle}{\langle \Omega, [\mathcal{M}_1\Omega]\Omega \rangle} : \frac{\langle \Omega, [\mathcal{M}_0'\Omega]\Omega \rangle}{\langle \Omega, [\mathcal{M}_1'\Omega]\Omega \rangle} = [\mathcal{M}_1 : \mathcal{M}_0]$$

and we have succeeded in constructing from the inclusion $\mathcal{M}_0 \subset \mathcal{M}_1$ with index κ another inclusion $\mathcal{M}_1 \subset \mathcal{M}_2$ with the same index. The above construction of Jones may be repeated and to produce a sequence $\mathcal{M}_0 \subset \mathcal{M}_1 \subset \ldots$ of finite factors and a sequence $E_0, E_1, \ldots$ of projections such that

(i) $\mathcal{M}_{n+2}$ is generated by E_n and $\mathcal{M}_{n+1}$; moreover, $\mathcal{M}_{n+2} = J_{n+1}\mathcal{M}_n' J_{n+1}$,

(ii) $[\mathcal{M}_{n+1} : \mathcal{M}_n]$ is independent of n.

Note that when repeating the basic construction we are working on larger and larger Hilbert spaces and obtain the projections E_0, E_1,... on the inductive limit. ($L^2(\mathcal{M}_n)$ is always identified with $E_{n-1}L^2(\mathcal{M}_{n+1})$.) It is not difficult to verify that E_0, E_1,... satisfy the relations (17.10) and (17.11). Indeed, $E_n \in \mathcal{M}_{n+2}$, and $\mathcal{M}_{n+3}$ consists of those elements of $\mathcal{M}_{n+4}$ which commute with E_{n+2}. This yields (17.11). The functional $a \mapsto \kappa\tau_2(E_0 a)$ is a tracial state on $\mathcal{M}_0$, hence it must be τ_0. Therefore,

$$\tau(aE_{n-1}) = \kappa^{-1}\tau(a) \quad \text{for} \quad a \in \mathcal{M}_n. \tag{17.18}$$

This implies $\mathcal{E}_n(E_{n-1}) = \kappa^{-1}I$ and we have $E_n E_{n-1} E_n = \kappa^{-1}E_n$, which is a part of (17.10). The other part $E_n E_{n+1} E_n = \kappa^{-1}E_n$ must be verified on $L^2(\mathcal{M}_{n+2})$.

Now Proposition 17.1 tells us that $[\mathcal{M}_1 : \mathcal{M}_0]$ is in **Ind**, and the other half of Jones' result consists in the construction of the inductive limit of finite dimensional algebras in order to show that every number in **Ind** is the index of a subfactor of a finite factor.

Let us consider a C*-algebra TL(κ, m) that is generated by the projections $e_1, e_2, \ldots, e_{m-1}$ satisfying the Temperley–Lieb relations (17.15) with a $\kappa > 4$. Such an algebra has a well-determined structure explained below. TL(κ, m) is the direct sum of full matrix algebras Q_j^m where

$$j = 1, 2, \ldots, [m/2] \quad \text{and} \quad Q_j^m \text{ is of order } \binom{m}{j} - \binom{m}{j-1} \equiv \left\{ \begin{matrix} m \\ j \end{matrix} \right\}.$$

The linear dimension of TL(κ, m) is

$$\sum_{j=0}^{[m/2]} \left(\binom{m}{j} - \binom{m}{j-1} \right)^2 = \frac{1}{m+1}\binom{2m}{m},$$

and its tracial dimension is

$$\sum_{j=0}^{[m/2]} \binom{m}{j} - \binom{m}{j-1} = \binom{m}{[m/2]}.$$

Since TL(κ, m) is not a factor, it has several tracial states. Among them the most important is given by a sequence $P_n(t)$ of polynomials. Let

$$P_n(t) = \frac{p(t)^{n+1} - q(t)^{n+1}}{p(t) - q(t)} \quad \text{where}$$

$$p(t) = \frac{1 + \sqrt{1-4t}}{2} \quad \text{and} \quad q(t) = \frac{1 - \sqrt{1-4t}}{2}. \tag{17.19}$$

(So $p(t)q(t) = t$ and $(p(t), q(t))$ is a probability distribution for $0 \leq t \leq 1/4$.) For a minimal projection $p_j \in Q_j^m$ set

$$\tau_m(p_j) = \kappa^{-j} P_{m-2j}(\kappa^{-1}) \quad \text{for} \quad j = 0, 1, \ldots, [m/2]. \tag{17.20}$$

The importance of τ_m will become more clear if we look at the embedding of TL (κ, m) into TL (κ, n) $(m < n)$. Let TL (κ, n) be generated by projections $e'_1, e'_2, \ldots, e'_{n-1}$ (satisfying the Temperley–Lieb relations) and consider TL (κ, m) as a subset of TL (κ, n) under the identification of e_j with e'_j for $j = 1, 2, \ldots, m - 1$. Then

$$\tau_n | \mathrm{TL}\,(\kappa, m) = \tau_m$$

but much more is true, namely, that

$$\tau_n(w\, e'_j) = \kappa^{-1}\tau_m(w) \quad \text{for} \quad w \in \mathrm{TL}\,(\kappa, m) \quad \text{and} \quad m \le j \le n-1. \tag{17.21}$$

More generally, if $w \in \mathrm{TL}\,(\kappa, m)$ and u is in the algebra generated by $e_m, e_{m+1}, \ldots, e_{n-1}$, then

$$\tau_n(w\, u) = \tau_m(w)\tau_m(u)\,. \tag{17.22}$$

Writing E_m for the conditional expectation of TL (κ, n) onto TL (κ, m) that preserves τ_n the above relation is seen to be equivalent to

$$E_m(u) = \tau_n(u)I\,. \tag{17.23}$$

Let the C*-algebra $\mathcal{A}$ be generated by an infinite sequence $e_1, e_2, \ldots$ of projections which satisfy the Temperley–Lieb relations (17.15) and let TL (κ, m) stand for the subalgebra spanned by $e_1, e_2, \ldots e_{m-1}$. Assume that the union of these subalgebras is norm dense in $\mathcal{A}$. (So $\mathcal{A}$ is an AF-algebra.)

Proposition 17.2 Let $\mathcal{A}$ be the AF-algebra constructed above as the limit of the algebras TL (κ, m) with $\kappa > 4$. Then there exists an endomorphism γ of the C*-algebra $\mathcal{A}$ so that $\gamma(e_i) = e_{i+1}$. The dynamical entropy of γ with respect to the tracial state τ of $\mathcal{A}$ is equal to $\eta(p) + \eta(q)$, where

$$p = \tfrac{1}{2}\big(1 + \sqrt{1 - 4\kappa^{-1}}\big), \quad q = 1 - p \quad \text{and} \quad \eta(x) = -x\log x\,.$$

Proof. The mean entropy

$$\lim_{m\to\infty} \frac{S(\tau | \mathrm{TL}\,(\kappa, 2m))}{2m}$$

will be shown to be equal to $\eta(p) + \eta(q)$, and it will be proved that the dynamical entropy must coincide with the mean one.

We shall need the values of τ on the minimal projections of TL $(\kappa, 2m)$. These values are

$$\kappa^{-j}\frac{p(\kappa^{-1})^{2m-2j+1} - q(\kappa^{-1})^{2m-2j+1}}{p(\kappa^{-1}) - q(\kappa^{-1})} \tag{17.24}$$

according to (17.19) and (17.20). Here $j = 0, 1, \ldots m$. Using the abbreviations

$$p(\kappa^{-1}) \equiv p, \quad q(\kappa^{-1}) \equiv q, \quad r \equiv q/p$$

(17.24) is rewritten as

$$q^j p^{2m-j}(1 - r^{2m-2j+1})(1-r)^{-1}.$$

Hence

$$\frac{S(\tau|\mathrm{TL}\,(\kappa, 2m))}{2m} = -(1-r)^{-1} \sum_{j=0}^{m} \left\{ \begin{matrix} 2m \\ j \end{matrix} \right\} q^j p^{2m-j}(1 - r^{2m-2j+1})$$

$$\times \left(\frac{\log(1 - r^{2m-2j+1}) - \log(1-r)}{2m} + \frac{j \log q + (2m - j)\log p}{2m} \right).$$

First we note that the factor

$$\frac{\log(1 - r^{2m-2j+1}) - \log(1-r)}{2m} + \frac{j \log q + (2m - j)\log p}{2m} \tag{17.25}$$

is bounded. We show that

$$\sum_{j=0}^{m} \left\{ \begin{matrix} 2m \\ j \end{matrix} \right\} q^j p^{2m-j} r^{2m-2j+1} \tag{17.26}$$

converges to 0 as $m \to \infty$ and so this term does not contribute to the limit. We have

$$\sum_{j=0}^{m} \left\{ \begin{matrix} 2m \\ j \end{matrix} \right\} q^j p^{2m-j} r^{2m-2j+1} = r \sum_{j=0}^{m} \left\{ \begin{matrix} 2m \\ j \end{matrix} \right\} q^{2m-j} p^j$$

$$= r \sum_{j=0}^{m} \left(\begin{matrix} 2m \\ j \end{matrix} \right) q^{2m-j} p^j - \sum_{j=0}^{m-1} \left(\begin{matrix} 2m \\ j \end{matrix} \right) q^{2m-j} p^j$$

and the limit is obtained from a probabilistic interpretation. The number

$$\sum_{j=0}^{m} \left(\begin{matrix} 2m \\ j \end{matrix} \right) q^{2m-j} p^j \tag{17.27}$$

is the probability of at most m successes in the course of $2m$ Bernoulli trials with success probability p. Since $p > 1/2$, the probability (17.27) goes to 0 and so does (17.26).

Since the first term of (17.25) converges to 0 we must determine the limit of the expression

$$-(1-r)^{-1} \sum_{j=0}^{m} \left\{ \begin{matrix} 2m \\ j \end{matrix} \right\} q^j p^{2m-j} \frac{j \log q + (2m - j)\log p}{2m}. \tag{17.28}$$

One shows by elementary methods that

$$\lim_{m\to\infty} \sum_{j=0}^{m} \left\{ \begin{matrix} 2m \\ j \end{matrix} \right\} q^j p^{2m-j} \frac{j}{2m} = q - rq$$

$$\lim_{m\to\infty} \sum_{j=0}^{m} \left\{ \begin{matrix} 2m \\ j \end{matrix} \right\} q^j p^{2m-j} \frac{2m - j}{2m} = p - rp$$

and these imply that the mean entropy (that is, the limit of (17.28)) is really $\eta(p) + \eta(q)$.

Now we have to show that the mean entropy coincides with $h_\tau(\gamma)$. First of all, we note that

$$h_\tau(\gamma) = \lim_{m \to \infty} h_\tau(\gamma\,; \mathrm{TL}\,(\kappa, 2m)) \tag{17.29}$$

due to the norm approximating net theorem (see Theorem 10.8 or rather Corollary 10.9). From the monotonicity of the multiple-subalgebra entropy, we have

$$H_\tau\big(\mathrm{TL}\,(\kappa, 2m), \gamma(\mathrm{TL}\,(\kappa, 2m)), \ldots, \gamma^n(\mathrm{TL}\,(\kappa, 2m))\big)$$
$$\leq H_\tau(\mathrm{TL}\,(\kappa, n+2m)) = S(\tau|\mathrm{TL}\,(\kappa, n+2m)),$$

which yields

$$h_\tau(\gamma\,; \mathrm{TL}\,(\kappa, 2m)) \leq \lim_{n \to \infty} \frac{1}{n} S(\tau|\mathrm{TL}\,(\kappa, n+2m)) = \eta(p) + \eta(q). \tag{17.30}$$

The converse estimate is based on the independence property understood as follows. The subalgebras

$$\mathrm{TL}\,(\kappa, 2m), \quad \gamma^{2m+1}(\mathrm{TL}\,(\kappa, 2m)), \quad \gamma^{2(2m+1)}(\mathrm{TL}\,(\kappa, 2m)), \quad \ldots$$

are independent under τ according to (17.22), and they commute pairwise due to the Temperley–Lieb relations. These facts allow for a simple estimate of the entropy

$$H_\tau\big(\mathrm{TL}\,(\kappa, 2m), \gamma^{2m+1}(\mathrm{TL}\,(\kappa, 2m)), \ldots, \gamma^{n(2m+1)}(\mathrm{TL}\,(\kappa, 2m))\big). \tag{17.31}$$

For $0 \leq j \leq n$ let $\mathcal{C}_j$ be a maximal abelian subalgebra of $\gamma^{j(2m+1)}(\mathrm{TL}\,(\kappa, 2m))$ that contains the density matrix of the restriction of τ. Furthermore, let $\mathcal{C}$ be the (finite dimensional abelian) subalgebra generated by $\mathcal{C}_0, \mathcal{C}_1, \ldots, \mathcal{C}_n$. So the triple $((\mathcal{C}_j)_{j=0}^n, \mathcal{C}, \tau|\mathcal{C})$ is an abelian model for the subalgebras $\mathrm{TL}\,(\kappa, 2m)$, $\gamma^{2m+1}(\mathrm{TL}\,(\kappa, 2m)), \ldots, \gamma^{n(2m+4)}(\mathrm{TL}\,(\kappa, 2m))$. The entropy with respect to this abelian model is easily computed. Namely, it is $S(\tau|\mathcal{C})$, because all the correction terms in (10.2) vanish. On the other hand,

$$S(\tau|\mathcal{C}) = (n+1)S(\tau|\mathcal{C}_0) = S(\tau|\mathrm{TL}\,(\kappa, 2m)).$$

Therefore, we arrive at

$$H_\tau\big(\mathrm{TL}\,(\kappa, 2m), \gamma^{2m+1}(\mathrm{TL}\,(\kappa, 2m)), \ldots, \gamma^{n(2m+1)}(\mathrm{TL}\,(\kappa, 2m))\big)$$
$$\geq (n+1)S(\tau|\mathrm{TL}\,(\kappa, 2m)). \tag{17.32}$$

The rest of our argumentation is rather standard:

$$H_\tau\big(\mathrm{TL}\,(\kappa, 2m), \gamma(\mathrm{TL}\,(\kappa, 2m)), \ldots, \gamma^{n(2m+1)}(\mathrm{TL}\,(\kappa, 2m))\big)$$
$$\geq H_\tau\big(\mathrm{TL}\,(\kappa, 2m), \gamma^{2m+1}(\mathrm{TL}\,(\kappa, 2m)), \ldots, \gamma^{n(2m+1)}(\mathrm{TL}\,(\kappa, 2m))\big)$$
$$\geq (n+1)S(\tau|\mathrm{TL}\,(\kappa, 2m))$$

and

$$h_\tau\left(\gamma\,;\mathrm{TL}\left(\kappa,2m\right)\right) \geq \lim_{n\to\infty} \frac{n+1}{n(2m+1)} S(\tau|\mathrm{TL}\left(\kappa,2m\right))$$

$$= \frac{1}{2m+1} S(\tau|\mathrm{TL}\left(\kappa,2m\right)). \tag{17.33}$$

Finally, reference to (17.29) makes the proof complete. □

It is striking that in the previous proposition the entropy of γ is the same as the entropy of an independent shift with distribution (p,q). To explain the reason behind, we consider the infinite tensor product of copies of $M_2(\mathbb{C})$. Let

$$\mathcal{B} = \otimes_{i\in\mathbb{N}}\mathcal{B}_i, \quad \mathcal{B}_i = M_2(\mathbb{C}) \quad \text{and} \quad \mathcal{B}_{[n,m]} = \otimes_{i=n}^m \mathcal{B}_i \subset \mathcal{B}.$$

We write $e_{ij}^{(n)}$ for the matrix units in $\mathcal{B}_n$ ($n \in \mathbb{N}$, $1 \leq i,j \leq 2$). Set $e_n \in \mathcal{B}_{[n,n+1]}$ for $n \in \mathbb{N}$ as

$$e_n = (1-t)e_{11}^{(n)} \otimes e_{22}^{(n+1)} + te_{22}^{(n+1)} \otimes e_{11}^{(n+1)}$$
$$+ \sqrt{t(1-t)}e_{12}^{(n)} \otimes e_{21}^{(n+1)} + \sqrt{t(1-t)}e_{21}^{(n)} \otimes e_{12}^{(n+1)} \tag{17.34}$$

with $t = \kappa^{-1}$. So the matrix of e_n is

$$\begin{pmatrix} 0 & 0 & 0 & 0 \\ 0 & 1-t & \sqrt{t(1-t)} & 0 \\ 0 & \sqrt{t(1-t)} & t & 0 \\ 0 & 0 & 0 & 0 \end{pmatrix}$$

and e_n is a projection. It is a matter of verification that the projections $e_1, e_2, \ldots$ satisfy the Temperley–Lieb relations (17.15). By construction the right shift on $\mathcal{B}$ sends e_n into e_{n+1}. Therefore the AF-algebra from Proposition 17.2 can be realized inside of the infinite tensor product, and γ is the restriction of the right shift β. On $\mathcal{B}$ we take the (shift-invariant) product state φ which has density $pe_{11}^{(n)} + qe_{22}^{(n)}$ in $\mathcal{B}_n$. Performing the GNS-construction we obtain $(\pi, \bar{\varphi}, \bar{\beta})$, where π is the representation, $\bar{\varphi}$ is a state of $\pi(\mathcal{B})''$ and $\bar{\beta}$ is its endomorphism. One can prove that $\pi(\mathcal{A})''$ is just the centralizer $\mathcal{Z}$ of $\bar{\varphi}$. This shows that γ is very close to an independent shift. Its dynamical entropy can be computed also in this way. We know from (10.16) that

$$h_\varphi(\beta) = \eta(p) + \eta(q) \quad \text{and} \quad h_\varphi(\beta) = h_{\bar{\varphi}}(\bar{\beta}) \tag{17.35}$$

is provided by Proposition 10.12. Since $h_{\bar{\varphi}|\mathcal{Z}}(\bar{\beta}|\mathcal{Z}) \leq h_{\bar{\varphi}}(\bar{\beta})$ by the existence of the conditional expectation onto a centralizer, in order to obtain another proof of the entropy formula one needs

$$\eta(p) + \eta(q) \leq h_{\bar{\varphi}|\mathcal{Z}}(\bar{\beta}|\mathcal{Z}), \tag{17.36}$$

which can be obtained from Lemma 7.12, for example. Neither this argument nor the above proof of Proposition 17.2 work in the case $\kappa = 4$. Nevertheless, for $\kappa = 4$ we have $h_\tau(\gamma) = \log 2$. Taking a closer inspection of the proof of Proposition 17.2 we observe that in the second part of the proof

$$h_\tau(\gamma) = \lim_{m \to \infty} \frac{1}{m} S(\tau | \mathrm{TL}(\kappa, m)) \tag{17.37}$$

was achieved without using the condition $\kappa > 4$. The assumption on κ was essential in the first part in which the concrete structure of the Temperley–Lieb algebras $\mathrm{TL}(\kappa, m))$ was required. For $\kappa \leq 4$ the structure of $\mathrm{TL}(\kappa, m))$ is not easily available, but the mean entropy can still be obtained.

Our next aim is the analogue of Proposition 17.2 in the case $\kappa < 4$. Let $\mathrm{TL}(\kappa, m)$ again be the direct sum of full matrix algebras Q_j^m. Opposite to the case $\kappa > 4$ the dimension of the centre of $\mathrm{TL}(\kappa, m)$ does not increase with m but stabilizes for large enough m. For $m = m_0 + 2k$ we have

$$\mathrm{TL}(\kappa, m) = \oplus_{j=1}^N Q_j^m \quad \text{and} \quad Q_j^m = M_{r(m,j)}(\mathbb{C}). \tag{17.38}$$

Moreover, if τ is the tracial state satisfying (17.23) and τ has the value $\varepsilon(m, j)$ on the minimal projections in the summand Q_j^m, then

$$\varepsilon(m_0 + 2k + 2, j) = \kappa^{-1}\varepsilon(m_0 + 2k, j) \tag{17.39}$$

(see [Jones 1983] for proofs).

Proposition 17.3 Let the AF-algebra $\mathcal{A}$ be the limit of the algebras $\mathrm{TL}(\kappa, m)$ with $\kappa < 4$ and let γ be its endomorphism which sends e_i into e_{i+1} ($i \in \mathbb{N}$). The dynamical entropy of γ with respect to the tracial state τ satisfying (17.22) equals to $\frac{1}{2} \log \kappa$.

Proof. It was shown in the proof of the previous proposition that the dynamical entropy is the same as

$$\lim_{k \to \infty} \frac{S(\tau | \mathrm{TL}(\kappa, m_0 + 2k))}{2k} \tag{17.40}$$

and we compute this limit using relations (17.38) and (17.39). Since

$$\sum_{j=1}^N r(m_0 + 2k, j)\varepsilon(m_0 + 2k, j) = 1,$$

we have

$$S(\tau | \mathrm{TL}(\kappa, m_0 + 2k)) = -\sum_{j=1}^N r(m_0 + 2k, j)\varepsilon(m_0 + 2k, j) \log \kappa^{-k}\varepsilon(m_0, j)$$

$$= -\sum_{j=1}^N r(m_0 + 2k, j)\varepsilon(m_0 + 2k, j)\varepsilon(m_0, j) + k \log \kappa.$$

Divided by $2k$, the first term tends to 0 so the limit (17.40) is $\frac{1}{2} \log \kappa$. □

Let us recall the definition of a stationary Markov processes. A process $(\mathcal{A}, \mathcal{A}_0, \gamma, \tau)$ is called a stationary Markov process if

(i) $\tau \circ \gamma = \tau$,

(ii) the τ-preserving conditional expectation E_J of $\mathcal{A}$ onto $\mathcal{A}_J$ exists for every interval $J = [n, k]$,

(iii) $E_{[n,k]}(\mathcal{A}_{[k,m]}) = \mathcal{A}_k$ for every $n < k < m$.

Choose $\mathcal{C}$ as the (unital) C*-algebra generated by the projections $e_1, e_2, \ldots$ satisfying the Temperley–Lieb relations (17.15), and let $\mathcal{A}_0$ be the (unital) subalgebra generated by e_1, e_2. If γ is an endomorphism of $\mathcal{A}$ so that $\gamma(e_i) = e_{i+1}$. If τ is a tracial state with property

$$\kappa \tau(w\, e_j) = \tau(w). \tag{17.41}$$

whenever w is in the subalgebra generated by the projections $\{e_1, e_2, \ldots, e_{j-1}\}$, then $(\mathcal{A}, \mathcal{A}_0, \gamma, \tau)$ is a Markov process. This fact is not completely obvious.

Notes and Remarks. The relations (17.10) and (17.11) appeared in [Temperley and Lieb 1971], see also [Baxter 1982]. Concerning the Potts model [Wu 1982] is suggested. There exists much activity in the mathematical theory of knots and links using methods of statistical mechanics. This is totally out of the scope of the book, see for example [Jones 1989].

The coupling constant was introduced by von Neumann (see [Strătilă and Zsidó 1979]) but it received more attention only after [Jones 1983], where the index of a subfactor was defined and the Temperley–Lieb relations showed up again. It was recognized in [Evans 1985] that projections with the same commutation relations had appeared in two different contexts. The set **Ind** was already met by Kronecker in connection with the possible norm of matrices with integer entries. The monograph [Goodman, de la Harpe and Jones 1989] contains a lot of information on the Temperley–Lieb algebras from a rather algebraic viewpoint. Proposition 17.1 is due to [Wenzl 1987].

The entropy of the Temperley–Lieb shift was computed in [Pimsner and Popa 1986], except for the case $\kappa = 4$ which was settled in [Yin 1990]. We benefited from both papers. The last statement concerning the Markov property is Example 4.2.9 in [Goodman, de la Harpe and Jones 1989].

[Kümmerer 1988] is a survey on Markov processes. The conecpt of index was extended to conditional expectations by [Kosaki 1986b]. Dynamical entropy was computed for related shifts in in [Choda and Hiai 1991]. All these recent developments have used modular theory rather deeply.

18 Optical Communication Processes

In communication processes we suffer the loss of information in the course of information transmission. In this chapter, we apply the theory of channel and mutual entropy to optical communication processes. We construct a concrete model of the channel by taking into account this attenuation of the information. We also discuss the efficiency of optical modulations with the quantum states by using the mutual entropy.

Let $\mathcal{H}$ be a Hilbert space. We generally denote the set of all normal states on the algebra $B(\mathcal{H})$ by $\mathfrak{S}_*(\mathcal{H})$. As always, $\mathfrak{S}_*(\mathcal{H})$ is identified with the set of all density operators on the Hilbert space $\mathcal{H}$. A quantum mechanical channel α is a completely positive unital mapping from $B(\mathcal{H}_2)$ to $B(\mathcal{H}_1)$, that is

$$\sum_{i=1}^{n} \sum_{j=1}^{n} A_i^* \alpha(B_i^* B_j) A_j \geq 0$$

for any $A_i \in B(\mathcal{H}_1)$, $B_j \in B(\mathcal{H}_2)$ and $n \in \mathbb{N}$. We restrict ourselves to normal channels. So we assume that $\alpha \circ \varphi \equiv \alpha_*(\varphi)$ belongs to $\mathfrak{S}_*(\mathcal{H}_2)$ whenever $\varphi \in \mathfrak{S}_*(\mathcal{H}_1)$. The channelling transformation will be regarded as state transformation in this chapter and we shall prefer to work with α_* that is usual in the context of communication theory.

An example of quantum mechanical channel α_* is given by

$$\alpha_*(\varrho) \equiv \mathrm{Tr}_{\mathcal{K}_2} \pi_*(\varrho \otimes \xi) \qquad \varrho \in \mathfrak{S}_*(\mathcal{H}_1), \tag{18.1}$$

where ξ is a state described by a noise and π_* is a channel mapping from $\mathfrak{S}_*(\mathcal{H}_1 \otimes \mathcal{K}_1)$ to $\mathfrak{S}_*(\mathcal{H}_2 \otimes \mathcal{K}_2)$. Moreover, $\mathrm{Tr}_{\mathcal{K}_2}$ is the partial trace:

$$\langle f, (\mathrm{Tr}_{\mathcal{K}_2} Q) g \rangle \equiv \sum_i \langle f \otimes h_i, Q(g \otimes h_i) \rangle$$

for any $Q \in B(\mathcal{H}_2 \otimes \mathcal{K}_2)$, any $f, g \in \mathcal{H}_2$, and for any orthonormal basis $(h_i)_i$ of $\mathcal{K}_2$.

Using (18.1) we give a general expression for an *attenuation process* by means of the Hamiltonian of each system. When we treat optical communication, each quantum system composed of photons is described by the Hamiltonian $H = a^* a + 1/2$, where a^* and a are creation and annihilation operators of a photon, respectively. By solving the Schrödinger equation $Hx(q) = Ex(q)$, we can easily get the eigenvalues $E_n = n + 1/2$ $\quad (n = 0, 1, 2, \ldots)$ and the eigenvectors are

$$x_n(q) = \sqrt{\frac{1}{\sqrt{\pi}n!}} H_n\left(\sqrt{2q}\right) \exp\left(-\frac{q^2}{2}\right),$$

where $H_n(q)$ is the nth Hermite function. The Hilbert space of each system is the closed linear span of the linear combinations of the functions $x_n(q)$ $\quad (n = 0, 1, 2, \ldots)$. Then a model for an optical attenuation process is considered as follows: When n_1 photons are transmitted from the input system, n_2 photons from the noise system add to the signal. Then m_1 photons are lost to the loss system through the channel, and m_2 photons are detected in the output system. According to the conservation of energy $n_1 + n_2 = m_1 + m_2$ holds. We take the following linear transformation among the coordinates q_1, t_1, q_2, t_2 of the input, noise, output, and loss systems, respectively:

$$\begin{cases} q_2 = cq_1 + dt_1 \\ t_2 = -dq_1 + ct_1 \end{cases} \qquad (c^2 + d^2 = 1).$$

By using this linear transformation, we define the mapping $\pi = U(\,.\,)U^*$ by

$$U(x_{n_1}^{(1)} \otimes y_{n_2}^{(1)})(q_2, t_2) = x_{n_1}^{(2)} \otimes y_{n_2}^{(2)}(cq_2 - dt_2, dq_2 + ct_2)$$

$$= \sum_{j=0}^{n_1+n_2} C_j^{n_1,n_2} x_j^{(2)} \otimes y_{n_1+n_2-j}^{(2)}(q_2, t_2), \qquad (18.2)$$

where $C_j^{n_1,n_2}$ is given by

$$C_j^{n_1,n_2} = \int\!\!\int x_{n_1}^{(2)} \otimes y_{n_2}^{(2)}(cq_2 - dt_2, dq_2 + ct_2)\overline{x_j^{(2)} \otimes y_{n_1+n_2-j}^{(2)}(q_2, t_2)}\, dq_2\, dt_2$$

$$= \sum_{r=L}^{K} (-1)^r \frac{\sqrt{n_1! n_2! j! (n_1 + n_2 - j)!}}{r!(n_1 - r)!(j - r)!(n_2 - j + r)!} c^{n_2-j+2r} d^{n_1+j-2r},$$

where $K = \min\{j, n_1\}$, $L = \max\{j - n_2, 0\}$. Then the channel α_* is expressed as

$$\alpha_* \varrho = \mathrm{Tr}_{\mathcal{K}_2} U(\varrho \otimes \xi)U^*. \qquad (18.3)$$

Here note that c^2 can be regarded as the transmission efficiency η for the channel α_*.

Now we introduce the new concept "lifting" and we apply it to the expression for an attenuation process.

(i) Let $\mathcal{H}, \mathcal{K}$ be Hilbert spaces and let $\mathcal{H} \otimes \mathcal{K}$ be a fixed tensor product of $\mathcal{H}$ and $\mathcal{K}$. A lifting $\mathcal{E}_*$ from $\mathcal{H}$ to $\mathcal{H} \otimes \mathcal{K}$ is a continuous map $\mathcal{E}_* : \mathfrak{S}_*(\mathcal{H}) \to \mathfrak{S}_*(\mathcal{H} \otimes \mathcal{K})$.

(ii) A lifting from $\mathcal{H}$ to $\mathcal{H} \otimes \mathcal{K}$ is called nondemolition for a state $\varphi \in \mathfrak{S}_*(\mathcal{H})$ if $(\mathcal{E}_*\varphi)(a \otimes 1) = \varphi(a)$ for any $a \in B(\mathcal{H})$.

If $\mathcal{E}_*$ is affine, we call it a linear lifting if it maps pure states into pure states, we call it pure. The notion of lifting is important to study quantum measuring processes. An example of the *nondemolition lifting* is a compound state (see Chapter 8). When we take $\mathcal{H} = \mathcal{H}_1 = \mathcal{H}_2$, $\mathcal{K} = \mathcal{K}_1 = \mathcal{K}_2$ and ξ be a vacuum state, that is, $\xi = |0\rangle\langle0| \in \mathfrak{S}_*(\mathcal{K}_1)$ is a noise state due to the "zero point fluctuation" of electromagnetic field ($y_0^{(1)}$ is a vacuum state vector in $\mathcal{K}_1$).

$$\mathcal{E}_* : \varrho \in \mathfrak{S}_*(\mathcal{H}) \to \pi_*(\varrho \otimes \xi) \in \mathfrak{S}_*(\mathcal{H} \otimes \mathcal{K})$$

is a lifting, and we can rewrite the channel:

$$\alpha_* \varrho = \mathrm{Tr}_\mathcal{K} \mathcal{E}_* \varrho. \tag{18.4}$$

By using the lifting, we can define a mapping V from $\mathcal{H}$ to $\mathcal{H} \otimes \mathcal{K}$ as

$$V|\theta\rangle = |c\theta\rangle \otimes |d\theta\rangle$$

where $|\theta\rangle$ represents a coherent vector (i.e.,vector state of a $a|\theta\rangle = \theta|\theta\rangle$). Now, let us show the equivalence of the above operator V and the operator U in the conventional expression.

$$V|\theta\rangle = |c\theta\rangle \otimes |d\theta\rangle$$

$$= \exp\left(-\tfrac{1}{2}|c\theta|^2\right) \sum_n \frac{(c\theta)^n}{\sqrt{n!}}|n\rangle \otimes \exp\left(-\tfrac{1}{2}|d\theta|^2\right) \sum_m \frac{(d\theta)^m}{\sqrt{m!}}|m\rangle$$

$$= \exp\left(-\tfrac{1}{2}|\theta|^2\right) \sum_n \sum_m \frac{(c\theta)^n (d\theta)^m}{\sqrt{n!m!}}|n\rangle \otimes |m\rangle$$

$$= \exp\left(-\tfrac{1}{2}|\theta|^2\right) \sum_{N=0}^{\infty} \frac{\theta^N}{\sqrt{N!}} \left(\sum_{n=0}^{\infty} c^n d^{N-n} \sqrt{\frac{N!}{n!(N-n)!}}|n\rangle \otimes |N-n\rangle\right),$$

which implies, for any nonnegative integer N,

$$V|N\rangle = \sum_{n=0}^{N} c^n d^{N-n} \sqrt{\frac{N!}{n!(N-n)!}}|n\rangle \otimes |N-n\rangle.$$

Thus U when $n_2 = 0$ in (18.2) equals to V by replacing d with $-d$. Therefore the attenuation channel can be rewritten as

$$\alpha_* \varrho = \mathrm{Tr}_\mathcal{K} V \varrho V^*. \tag{18.5}$$

We derive error probabilities for some important communication systems such as PPM (explained later) and BPSK (Binary Phase Shift Keying) homodyne detection for the above expression (18.5). In particular, each error probability for a squeezed state taken as an input state is calculated.

Here we briefly review the basic facts for quantum coding, BPSK, homodyne detection and a squeezed state.

Suppose that, by some procedure, we encode an information representing it by a sequence of letters $c^{(1)}, \ldots, c^{(n)}, \ldots$, where $c^{(k)}$ is an element in a set

$\mathcal{C}$ of symbols called the alphabet. A quantum code is a map which associates to each symbol (or sequence of symbols) in $\mathcal{C}$ a quantum state, representing an optical signal. Sometimes one uses a state as two codes: one for input and one for output. In the sequel we shall only consider a two symbols alphabet:

$$\mathcal{C} = \{0,1\} \iff \Xi = \{\varrho_0, \varrho_1\}.$$

One example of quantum code $\Xi = \{\varrho_0, \varrho_1\}$ where ϱ_i is the quantum state corresponding to the symbol $c_i \in \mathcal{C}$, is obtained by choosing ϱ_0 as the vacuum state and ϱ_1 another state such as a coherent or a squeezed state of a one-mode field (OOK (On Off Keying), Pulse Modulation etc). Two states (quantum codes) $\varrho_0^{(1)}$ and $\varrho_1^{(1)}$ in the input system are transmitted to the output system through a channel α_*. If we here assume a Z-type signal transmission, namely that the input signal "0", represented by the state $\varrho_0^{(1)}$, is error free in the sense that it goes always to the output signal "0" represented by $\varrho_0^{(2)}$, while the input signal "1", represented by the state $\varrho_1^{(1)}$, is not error free in the sense that its output can give rise to both states $\varrho_0^{(2)}$ or $\varrho_1^{(2)}$ with different probabilities. On the other hand, in the case that the information is carried by the amplitude, frequency or phase of the input state, regardless of the noise state ξ, both of transmitted input signals "0" and "1" have a possibility to be recognized by mistake in the output system. We call this channel X-type channel. For example, BPSK is one of the modulations for X-type channel so that the information is represented by phases of an input state in the sense that ϱ_0 represents a state with phase 0 and ϱ_1 does a state with phase π. Therefore in the output system it is enough to detect the real part of complex amplitude of a received state. This is homodyne detection. Next, we give the definition of a squeezed state for the sequel discussion. A squeezed state is made by a eigenvector $|\theta; \mu, \nu\rangle$ of a new annihilation operator such as

$$b \equiv \mu a + \nu a^*, \qquad b|\theta; \mu, \nu\rangle = (\mu\theta + \nu\bar{\theta})|\theta; \mu, \nu\rangle$$

$$|\mu|^2 - |\nu|^2 = 1, \quad \mu = \cosh z, \quad \nu = \exp(i\phi)\sinh z.$$

The squeezed vector state $|\theta; \mu, \nu\rangle$ can be represented as

$$U(z)|\theta\rangle \equiv |\theta; \mu, \nu\rangle,$$

where $U(z)$ is defined by

$$U(z) \equiv \exp(\tfrac{1}{2}(z^2 a - \bar{z}a^{*2})) \qquad (z \in \mathbb{C}).$$

Then positive operator-valued measure is a useful tool to describe quantum measurement processes. Therefore we apply the attenuation channel and each positive operator-valued measure expression to the derivation of error probability for a squeezed input state. Direct detection is a measurement of photons in a transmitted state, so that the positive operator-valued measure for the direct detection is given by

$$E_{\mathrm{DD}}(n) = |n\rangle\langle n| \tag{18.6}$$

where $|n\rangle$ is the n-th number photon vector state in $\mathcal{H}_2$. Especially in digital modulation direct detection we usually encode the information by OOK. Namely, ϱ_0 is a vacuum state and ϱ_1 is another state in the input space $\mathcal{H}_1$. Therefore we have only to consider the case that the signal "1" is recognized as the signal "0" by a certain mistake because the noise state in $\mathcal{K}_1$ is assumed to be vacuum. That is, we consider only Z-type channel. Therefore, when the input state ϱ_1 is transmitted to an output state $\varrho_1 \circ \alpha$, the error probability q_e that the state $\varrho_1 \circ \alpha$ is recognized as a vacuum state by mistake is given by :

$$q_e = \mathrm{Tr}_{\mathcal{H}_2} \varrho_1 \alpha(E_{\mathrm{DD}}(0))$$
$$= \mathrm{Tr}_{\mathcal{H}_2} \mathrm{Tr}_{\mathcal{K}_2} V \varrho_1 V^* E_{\mathrm{DD}}(0). \tag{18.7}$$

In the case of PPM, since each symbol pulse is used for each quantum code, the error probability P_e^{PPM} becomes

$$P_e^{\mathrm{PPM}} = q_e.$$

A squeezed state can be expressed by a unitary operator $U(z)$ $(z \in \mathbb{C})$ such that

$$\varrho_1 = U(z)|\theta\rangle\langle\theta|U(z)^*$$

where $|\theta\rangle$ is a certain coherent vector. Then, from (18.6) and (18.7), the error probability $P_{e(SQ)}^{\mathrm{PPM}}$ is given by

$$q_e = \mathrm{Tr}_{\mathcal{H}_2}\left(\mathrm{Tr}_{\mathcal{K}_2} V U(z)|\theta\rangle\langle\theta|U(z)^* V^*\right)|0\rangle\langle 0|$$
$$= \mathrm{Tr}_{\mathcal{H}_1} U(z)|\theta\rangle\langle\theta|U(z)^*\left(V^*(|0\rangle\langle 0| \otimes I)V\right)$$
$$= \langle U(z)\theta, V^*(|0\rangle\langle 0| \otimes I)V U(z)\theta\rangle$$
$$= \frac{1}{\pi^2} \iint \langle U(z)\theta, w\rangle\langle cw, 0\rangle\langle dw, dv\rangle\langle 0, cv\rangle\langle v, U(z)\theta\rangle \, d^2v \, d^2w.$$

This can be computed by the following Gaussian type integration:

$$\frac{1}{\pi} \int \exp(-|w|^2 + aw + b\overline{w} + cw^2 + d\overline{w}^2) \, d^2w$$
$$= \frac{1}{\sqrt{1 - 4cd}} \exp\left(\frac{a^2 d + ab + b^2 c}{1 - 4cd}\right).$$

The result is

$$P_{e(SQ)}^{\mathrm{PPM}} = \sqrt{\tau} \exp\left(((1-\eta)\tau - 1)|\theta|^2 + \left(1 - (1-\eta)^2\tau\right)\left(\frac{\overline{\nu}\theta^2}{2\mu} + \frac{\nu\overline{\theta}^2}{2\overline{\mu}}\right)\right),$$

where $\tau = (|\mu|^2 - (1-\eta)^2|\nu|^2)^{-1}$, μ and ν are complex numbers which satisfy $|\mu|^2 - |\nu|^2 = 1$. On the other hand, homodyne detection is a measurement of the real part of the complex amplitude of a transmitted state. Therefore the positive operator-valued measure E_{HO} for homodyne detection is given by

$$E_{HO}(\Delta^{HO}) = \int_{\Delta^{HO}} |\theta_x\rangle\langle\theta_x| \, d\theta_x \,,$$

where $|\theta_x\rangle$ is the eigenvector of the operator $a_x = (a + a^*)/2$, a is the annihilation operator of photon, and Δ^{HO} is the set of real variables θ_x. So the infinitesimal nonnegative definite hermitian operator $dE_{HO}(\theta_x)$ is given by

$$dE_{HO}(\theta_x) = |\theta_x\rangle\langle\theta_x| \, d\theta_x \,.$$

The probability density function $p^{HO}(\theta_x)$ of the outcomes is

$$
\begin{aligned}
p^{HO}(\theta_x)d\theta_x &= \mathrm{Tr}_{\mathcal{H}_2}\alpha_* \varrho dE_{HO}(\theta_x) \\
&= \mathrm{Tr}_{\mathcal{H}_2}\alpha_* \varrho |\theta_x\rangle\langle\theta_x| d\theta_x,
\end{aligned}
$$

so that the probability density function $p^{HO}(\theta_x)$ is

$$p^{HO}(\theta_x) = \mathrm{Tr}_{\mathcal{H}_2}\alpha_* \varrho |\theta_x\rangle\langle\theta_x| \,. \tag{18.8}$$

Here we derive the probability density function $p_{SQ}^{HO}(\theta_x)$ for a squeezed input state.

$$
\begin{aligned}
p_{SQ}^{HO}(\theta_x) &= \mathrm{Tr}_{\mathcal{H}_2}(\mathrm{Tr}_{\mathcal{K}_2}(VU(z)|\theta\rangle\langle\theta|U(z)^*V^*)|\theta_x\rangle\langle\theta_x| \\
&= \mathrm{Tr}_{\mathcal{H}_1}U(z)|\theta\rangle\langle\theta|U(z)^* \, (V^*(|\theta_x\rangle\langle\theta_x| \otimes I)V) \\
&= \frac{1}{\pi^2}\int\int \langle U(z)\theta, w\rangle \langle cw, \theta_x\rangle \langle dw, dv\rangle \langle\theta_x, cv\rangle \langle v, U(z)\theta\rangle \, d^2v \, d^2w \\
&= \frac{1}{\sqrt{\frac{\pi}{2}\eta|\mu - \nu|^2 + \frac{\pi}{2}(1 - \eta)}} \\
&\quad \times \exp\left(-\frac{(\theta_x - c\,\mathrm{Re}\,((\overline{\mu} - \overline{\nu})\theta))^2}{\frac{1}{2}\eta|\mu - \nu|^2 + \frac{1}{2}(1 - \eta)}\right).
\end{aligned}
$$

This probability density function $p_{SQ}^{HO}(\theta_x)$ is a Gaussian type. Then m_{SQ} and σ_{SQ}^2, the average and the variance of this distribution (18.8), are calculated as

$$m_{SQ} = c\,\mathrm{Re}\,((\overline{\mu} - \overline{\nu})\theta), \quad \sigma_{SQ}^2 = \tfrac{1}{4}\eta|\mu - \nu|^2 + \tfrac{1}{4}(1 - \eta).$$

In BPSK, ϱ_0 is a state with phase 0 and ϱ_1 is a state with phase π. Therefore the probability density functions $p_{0(SQ)}^{HO}(\theta_x)$ and $p_{1(SQ)}^{HO}(\theta_x)$ for the signal "0" and "1" are respectively obtained from (18.8)

$$
\begin{aligned}
p_{0(SQ)}^{HO}(\theta_x) &= \frac{1}{\sqrt{2\pi(\frac{1}{4}\eta|\mu - \nu|^2 + \frac{1}{4}(1 - \eta))}} \\
&\quad \times \exp\left(-\frac{(\theta_x - c\,\mathrm{Re}\,(\overline{\mu} - \overline{\nu})|\theta|)^2}{2\{\frac{1}{4}\eta|\mu - \nu|^2 + \frac{1}{4}(1 - \eta)\}}\right) \\
p_{1(SQ)}^{HO}(\theta_x) &= \frac{1}{\sqrt{2\pi(\frac{1}{4}\eta|\mu - \nu|^2 + \frac{1}{4}(1 - \eta))}} \\
&\quad \times \exp\left(-\frac{(\theta_x + c\,\mathrm{Re}\,(\overline{\mu} - \overline{\nu})|\theta|)^2}{2(\frac{1}{4}\eta|\mu - \nu|^2 + \frac{1}{4}(1 - \eta))}\right).
\end{aligned}
$$

The error probability of each signal for BPSK is equal, that is, so

$$P_{e0(CO)}^{BPSK-HO} = P_{e1(SQ)}^{BPSK-HO} = \int_0^\infty p_{1(SQ)}^{HO}(\theta_x)\, d\theta_x \,.$$

Then the error probability $P_{e(SQ)}^{BPSK-HO}$ is given by

$$P_{e(SQ)}^{BPSK-HO} = \frac{1}{2}\mathrm{erfc}\left(\frac{\sqrt{2\eta}|\theta|\,\mathrm{Re}\,(\bar{\mu}-\bar{\nu})}{\sqrt{\eta|\mu-\nu|^2+(1-\eta)}}\right)\,.$$

In optical communication, an electric current or an electric wave is generally used as a carrier wave (a medium of an information transmission). Optical communication using a photon (laser beam) as a carrier wave has been widely used recently. In order to send information by optical device, we have to modulate the signal by a proper way. We now discuss the efficiency of such optical modulations with the quantum states by using the mutual entropy.

Let $\mathcal{H}_1$, $\mathcal{K}_1$, $\mathcal{H}_2$ and $\mathcal{K}_2$ be the Hilbert spaces of input, noise, output and loss systems with orthonormal basises $\{x_{n_1}^{(1)}\}$, $\{y_{m_1}^{(1)}\}$, $\{x_{n_2}^{(2)}\}$ and $\{y_{m_2}^{(2)}\}$, respectively. We here discuss about the efficiency of optical pulse modulations (PAM, PPM, PWM, PCM) for the photon number states. Optical pulse modulations are methods to carry the information on the carrier wave (optical pulse) and we usually use the following methods.

There are four main ways, called pulse modulation, to code the symbols of the alphabet $\mathcal{C}$. We briefly explain them for completeness of this book. A pulse is an optical signal, represented by a non vacuum state of the electro-magnetic field; its energy is here called the height of the pulse. To a single symbol of the alphabet $\mathcal{C}$ one associates one or more pulses. Time is discretized and each time interval between t_k and t_{k+1} has length τ. Each time interval corresponds to a single symbol of the alphabet. s

(1) PAM (Pulse Amplitude Modulation when Standard Photon Number is d): To the kth symbol a_k of the input sequence, one associates the optical pulse having $k*d$-photon numbers.

(2) PPM (Pulse Position Modulation): In this case there is only one non vacuum pulse in each time interval of length τ. The ordered set of these pulses is denoted x_k. The code x_k expressing a signal a_k is determined by the position of the non vacuum pulse, so that we need M slots (sites) in each time interval in order to express M signals. For instance, for the alphabet $\{a_1, a_3\}$, for $N = 5$ and choosing the elementary pulses to be the vacuum (i.e., no pulse) denoted 0, and another fixed pulse, e.g., a coherent state, denoted 1, the code x_k corresponding to a_k is determined by $x_1 = (1, 0, 0, 0, 0), x_3 = (0, 0, 1, 0, 0)$.

(3) PWM (Pulse Width Modulation): In this case there is some non vacuum pulse in each time interval of length τ. The code x_k expressing a signal a_k is determined by the width (interval) of the non vacuum pulse, so that we need M slots (sites) in each time interval in order

to express M signals. For instance, in the same notations as above, $x_1 = (1, 0, 0, 0, 0), x_3 = (1, 1, 1, 0, 0)$.

(4) PCM (Pulse Code Modulation): To the k-th symbol a_k of the input sequence, one associates N pulses starting at a time t_k. For instance in the same notations as above, $x_1 = (1, 0, 0), x_3 = (1, 1, 0)$ and so on. For this modulation, we need N slots (sites) in one time interval (e.g., between t_k and t_{k+1}) to fully represent all M signals; $2^{N-1} < M \leq 2^N$.

The optical signal modulated by an input signal corresponds to an input state ϱ for the quantum communication process discussed before, and the optical channel is a kind of quantum mechanical channel α. Moreover, each input state ϱ_M is given by each optical modulation M. The mutual entropy $I(\varrho_M; \alpha)$ with the input state ϱ_M and the channel α represents the amount of information transmitted through the channel α to the output system using this optical modulation M. When the channel is fixed, it is natural for us to consider the efficiency of the modulation by the entropy ratio

$$r(\varrho_M; \alpha) = \frac{I(\varrho_M; \alpha)}{S(\varrho_M)}.$$

In the sequel, we write the input state corresponding to each optical modulation as follows: ϱ_{PAM} for the pulse amplitude modulation, ϱ_{PPM} for the pulse position modulation, ϱ_{PWM} for the pulse width modulation and ϱ_{PCM} for the pulse code modulation. Each input signal is represented as follows:

(1) PAM (with a standard photon number d)

$$\text{Input signal } 1 \rightarrow E_{1d} = |x_{1d}^{(1)}\rangle\langle x_{1d}^{(1)}|$$
$$\text{Input signal } 2 \rightarrow E_{2d} = |x_{2d}^{(1)}\rangle\langle x_{2d}^{(1)}|$$

$$\vdots \qquad\qquad \vdots$$

$$\text{Input signal } N \rightarrow E_{Nd} = |x_{Nd}^{(1)}\rangle\langle x_{Nd}^{(1)}|$$

$$\varrho_{PAM}^{(d)} = \sum_{n=1}^{N} \lambda_n E_{nd}, \qquad \left(\sum_{n=1}^{N} \lambda_n = 1, \ \lambda_n \neq \lambda_m, \ n \neq m\right).$$

$$S(\varrho_{PAM}^{(d)}) = -\sum_{n=1}^{N} \lambda_n \log \lambda_n.$$

$$I(\varrho_{PAM}^{(d)}; \alpha) = -\sum_{j=0}^{Nd} \sum_{n=J}^{N} \lambda_n |C_j^{nd}|^2 \log \frac{|C_j^{nd}|^2}{\sum_{k=J}^{N} \lambda_k |C_j^{kd}|^2},$$

where $J = \max\{1, j\}$.

(2) PPM (with a photon number p)

$$\text{Input signal } 1 \rightarrow \bar{E}_1 = E_p \otimes E_0 \otimes \ldots \otimes E_0$$
$$\text{Input signal } 2 \rightarrow \bar{E}_2 = E_0 \otimes E_p \otimes \ldots \otimes E_0$$

$$\vdots \qquad\qquad \vdots$$

Input signal $N \rightarrow \bar{E}_N = E_0 \otimes E_0 \otimes \ldots \otimes E_p$

$$\varrho_{\text{PPM}}^{(p)} = \sum_{n=1}^{N} \lambda_n \bar{E}_n, \qquad \left(\sum_{n=1}^{N} \lambda_n = 1, \; \lambda_n \neq \lambda_m, \; n \neq m \right).$$

$$S(\varrho_{\text{PPM}}^{(p)}) = -\sum_{n=1}^{N} \lambda_n \log \lambda_n.$$

$$I(\varrho_{\text{PPM}}^{(p)}; \; \alpha) = \left(1 - (1-\eta)^p\right) \left(-\sum_{n=1}^{N} \lambda_n \log \lambda_n \right).$$

(3) PWM (with a photon number q)

Input signal $1 \rightarrow \hat{E}_1 = E_q \otimes E_0 \otimes \ldots \otimes E_0$

Input signal $2 \rightarrow \hat{E}_2 = E_q \otimes E_q \otimes \ldots \otimes E_0$

$$\vdots \qquad\qquad \vdots$$

Input signal $N \rightarrow \hat{E}_N = E_q \otimes E_q \otimes \ldots \otimes E_q$

$$S(\varrho_{\text{PWM}}^{(q)}) = -\sum_{n=1}^{N} \lambda_n \log \lambda_n, \qquad \left(\sum_{n=1}^{N} \lambda_n = 1, \; \lambda_n \neq \lambda_m, \; n \neq m \right).$$

$$S(\varrho_{\text{PWM}}^{(q)}) = -\sum_{n=1}^{N} \lambda_n \log \lambda_n.$$

$$I(\varrho_{\text{PWM}}^{(q)}; \; \alpha) = -\sum_{n=1}^{N} \sum_{i=1}^{n} \lambda_n (1-\eta)^{(n-i)d} \left(\eta \sum_{k=0}^{d-1} (1-\eta)^k \right)^{L_i}$$

$$\log \frac{\sum_{j=i}^{N} \lambda_j (1-\eta)^{(j-i)d}}{(1-\eta)^{(n-i)d}},$$

where $L_i = \min\{1, i - 1\}$.

(4) PCM (of binary code with a photon number r)

	Binary code	Binary coded pulse sequence

Input signal $1 \rightarrow b(1) = (0, 0, \ldots, 0, 1) \rightarrow \tilde{E}_1 = E_0 \otimes \ldots \otimes E_0 \otimes E_r$

Input signal $2 \rightarrow b(2) = (0, 0, \ldots, 1, 0) \rightarrow \tilde{E}_2 = E_0 \otimes \ldots \otimes E_r \otimes E_0$

$$\vdots \qquad\qquad \vdots$$

Input signal $N \rightarrow b(N) = (1, 0, \ldots, 0, 1) \rightarrow \tilde{E}_N = E_r \otimes \ldots \otimes E_0 \otimes E_r$

$$\varrho_{\text{PCM}}^{(r)} = \sum_{n=1}^{N} \lambda_n \tilde{E}_n, \qquad \left(\sum_{n=1}^{N} \lambda_n = 1, \; \lambda_n \neq \lambda_m, \; n \neq m \right).$$

$$S(\varrho_{PCM}^{(r)}) = -\sum_{n=1}^{N} \lambda_n \log \lambda_n.$$

For given N, we put

$$\Omega = \{1, 2, \ldots, N\}, \quad X = \{0, 1\}, \quad M = \min\{n; \; 2^n \geq N\},$$

and we define a binary transformation b from Ω to X^M such that

$$b(n) = (n_M, \ldots, n_1), \quad n = \sum_{i=1}^{M} n_i 2^{i-1}.$$

Moreover we put

$$W_n^0 = \{i; n_i \neq 0\}, \quad e_m = \min W_n^{m-1} \quad (1 \leq m \leq \|b(n)\|_1),$$

$$W_n^m = W_n^{m-1} \setminus \{e_m\}, \quad n^{(m)} = \sum_{i=1}^{m} n_{e_i} 2^{e_i - 1}.$$

Then the mutual entropy $I(\varrho_{PCM}^{(r)}; \alpha)$ is represented by

$$I(\varrho_{PCM}^{(r)}; \alpha) = \sum_{n=1}^{N} \sum_{m=1}^{\|b(n)\|_1} \lambda_n (1-\eta)^{(\|b(n)\|_1 - m)r} \left(\eta \sum_{k=0}^{r-1} (1-\eta)^k \right)^{L_m}$$

$$\times \log \frac{\sum_{j \in R} \lambda_j (1-\eta)^{(\|b(j)\|_1 - \|b(n^{(m)})\|_1)r}}{(1-\eta)^{(\|b(n)\|_1 - m)r}}$$

where $R = \{j \in \Omega; \; \langle b(n^{(m)}), b(j) \rangle = \|b(n^{(m)})\|_1\}$.

For the above modulations, we obtain the following results:

Theorem 18.1 (1) Comparison of PPM (of a photon number p) with PAM (of a standard photon number d): If $p \geq Nd$, then

$$r(\varrho_{PPM}^{(p)}; \alpha) \geq r(\varrho_{PAM}^{(d)}; \alpha).$$

(2) Comparison of PPM (of a photon number p) with PWM (of a photon number q): If $p \geq (N-1)q$, then

$$r(\varrho_{PPM}^{(p)}; \alpha) \geq r(\varrho_{PWM}^{(q)}; \alpha).$$

(3) Comparison of PPM (of a photon number p_1) with PPM (of a photon number p_2): If $p_1 \geq p_2$, then

$$r(\varrho_{PPM}^{(p_1)}; \alpha) \geq r(\varrho_{PPM}^{(p_2)}; \alpha).$$

(4) Comparison of PPM (of a transmition ratio η_1) with PAM (of a transmition ratio η_2): If $\eta_1 \geq \eta_2$, then

$$r(\varrho_{PPM}^{(p; \eta_1)}; \alpha) \geq r(\varrho_{PPM}^{(p; \eta_2)}; \alpha).$$

(5) Comparison of PPM (of a photon number p) with PCM (of binary code with a photon number r): If $N = 3$ and $p = r = 1$, then

$$r(\varrho_{PPM}^{(1)}; \alpha) \geq r(\varrho_{PCM}^{(1)}; \alpha).$$

Proof. Since we have

$$S(\varrho_{\mathrm{PPM}}^{(p)};\ \alpha) = S(\varrho_{\mathrm{PAM}}^{(d)};\ \alpha),$$

in order to conclude (1) we have only to prove

$$I(\varrho_{\mathrm{PPM}}^{(p)};\ \alpha) - I(\varrho_{\mathrm{PAM}}^{(d)};\ \alpha) \geq 0. \tag{18.9}$$

The left hand side of (18.9) equals to the following expression:

$$-\sum_{n=1}^{N} \lambda_n |C_0^{nd}|^2 \log \frac{|C_0^{nd}|^2 \lambda_n \left(\sum_{j=1}^{p-nd} |C_j^{p-nd}|^2\right)}{\sum_{k=1}^{N} \lambda_k |C_0^{kd}|^2}$$

$$-\sum_{j=1}^{Nd} \sum_{n=J}^{N} \lambda_n |C_j^{nd}|^2 \log \frac{\lambda_n |C_j^{nd}|^2}{\sum_{k=J}^{N} \lambda_k |C_j^{kd}|^2}.$$

By using the inequality

$$\lambda^t \leq 1 - (1-\lambda)t \qquad \text{(for any } \lambda \text{ and } t \text{ satisfying } 0 \leq \lambda, t \leq 1),$$

we obtain

$$I(\varrho_{\mathrm{PPM}}^{(p)};\ \alpha) - I(\varrho_{\mathrm{PAM}}^{(d)};\ \alpha)$$

$$\geq -\sum_{j=0}^{Nd} \sum_{n=J}^{N} \lambda_n |C_j^{nd}|^2 \log \frac{|C_0^{nd}|^2 (1 - (1-\lambda_n) \sum_{j=1}^{p-nd} |C_j^{p-nd}|^2)}{\sum_{k=1}^{N} \lambda_k |C_0^{kd}|^2}$$

$$-\sum_{j=1}^{Nd} \sum_{n=J}^{N} \lambda_n |C_j^{nd}|^2 \log \frac{\lambda_n |C_j^{nd}|^2}{\sum_{k=J}^{N} \lambda_k |C_j^{kd}|^2}$$

$$= -\sum_{j=0}^{Nd} \sum_{n=J}^{N} \lambda_n |C_j^{nd}|^2 \log \frac{\lambda_n |C_0^{nd}|^2 + \sum_{m\neq n}^{N} \lambda_m |C_0^{p}|^2}{\lambda_n |C_0^{nd}|^2 + \sum_{k\neq n}^{N} \lambda_k |C_0^{kd}|^2}$$

$$-\sum_{j=1}^{Nd} \sum_{n=J}^{N} \lambda_n |C_j^{nd}|^2 \log \frac{\lambda_n |C_j^{nd}|^2}{\sum_{k=J}^{N} \lambda_k |C_j^{kd}|^2}$$

$$\geq 0.$$

In this way we have arrived at (18.9) and the proof of (1) is complete. The other statements can be proved in the same way.   □

Notes and Remarks. The "classical" source on quantum detection and estimation theory is the monograph [Helstrom 1976]. Concerning coherent states we refer to [Klauder and Sudarshan 1968] but [Louisell 1973] also contains a review. The chapter is based on the papers [Ohya 1991a] and [Watanabe 1991b].

Bibliography

Accardi, L. (1975): "A noncommutative Markov property" (in Russian), Funkcional. Anal. i Prilozen. **9**, pp. 1–8

Accardi, L. (1981): "Topics in quantum probability", Phys. Rep. **71**, pp. 169–192

Accardi, L. (1984): "Some trends and problems in quantum probability", in *Quantum Probability and Applications to the Quantum Theory of Irreversible Processes*, ed. L. Accardi, A. Frigerio and V. Gorini (Springer-Verlag, Berlin, Heidelberg, New York) pp. 1–19

Accardi, L. and Bach, A. (1989): "Central limits of squeezing operators", in *Quantum Probability and Applications IV*, ed. L. Accardi, W. von Waldenfels (Lecture Notes in Math. **1396**, Springer-Verlag, Berlin, Heidelberg, New York) pp. 7–19

Accardi, L. and Cecchini, C. (1982): "Conditional expectations in von Neumann algebras and a theorem of Takesaki", J. Functional. Anal. **45**, pp. 245–273

Accardi, L. and Frigerio, A. (1983): "Markovian cocycles", Proc. R. Ir. Acad. **83A**, pp. 251–263

Accardi, L., Frigerio, A. and Lewis, J. (1982): "Quantum stochastic processes", Publ. RIMS Kyoto Univ. **18**, pp. 97–133

* Accardi, L. and Liebscher, V. (1999): "Markovian KMS-states for one-dimensional spin chains", Infin. Dimens. Anal. Quantum Probab. Relat. Top. **2**, pp. 645–661

* Accardi, L. and Ohya, M. (1999): "Compound channels, transition expectations and liftings", Appl. Math. Optim. **39**, pp. 33–59

Accardi, L. and Pistone, G. (1982): "De Finetti's theorem, sufficiency, and Dobrushin's theory", in *Exchangeability in probability and statistics*, ed. G. Koch and F. Spizzichino (North-Holland, Amsterdam) pp. 125–156

Aczél, J. and Daróczy, Z. (1975): *On measures of information and their characterizations* (Academic Press, New York)

Akashi, S. (1990): "The asymptotic behaviour of ε-entropy of a compact positive operator", J. Math. Anal. Appl. **153**, pp. 250–257

* Akashi, S. (1992): "A relation between Kolmogorov-Prokhorov's condition and Ohya's fractal dimensions", IEEE Transaction on Inf. Theory, **38**, pp. 1567–1570

Alberti, P.M. and Uhlmann, A. (1981): *Stochasticity and partial order. Doubly stochastic maps and unitary mixing* (VEB Deutscher Verlag Wiss., Berlin)

* Alicki, R. and Fannes, M. (1994): "Defining quantum dynamical entropy", Lett. Math. Phys. **32**, pp. 75–82

Amann, A. (1990): "Perturbation theory of boson dynamical systems", J. Phys. A **23**, pp. L783–L788

Amann, A. (1991): "Perturbation theory of C*-systems without norm continuity properties", J. Math. Phys. **32**, pp. 2875–2879

Amari, S. (1985): *Differential-geometrical methods in statistics*, Lecture Notes in Stat. **28** (Springer, Berlin, Heidelberg, New York)

Ando, T. (1982): *Majorization, double stochastic matrices and comparison of eigenvalues* Lecture Notes (Hokkaido University, Sapporo)

Araki, H. (1969): "Gibbs states of a one dimensional quantum lattice", Commun. Math. Phys. **14**, pp. 120–157

Araki. H. (1973a): "Relative Hamiltonian for faithful normal states a von Neumann algebra", Publ. RIMS Kyoto Univ. **9**, pp. 165–209

Araki, H. (1973b): "Golden-Thompson and Peierls-Bogoliubov inequalities for a general von Neumann algebra", Commun. Math. Phys. **34**, pp. 167–178

Araki, H. (1974): "On the equivalence of the KMS condition and the variational principle for quantum lattice systems", Commun. Math. Phys. **38**, pp. 1–10

Araki, H. (1975): "On uniqueness of KMS states of one-dimensional quantum lattice systems", Commun. Math. Phys. **44**, pp. 1–7

Araki, H. (1976): "Relative entropy for states of von Neumann algebras", Publ. RIMS Kyoto Univ. **11**, pp. 809–833

Araki, H. (1977): "Relative entropy of states of von Neumann algebras II", Publ. RIMS Kyoto Univ. **13**, pp. 173–192

Araki, H. (1978): "On KMS states of a C*-dynamical system", in *C*-algebras and Applications to Physics* ed. by H. Araki and R.V. Kadison (Lecture Notes in Math. **650**, Springer), pp. 66–84

Araki, H. (1987a): "Recent progress on entropy and relative entropy", in *VIIIth Int. Cong. on Math. Phys.* (World Sci. Publishing, Singapore) pp. 354–365

Araki, H. (1987b): "On subadditivity of relative entropy", in *Wandering in the Fields*, ed. by K. Kawarabayashi and P. Ukawa (World Scientific, Singapore) pp. 215–226

Araki, H. (1990): "On an inequality of Lieb and Thirring", Lett. Math. Phys. **19**, pp. 167–170

Araki, H. and Lieb, E.H. (1970): "Entropy inequalities", Commun. Math. Phys. **18**, pp. 160–170

Araki, H. and Masuda, T. (1982): "Positive cones and L_p-spaces for von Neumann algebras", Publ. RIMS Kyoto Univ. **18**, pp. 339–411

Araki, H. and Sewell, G.L. (1977): "KMS conditions and local thermodynamical stability of quantum lattice systems", Commun. Math. Phys. **52**, pp. 103–109

* Barnum, H., Caves, C.M., Fuchs, C.A., Jozsa, R. and Schumacher, B.W. (2001): "On quantum coding for ensembles of mixed states", J. Phys. A **34**, pp. 6767–6785

Barron, A.R. (1986): "Entropy and the central limit theorem", Ann. Prob. **14**, pp. 336–342

Baxter, R.J. (1982): *Exactly solved models in statistical mechanics* (Academic Press, New York)

Beckner, W. (1975): "Inequalities in Fourier analysis", Ann. Math. **102**, pp. 159–182

* Belavkin, V.P. and Ohya, M. (2001): "Quantum entropy and information in discrete entangled states", Infin. Dimens. Anal. Quantum Probab. Relat. Top. **4**, pp. 137–160

* Belavkin, V.P. and Ohya, M. (2002): "Entanglement, quantum entropy and mutual information", R. Soc. Lond. Proc. Ser. A Math. Phys. Eng. Sci. **458**, pp. 209–231

Belavkin, V.P. and Staszewski, P. (1982): "C*-algebraic generalization of relative entropy and entropy", Ann. Inst. Henri Poincaré, Sec. A **37**, pp. 51–58.

Beltrametti, E. and Cassinelli, G. (1981): *The logic of quantum mechanics* (Addison-Wesley, Reading)

Benatti, F. and Narnhofer, H. (1988a): "Entropic behaviour under completely positive maps", Lett. Math. Phys. **15**, pp. 325–334

Benatti, F. and Narnhofer, H. (1988b): "Entropic dimension for completely positive maps", J. Stat. Phys. **53**, pp. 1273–1298

Benatti, F. and Narnhofer, H. (1991): "Strong asymptotic abelienes for entropic K-systems", Commun. Math. Phys. **136**, pp. 231–250

Benatti, F., Narnhofer, H. and Sewell, G.L. (1991): "A noncommutative version of the Arnold cat map", Lett. Math. Phys. **21**, pp. 325–334

Benjaballah, C. and Charbit, M. (1989): "Quantum communication with coherent states", IEEE Trans. Inform. Theory **35**, pp. 1114–1123

* Bennett, C.H., Shor, P.W., Smolin, J.A. and Thapliyal, A.V. (2002): "Entanglement-assisted capacity of a quantum channel and the reverse Shannon theorem", IEEE Trans. Inform. Theory **48**, pp. 2637–2655

Benoist, R.W., Merchand, J.P. and Wyss, W. (1979): "A note on the relative entropy", Lett. Math. Phys. **3**, pp. 169–173

Berezin, F.A. (1972): "Covariant and contravariant symbols of operators", Izv. Akad. SSSR Ser. Mat. **6**, pp. 1134–1167

Bernstein, D.S. (1988): "Inequalities for trace of matrix exponentials", SIAM J. Matrix Anal. Appl. **9**, pp. 156–158

Bersohn, M. (1966): "Uncertainty principle and normal operators", Am. J. Phys. **34**, pp. 62–63

Besson, O. (1981): "On entropy in II_1 von Neumann algebras", Ergodic Theory and Dyn. Syst. **1**, pp. 419–429

Besson, O. (1985): "The entropy of the quantum Markov states", in *Quantum Probability and Applications II* ed. by L. Accardi and W. von Waldenfels (Lecture Notes in Math. **1136**, Springer), pp. 81–89

Bialynicki-Birula, I. (1985): "Entropic uncertainty relations in quantum mechanics", in *Quantum Probability and Applications II* ed. by L. Accardi and W. von Waldenfels (Lecture Notes in Math. **1136**, Springer), pp. 90–103

Bialyniczki-Birula, I. and Mycielski, I. (1975): "Uncertainty relations for information entropy in wave mechanics", Commun. Math. Phys. **44**, pp. 129–132

Blahut, R.E. (1987): *Principles and practice of information theory* (Addison–Wesley, Reading)

Boltzmann, L. (1877): "Beziehung zwischen dem zweiten Hauptsatze der mechanischen Wärmetheorie und der Wahrscheinlichkeitsrechnung respektive den Sätzen über das Wärmegleichgewicht", Wien. Ber. **76**, pp. 373–435

Bóna, P. (1989): "Equlibrium states of a class of mean field theories", J. Math. Phys. **30**, pp. 2994–3007

Bratteli, O. and Robinson, D.W. (1979): *Operator algebras and quantum statistical mechanics I* (Springer, New York, Berlin, Heidelberg)

Bratteli, O. and Robinson, D.W. (1981): *Operator algebras and quantum statistical mechanics II*, (Springer, New York, Berlin, Heidelberg)

Breitenbecker, M. and Grümm, H.R. (1972): "Note on trace inequalities", Commun. Math. Phys. **26**, pp. 276–279

Breitenberger, E. (1985): "Uncertainty measures and uncertainty relations for angle observables", Found. Phys. **15**, pp. 353–364

Busch, P. (1986): "Unsharp reality and joint measurements for spin observables", Phys. Rev. D **33**, pp. 2253–2261

Busch, P., Grabowski, M. and Lahti, P.J. (1989): "Some remarks on effects, operations, and unsharp measurements", Found. Phys. Lett. **2**, pp. 331–345

Campbell, L.L. (1985): "The relation between information theory and differential geometry approach to statistics", Inf. Sciences, **35**, pp. 199–210

Carlen, E.A. (1991): "Some integral identities and inequalities for entire functions and theor application to the coherent state transform", J. Functional Anal. **97**, pp. 231–249

Carlen, E.A. and Soffer, A. (1991): "Entropy production by block variable summation and central limit theorems", Commun. Math. Phys. **140**, pp. 339–371

* Cecchini, C. (1992): "Markovianity for states on von Neumann algebras", *Quantum Probability and Related Topics VII*, (World Scientific, Singapore) pp. 93–107

Cecchini, C. and Petz, D. (1989a): "On the fixed point algebras for φ-conditional expectations in von Neumann algebras", Studia Sci. Math. Hungar. **24**, pp. 133–157

Cecchini, C. and Petz, D. (1989b): "State extensions and a Radon-Nikodym theorem for conditional expectations on von Neumann algebras", Pacific J. Math. **138**, pp. 9–24

Cencov, N.N. (1982): *Statistical decision rules and optimal inferences* (Amer. Math. Society, Providence)

Chayes, J.T., Chayes, L. and Ruskai, M.B. (1985): "Density functional approach to quantum lattice systems", J. Stat. Phys. **38**, pp. 497–518

Chen, E. (1977): "Entropy and the superposition principle", Rep. Math. Phys. **11**, pp. 189–195

* Choda, M. (1996): "Entropy for extensions of Bernoulli shifts" Ergodic Theory Dynam. Systems **16**, pp. 1197–1206

* Choda, M. (1999): "Entropy of Cuntz's canonical endomorphism" Pacific J. Math. **190**, pp. 235–245

* Choda, M. (2000): "A C*-dynamical entropy and applications to canonical endomorphisms", J. Funct. Anal. **173**, pp. 453–480

* Choda, M. (2003): "Dynamical entropy for automorphisms of exact C*-algebras", J. Funct. Anal. **198**, pp. 481–498

Choda, M. and Hiai, F. (1991): "Entropy for canonical shifts. II", Publ. RIMS Kyoto Uni. **27**, pp. 461–489

Choi, M.D. (1974): "A Schwarz inequality for positive linear maps on C*-algebras", Illinois J. Math. **18**, pp. 565–574

Choi, M.D. and Effros, E. (1978): "Nuclear C*-algebras and the approximation property", Amer. J. Math. **100**, pp. 61–79

Cohen, J.E., Friedland, S., Kato, T. and Kelly, F.P. (1982): "Eigenvalue inequalities for products of matrix exponentials", Linear Alg. Appl. **45**, pp. 55–95

Connes, A. (1973): "Sur la théorème de Radon-Nikodym pour les poids normaux fidèles semifinis", Bull. Sci. Math. Sec. II. **97**, pp. 253–258

Connes, A. (1980): "On a spatial theory of von Neumann algebras", J. Functional Anal. **35**, pp. 153–164

Connes, A. (1985): "Entropie de Kolmogoroff-Sinai et mècanique statistique quantique", C.R. Acad. Sci. Paris Ser. I. Math. **301**, pp. 1–6

Connes, A., Narnhofer, H. and Thirring, W. (1987a): "Dynamical entropy of C*-algebras and von Neumann algebras", Commun. Math. Phys. **112**, pp. 691–719

Connes, A., Narnhofer, H. and Thirring, W. (1987b): "The dynamical entropy of quantum systems", in *Recent developments in mathematical physics* ed. by H. Mitter and L. Pittner (Springer, New York, Berlin, Heidelberg) pp. 102–136

Connes, A. and Størmer, E. (1975): "Entropy for automorphisms of II₁ von Neumann algebras", Acta Math. **134**, pp. 289–306

Connes, A. and Størmer, E. (1978): "Homogeneity of the state space of factors of type III₁", J. Functional Anal. **28**, pp. 187-196

Connes, A. and Størmer, E. (1983): "A connection between the classical and the quantum mechanical entropies", in *Operator Algebras and Group Representations*, pp. 113–123

Csiszár, I. (1967): "Information type measure of difference of probability distributions and indirect observations", Studia Sci. Math. Hungar. **2**, pp. 299–318

Csiszár, I. (1975): "I-divergence geometry of probability distributions and minimization problems", Ann. Prob. **3**, pp. 146–158

Csiszár, I. and Körner, J. (1981): *Information theory. Coding theorems for discrete memoryless systems* (Akadémiai Kiadó, Budapest)

Cushen, C.D. and Hudson, R. (1971): "A quantum mechanical central limit theorem", J. Appl. Prob. **8**, pp. 454–469

Davies, E.B. (1976): *Quantum Theory of Open Systems* (Academic Press, London, New York)

Davies, E.B. and Lewis, J.T. (1970): "An operational approach to quantum probability", Commun. Math. Phys. **17**, pp. 239–260

Dembo, A., Cover, T.M. and Thomas, J.A. (19991): "Information theoretic inequalities", IEEE Trans. Inf. Theory, **37**, pp. 1501–1518

Deutsch, D. (1983): "Uncertainty in quantum measurements", Phys. Rev. Lett. **50**, pp. 631–633

* Dold, J.L. and Nielsen, M.A.: "A simple operational interpretation of fidelity", arXiv e-print quant-ph/0111053

Donald, M.J. (1985): "On the relative entropy", Commun. Math. Phys. **105**, pp. 13–34

Donald, M.J. (1987a): "Further results on the relative entropy", Math. Proc. Camb. Phil. Soc. **101**, pp. 363-373

Donald, M.J. (1987b): "Free energy and relative entropy", J. Stat. Phys. **49**, pp. 81-87

Donald, M.J. (1990): "Relative Hamiltonian which are not bounded from above", J. Functional Anal. **91**, pp. 143–173

Donald, M.J. (1991): "Continuity of relative hamiltonians", Commun. Math. Phys. **136**, pp. 625–632

* Donald, M.J. (1992): "A priori probability and localized observers", Found. Phys. **22**, pp. 1111–1172

* Donald, M.J., Horodecki, M. and Rudolph, O. (2002): "The uniqueness theorem for entanglement measures. Quantum information theory", J. Math. Phys. **43**, pp. 4252–4272

Donogue, Jr.W. (1974): *Monotone matrix functions and analytic continuations* (Springer-Verlag, Berlin, Heidelberg, New York)

Effros, E. (1978): "Aspects of non-commutative order", in *C*-algebras and Applications to Physics* ed. by H. Araki and R.V. Kadison (Lecture Notes in Math. **650**, Springer), pp. 1–40

Effros, E. and Lance, E.C. (1977): "Tensor products of operator algebras", Advances Math. **25**, pp. 1–34

Ekeland, I. and Temam, R. (1976): *Convex analysis and variational problems* (North-Holland, Amsterdam)

Ellis, R.S. (1985): *Entropy, large deviations, and statistical mechanics* (Springer, Berlin, Heidelberg, New York)

Emch, G.G. (1974): *Algebraic Methods in Statistical mechanics and Quantum Field Theory* (Wiley, New York)

Emch, G.G. (1974): "Positivity of the K-entropy on non-abelian K-flows", Z. Wahrscheinlichkeitstheory verw. Gebiete **29**, pp. 241–252

* Emch, G.G. (1992): "Kolmogorov flows, dynamical entropies and mechanics", *Quantum Probability and Related Topics VII* (World Scientific, Singapore) pp. 125–137

Emch, G.G. and Wolfe, F.C. (1974): "C*-algebraic formalism for coarse graining. I. General theory, J. Math. Phys. **105**, pp. 1343–1347

Epstein, H. (1973): "Remarks on two theorems of E. Lieb", Comm. Math. Phys. **31**, pp. 317–325

Evans, D.E. (1982): "Entropy of Automorphisms of AF algebras", Publ. RIMS Kyoto Univ. **18**, pp. 1045–1051

Evans, D.E. (1985): "The C*-algebras of the two-dimensional Ising model", in *Quantum Probability and Applications II* ed. by L. Accardi and W. von Waldenfels (Lecture Notes in Math. **1136**, Springer), pp. 162–176

Evans, D.E. (1989): "An algebraic approach to critical phenomena in two dimensional lattice models", in *Mathematical Methods in Statistical Mechanics* ed. by M. Fannes and A. Verbeure (Leuven University Press, Leuven), pp. 15–30

Evans, D.E., and Lewis, F.T. (1977): "Dilations of irreversible evolutions in algebraic quantum theory" (Dublin Institute for Advanced Studies)

Fack, T. (1982): "Sur la notion de valeur caractéristique", J. Operator Theory **7**, pp. 307–334.

Fack, T. and Kosaki, H. (1986): "Generalized s-numbers of τ-measurable operators", Pacific J. Math. **123**, pp. 269–300

Fannes, M. (1973a): "The entropy density of quasi free states", Commun. Math. Phys. **31**, pp. 279–290

Fannes, M. (1973b): "A continuity property of the entropy density for spin lattice systems", Commun. Math. Phys. **31**, pp. 291–294

Fannes, M., Kossakowski, A. and Verbeure, A. (1991): "Critical fluctuations for quantum mean field models", J. Stat. Phys. **65**, pp. 801–811

Fannes, M., Nachtergaele, B. and Werner, R.F. (1992a): "Finitely correlated states of quantum spin chains", Commun. Math. Phys. **144**, pp. 443–490

* Fannes, M., Nachtergaele, B. and Werner, R.F. (1992b): "Entropy estimates for finitely correlated states", Ann. Inst. H. Poincaré Phys. Théor. **57**, pp. 259–277

* Fannes, M. Nachtergaele, B. and Werner, R.F. (1992c): "Finitely correlated states on quantum spin chains", Comm. Math. Phys. **144**, pp. 443–490

Fannes, M. and Quaegebeur, J. (1983): "Product mappings between CAR algebras", Publ. RIMS Kyoto Univ. **19**, pp. 469–491

Fannes, M., Spohn, H. and Verbeure, A. (1980): "Equilibrium states of mean field models", J. Math. Phys. **21**, pp. 355–358

* Fannes, M. and Tuyls, P. (2003): "On quantum dynamical entropy", to appear in Periodica Math. Hungar.

Faris, W.G. (1975): *Selfadjoint operators*, Lecture Notes in Math. **433** (Springer, Berlin, Heidelberg, New York)

Feller, W. (1966): *An introduction to probability theory and its applications II* (John Wiley, New York, London, Sydney)

* Friedland, S. and So, W. (1994): "On the product of matrix exponentials", Linear Algebra Appl. **196**, pp. 193–205

Fröhlich, J. (1978): "The pure phases (harmonic functions) of generalized processes or: mathematical physics of phase transitions and symmetry breaking", Bull. Amer. Math. Soc. **84**, pp. 165–193

Fröhlich, J. and Pfister, C. (1981): " On the absence of spontaneous symmetry breaking and of cristalline ordering in two-dimensional systems", Commun. Math. Phys. **81**, pp. 277–298

Fujii, J.I. and Kamei, E. (1989): "Relative operator entropy in noncommutative information theory", Math. Japon. **34**, pp. 341–348

Georgii, H.O. (1979): *Canonical Gibbs measures*, Lecture Notes in Math. **760** (Springer, Berlin, Heidelberg, New York)

Gerisch, Th. and Rieckers, A. (1990): "The quantum statistical free energy minimum principle for multy-lattice mean field theories", Z. Naturforsch. **45a**, pp. 931–945

Giri, N. and Waldenfels, W. von (1978): "An algebraic version of the central limit theorem", Z. Wahrsheinlichkeitstheorie Verw. Gebiete **42**, pp. 129–134

Goderis, D., Verbeure, A. and Vets, P. (1989a): "Non-commutative central limits", Probability Theor. Rel. Fields **82**, pp. 527–544

Goderis, D., Verbeure, A. and Vets, P. (1989b): "Theory of quantum fluctuations and the Onsager relations", J. Stat. Phys. **56**, pp. 721–746

Goderis, D., Verbeure, A. and Vets, P. (1990): "Quantum central limit and coarse graining", in *Quantum probability and applications V* ed. by L. Accardi and W. von Waldenfels (Lecture Notes in Math. **1442**, Springer), pp. 178–193

Golodets, V.Ya. and Zholtkevich, G.N. (1983): "Markovian Kubo-Martin-Schwinger states" (in Russian), Teoret. Mat. Fiz. **56**, pp. 80–86

* Golodets, V.Ya. and Størmer, E. (1999): "Generators and comparison of entropies of automorphisms of finite von Neumann algebras", J. Funct. Anal. **164**, pp. 110–133

Golden, S. (1965): "Lower bounds for the Helmholtz function", Phys. Rev. **137**, pp. B1127-B1128

Goodman, F.M., de la Harpe, P. and Jones, V.F.R. (1989): *Coxeter graphs and towers of algebras* (Springer, Berlin, Heidelberg, New York)

Gudder, S.P. (1979): *Stochastic Methods in Quantum Mechanics* (North Holand, New York, Oxford)

Gudder, S., Merchand, J.P. and Wyss, W. (1979): "Bures distance and relative entropy", J. Math. Phys. **20**, pp. 1963–1966

Haagerup, U. (1975): "The standard form of von Neumann algebras", Math. Scand. **37**, pp. 271–283

Haagerup, U. (1979): "Operator valued weights in von Neumann algebras I", J. Functinal Anal. **32**, pp. 175–206

Haagerup, U. (1985): "A new proof of the equivalence of injectivity and hyper-finiteness for factors on a separable Hilbert space", J. Functinal Anal. **62**, pp. 160–201

* Haagerup, U. and Størmer, E. (1998): "Maximality of entropy in finite von Neumann algebras", Invent. Math. **132**, pp. 433–455

Hansen, F. and Pedersen, G.K. (1982): "Jensen's inequality for operator and Löwner's theorem", Math. Anal. **258**, pp. 229–241

Hegerfeld, G.C. (1985): "Noncommutative analogues of probabilistic notions and results", J. Functional Anal. **64**, pp. 436–456

* Hayden, P., Jozsa, R., Petz, D. and Winter, A. (2003): "Structure of states which satisfy strong subadditivity of quantum entropy with equality", to appear in Comm. Math. Phys.

Heisenberg, W. (1927): "Über den anschaulichen Inhalt der quantummechanischen Kinematik und Mechanik", Zeitschrift für Physik, **43**, pp. 172–198

Helstrom, C.W. (1976): *Quantum detection and estimation theory*, (Academic Press, New York)

Heyer, H. (1982): *Theory of statistical experiments* (Springer-Verlag, New York, Heidelberg, Berlin)

Hiai, F. (1984): "Martingale-type convergence of modular automorphism groups on von Neumann algebras", J. Functional Anal. **56**, pp. 265–278

Hiai, F. (1991a): "Minimizing indeces of conditional expectations onto a subfactor", Publ. RIMS Kyoto Univ. **24**, pp. 673–678

Hiai, F. (1991b): "Minimum index for subfactors and entropy II", J. Math. Soc. Japan **43**, pp. 347–379

* Hiai, F. (1993): "Some remarks on the trace operator logarithm and relative entropy", *Quantum Probability and Related Topics VIII* (World Scientific, Singapore) pp. 223–236

* Hiai, F., Ohya, M. and Petz, D. (1995): "McMillan type convergence for quantum Gibbs states", Arch. der Math. **65**, 154–158

Hiai, F., Ohya, M. and Tsukada, M. (1981): "Sufficiency, KMS condition a relative entropy in von Neumann algebras", Pacific J. Math. **96**, pp. 99–109

Hiai, F., Ohya, M. and Tsukada, M. (1983): "Sufficiency and relative entropy in *-algebras with applications to quantum systems", Pacific J. Math. **107**, pp. 117–140

Hiai, F. and Petz, D. (1991): "The proper formula for relative entropy and its asymptotics in quantum probability", Comm. Math. Phys. **143**, pp. 99–114

* Hiai, F. and Petz, D. (1993): "The Golden-Thompson trace inequality is complemented", Linear Alg. Appl. **181**, pp. 153–185

* Hiai, F. and Petz, D. (1994): "Entropy densities for algebraic states", J. Funct. Anal. **125**, pp. 287–308

Hiai, F. and Tsukada, M. (1984): "Strong martingale convergence of generalized conditional expectations on von Neumann algebras", Trans. Amer. Math. Soc. **282**, pp. 791–798

Hilgevoord, J. and Uffink, J. (1988): "The mathematical expression of the uncertainty principle", in *Proceedings of the International Conference on Microphysical Reality and Quantum Description*, ed. by F. Selleri, A. von der Meerwe and G. Tarozzi (Reidel, Dordrecht) pp. 91–114

Hirota, H., Ohya, M., Nakagawa, M. and Uyematsu, K. (1987): "Effect of error correcting code in photon communications with energy loss", Trans. IECE **E70**, pp. 689–692

Hirota, H., Ohya, M., Nakagawa, M. and Yamazaki, K. (1986): "Properties of error correcting code using photon pulse", Trans. IECE, **E69**, pp. 917–919

Hirota, H., Ohya, M. and Yoshimi, H. (1988): "Rigorous derivation of error probability in quantum control communication processes", IEICE of Japan, **J71-B**, pp. 533–539

Hirschman, I.I.Jr. (1957): "A note on entropy", Am. J. Math. **79**, pp. 152–156

Holevo, A.S. (1972): "An analogue of the theory of statistical decisions in noncommutative probability theory", Trans. Moscow Math. Soc. **26**, pp. 133–149

Holevo, A.S. (1973a): "Statistical problems in quantum physics" in *Proc. Second Japan-USSR Sympos. Probability Theory* Lecture Notes in Math. **330**, pp. 104–119.

Holevo, A.S. (1973b): "Some estimates for the amount of information transmittable by a quantum communication channel" (in Russian), Problemy Peredachi Informacii, **9**, pp. 3–11

Holevo, A.S. (1973c): "Informational aspects of quantum measurement" (in Russian), Problemy Peredachi Informacii, **9**, pp. 31–42

Holevo, A.S. (1978): *Investigations in the general theory of statistical decisions* Proc. Steklov Inst. of Math. **124** (Amer. Math. Soc., Providence)

Holevo, A.S. (1979): "Capacity of a quantum communication channel" (in Russian), Problemy Peredachi Informatsii, **15**, pp. 3–11; English translation: Problems Inform. Transmission **15**, pp. 247–253

Holevo, A.S. (1982): *Probabilistic and statistical aspects of quantum theory* (North-Holland, Amsterdam)

Holevo, A.S. (1991): "Quantum probability and quantum statistics" (in Russian), Itogi Nauki i Tehniki **83** (Viniti, Moscow)

* Holevo, A.S. (1998a): "Quantum coding theorems", Russian Math. Surveys, **53**, pp. 1295–1331

* Holevo, A.S. (1998b): "The capacity of the quantum channel with general signal states", IEEE Trans. Inform. Theory **44**, pp. 269–273.

* Holevo, A.S. (2002): "On entanglement-assisted classical capacity. Quantum information theory", J. Math. Phys. **43**, pp. 4326–4333

Honegger, R. (1990): "Unbounded perturbations of Boson equilibrium states in their GNS representation", Helvetica Phys. Acta **63**, pp. 139–155

* Horodecki, M., Horodecki, P., Horodecki, R., Leung, D.W. and Terhal, B.M. (2001): "Classical capacity of a noiseless quantum channel assisted by noisy entanglement", Quantum Inf. Comput. **1**, pp. 70–78

Hudetz, T. (1988): "Spacetime dynamical entropy of quantum systems", Lett. Math. Phys. **16**, pp. 151–161

Hudetz, T. (1991): "Dynamical entropy for infinite quantum systems", in *Information Dynamics*, ed. by H. Atmanspacher and H. Scheingraber (Plenum Press, New York), pp. 279–287

* Hudetz, T. (1993): "Quantum topological entropy: First steps of a "pedestrian" approach", *Quantum Probability and Related Topics VIII*, (World Scientific, Singapore) pp. 237–261

Hudson, R.L. (1973): "A quantum-mechanical central limit theorem for anti-commuting observables", J. Appl. Prob. **10**, pp. 502–509

Ingarden, R.S. (1976): "Quantum information theory", Rep. Math. Phys. **10**, pp. 43–72

Ingarden, R.S., Janyszek, H., Kossakowski, A. and Kawaguchi, T. (1979): "Information geometry of quantum statistical systems", Tensor, **33**, pp. 347–353

* Ingarden, R.S., Kossakowski, A. and Ohya M. (1997): *Information dynamics and open systems. Classical and quantum approach* (Kluwer Academic Publishers, Dordrecht)

Ingarden, R.S. and Urbanik K. (1962): "Quantum information thermodynamics", Acta Phys. Polon. **21**. pp. 281–304

Israel, R.B. (1979): *Convexity in the theory of lattice gases* (Princeton University Press, Princeton)

Jajte, R. (1985): *Strong limit theorems in non-commutative probability* Lecture Notes in Math. **1110** (Springer, Berlin, Heidelberg, New York)

* Jajte, R. (1997): "On some versions of Jensen's inequality on operator algebras", Math. Slovaca **47**, pp. 303–311

Janyszek, H. (1986): "Geometrical structure of quantum Gibbs states", Rep. Math. Phys. **24**, pp. 11–19

Jaynes, E.T. (1956): "Information theory and statistical mechanics. II", Phys. Rev. **108**, pp. 171–190

Jones, V.F.R. (1983): "Index for subfactors", Invent. Math. **134**, pp. 1–25

Jones, V.F.R. (1989): "On knot invariants related to some statistical mechanical models", Pacific J. Math. **137**, pp. 311–334

Jones, V.F.R. (1990): "Notes on subfactors and statistical mechanics", in *Braid group, knot theory and statistical mechanics*, ed. C.N. Yang and M.L. Ge (World Scientific, Singapore) pp. 1–25

* Jozsa, R. and Schlienz, J. (2000): "Distinguishability of states and von Neumann entropy", Phys. Rev. A (3) **62**, pp. 012301-1–012301-11

* Jozsa, R. and Schumacher, B. (1994): "A new proof of the quantum noiseless coding theorem", J. Modern Opt. **41**, pp. 2343–2349

Kadison, R.V. and Ringrose, J.R. (1986): *Fundamentals of the theory of operator algebras. Volume II. Advanced thory* (Academic Press, New York)

Kamei, E. (1984): "Double stochasticity in finite factors", Math. Japon. **29**, pp. 903–907

Kamei, E. (1985): "An order on statistical operators implicitly introduced by von Neumann", Math. Japon. **30**, pp. 891–895

Kato, T. (1966): *Perturbation theory for linear operators* (Springer, New York, Berlin, Heidelberg)

Kawakami, S. and Yoshida, H. (1987): "Actions of finite groups on finite von Neumann algebras and the relative entropy", J. Math. Soc. Japan **39**, pp. 609–626

Kennard, E.H. (1927): "Zur Quantenmechanik einfacher Bewegungstypen", Zeit-schrift für Physik **44**, pp. 1–25

* King, C. and Ruskai, M.B. (2001): "Minimal entropy of states emerging from noisy quantum channels", IEEE Trans. Inform. Theory **47**, pp. 192–209

* King, C. (2002): "Maximization of capacity and l_p norms for some product chan-nels", J. Math. Phys. **43**, pp. 1247–1260

* King, C. and Ruskai, M.B. (2001): "Minimal entropy of states emerging from noisy quantum channels", IEEE Trans. Inform. Theory **47**, No. 1, pp. 192–209

Klimek, S. and Lesniewski, A. (1991): "A Golden-Thompson inequality in super-symmetric quantum mechanics", Lett. Math. Phys. **21**, 237–244

Klauder, J.R. and Sudarshan, E.C.G. (1968): *Fundamentals of Quantum Optics* (Benjamin, New York)

Kosaki, H. (1982): "Interpolation theory and the Wigner-Yanase-Dyson-Lieb con-cavity", Commun. Math. Phys. **87**, pp. 315–329

Kosaki, H. (1986a): "Relative entropy for states: a variational expression", J. Operator Theory **16**, pp. 335–348

Kosaki, H. (1986b): "Extension of Jones' theory on index to arbitrary factors", J. Functional Anal. **66**, pp. 123–140

Kosaki, H. (1992): "An inequality of Araki-Lieb-Thirring (von Neumann algebra case)", Proc. Amer. Math. Soc. **114**, pp. 477–481

* Kossakowski, A., Ohya, M. and Watanabe, N. (199):"Quantum dynamical en-tropy for completely positive map", Infin. Dimens. Anal. Quantum Probab. Relat. Top. **2**, pp. 267–282.

Kovács, I. and Szűcs, J. (1966): "Ergodic type theorems in von Neumann alge-bras", Acta Sci. Math. **27**, pp. 233–246

Kraus, K. (1983): *States, effects and operations* (Springer-Verlag, Berlin)

Kraus, K. (1987): "Complementarity and uncertainty relations", Phys. Rev D. **35**, pp. 3070–3075

Krengel, U. (1985): *Ergodic theorems* (de Greuter, Berlin, New York)

Kubo, R. (1957): "Statistical mechanical theory of irreversible processes I", J. Phys. Soc. Japan, **12**, pp. 570–586

Kullback, S. and Leibler, R. (1951): "On information and sufficiency", Ann. Math. Stat. **22**, pp. 79–86

Kümmerer, B. (1984): "Examples of Markov dilation over the 2×2 matrices", in *Quantum Probability and Applications to the Quantum Theory of Irreversible Processes*, ed. L. Accardi, A. Frigerio and V. Gorini (Springer-Verlag, Berlin, Heidelberg, New York) pp. 228–244

Kümmerer, B. (1985): "Markov dilations on W*-algebras", J. Functional Anal-ysis, **63**, pp. 139–177

Kümmerer, B. (1986): "Construction and structure of Markov dilations on W*-algebras", Habilitationsschrift, Tübingen, 1986

Kümmerer, B. (1988): "Survey on a theory of non-commutative stationary Markov processes" in *Quantum Probability and Applications III*, ed. L. Ac-cardi and W. von Waldenfels (Springer-Verlag, Berlin, Heidelberg, New York) pp. 228–244

Kümmerer, B. and Maasen, H. (1987): "The essentially commutative dilations of dynamical semigroups on M_n", Commun. Math. Phys. **109**, pp. 1–22

Kümmerer, B. and Nagel, R. (1979): "Mean ergodic semigroups on W*-algebras", Acta Sci. Math. **41**, pp. 151–159

Lance, E.Ch. (1982): " Tensor products and nuclear C*-algebras", in *Proc. Symposia Pure Math.* **38** ed. by R.V. Kadison pp. 379–399

Landford III, O.E. (1973): " Entropy and equilibrium states in classical statistical mechanics", in *Statistical Mechanics and Mathematical Problems* ed. A. Lenard, Lect. Notes in Phys. **20**, pp. 1–113

Landford, O.E. and Robinson, D.W. (1968): "Mean entropy of states in quantum statistical mechanics", J. Math. Phys. **9**, pp. 1120–1125

Lenard, A. (1971): "Generalization of the Golden-Thompson inequality $\mathrm{Tr}e^A e^B$ $\geq \mathrm{Tr}\, e^{A+B}$", Indiana Univ. Math. J. **21**, pp. 457–467

Lévy-Leblond, J-M. (1976): "Who is afraid of nonhermitean operators? A quantum description of angle and phase", Ann. Phys. **101**, pp. 319–341

Lewis, J.T. (1986): "The large deviation principle in statistical mechanics: an expository account", in *Stochastics mechanics and stochastic processes*, ed. A. Truman and I.H. Davies (Springer-Verlag, Berlin, Heidelberg, New York) pp. 141–155

Lewis, J.T. (1988): "Probabilistic aspects of statistical mechanics", in *Mark Kac seminar on probability and physics, Syllabus 1985-1987*, ed. F. den Hollander and H. Maasen (Centrum voor Wiskunde & Informatica, Amsterdam) pp. 85–146

Lieb, E.H. (1973a): "Convex trace functions and the Wigner-Yanase-Dyson conjecture", Advances in Math. **11**, pp. 267–288

Lieb, E.H. (1973b): "The classical limit of quantum systems", Commun. Math. Phys. . **31**, pp. 327–340

Lieb, E.H. (1975): "Some convexity and subadditivity properties of entropy", Bull. Amer. Math. Soc. **81**, pp. 1–14

Lieb, E.H. (1978): "Proof of an entropy conjecture of Wehrl", Commun. Math. Phys. **62**, pp. 35–41

Lieb, E.H. (1990a): "Gaussian kernels have only Gaussian maximizers", Invent. Math. **102**, pp. 179–208

Lieb, E.H. (1990b): "Integral bounds for radar ambiguity functions and Wigner distributions", J. Math. Phys. **31**, pp. 594–599

* Lieb, E.H. and Pedersen, G.K. (2002): "Convex multivariable trace functions", Rev. Math. Phys. **14**, pp. 631–648

Lieb, E.H. and Ruskai, M.B. (1973a): "Proof of the strong subadditivity of quantum mechanical entropy", J. Math, Phys. **14**, pp. 1938–1941

Lieb, E.H. and Ruskai, M.B. (1973b): "A fundamental property of quantum mechanical entropy", Phys. Rev. Lett. **30**, pp. 434–436

Lieb, E.H. and Ruskai, M.B. (1974): "Some operator inequalities of the Schwarz type", Adv. Math. **12**, pp. 269–273

Lieb, E.H. and Thirring, W. (1976): "Inequalities for the moments of the eigenvalues of the Schrödinger Hamiltonian and their relation to Sobolev inequalities", in *Essays in Honor of Valentine Bargmann*, ed. by E.H. Lieb, B. Simon and A.S. Wightmann, pp. 301–302 (Princeton University Press, Princeton)

Lindblad, G. (1973): "Entropy, information and quantum measurement", Commun. Math. Phys. **33**, pp. 305–322

Lindblad, G. (1974): "Expectations and entropy inequalities for finite quantum systems", Commun. Math. Phys. **39**, pp. 111–119

Lindblad, G. (1975): "Completely positive maps and entropy inequalities", Commun. Math. Phys. **40**, pp. 147–151

Lindblad, G. (1979a): "Non-Markovian stochastic processes and their entropy", Commun. Math. Phys. **65**, pp. 281–294

Lindblad, G. (1979b): "Gaussian quantum stochastic processes on the CCR algebra", J. Math. Phys. **20**, pp. 2081–2087

Lindblad, G. (1983): *Nonequlibrium entropy and irreversibility* (Reidel, Dordrecht-Boston)

Lindblad, G. (1986): "Quantum ergodicity and chaos", in *Fundamental Aspects of Quantum Theory* ed. by A. Frigerio and V. Gorini (Plenum Press, New York), pp. 199–208

Lindblad, G. (1988): "Dynamical entropy for quantum probability", *Quantum Probability and Application III* ed. by L. Accardi and W. von Waldenfels (Lecture Notes in Math. **1303**, Springer), pp. 183–191.

Lindblad, G. (1991): "Quantum entropy and quantum measurements", in *Quantum Aspects of Optical Communication* ed. by C. Benjaballah, O. Hirota, S. Reynaud (Lecture Notes in Physics **378**, Springer), pp. 199–208

Louisell, W.H. (1973): *Quantum Statistical Properties of Radiation* (John Wiley and Sons, New York)

Maasen, H. (1990): "A discrete entropic uncertainty relation", in *Quantum probability and applications V* ed. by L. Accardi and W. von Waldenfels (Lecture Notes in Math. **1442**, Springer), pp. 263–266

Maasen, H. (1992): "Addition of freely independent random variables", J. Funct. Anal. **106**, pp. 409–438.

Maasen, H. and Uffink, I. (1988): "Generalized entropic uncertainty relations", Phys. Rev. Lett. **60**, pp. 1103–1106

Mackey, M.C. (1989): "The dynamical origin of increasing entropy", Rev. Mod. Phys. **61**, pp. 981–1015

Majewski, W.A. and Kuma, M. (1990): "An example of an infinite system with zero quantum dynamical entropy", Lett. Math. Phys. **20**, pp. 337–341

Manuceau, J., Naudts J. and Verbeure, A. (1972): "Entropy of normal states", Commun. Math. Phys. **27**, pp. 327–338

Manuceau, J., Sirugue, M., Testard, D. and Verbeure, A. (1973): "The smallest C*-algebra for canonical commutation relations", Commun. Math. Phys. **32**, pp. 231–243

Marshall, A.W. and Olkin, I. (1979): *Inequalities: Theory of majorization and its applications* (Academic Press, New York)

Marshall, A.W. and Olkin, I. (1985): "Inequalities for the trace function", Aequations Math. **29**, pp. 36–39

Martin, N.F.G. and England, J.W. (1981): *Mathematical theory of entropy* (Addison-Wesley, Reading)

* Mosonyi, M. and Petz, D. (2003): 'Structure of sufficient coarse-grainings", preprint

Muraki, N. (1991): "Remarks on continuity of entropy of states on C*-dynamical systems, J. Math. Phys. **32**, pp. 1796–1798

* Muraki, N. and Ohya, M. (1992): "Note on continuity of information rate", Illinois J. of Math. **36**, pp. 529–550

Muraki, N., Ohya, M. and Petz, D. (1992): "Note on entropy of general quantum systems", Open Syst. Inf. Dynamics **1**, pp. 43–56.

Müller, E.E. (1985): "Note on the relative entropy and thermodynamical limit", Helvetica Phys. Acta, **58**, pp. 622–632

Nachtergaele, B. (1990): "Working with quantum Markov states and their classical analogues", in *Quantum probability and applications V* ed. by L. Accardi and W. von Waldenfels (Lecture Notes in Math. **1442**, Springer), pp. 267–285

Nagaoka, H. (1989a): "A new approach to Cramer-Rao bounds for quantum state estimation", IEICE Technical Report, **89**, IT89-42, pp. 9–14

Nagaoka, H. (1989b): "On the parameter estimation problem for quantum statistical models", in *Proc. of the 12th Symposium on Information Theory and Its Applications (held in Inuyama, Japan)*, (Soc. of Information Theory and Its Appl., Tokyo) vol. **2**, pp. 577–582

Nakamura, M. and Umegaki, H. (1961): "A note on the entropy on operator algebras", Proc. Jap. Acad. **37**, pp. 149–154

Narnhofer, H. (1988): "Free energy and the dynamical entropy of space translations", Rep. Math. Phys. **25**, pp. 345–356

Narnhofer, H. (1989): "Dynamical entropy in quantum theory", in *XIth Int. Congress on Math. Phys.* ed. by B. Simon, A. Truman, I.M. Davies (Adam Hilger, Bristol and New York) pp. 64–76

Narnhofer, H. (1990a): "Dynamical entropy, quantum K-systems and clustering", in *Quantum probability and applications V* ed. by L. Accardi and W. von Waldenfels (Lecture Notes in Math. **1442**, Springer), pp. 286–295

Narnhofer, H. (1990b): "Beispiele für Algebren mit gleicher dynamischer Entropie", Ann. Physik **47**, pp. 166–176

Narnhofer, H. (1990c): "The concept of K-automorphism in quantum field theory and the Jones index", in *Selected topics in quantum field theory and mathematical physics* ed. by J. Niederle and J. Fischer (World Scientific, Singapore) pp. 91–103

Narnhofer, H. (1990d): "K-automorphisms in quantum theory in nonlinear dynamics and quantum dynamical systems", Mathematical Research **59**, ed. by G.A. Leonov, V. Reitmann, W. Timmermann (Akademie Verlag, Berlin) pp. 86–95

Narnhofer, H., Pflug, A. and Thirring, W. (1989): " Mixing and entropy increase in quantum systems", in *Symmetry and Nature*, (Scoula Normale Superiore, Pisa) pp. 597–626

* Narnhofer, H., Størmer, E. and Thirring, W. (1995): "C*-dynamical systems for which the tensor product formula for entropy fais", Ergodic Th. and Dynam. Sys. **15**, pp. 961–968

Narnhofer, H. and Thirring, W. (1985): "From relative entropy to entropy", Fizika **17**, pp. 257–265

Narnhofer, H. and Thirring, W. (1987): "Dynamical entropy of quasifree automorphisms", Lett. Math. Phys. **14**, pp. 89–96

Narnhofer, H. and Thirring, W. (1988): "Dynamical entropy and the third law of thermodynamics", Lett. Math. Phys. **15**, pp. 261–273

Narnhofer, H. and Thirring, W. (1989): "Quantum K-systems", Commun. Math. Phys. **125**, pp. 565–567

Narnhofer, H. and Thirring, W. (1990): "Algebraic K-systems", Rep. Math. Phys. **20**, pp. 231–250

* Narnhofer, H., Thirring, W. and Størmer, E. (1995): "C*-dynamical systems for which the tensor product formula for entropy fails", Ergodic Theory Dynam. Systems **15**, pp. 961–968

Naudts, J. (1974): "A generalized entropy function", Comm. Math. Phys. **37**, pp. 175–182

Naudts, J., Verbeure, A. and Weder, R. (1975): "Linear response theory and the KMS condition", Commun. Math. Phys. **44**, pp. 87–99

* Neshveyev, S. and Størmer, E. (2001): "Entropy of type I algebras", Pacific J. Math. **201**, pp. 421–428

* Neshveyev, S. and Størmer, E. (2002): "The McMillan theorem for a class of asymptotically abelian C*-algebras", Ergodic Theory Dynam. Systems **22**, pp. 889–897

Neumann, J. von (1927): "Thermodynamik quantenmechanischer Gesamtheiten", Gött. Nachr. pp. 273–291

Neumann, J. von (1932): *Mathematische Grundlagen der Quantenmechanik* (Springer, Berlin)

* Nielsen, M.A. and Chuang, I.L. (2000): *Quantum Computation and Quantum Information*, Cambridge University Press.

Ochs, W. (1975): "A new characterization of von Neumann entropy", Rep. Math. Phys. **8**, pp. 109–120

* Ogawa, T. and Nagaoka, H. (2000): "Strong converse and Stein's lemma in quantum hypothesis testing", IEEE Trans. **IT-46**, pp. 2428–2433

Ohya, M. (1981): "Quantum ergodic channels in operator algebras", J. Math. Anal. Appl. **84**, pp. 318–327

Ohya, M. (1983a): "On compound state and mutual information in quantum information theory", IEEE Trans. Information Theory, **29**, pp. 770–777

Ohya, M (1983b): "Note on quantum probability", L. Nuovo Cimento, **38**, pp. 402–406

Ohya, M. (1984): "Entropy transmission in C*-dynamical systems", J. Math. Anal. Appl. **100**, pp. 222–235

Ohya, M. (1985): "State change and entropies in quantum dynamical systems", in *Quantum Probability and Applications II* ed. by L. Accardi and W. von Waldenfels (Lecture Notes in Math. **1136**, Springer), pp. 397–408

Ohya, M. (1986a): "Entropy operators and McMillan type convergence theorems in a noncommutative dynamical system", in *Probability Theory and Mathematical Statistics* ed. by S. Watanabe, Yu. V. Prokhorov (Lecture Notes in Math. **1299**, Springer), pp. 384–390

Ohya, M. (1986b): "Entropy change in linear response dynamics", Nuovo Cimento, **91B**, pp. 25–30

Ohya, M. (1987): "A mathematical formulation of channel and mutual entropy in quantum systems", Proc. Intern. Conf. Opt. Commun. Theory, pp. 31–34

Ohya, M. (1989): "Some aspects of quantum information theory and their applications to irreversible processes", Rep. Math. Phys. **27**, pp. 19–47

Ohya, M. (1991a): "Information dynamics and its application to optical communication processes", in *Quantum Aspects of Optical Communication* ed. by C. Benjaballah, O. Hirota, S. Reynaud (Lecture Notes in Physics **378**, Springer), pp. 81–92

Ohya, M. (1991b): "Fractal dimensions of states", in *Quantum Probability and Related Topics VI* (World Scientific, Singapore) pp. 359–369

* Ohya, M. (1999): "Fundamentals of quantum mutual entropy and capacity", Open Syst. Inf. Dyn. **6**, pp. 69–78

Ohya, M. and Matsuoka, T. (1986): "Continuity of entropy and mutual entropy in C*-dynamical systems", J. Math. Phys. **27**, pp. 2076–2079

* Ohya, M. and Muraki, N. (1992): "Note on continuity of information rate", Illinois J. Math. **36**, pp. 529–550

* Ohya, M. and Petz, D. (1996): "Notes on quantum entropy", Studia Sci. Math. Hungar. **31**, pp. 423–430

* Ohya, M., Petz, D. and Watanabe, N. (1997): "On capacities of quantum channels", Prob. Math. Stat. **17**, pp. 179–196

Ohya, M. and Suyari, H. (1990): "Optimization of error probability in quantum control communication processes", IEICE of Japan, **J73-B-I**, pp. 200–207

Ohya, M. and Suyari, H. (1991): "Rigorous derivation of error probability in coherent optical communication", in *Quantum Aspects of Optical Communication* ed. by C. Benjaballah, O. Hirota, S. Reynaud (Lecture Notes in Physics **378**, Springer), pp. 203–212

Ohya, M., Tsukada M. and Umegaki, H. (1987): "A formulation of noncommutative McMillan theorem", Proc. Japan Academy Ser. A, **63**, pp. 50–53

Ohya, M. and Watanabe, N. (1985): "Construction and analysis of a mathematical model in quantum communication processes", Scripta Thechnica Inc. Elect. Commun. Japan, **68**, pp. 29–34

Ohya, M. and Watanabe, N., (1986): "A new treatment of communication processes with gaussian channels", Japan. J. Appl. Math. **3**, pp. 197–206

Ojima, I. (1989): "Entropy production and nonequilibrium stationarity in quantum dynamical systems. Physical meaning of van Hove limit", J. Stat. Phys. **56**, pp. 203–226

Ojima, I. (1991): "Entropy production and nonequilibrium stationarity in quantum dynamical systems", in *Quantum Aspects of Optical Communication* ed. by C. Benjaballah, O. Hirota, S. Reynaud (Lecture Notes in Physics **378**, Springer), pp. 164–178

Ojima, I., Hasegawa, H. and Ichiyanagi, M. (1988): "Entropy production and its positivity in nonlinear response theory of quantum dynamical systems", J. Stat. Phys. . **50**, pp. 633–655

Ovchinnikov, V.I. (1970): "s–numbers of measurable operators", F Anal. Appl. **4**, pp. 236–242.

Ozawa, M. (19860: "On information gain by quantum measurements of continuous observables", J. Math. Phys. **27**, pp. 759–763

Padmanabhan, A.R. (1979): "Probabilistic aspects of von Neumann algebras", J. Functional Anal. **31**, pp. 139–149

* Park, Y.M. (1994): "Dynamical entropy of generalized quantum Markov chains", Lett. Math. Phys. **32**, pp. 63–74

Partovy, M.H. (1983): "Entropic formulation of uncertainty in quantu measurements", Phys. Rev. Lett. **50**, pp. 1883-1885

Paulsen, V.I. (1986): *Completely bounded maps and dilations* Research Notes in Mathematics **146** (Pitman, Boston, London, Melbourne)

Perelomov, A. (1986): *Generalized coherent states and their applications* (Springer, New York, Berlin, Heidelberg)

Petz, D. (1984): "A dual in von Neumann algebras", Quart. J. Math. Oxford **35**, pp. 475–483

Petz, D. (1985a): "Properties of quantum entropy", in *Quantum Probability and Applications II* ed. by L. Accardi and W. von Waldenfels (Lecture Notes in Math. **1136**, Springer), pp. 428–441

Petz, D. (1985b): "Quasi-entropies for states of a von Neumann algebebras", Publ. RIMS Kyoto Univ. **21**, pp. 787–800

Petz, D. (1985c): "Spectral scale of selfadjoint operators and trace inequalities", Math. Anal. Appl. **109**, pp. 74–82

Petz, D. (1986a): "Properties of the relative entropy of states of a von Neumann algebra", Acta Math. Hung. **47**, pp. 65–72

Petz, D. (1986b): "Quasi-entropies for finite quantum systems", Rep. Math. Phys. **21**, pp. 57–65

Petz, D. (1986c): "Sufficient subalgebras and the relative entropy of states on a von Neumann algebra", Commun. Math. Phys. **105**, pp. 123–131

Petz, D. (1987): "Jensen's inequality for contractions of operator algebras", Proc. Amer. Math. Soc. **99**, pp. 273–277

Petz, D. (1988a): "Conditional expectation in quantum probability", *Quantum Probability and Application III* ed. by L. Accardi and W. von Waldenfels (Lecture Notes in Math. **1303**, Springer), pp. 251–260.

Petz, D. (1988b): "Sufficiency of channels over von Neumann algebras", Quart. J. Math. Oxford **39**, pp. 907–108

Petz, D. (1988c): "A variational expression for the relative entropy", Commun. Math. Phys. **114**, pp. 345–348

Petz, D. (1989): "Characterization of sufficient observation channels", in *Mathematical Methods in Statistical Mechanics* ed. by M. Fannes and A. Verbeure (Leuven University Press, Leuven), pp. 167–178

Petz, D. (1990a): *The algebra of the canonical commutation relation* (Leuven University Press, Leuven)

Petz, D. (1990b): "First steps towards a Donsker and Varadhan theory in operator algebras", in *Quantum probability and applications V* ed. by L. Accardi and W. von Waldenfels (Lecture Notes in Math. **1442**, Springer), pp. 311–319

Petz, D. (1991): "On certain properties of the relative entropy of states of operator algebras", Math. Zeitsch. **206**, pp. 351–361

* Petz, D. (1992a): "Characterization of the relative entropy of states of operator algebras", Acta Math. Hungar. **59**, pp. 449–455

Petz, D. (1992b): "Entropy and central limit on the algebra of the canonical commutation relation", Lett. Math. Phys. **24**, pp. 211–220

* Petz, D. (1992c): "Entropy in quantum probability I", *Quantum Probability and Related Topics VII* (World Scientific, Singapore) pp. 275–297

Petz, D. (1992d): "Characterization of the relative entropy of state of matrix algebras", Acta Math. Hungar. **59**, pp. 449–455

* Petz, D. (1994a): "Discrimination between states of quantum systems by observations and an application to injective von Neumann algebras", J. Funct. Anal. **120**, pp. 82–97

* Petz, D. (1994b): "On entropy functionals of states of operator algebras", Acta Math. Hungar. **64**, pp. 333–340

* Petz, D. (2001): Entropy, von Neumann and the von Neumann entropy, in *John von Neumann and the Foundations of Quantum Physics*, eds. M. Rédei and M. Stöltzner, Kluwer.

* Petz, D. and Mosonyi, M. (2001): "Stationary quantum source coding", J. Math. Phys. **42**, pp. 4857–4864

* Petz, D. (2002): "Covariance and Fisher information in quantum mechanics", J. Phys. A: Math. Gen. **35**, pp. 929–939

* Petz, D. (2003): "Monotonicity of quantum relative entropy revisited", Rev. Math. Phys. **15**, pp. 79–91

Petz, D., Raggio, G.A. and Verbeure, A (1989): "Asymptotics of Varadhan-type and the Gibbs variational principle", Commun. Math. Phys. **121**, pp. 271–282

Petz, D. and Zemánek, J. (1988): "Characterizations of the trace", Linear Alg. Appl. **111**, pp. 43–52

Pimsner, M. and Popa, S. (1986): "Entropy and index for subfactors", Ann. Sci. École. Norm. Sup. **19**, pp. 289–306

Powers, R.T (1967): "Representation of uniformly hyperfinite algebras and their associated von Neumann algebras", Ann. of Math. **86**, pp. 138–171

Pusz, W. and Woronowicz, S.L. (1975): "Functional calculus for sesquilinear forms and the purification map", Rep. Math. Phys. **8**, pp. 159–170

Pusz, W. and Woronowicz, S.L. (1978): "From convex functions and the Wydl and other inequalities", Lett. Math. Phys. **2**, pp. 505–512

Raggio, G.A. (1982): "Comparison of Uhmann's transition probability with one induced by the natural positive cone of a von Neumann algebra", Lett. Math. Phys. **6**, pp. 233–236

Raggio, G.A. (1984): "Generalized transition probabilities and applications", in *Quantum Probability and Applications to the Quantum Theory of Irreversible Processes*, ed. L. Accardi, A. Frigeris and V. Gorini (Springer-Verlag, Berlin, Heidelberg, New York) pp. 327–335.

Raggio, G.A. and Werner, R.F. (1989): "Quantum statistical mechanics of general mean field systems", Helvetica Phys. Acta **62**, pp. 980–1003

Raggio, G.A. and Werner, R.F. (1990): "Minimizing entropy in a face", Lett. Math. Phys. **19**, pp. 7–14

Raggio, G.A. and Werner, R.F. (1991): "The Gibbs variational principle for inhomogenous mean-field systems", Helvetica Phys. Acta **64**, pp. 633–667

Rédei, M. (1987): "Reformulation of the hidden variable problem entropic measure of uncertainty", Synthese **73**, pp. 371–379

Reed, M. and Simon, B. (1972): *Methods of modern mathematical physics, I: Functional analysis* (Academic Press, New York)

Reed, M. and Simon, B. (1975): *Methods of modern mathematical physics, II: Fourier analysis, selfadjointness* (Academic Press, New York)

Rényi, A. (1962): *Wahrscheinlichkeitsrechnung* (VEB Deutsch. Ver. der Wiss., Berlin)

Rényi, A. (1965): "On the foundations of information theory", Rev. Int. Stat. Inst. **33**, pp. 1–14

Robertson, H.P. (1929): "The uncertainty principle", Phys. Rev. **34**, pp. 163–164

Roos, H. (1970): "Independence of local algebras in quantum field theory", Commun. Math. Phys. **16**, pp. 238–246

Ruelle, D. (1969): *Statistical mechanics. Rigorous results* (Benjamin, New York-Amsterdam)

Ruelle, D. (1978): *Thermodynamic formalism* (Addison-Wesley, Reading)

Ruskai, M.B. (1972): "Inequalities for traces on von Neumann algebras", Commun. Math. Phys. **26**, pp. 280–289

Ruskai, M.B. (1973): "A generalization of entropy using traces on von Neumann algebras", Ann. Inst. Henri Poincaré, Sec. A **19**, pp. 357–373.

Ruskai, M.B. (1987): "Entropy of reduced density matrices", in *Density Matrices and Density Functionals*, ed. R. Erdahl and V. Smith (Reidel, Dordrecht) pp. 213–230

Ruskai, M.B. (1988): "Extremal properties of relative entropy in quantum statistical mechanics", Rep. Math. Phys. **26**, pp. 143–150

* Ruskai, M.B. (2002): "Inequalities for quantum entropy: a review with conditions for equality. Quantum information theory", J. Math. Phys. **43**, pp. 4358–4375

Ruskai, M.B. and Stillinger, F.K. (1990): "Convexity inequalities for estimating free energy and relative entropy", J. Phys. A: Math. Gen. **23**, pp. 2421–2437

Sakai, S. (1985): "Bounded perturbations and KMS states in C*-algebras", Proc. Inst. Nat. Sciences, Nikon Univ, **20**, pp. 15–30

Sakai, S. (1987): "Perturbations and KMS states in C*-dynamical systems (Generalization of the absence theorem of phase transition to continuous quantum systems)", Contemp. Math. **62**, pp. 187–217

Sakai, S. (1991): *Operator algebras in dynamical systems* (Cambridge University Press, Cambridge)

Sauvageot, J-L. and Thouvenot, J-P. (1992): "One nouvelle définition de l'entropie dynamique des systemes non commutatifs", Commun. Math. Phys. **145**, pp. 411–423

Schmetterer, L. (1974): *Introduction to mathematical statistics* (Springer-Verlag, Berlin, Heidelberg, New York)

Schmüdgen, K. (1983a): "On the Heisenberg commutation relation I", J. Functional Anal. **50**, pp. 8–49

Schmüdgen, K. (1983b): "On the Heisenberg commutation relation II", Publ. RIMS Kyoto Univ. **19**, pp. 601–671

Schroeck, F.E. (1989): "On the entropic formulation of uncertainty for quantum measurement", J. Math. Phys. **30**, pp. 2078–2082

* Schumacher, B. (1995): Quantum coding, Phys. Rev. A **51**, pp. 2738–2747

* Schumacher, B. and Westmoreland, M.D. (1997): "Sending classical information via noisy quantum channels", Phys. Rev. A**51**, pp. 131–138

* Schupp, P. (1999): " On Lieb's conjecture for the Wehrl entropy of Bloch coherent states", Comm. Math. Phys. **207**, pp. 481–493

Schürmann, M. (1990): "Gaussian states on bialgebras", in *Quantum probability and applications V* ed. by L. Accardi and W. von Waldenfels (Lecture Notes in Math. **1442**, Springer), pp. 347–367

Segal, I.E. (1960): "A note on the concept of entropy", J. Math. Mech. **9**, pp. 623–629

Sewell, G.L. (1980): "Stability, equilibrium and metastability in statistical mechanics", Physics Rep. **5**, pp. 307–342

Sewell, G.L. (1986): *Quantum theory of collective phenomena* (Clarendon Press, New York)

Sewell, G.L. (1988): "Entropy, observability and the generalized second law of thermodynamics", *Quantum Probability and Application III* ed. by L. Accardi and W. von Waldenfels (Lecture Notes in Math. **1303**, Springer), pp. 319–328

Shannon, C.E. (1948): "Mathematical theory of communication", Bell System Tech. J. **27**, pp. 379–423

Shields, P. (1973): *The theory of Bernoulli shifts* (The University of Chicago Press, Chicago and London)

* Shor, P. (2000): "Quantum information theory: results and open problems", Geom. Funct. Anal. Special Volume, Part II, pp. 816–838

* Shor, P.W. (2002): "Additivity of the classical capacity of entanglement-breaking quantum channels. Quantum information theory", J. Math. Phys. **43**, pp. 4334–4340

Shukhov, A.G. (1980): "Entropy of diagonalizable states of von Neumann algebras" (in Russian), Funktional. Analiz. i Prilozhen. **14**, pp. 95–96

Shukhov, A.G. (1981): "The entropy invariant of automorphisms of W*-algebras" (in Russian), Funktional. Analiz. i Prilozhen. **15**, pp. 94–95

Shukhov, A.G., Stepin, A.M. (1982): "The centralizer of diagonalizable states and the entropy of automorphisms of W*-algebras" (in Russian), Izv. Vyssh. Uchebn. Zaved. Mat. **8**, pp. 52–60

Simon, B. (1979): *Trace ideals and their applications* (Cambridge University Press, Cambridge)

Simon, B. (1980): "The classical limit of quantum partition functions", Commun. Math. Phys. **71**, pp. 247–276

Slawny, F. (1971): "On factor representations and the C*-algebra of canonical commutation relation", Commun. Math. Phys. **24**, pp. 151–170

* So, W. (1992): "Equality cases in matrix exponential inequalities", SIAM J. Matrix Anal. Appl. **13**, pp. 1154–1158

Speicher, R. (1990): "A new example of independence and white noise", Probab. Th. Rel. Fields **84**, pp. 141–159

Speicher, R. (1992): "A non-commutative central limit theorem", Math. Zeitsch. **209**, pp. 55–66

Størmer, E. (1969): "Symmetric states of infinite tensor product C*-algebras". J. Functional Anal. **48**, pp. 48–68

Størmer, E. (1976): "Entropy in finite von Neumann algebras", in *Symposia Mathematica* **XX**, pp. 197–205

* Størmer, E. (1992): "Entropy of some automorphisms of the II$_1$-factor of the free group in infinite number of generators", Invent. Math. **110**, pp. 63–73

* Størmer, E. (2000): "Entropy of endomorphisms and relative entropy in finite von Neumann algebras", J. Funct. Anal. **171**, pp. 34–52

Størmer, E. and Voiculescu, D. (1990): "Entropy of Bogoliubov automorphisms of the canonical anticonnutation relation", Commun. Math. Phys. **133**, pp. 511–542

Strătilă, S. (1981): *Modular theory in operator algebras* (Abacuss Press, Tunbridge Wells)

Strătilă, S. and Zsidó, L. (1979): *Lectures on von Neumann algebras* (Abacuss Press, Tunbridge Wells)

Streater, R.F. (1985): "Convergence of the quantum Boltzmann map", Commun. Math. Phys. **98**, pp. 177–185

Streater, R.F. (1987a): "Entropy and the central limit theorem in quantum mechanics", J. Phys. A **20**, pp. 4321–4330

Streater, R.F. (1987b): "A Boltzmann map for quantum oscillators", J. Statist. Phys. **48**, 753–767

Summers, S.J. (1990): "On the independence of local algebras in quantum field theory", Rev. Math.Phys. **2**, pp. 201–247

* Suyari, H. (2002): "On the most concise set of axioms and the uniqueness theorem for Tsallis entropy", J. Phys. **A35**, pp. 10731–10738

Quaegebeur, J. (1984): "A noncommutative central limit theorem for CCR-algebras", J. Functional Anal. **57**, pp. 1–20

Quaegebeur, J. and Verbeure, A. (1980): "Stability for mean field models", Ann. Inst. Henri Poincaré **32**, pp. 343–349

Takesaki, M. (1970): *Tomita's theory of modular Hilbert algebras and its applications*, Lecture Notes in Math. **128** (Springer, Berlin, New York)

Takesaki, M. (1972): "Conditional expectation in von Neumann algebra J. Functional Anal. **9**, pp. 306–321

Takesaki, M. (1979): *Theory of operator algebras I* (Springer, New York, Berlin, Heidelberg)

Temperly, H.N.B. and Lieb, E.H. (1971): "Relations between the percolation and colouring problem and other graph-theoretical problems assocated with regular planar lattices: some exact results for the percolation problem", Proc. Roy. Soc. (London) Ser. A **322**, pp. 251–280

Thirring, W. (1983): *A Course in Mathematical Physics. 4. Quantum Mechanics of Large Systems* (Springer, New York, Berlin, Heidelberg)

Thirring, W. and Wehrl, A. (1967): "On the mathematical structure of the B.C.S. model", Commun. Math. Phys. **4**. pp. 303–314

Thompson, C.J. (1965): "Inequality with applications in statistical mechanics", J. Math. Phys. **6**, pp. 1812–1813

Thompson, C.J. (1971): "Inequalities and partial orders on matrix spaces", Indiana Univ. Math. J. **21**, pp. 469–480

Triebel, H. (1978): *Interpolation theory, function spaces, differential operators* (VEB Deutscher Verlag der Wiss., Berlin)

* Tsallis, C. (1988): "Possible generalization of Boltzmann-Gibbs statistics", J. Statist. Phys. **52**, pp. 479–487

Uffink, J. (1990): "Measures of uncertainty and the uncertainty principle", Thesis, University of Utrecht

Uhlmann, A. (1970): "On the Shannon entropy and related functionals on convex sets", Rep. Math. Phys. **1**, pp. 147–159

Uhlmann, A. (1977): "Relative entropy and the Wigner-Yanase-Dyson-Lieb concavity in an interpolation theory", Commun. Math. Phys. **54**, pp. 21–32

Uhlmann, A. (1985): "The transition probability for states of *-algebras", Ann. Physik, **42**, pp. 524–532

Umegaki, H. (1954): "Conditional expectations in an operator algebra", Tohoku Math. J. **6**, pp. 177–181

Umegaki, H. (1956): "Conditional expectations in an operator algebra II", Tohoku Math. J. **8**, pp. 86–100

Umegaki, H. (1959): "Conditional expectations in an operator algebra III", Kodai Math. Sem. Rep. **11**, pp. 51–64

Umegaki, H. (1962): "Conditional expectations in an operator algebra IV (entropy and information)", Kodai Math. Sem. Rep. **14**, pp. 59–85

Varadhan, S.R.S. (1984): *Large deviation and applications* (Soc. for Indust. and Appl. Math., Philadelphia)

* Vedral, V., Plenio, M.B., Rippin, M.A. and Knight, P.L. (1998a): "Quantifying entanglement", Phys. Rev.Lett. **78**, pp. 2275–2279

* V. Vedral, V. and Plenio, M.B. (1998b): "Entanglement measures and purification procedures", Phys. Rev. **A57**, pp. 1619–1633

Verbeure, A. (1988): "Detailed balance and critical slowing down", in *Quantum Probability and Application III* ed. by L. Accardi and W. von Waldenfels (Lecture Notes in Math. **1303**, Springer), pp. 354–362

Verbeure, A. (1989): "Equilibrium states of quantum systems and applications", in *Stochastic Methods in Mathematical Physics*, ed. by R. Gielerak and W. Karwowski (World Scientific, Singapore) pp. 134–170

Voiculescu, D. (1986): "Addition of non-commuting random variables", J. Functional Anal. **66**, pp. 323–346

Voiculescu, D. (1991): "Entropy of dynamical systems and perturbations of operators", Ergodic Th. Dyn. Syst. **11**, pp. 779–786

Waldenfels, W. von (1978): "An algebraic central limit theorem in the anticommuting case", Z. Wahrsheinlichkeitstheorie Verw. Gebiete **42**, pp. 135–140

Waldenfels, W. von (1990): "Illustration of the quantum central limit theorem by independent addition of spins", in *Séminarie de probabilités XXIV*, ed. J. Azéma, P.A. Meyer, M. Yor (Lect. Notes in Math. **1426**, Springer, Berlin) pp. 343–356

Walters, P. (1982): *An introduction to ergodic theory* (Springer, New York, Berlin, Heidelberg)

Watanabe, N. (1986): "Noncommutative extension of an integral representation theorem of entropy", Kodai Math. J. **9**, pp. 165–169

Watanabe, N. (1991a): "Efficiency of optical modulations with coherent states", in *Quantum Probability and Related Topics VI* (World Scientific, Singapore) pp. 489–498

Watanabe, N. (1991b): "Efficiency of optical modulations for photon number states", in *Quantum Aspects of Optical Communication* ed. by C. Benjaballah, O. Hirota, S. Reynaud (Lecture Notes in Physics **378**, Springer), pp. 350–360

Wehrl, A. (1974): "How chaotic is a state of a quantum system", Rep. Math. Phys. **6**, pp. 15–28

Wehrl, A. (1978): "General properties of entropy", Rev. Mod. Physics **50**, pp. 221–260

Wehrl, A. (1979a): "On the relation between classical and quantum-mechanical entropy", Rep. Math. Phys. **16**, pp. 353–358

Wehrl, A. (1979b): "A remark on the concavity of entropy", Found. Phys. **9**, pp. 939–946

* Wehrl, A. (1991): "The many facets of entropy", Rep. Math. Phys. **30**, pp. 119–129

Wenzl, H. (1987): "On sequences of projections", C.R. Math. Rep. Acad. Sci. Canada, **9**, pp. 5–9

* Werner R.F. (1992): "Large deviations and mean field quantum sytems", *Quantum Probability and Related Topics VII* (World Scientific, Singapore) pp. 349–381

* Werner, R.F. and Holevo, A.S. (2002): "Counterexample to an additivity conjecture for output purity of quantum channels. Quantum information theory", J. Math. Phys. **43**, pp. 4353–4357

Wigner, E.P. and Yanase, M.M. (1963): "Information content of distributions", Proc. Nat. Acad. Sci. USA **49**, pp. 910-918

* Winter, A. and Massar, S. (2001): "Compression of quantum measurement operations", Phys. Rev. A **64**, 012311

Wu, F.Y. (1982): "The Potts model", Rev. Mod. Phys. **54**, pp. 235–268

Yin, H.S. (1990): "Entropy of certain noncommutative shifts", Rocky Mount. J. Math. **20**, pp. 651–656

Index

Texts and Monographs in Physics

Series Editors: R. Balian   W. Beiglböck   H. Grosse   E. H. Lieb
N. Reshetikhin   H. Spohn   W. Thirring

Texts and Monographs in Physics

Series Editors: R. Balian   W. Beiglböck   H. Grosse   E. H. Lieb
N. Reshetikhin   H. Spohn   W. Thirring

The Statistical Mechanics of Financial Markets 2nd Edition By J. Voit

Statistical Mechanics A Short Treatise
By G. Gallavotti

Statistical Physics of Fluids
Basic Concepts and Applications
By V. I. Kalikmanov

Many-Body Problems and Quantum Field Theory An Introduction 2nd Edition
By Ph. A. Martin and F. Rothen

Foundations of Fluid Dynamics
By G. Gallavotti

High-Energy Particle Diffraction
By E. Barone and V. Predazzi

Physics of Neutrinos
and Applications to Astrophysics
By M. Fukugita and T. Yanagida

Relativistic Quantum Mechanics
By H. M. Pilkuhn

The Geometric Phase in Quantum Systems
Foundations, Mathematical Concepts,
and Applications in Molecular
and Condensed Matter Physics
By A. Bohm, A. Mostafazadeh, H. Koizumi,
Q. Niu and J. Zwanziger

The Atomic Nucleus as a Relativistic System
By L.N. Savushkin and H. Toki

The Frenkel–Kontorova Model
Concepts, Methods, and Applications
By O. M. Braun and Y. S. Kivshar

**Aspects of Ergodic, Qualitative
and Statistical Theory of Motions**
By G. Gallavotti, F. Bonetto and G. Gentile

Quantum Entropy and Its Use
By M. Ohya and D. Petz

The New Springer Global Website

Be the first to know

▸ Benefit from new practice-driven features.

▸ Search all books and journals –
now faster and easier than ever before.

▸ Enjoy big savings through online sales.

springeronline.com – the innovative website
with you in focus.

springeronline.com

The interactive website for all Springer books and journals

010048x

 Springer